TAMING THE YELLOW RIVER: SILT AND FLOODS

The GeoJournal Library

Series Editor: WOLF TIETZE, Helmstedt, FR Germany

The titles published in this series are listed at the end of this volume.

Taming the Yellow River: Silt and Floods

Proceedings of a Bilateral Seminar on Problems in the Lower Reaches of the Yellow River, China

edited by

Lucien M. Brush and M. Gordon Wolman

The Johns Hopkins University, Baltimore, Maryland, U.S.A.

and

Huang Bing-Wei

Chinese Academy of Sciences, Beijing, People's Republic of China

KLUWER ACADEMIC PUBLISHERS
DORDRECHT / BOSTON / LONDON

Library of Congress Cataloging in Publication Data

Bilateral Seminar on Problems in the Lower Reaches of the Yellow River, China (1987 : Peking, China?)
Taming the Yellow River : silt and floods / edited by Lucien M. Brush and M. Gordon Wolman and Huang Bing-Wei.
p. cm. -- (The GeoJournal library ; 13)
"Proceedings of a Bilateral Seminar on Problems in the Lower Reaches of the Yellow River, China."
Conference held in October 1987.
Includes bibliographical references.
ISBN 0-7923-0416-0
1. Water resources development--China--Yellow River Watershed--Congresses. 2. Silt--China--Yellow River Watershed--Congresses. 3. Floodplain management--China--Yellow River Watershed--Congresses. I. Brush, Lucien M. II. Wolman, M. Gordon (Markley Gordon), 1924- . III. Huang, Bing-Wei, 1913- IV. Title. V. Series: GeoJournal library ; v. 13.
TC502.Y44B55 1987
627'.12--dc20 89-15618

ISBN 0-7923-0416-0

Published by Kluwer Academic Publishers,
P.O. Box 17, 3300 AA Dordrecht, The Netherlands.

Kluwer Academic Publishers incorporates
the publishing programmes of
D. Reidel, Martinus Nijhoff, Dr W. Junk and MTP Press.

Sold and distributed in the U.S.A. and Canada
by Kluwer Academic Publishers,
101 Philip Drive, Norwell, MA 02061, U.S.A.

In all other countries, sold and distributed
by Kluwer Academic Publishers Group,
P.O. Box 322, 3300 AH Dordrecht, The Netherlands.

printed on acid free paper

Printed in The Netherlands

TABLE OF CONTENTS

Preface ix

American soil and water conservation policy 1
Sandra S. Batie

Measuring the benefits of water quality improvement: recent U.S. practice 21
John J. Boland

Wash load, autosuspensions and hyperconcentrations 31
Lucien M. Brush

An outline for planning development of river basins: a regional approach 45
Nathan Buras

New research directions in agricultural hydrology and erosion 73
Klaus W. Flach and John M. Laflen

Expert systems and river basin management 85
Mark H. Houck

Sediment yield variations as a function of incised channel evolution 99
S.A. Schumm and Allen C. Gellis

Synthesis of data on bed configurations in alluvial channels, and the effect of water temperature and suspended-load concentration 111
John B. Southard

Erosion control and reservoir deposition 139
M. Gordon Wolman

Flood forecasting and its reliability 163
Ben Chie Yen

Remarks 197
Huang Bingwei

The importance of Yellow River safety and the urgency of research for its control 201
Zuo Da-Kang

Temporal and spatial distribution of heavy rain in the middle and lower reaches of the Huanghe river 205
Shen Jianzhu, Guo Yimei and Xu Shuying

Some problems of flooding and its prevention in the lower Yellow River 223
Liu Changming and Liang Jiyang

Simulation of flood propagation in channel of the Yellow River by convection-diffusion equation 243
Liang Jiyang and Liu Changming

A general survey of problems in flood study on the lower reach of the Yellow River 257
Hua Shi Qian

Fluvial processes of the lower Yellow River and estimation of flood conditions 261
Ye Qingchao

The temporal and spatial variation of erosion and sediment yield on the Loess Plateau 275
Chen Yongzong, Jing Ke, Lu Jinfa and Zhang Xunchang

Study of counter measures for the Yellow River regulation and flood prevention 289
Wang Wenkai, Wang Tingwu and Ma Min

A story of soil loss and soil degradation on the Loess Plateau in China 299
Tang Keli, Zhang Zhong-zi and Kong Xiao-Ling

On the strategy of harnessing the Yellow River 315
Feng Yin

The prospects of flood control measures on the lower Yellow River 331
Gu Wenshu

Tentative ideas for controlling the aggradation of the lower Yellow River in the near and remote future 339
Xie Jianheng

The proper measures for flood management and water resources exploitation on the Yellow River 355
Chen Jiaqi

Yellow River sediment problems and synopsis on the strategy of harnessing the river 363
Dai Yingshen

Discussion of the lifespan of the present channel of the lower Yellow River 377
Zhang Ren and Xie Shu-nan

Study on and assessment of flood prevention measures on the lower Yellow River 387
Wu Zhi Yao

Flood characteristics and design flood for flood protection in lower Yellow River 407
Shi Fucheng and Wang Guoan

Flood forecasting and flood warning system in the lower Yellow River 425
Chen Zanting

Ice floods on the lower reach of the Yellow River and measures for ice flood prevention 451
Cai Lin

Current utilization and future prospects of water resources of the Yellow River 469
Bai Zhaoxi

Sedimentation in the lower reaches of the Yellow River and its basic laws 477
Zhao Yean, Pan Xiandi, Fan Zuoying and Han Shaofa

Harnessing and development of the Loess Plateau in the middle reaches of the Yellow River 517
Hua Shaozu and Mou Jinze

Assessment of sediment reduction due to upland water and soil conservation works in the Yellow River 529
Xiong Guishu, Xu Jianhua, Gu Bisheng and Dong Xuena

Evolution of the course of the lower Yellow River in past history 545
Xu Fuling

Review of measures for reducing sedimentation of the lower Yellow River 557
Chen Zhilin

Estimate of reduction in deposition in lower Yellow River by warping on floodplains 585
Hong Shangchi and Chen Zhilin

Effects and significance of water and sediment regulation by reservoirs in the middle Yellow River 597
Qian Yiying

Effects of river training works on flood control 617
Hu Yisan and Xu Fuling

Improvement of the mouth of the Yellow River and sediment disposal 637
Ding Liuyi

Sediment problems in diverting water-to-water supply in the lower Yellow River 657
Tong Erxun

Synopsis of the west-route project of south-to-north water transfer from the Yangtze to the Yellow River 677
Zhang Weiming

Postface 687

PREFACE

About four years ago Dr. Gilbert White visited China and sowed the seeds of this project through conversations with Drs. Huang and Gong of the Chinese Academy of Sciences, and Mr. Long of the Yellow River Conservancy Commission. After some additional rounds of communications by letter, the plan for a workshop evolved and Drs. Wolman and Brush visited with Dr. Sabadell of the National Science Foundation to begin the initial planning. In March 1987 Dr. Brush visited China and the details were worked out for the October 1987 workshop.

At the outset it was recognized that the 10 American scientists and engineers had very little knowledge of the Yellow River and none had ever seen it. Therefore, it became important that field trips be scheduled before the workshop to better set the stage for fruitful discussions. It was also acknowledged that the American participants could not present papers about the Yellow River per se so their offerings reflected their general knowledge of rivers using other rivers as examples. On the other hand the Chinese participants were all well into the difficult problems of harnessing the Yellow River and made their presentations accordingly. Despite these differences the subject matter was the unifying thread and cross communication was excellent.

More than just a workshop this project was designed to serve a number of objectives. One, of course, was bringing together scientists and engineers who may have heard of one another but who never had a chance to enter into meaningful discussions on a vital subject of mutual interest. The main driving force behind the workshop was the hope that it would produce ideas and projects which would lead to future cooperative research between the People's Republic of China and the United States of America. Some ideas along these lines are currently being pursued. In effect, three agencies--The Chinese Academy of Science, the Yellow River Conservancy Commission, and the National Science Foundation--together with participating scientists and engineers are attempting to widen the scientific horizons now so often relegated to provincial problems.

Hopefully, an additional by-product will emerge. Publication of the proceedings in both Chinese and English, by providing information to colleagues in both countries, will enlarge the understanding of the fundamental and complex issues involved on the Yellow River and stimulate contributions to these problems among a broad range of scientists and engineers.

Editing of these papers was not easy, and some thoughts may be blurred through translation, but we hope that enough of the excellent work that is going on can be seen in this volume.

L. M. Brush
M. G. Wolman
Huang B.W.

The editors wish to thank all of the participants for their printed contributions and their informative discussions during the workshop. Special thanks are extended to Mr. Long and Mr. Liu for their help in making the workshop and field trips successful and for their follow-up efforts in obtaining the manuscript from the individual authors. We would also like to acknowledge Dr. Sabadell for her constant help and encouragement from the planning stages to the completion of this project. Throughout the proceedings and publication of the manuscripts, translators in China performed yeoman services. Too many to single out, all of the participants are in their debt. In addition, the efforts of Sue McLaughlin, Judy Black and Lane England in helping to produce the manuscript are greatly appreciated.

AMERICAN SOIL AND WATER CONSERVATION POLICY

Sandra S. Batie
Department of Agricultural Economics
Virginia Polytechnic Institute
and State University
Blacksburg, Virginia 24060, U.S.A.

ABSTRACT. The early history of United States conservation policy demonstrates U.S. policies were governed by a commitment to support agriculture's expansion and to create a more equitable economic base for the nation[1]. Institutions such as The Tennessee Valley Authority (TVA), enacted in 1933, embodied and reflected this commitment. Until recently traditional conservation programs were immune from criticism. However, changes in public attitudes are challenging these programs. The current policy encourages the design of natural resource programs in such a manner as to be more sensitive to public goals of improved environmental quality.

Introduction

For the early history of the United States of America, responsibility for water and soil was mostly in the private sector, although there was some individual involvement of States with water management. These roles changed with the Great Depression of the 1930s. In the 1930s over one-quarter of the U.S. labor force was unemployed. Per capita income fell 15 percent in 1931, and 24 percent in 1932 [Chandler, 1984]. Many of the U.S. traditional natural resource programs of today were borne in response to the perceived need for relief, recovery, and reform in the agricultural sector from such hardships [Paarlberg, 1983].

However, the natural resource programs of this "New Deal" period had an intellectual base extending back to the Progressive Conservation movement of the early 1900s. Indeed, the Depression created the social context which permitted final acceptance of the basic arguments about resource policy made by Progressive Conservation movement leaders such as Theodore Roosevelt, Gifford Pinchot, and Frank Newell. The Depression, they argued, demonstrated that market processes could not be solely relied on to provide for equitable growth.

L. M. Brush et al. (eds.), Taming the Yellow River: Silt and Floods, 1–19.

They articulated a basic rationale for public sector (largely federal) intervention into the market economy to manage the nation's natural resource base, in part to promote the development of agriculture.

The Progressive Conservation movement was an amalgamation of beliefs about both resource policy and social policy. Three general themes can be identified. First, there was a belief that natural resources were the source of the nation's material welfare. At the 1908 Governors' Conference on Conservation, President Theodore Roosevelt commented that "...[I]t is safe to say that the prosperity of our people depends directly on the energy and intelligence with which our natural resources are used. It is equally clear that these resources are the final basis of national power and perpetuity" [as quoted in Nash, 1976, p. 49]. In 1908 the U.S. was a largely agrarian nation, but one in transition to an industrial base. The early conservationists openly expressed alarm over the industrial sector's focus on profits rather than the long-term sustainability of the nation's natural wealth. The industrial transformation was seen as threatening the resource base of the nation and the role of the farmer as a manager of the land -- a manager who, in Roosevelt's words, "...leaves his farm more valuable at the end of his life than it was when he took hold of it" [as quoted in Nash, p. 51].

Second, the industrial developments of the late 19th century were seen as a challenge to the maintenance of democratic values and a fair distribution of the nation's wealth -- a distribution tied directly to the distribution of property rights in the resource base. The industrial sector was viewed as exercising monopoly power to the detriment of the general welfare, and it was the public sector's duty to develop and extend resources and their ownership as widely as possible. Gifford Pinchot, Theodore Roosevelt's Director of the Bureau of Forests, defended this viewpoint as follows:

> The natural resources must be developed and preserved for the many, and not merely the profit of a few. We are coming to understand in this country that public action for the public benefit has a very much wider field to cover and a much larger part to play than was the case when there were resources enough for everyone...
>
> ...[By] reason of the XIVth Amendment to the Constitution, property rights in the United States occupy a...[strong] position...It becomes then a matter of multiplied importance...to see that they [property rights] shall be so granted that the people shall get their fair share of the benefit which comes from the development of resources that belong to us all. The time to do that is now. By so doing we shall avoid the

> difficulties and conflicts which will surely arise if we allow vested rights to accrue outside the possibility of governmental and popular control [as quoted in Nash, 1976, pp. 60-61].

A third belief of the Progressive movement was faith in science as the path to human betterment. Products of science were accepted as positive contributions to human welfare. In such a setting, the creation of modern scientific agriculture proceeded with few people questioning the wisdom of the changes in production practices it fostered.

These three beliefs worked together to affect the way the public sector responded to natural resource issues affecting agriculture from the early 1900s until the late 1960s.

Water Resources

Federal and state policies and programs toward water resources from the 1900s to the late 1960s were governed by a commitment to support agriculture's expansion and the creation of a more equitable economic base for the nation. The early reclamation programs'rationale is described by Frank Newell, Director of the United States Reclamation Service during the administration of Theodore Roosevelt:

> The building of large structures for water conservation and for the reclamation of land is not, however, the ultimate object. These works in themselves are notable, but their importance to the nation comes from the fact that they make possible opportunities for the creation of small farms and building of homes for an independent citizenship. The question of prime interest to the general public is not so much that of how far these works may be extended, but what may be accomplished through them by the reclamation of lands otherwise valueless [as quoted in Nash, 1976, p. 55].

Water development was seen as a way to provide jobs as well as irrigation water, flood control and power. Federal development of water sources was achieved during this period of 1930 to 1970 mainly by federal agencies working directly with and providing funds to local sponsors -- for the most part ignoring state governments [North, 1983]. The historical commitment to these programs is illustrated by the work of Pavelis [1977]. He reviewed U.S. investments in irrigation, drainage and soil and water conservation (IDC) from 1900 to 1975 and estimated the net depreciated value of IDC facilities and equipment at $27.5 billion. This included $12.3 billion for irrigation, $9.7 billion for conservation and $5.5 billion for drainage. Of particular note, he found that "[t]he federal government, through direct construction or indirect cost sharing had created up to 1975 about 45 percent of the value of all IDC facilities in the United States -- about 8 percent for drainage, 52 percent for irrigation, and 55 percent for soil and water conservation" [p.1-2].

THE TENNESSEE VALLEY AUTHORITY

The Tennessee Valley Authority (TVA) enacted in 1933 embodies the spirit of these times (1930-1970). The Tennessee Valley Authority was created to manage the areas drained by the Tennessee River. The region includes parts of seven states: Alabama, Georgia, Kentucky, Mississippi, North Carolina, Tennessee and Virginia. The U.S. Congress intended TVA to "tame the river" with dams and to save the Tennessee River Valley citizens from economic despair by providing navigation, flood control, power generation, soil erosion control, and reforestation by producing and encouraging the use of fertilizer and economic development. It was also created to counter a feared monopolization of water power by private companies [Chandler, 1984].

President Franklin D. Roosevelt in asking for legislation to create the Tennessee Valley Authority stated,

> [T]he potential public usefulness of the entire Tennessee River...transcends mere power development; it enters the wide fields of flood control, soil erosion, afforestation, elimination from agricultural production use of marginal lands, and distribution and diversification of industry. In short, this power development of war days leads logically to rational planning for a complete watershed involving many states and the future lives and welfare of millions...I, therefore, suggest to the Congress legislation to create a Tennessee Valley Authority--a corporation clothed with the power of Government but possessed with the flexibility and initiative of a private enterprise. It should be charged with the broadest duty of planning for the proper use, conservation, and development of the natural resources of the Tennessee River drainage basin and its adjoining territory for the general social and economic welfare of the nation. [King, p. 268 cited in Chandler, 1984, p. 26]

> Many hard lessons have taught use the human waste that results from lack of planning. Here and there a few wise cities and counties have looked ahead and planned. But our Nation has "just grown." It is time to extend planning to a wider field, in this instance comprehending in one great project many States directly concerned with the basin of one of our greatest rivers. [Wirtz, p. 573,

as cited in Chandler, 1984, p. 27]

The TVA was created by the federal government as an independent authority--a totally unique social experiment. TVA's funding depends on appropriated funds from Congress and the revenue from power generation, fertilizer sales and consulting fees. It has historically received funds in a lump sum, and has therefore "enjoyed the extraordinary liberty of self-authorizing power" [Chandler, 1984, p. 5]. It has not been accountable to any voter, or subject to the usual legislative checks and balances found in American government.

The TVA region was provided with cheap electric power--residential rates averaged only half the national level for decades [Chandler, 1984]. It had large research programs in the use of fertilizer, crop rotation and soil conservation. TVA created over 13,000 demonstration farms in the Tennessee River Valley. Farmers got free assistance and fertilizer if they would participate in demonstrations of up to 5 years. Over 200,000 acres--80,940 hectares--were planted in trees. TVA encouraged terracing of and rotations on cropland.

The TVA's emphasis on protecting the nation's agrarian base and resisting private monopolies of resources typifies the natural resource policy attitudes dominating the 1930 to 1970 period.[2] For example, throughout the nation, as within the TVA region, federal investments in extending the navigation system were justified as being necessary competitors to the railroads. Railroads' rate structures were perceived as being excessively high due to the railroads' monopoly power over their customers, including farmers shipping goods to market. An extension of the "toll-free" waterway policy that originated in the early years of the nation was justified, in part, by the argument that the agricultural sector should be supported by the low-cost transportation offered by the waterways.

The promotion of social and economic change through the development of resources and the unsupervised planning and management by undisputed experts in TVA also typifies public programs throughout the nation during the 1930s to 1970.

Soil Conservation

Like water resource programs, federal soil conservation programs historically have been designed to support agricultural expansion and more equitable distribution of economic wealth. Soil conservation control was first authorized as a means of unemployment relief as part of the 1933 National Industrial Recovery Act (PL 73-67), and under the Act's authority the Soil Erosion Service was established within the Department of Interior. Public works funds were used to build check dams, fences, contour strips, terraces, to plant trees and to grass eroded areas [Morgan, 1968, p. 1]. However, the soil conservation programs soon changed so that they supported the goals of increasing farm income and improving agricultural productivity.

In 1933 Congress passed the first Agricultural Adjustment Act (AAA). The AAA was the Congressional response to President Franklin D. Roosevelt's request for legislation that would "increase the purchasing power of our farmers and the consumption of articles manufactured in our industrial communities" [Roosevelt, 1938, p. 74]. Because many policymakers thought that agriculture's main problem was overproduction, paying farmers not to plant in order to reduce surpluses was seen as a way of increasing farm income and providing emergency farm relief. Therefore the AAA allowed the government to "enter into agreements with farmers to control production by reducing acreages devoted to basic crops, to store crops on the farm and make payment advances on them and to enter into marketing agreements...to stabilize product prices and to levy processing taxes" [Cochrane, 1979, p.141]. When the AAA was ruled unconstitutional by the U.S. Supreme Court, Congress enacted the Soil Conservation and Domestic Allotment Act of 1936. This Act enabled farmers to receive soil conservation payments for reducing "soil depleting" crops, which were also the crops that were in surplus. Two goals--soil conservation and the maintenance of farm income--could thus be pursued simultaneously, and soil conservation legislation provided a legal vehicle for the pursuit of farm "relief" and "recovery." Thus, for the last 50 years the national farm programs and soil conservation programs have been the de facto national rural development policy.

The soil conservation programs have continued to be implemented since the 1930s with a primary focus of improving farm income. As a result, selection criteria for the receipt of cost-sharing funds for soil and water conservation have not been tied to the severity of the farmer's erosion problem. Rather program administration has spread cost-sharing funds widely so as to reach as many farmers as possible [Batie, 1983]. Eligible farmers can receive technical assistance and dollars to implement soil conserving practices--such as terracing planting cover crops, or using no-till agricultural practices. Payments also were made for numerous production enhancing practices such as farmers' purchases of fertilizers and lime.

Drainage of wet soils also received public attention. Public works projects, in conjunction with cost sharing and technical assistance programs for on-farm drainage projects, were justified as turning "wastelands" into valuable agricultural soils, thus extending the agricultural base of the nation.

A Subsidized and Intensive Modern Agriculture

The farm programs, the soil conservation programs and the water resource programs have worked together to encourage an expanding and more intensive modern agriculture. For example, cheap water and technical assistance plus high agricultural price supports have encouraged farmers to farm more intensively; to use wetlands and fragile soil-eroding lands despite chronic grain surpluses.

The U.S. has gone from hoes to herbicides in less than four decades[3] [Hallberg, 1986]. As farmers sought to reduce risks of crop failure and to take advantage of price supports they quickly adopted chemical inputs. Between 1964 and 1985, farmers' use of pesticides has more than tripled [USDA, 1985]. Nitrogen fertilizers, used to any degree before World War II, now have use-rates of 10 million metric tons per year [Hallberg, 1986]. In the Corn Belt--in the middle of the U.S.--average rates of fertilizer use on corn were 65 pounds per acre (73 kilograms per hectare) in 1965; they are now 135 pounds per acre (151 kilograms per hectare) [Hargett and Berry, 1983]. Today over 91 percent of the U.S. row crop acreage and 44 percent of U.S. small grain crop acreage have herbicides applied annually [The Conservation Foundation, 1987].

The programs have also encouraged specialization--the devoting of large acreages to single crops. Thus, fertilizer-saving and insect-controlling rotations are less likely to be used. Indeed, U.S. farm programs have historically been structured so that they penalize farmers who choose to undertake rotations and reward farmers that plow out pastures, level their fields, and plant corn year after year.

AGRICULTURE AS THE PROBLEM

Until recently these traditional federal programs have remained largely intact and immune from criticism. However, changes in public attitudes toward the agricultural production sector and changed perceptions about the relationships of agriculture to the environment are resulting in challenges to these programs as well as the development of new policies affecting farmers' access to, and use of, resources.

The questioning of the federal programs corresponding with the rise of the "environmental movement of in the 1970s." Although environmentalism did not represent a complete break with all the traditions of the progressive conservation movement, there were some fundamental differences which began to undermine the political rationale for the traditional natural resource programs.

Environmentalists questioned the development of resources

for profit, and they promoted a more harmonious relationship of people with the natural world. The environmental community wondered whether the nation's economic arrangements were actually a threat to the long run sustainability of the system. They criticized the structure of agriculture which had lost its "family farm orientation" and became a large farm, monoculture, input dependent agriculture. The image of agriculture projected by environmentalists was more akin to a factory than to a family farm. The critics asserted that modern science and technology may have created more problems than they solved. As the National Research Council of the National Academy of Sciences [1982] summarized:

> Modern agriculture technology, together with economic forces, has favored large contiguous fields devoted to single crops. The increasingly efficient drainage of lowlands, improved varieties of crops capable of growing on marginal soils, increased and more efficient use of fertilizer and pesticides, and development of irrigation have expanded cropland at the expense of natural ecosystems [p. 3].

The range of concerns about the harmful impacts of scientific contributions to improve productivity was widespread: for example, chemical contamination of the environment, groundwater and surface water depletion made possible by large irrigation systems, and the destruction of wetlands and the soil resource for the convenience of large machinery.

Chemical Contamination. The initial mainstream scientific thought with respect to increased chemical use by farmers is that it was harmless to mammals, and in any event the soil would filter chemicals and protect ground water [Hinkle, 1987]. Public doubt was first raised in 1962 when Rachel Carson completed Silent Spring. Her book alerted many Americans to the possibility of problems associated with widespread chemical use. Farm chemicals were indicted as possible human health risks, as catalysts in the evolution of pesticide resistant plants and insects, as destroyers of non-targeted species, and as creators of new pest infestations.

Still, it was not until 1972 that amendments changed the Federal Insecticide, Fungicide, and Rodenticide Act [FIFRA] from the previous United States Department of Agriculture (USDA) administered act which provided farmers protection from ineffective and dangerous pesticides. The amended FIFRA gave the Environmental Protection Agency (EPA) authority to directly control pesticides and to give adequate consideration to environmental risks [Aidala, 1986]. The concern was less for farmers' welfare and more for protection of the environment.

Groundwater quality policies are a product of the 1980s. It was not until then that there was scientific corroboration of groundwater contamination resulting from normal agricultural use of chemicals [Holden, 1986]. The 1979 discovery of aldicarb in the groundwater in Suffolk County, New York, caused many states to begin monitoring their own groundwater. Evidence of existing and increasing contamination of groundwater by agricultural pesticides and fertilizers is accumulating.

By 1984, the United States Geological Survey (USGS) had found over 20 percent of 124,000 sampled wells had nitrate concentrations high enough to assume man-caused contamination; 6 percent exceeded federal drinking standards. By 1985, the Environmental Protection Agency (EPA) had found 17 pesticides in the groundwater of 23 states. The pesticides were there as a result of normal farm use of chemicals following label instructions [USDA, 1985].

There is increasing evidence that chemicals persist in groundwater. Such findings are distressing because they indicate that some pesticides and herbicides, once thought to quickly decompose in the environment, are apparently long-lived. Furthermore, some of these chemicals are carcinogenic. Lasso, a widely used pesticide, for example, has been implicated in cancers in the lung, stomach, and thyroids of laboratory animals.

Chemical contamination concerns also encompass food safety. Recently, the National Food Marketing Institute polled grocery shoppers and asked their attitudes about food safety. Seventy-seven percent of the shoppers thought pesticide residues in food products were a serious health hazard. Another 18 percent thought these chemicals were "something of a hazard." This was by far the greatest concern expressed by the shoppers, out-distancing cholesterol, salt, additives, preservatives, sugar, and artificial coloring. The poll was conducted before an aldicarb poisoning of watermelons on the West Coast, and before ethylenedibromide (EDB) was found in groundwater in Florida.

While there is not a scientific consensus nor a political consensus on the nature and magnitude of these chemical residuals in our food system nor on many other environmental impacts, the resource debate is clearly focused on these issues. Those that fear that their future in the environment is being put at risk by current agriculture practices are taking the initiative in framing the debate. They are taking the initiative in proposing policy options. Increasingly, agencies which previously did not have much interest in agriculture find themselves putting agriculture as an important part of their agenda. The Environmental Protection Agency is considering banning alar for use on apples, peanuts, fruits and vegetables as a suspected carcinogen. The Food and

Drug Administration is reluctantly bowing to heavy pressure to provide closer oversight to animal drugs, whose residues might show up in meat and perhaps pose a health hazard. U.S. Secretary of the Interior Hodel recently reached a compromise with a group of 250 California farmers that meant they could continue to receive federal irrigation water, provided they could assure the Secretary that their drainage water would no longer go into the Kesterson Wildlife Refuge. The irrigation water was so laden with the trace element selenium that 20 percent of the wildfowl hatched at that refuge had deformed eyes, beaks, wings, legs, or skulls.

Depletion and Water Quality Degradation Due to Irrigation. Irrigated acreage has tripled in the United States since the 1940s; it now accounts for more than one-fourth of the value of the nation's crops. In some arid regions, such as California's deserts, irrigation has transformed seemingly worthless land into some of the world's most productive agricultural regions. Much of the irrigation water from surface sources has been provided to farmers by the federal government virtually cost-free. The irrigation of western lands is associated with two major environmental problems: depletion of surface groundwater and salinity.

Some 22 million acre-feet (the equivalent of 9 million hectares covered by one-third of a meter of water) of water are pumped from western aquifers yearly. The volume of water is the equivalent of considerably more than one entire annual flow of the Colorado River. Water tables are declining, particularly in California and the High Plains states of Nebraska, Kansas, Colorado, Oklahoma, and Texas. If groundwater pumping rates greatly exceed rates of water recharge, deeper and more expensive wells are required. In certain areas the impacts are severe. Some farmers already cannot afford deep-well irrigation; some regions will suffer economic losses if they return to dryland farming.

Conflicts between farmers and nonagriculturalists over water allocation seem certain to become increasingly severe. Western water supplies are already seriously strained by the demands of current users--almost 90 percent of whom are farmers. According to a 1978 U.S. Water Resources Council report, western water consumption for nonagricultural uses--chiefly for urban expansion--will increase 81 percent by the year 2000. Such increases will translate into intense competition for water, particularly because most surface water supplies are already severely limited. Already, in a year of average rainfall, the total water requirements for the lower Colorado River exceed stream flow by 225 percent. In the western states water requirements exceed stream flow in 48 of the 53 water subregions in a dry year (defined as having rainfall less than 80 percent of average).

In a majority of states, there is considerable political

pressure for obtaining additional water through interbasin water transfers. Such transfers are very controversial and frequently financially infeasible. For example, one proposal to bring Mississippi water to the High Plains of Texas would cost over $1000 an acre-foot (an acre-foot of water is the equivalent of 9 hectares covered with one-third of a meter). The water would be used to produce crops that translate into a value of the water at no more than $75 per acre-foot. But the demands for water persist, because farmers hope to obtain the water at highly subsidized rates and have the federal government pay for most of the costs [Kneese, 1985].

Salinity problems affect an estimated 35 percent of all western irrigated cropland. Moreover, salinity conditions appear to be worsening in all western river basins except the Columbia River in the Northwest region of the U.S. As irrigation water passes through soils, it dissolves naturally occurring salts, which are then carried back into the irrigation canal or river. Increased salt concentrations in irrigation water may decrease plant yield significantly. Moreover, downstream users suffer from the accumulation of salt. Irrigation also causes salinization of the soil; as crops draw up irrigation water, dissolved salts travel with it to the surface, where they accumulate in the soils. Additional amounts of irrigation water are then necessary to flush out the salt from the crop root area. High salt concentrations in soil decrease crop yields, alter crop patterns, increase leaching and drainage requirements, and increase water management costs.

Nearly 40 percent of the salt load of the Colorado River can be attributed to irrigation waters. In this river system alone, annual damage from salinity was estimated at $104 million in 1980--and the cost is expected to rise to $165 million by the year 2000. Much of this damage is to crops, but salinity also increases water treatment costs for municipal and industrial users. In addition, there are the uncounted costs of wildlife habitat lost from saline contamination.

Irrigation causes other water pollution problems as well. In addition to salt, selenium, and heavy metals, other contaminants such as pesticides and fertilizers are returned to receiving waters--sometimes in high concentrations.

Wetlands Destruction. Between the mid-1950s and the mid-1970s, the nation lost 550,000 acres of wetlands, according to a 1984 study by the Congressional Office of Technology Assessment (OTA). Wetlands provide an important habitat for fish and wildlife, and they can also improve water quality and aid in flood and erosion control. Almost 80 percent of these losses, OTA reported, can be attributed to the draining and clearing of inland wetlands for agricultural purposes.

Over 90 percent of Nebraska's original wetlands has been cleared and planted. An estimated 80 percent of the bottomland hardwood forests in the Mississippi Valley has been cleared, drained, and planted. Only 56 percent of the original prairie pothole area of the Red River Valley of Minnesota and North Dakota remains.

While most of these conversions of wetlands to agricultural purposes were prompted by favorable prices for crops, federal farm programs and the tax code have also played a role. Between 1940 and 1977 USDA was authorized to provide technical and cost-sharing assistance to farmers for draining wetlands; farmers used this help in draining nearly 57 million acres during that period. Although USDA assistance for wetlands drainage has been eliminated for the most part, federal tax-code provisions can still reduce the costs of wetlands conversion by as much as 45 percent. Other federal programs, such as the farm commodity price and income support programs, have also provided financial incentives for wetland conversion by raising the price of crops and thus encouraging new planting. In the Mississippi Delta states, for example, crops can bring a farmer more income than the bottomland hardwood forest production that takes place on wetlands. As a result farmers clear, drain and level these wetlands for crop production.

Soil and Water Quality Degradation Due to Erosion. The 1977 Natural Resource Inventory (NRI) of the U.S. Department of Agriculture (USDA), provided for the first time a statistically valid picture of the magnitude and location of erosion in the United States. In 1982 the inventory was updated by an even more extensive survey of erosion problems. According to the 1982 NRI, 1.83 billion tons (1.66 billion metric tons) of soil on cropland were displaced by rainfall-caused erosion, and wind erosion displaced another 1.25 billion tons (1.14 billion metric tons). In all, over 3 billion tons of cropland soil erode annually; the total erosion rate, including nonfarm sources, is over 6 billion tons (5.443 billion metric tons) a year.

The 1982 NRI data also revealed that erosion of cropland is highly concentrated. Almost 56 percent of the cropland was eroding at rates that exceeded the tolerance levels established by USDA. These tolerance levels range from two to five tons per acre [4.5-11.2 metric tons per hectare] annually. These figures corroborated the 1977 inventory, which found that 5 percent of cropland acreage accounts for 36 percent of the total erosion and 66 percent of the excessive erosion. The states that have the greater share of the water-caused excessive erosion are Iowa, Illinois, Missouri, Nebraska, Kansas, Texas, Mississippi, Tennessee, Indiana, and North Dakota. These ten states account for 61 percent of the total water-caused erosion of U.S. cropland.

Soil erosion causes damages of two types: (1) on-farm losses in soil productivity, and (2) off-farm pollution of air and water. The two types of damages do not necessarily occur simultaneously, nor are they always correlated with high erosion rates. Consequently, high erosion rates occurring on rich deep soils far from water water may have less immediate impact than more moderate erosion rates on croplands with shallow topsoils or on croplands near important rivers, reservoirs, lakes, or bays. Unfortunately, precise information is not yet available on the location and magnitude of the damages caused by soil erosion.

Many studies have demonstrated the effect of soil erosion on crop yields. In a 1982 paper, for example, George W. Langdale and William Schrader presented the results of eight earlier studies (conducted between 1961 and 1977) on the effect that the removal of all topsoil would have on various crop yields. For corn the reduction ranged from 8 to 30 percent; for soybeans, from 20 to 40 percent; for cotton, from 12 to 20 percent; for small gains, from 11 to 24 percent; and for forages, from 5 to 17 percent. Different soils, however, have different propensities for erosion. These studies were conducted on deep, medium-textured soils; similar studies on shallow, medium-to-coarse-textured soils showed even greater declines in yield. The crops most affected by erosion are corn, soybeans and cotton; they tend to have higher average erosion rates than other crops. In 1977 annual erosion rates exceeded 5 tons per acre (11.2 metric tons per hectare) on 33 percent of the land planted in corn, on 44 percent of the land in soybeans, and on 34 percent of the land in cotton.

The ultimate impact of erosion on soil productivity is difficult to calculate, however, because farmers are able to compensate for erosion by applying more fertilizer and water and using higher-yielding crop varieties and different management practices. Uncertainties concerning future changes in food demand and technology advances also complicate predictions about the significance of reduced soil fertility. The impact of erosion-caused productivity losses would be heightened if the demand for crops expanded rapidly, if few or very highly cost technological substitutes for soil fertility were available (for example, improved crop hybrids and improved fertilizers), and if farming costs rose rapidly.

On the other hand, if current erosion rates continue but low-cost substitutes for soil fertility are available, the aggregate damages of soil erosion to soil productivity may not be great. Agronomists William E. Larson, Frank J. Pierce, and Robert H. Dowdy recently estimated that a continuation of current erosion rates, in combination with today's farming technology, would reduce national U.S. crop yields by less than 10 percent over 100 years. Moreover, even a modest rate of technological advance would reduce that impact. In terms of costs, on-farm productivity losses from soil erosion are estimated at approximately $40 million per year. However, in localized areas the damages could be very serious.

Soil erosion also affects air and water quality off the farm. Soil erosion causes the premature siltation of harbors, lakes, rivers and reservoirs, and the filling of irrigation and drainage ditches. An Illinois state survey, for example, found that sediment removal from roadside ditches cost the state $6.3 million in 1977.

In addition, as soil is carried off the cropland, it brings with it fertilizers, pesticides, and herbicides that can pollute drinking, recreational and industrial water. As municipalities and industries have cleaned up point sources such as sewage discharge and industrial effluents, nonpoint pollution from agriculture and urban runoff has emerged as a major cause of current water quality problems. Indeed, 33 percent of the oxygen-demand load, 66 percent of the phosphorus, and 75 percent of the nitrogen content in streams comes from dispersed agricultural sources--of which soil erosion is probably the largest contributor.

EPA's Office of Water Planning and Standards (1980) estimated that in 1980, 2.1 million fish were killed as a direct result of pesticide, fertilizer or manure-silage water contaminants. From 1961 to 1975 agricultural pesticides were responsible for nearly one-fourth of all reported fish kills for which the cause was known. Some individual incidents can be spectacular. In 1976 pesticide runoff killed 109,000 fish in an Alabama fish hatchery. In Louisiana the herbicide paraquat has killed crayfish and shrimp. Some commercial fishing in the Great Lakes was banned in the mid-1970s because unacceptable levels of pesticides accumulated in fish tissues.

The state of the art in assessing off-farm damages is far from advanced. Nonetheless, various studies, which have attempted to estimate costs as accurately as possible with the limited evidence available, suggest that the costs of off-farm erosion damage are much higher than the costs of on-farm lost productivity. A recent study by the Conservation Foundation (Clark, Haverkamp and Chapman, 1985) using what they term highly unsatisfactory information, estimated that cropland erosion costs $2.2 billion annually in water quality damage. This estimate did not include damages to aquatic ecosystems. There are no comparable estimates for erosion-caused air quality damages.

Effective techniques to combat erosion are available. Conservation tillage, contour planting, strip cropping, terracing, and other techniques, used singly or in combination, can reduce soil erosion rates by 60 to 95 percent. For a variety of reasons, related to both cost and farmers' perceptions and values, many farmers do not implement these measures, however.

In addition, soil erosion has been exacerbated by federal policies that encourage farmers to increase their production of subsidized crops such as corn and wheat. Often this has meant bringing marginal, highly erodible soils into cultivation. Tax policies that encourage farmland speculation can also lead to soil-depleting practices on the farm, even though such impacts are often unintended.

Change in Public Attitudes Toward Managing Resources

From these basic concerns there now appears to be an increased public support for mutually supportive agricultural and resource policies. At the most general level, there is a "shift in constraints." In the past many resource policies were constrained by a concern for how well they served the interests of agriculture. Increasingly the question is now asked: "How do agricultural programs and policies conform to goals of environmental protection and enhancement?"

The traditional resource programs have been altered in numerous ways. Federal water resources development programs have come to a comparative standstill, and future water project development and operation costs will fall increasingly upon project users as cost-sharing reform moves forward. The commitment is now to increasing the share of cost paid by project beneficiaries, including agriculture, and a general reluctance to support federal water development. Taken together these trends suggest increased control over farmers' use of groundwater for irrigation and an increased recognition of the values served by reserving in-stream water from diversion to agricultural and other uses [Dewsnup and Jensen, 1977]. Similarly, federal investments, cost-sharing and advisory assistance to promote the drainage of wet soils have been sharply curtailed in response to concern that such programs have promoted the destruction of wildlife habitat such as wetlands and bottomland hardwood forests.

Soil conservation programs have also been under increasing scrutiny. In 1977 a major evaluation effort was begun with the passage of the Soil and Water Resources Conservation Act (RCA). RCA requires USDA to provide a review of past programs effectiveness, to design improved programs, and to display the results to the public and Congress. Soil conservation under RCA directs USDA to focus on the conservation, protection and enhancement of soil, water and related resources on nonfederal lands--not on farmers' economic welfare [USDA, 1982].

Conclusion

American public opinion about agriculture and its use of resources is changing. At one time most Americans either lived on a farm or were only one generation removed from a farm. They saw a link between the farmer's welfare and the nation's welfare. They saw the farmer as the steward of the environment, and they had faith that science and technology would bring societal benefits. As citizens become further removed from farming enterprises, farmers are losing their special place in American society. Increasingly, they are finding their traditional programs and practices questioned.

The current policy setting is one of designing programs for the use of natural resources that is more sensitive to public goals of an improved environment.

Footnotes

[1] Much of this paper draws from Batie, Shabman and Kramer (1986); Batie (1985); and Batie (1987).

[2] The TVA still exists; it is the nation's largest power utility. It provides power to over 6 million people--although much of its regional development and conservation efforts have lanquished [Chandler, 1984]. Whether it has been a success is actively debated. There is no doubt that per capita income and employment rose dramatically in the valley. For example, per capita income rose from 44 percent of the national average in 1933 to 61 percent by 1953. The provision of running water in homes and rural electrification has been largely completed. Critics question, however, whether the TVA region might now have achieved these levels of welfare without the Tennessee Valley Authority, and they point to the success of similar regions outside the Tennessee Valley Authority boundaries. The critics question whether it was the TVA projects that lead to economic growth. Critics also note that TVA historically has engaged in destructive coal stripmining and the debilitation of the underground coal industry so important to this region [Chandler, 1984].

[3] There are now 1800 biologically active chemicals sold in 32,000 different formulations [Sampson, 1981].

References

Aidala, James. (Aug) 1986. "Pesticide Regulation: 1986 Amendments to FIFRA." Congressional Research Series 86-796 ENR. Washington, D.C.: Library of Congress.

Batie, Sandra S. 1983. Soil Erosion: Crisis in America's Croplands? Washington, D.C.: The Conservation Foundation.

Batie, Sandra S. 1985. "Environmental Limits: The New Constraints." Issues in Science and Technology. II(1):134-143. Washington, D.C.: National Academy of Science.

Batie, Sandra S., Leonard A. Shabman, and Randall A Kramer. 1986 "U.S. Agricultural and Natural Resource Policy." pp. 132-138, In The Future of the North American Granary: Politics, Economics, and Resource Constraints in American Agriculture, ed. C. Ford Runge, Ames, Iowa: Iowa State Press.

Batie, Sandra. 1987. "Institutions and Ground Water Quality," pp. 22-40. In Proceeding of a National Symposium on Agricultural Chemicals and Ground Water Pollution Control. March 26-27, 1987; Kansas City, Missouri.

Carson, Rachel. 1962. Silent Spring. Boston: Houghton Mifflin.

Chandler, William U. 1984. The Myth of TVA: Conservation and Development in the Tennessee Valley, 1933-1983. Cambridge, Mass: Ballinger Publishing Co.

Clark, Edwin H.,II, Jennifer A. Haverkamp, and William Chapman. 1985. Eroding Soils: The Off-Farm Impact. Washington, D.C.: Government Printing Office.

Cochrane, Willard W. 1974. The Development of American Agriculture: A Historical Analysis. Minneapolis, MN: University of Minnesota Press.

Dewsnup, Richard L., and Dallin W. Jensen. (Mar) 1977. State Laws and Instream Flows. U.S. Department of Interior: Office of Biological Services, Fish and Wildlife Service (FWS/OBS-77127).

Hallberg, George R. (Nov-Dec). 1986. "From Hoes to Herbicides: Agriculture and Groundwater Quality." Journal of Soil and Water Conservation. 41(6): 357-363.

Hargett, N. L. and J. T. Berry. 1983. "1982 Fertilizer Data

Use Summary Development Center," Tennessee Valley Authority. Muscle Shoals, Alabama.

Hargett, N. L. and J. T. Berry. 1983. "1982 Fertilizer Data Use Summary." Development Center. Tennessee Valley Authority. Alabama: Muscle Shoals.

Hinkle, Maureen. (Jan) 1987. "Comments on EPA's Proposed Decision to Reregister Alachlor." Washington, D.C.: The Audubon Society.

Holden, Patrick W. 1986. Pesticides and Groundwater Quality: Issues and Problems in Four States. Washington, D.C.: National Academy Press.

King, Judson. 1959. The Conservation Fight. Washington, D.C.: Public Affairs Press, p. 268.

Kneese, Allen U. (April 19) 1985. "Policy Considerations In Intra State, Interstate and Interbasin Transfers of Water." Presentation made to the Food and Agricultural Committee of the National Planning Association in Denver, Colorado.

Langdale, George W. and William Schrader. 1982. "Soil Erosion Effects on Soil Productivity of Cultivated Cropland," in Determinants of Soil Loss Tolerance, ASA Special Publication no. 45, ed. B. L. Schmidt, R. R. Allmaras, J. V. Mannering and R.J. Papendick. Madison, Wis.: American Society of Agronomy and Soil Science Society of America.

Larson, William E., Frank J. Pierce, and Robert H. Dowdy. 1983. "The Threat of Soil Erosion to Long-Term Crop Production," Science, 219: 458-65.

Morgan, Robert J. 1968. Governing Soil Conservation: Thirty Years of the New Decentralization. Baltimore, MD: Johns Hopkins Press.

Nash, Roderick. 1976. The American Environment: Readings in the History of Conservation. Readings, Mass.: Addison-Wesley Publishing Co.

National Research Council, National Academy of Sciences (NAS). 1982. Impacts of Emerging Agricultural Trends on Fish and Wildlife Habitat. Washington, D.C.: National Academy Press.

North, Ronald M. (Feb) 1983. "Roles and Responsibilities of Governments in Water Resources Management," p. 21-36. In Emerging Issues in Water Management and Policy, ed. Sandra S. Batie and J. Paxton Marshall. Mississippi State, Mississippi Southern Natural Resource Economics Publication 17.

Office of Technology Assessment. 1984. Wetlands: Their Use and Regulation. Washington, D.C.: U.S. Congress.

Paarlberg, Don. 1983. "Effects of New Deal Farm Programs on the Agricultural Agenda a Half Century Later and Prospects For the Future." American Journal of Agricultural Economics, 65:1163-67.

Pavelis, George A. 1978. Natural Resource Capital Stocks: Measurement and Significance for U.S. Agriculture Since 1900. U.S. Department of Agriculture: Economics, Statistics and Cooperatives Service, Washington, D.C. February 1978.

Roosevelt, Franklin D. 1938. The Public Papers and Addresses, Vol. 2, New York: Random House.

Sampson, R. Neil. 1981. Wasteland: A Time to Choose. Ammaus, PA: Rodale Press.

The Conservation Foundation. 1987. Groundwater Protection. Washington, D.C.: The Conservation Foundation.

U.S. Department of Agriculture (USDA). 1982. A National Program for Soil and Water Conservation: 1982 Final Report and Environmental Impact Statement. Washington, D.C.: U.S. Government Printing Office. 1982.

United States Dept. of Agriculture (USDA). 1985. "Inputs, Outlook, and Situation Report." Economic Research Service IOS-7. Washington, D.C.

U.S. Environmental Protection Agency, Office of Water Planning and Standards. 1980. Fish Kills Caused by Pollution, Fifteen Year Summary 1961-1975. Washington, D.C.: Government Printing Office.

U.S. Water Resources Council. 1978. The Nation's Water Resources 1975-2000, Second National Assessment. Washington, D.C.

Wirtz, Richard. (Sum) 1976 "The Legal Framework of the Tennessee Valley Authority," Tennessee Law Review, p. 573.

MEASURING THE BENEFITS OF WATER QUALITY IMPROVEMENT: RECENT U.S. PRACTICE

John J. Boland
Department of Geography and Environmental Engineering
The Johns Hopkins University
Baltimore, MD 21218, U.S.A.

ABSTRACT: Recent U.S. practice in measuring the benefits of water quality improvement is reviewed and interpreted. Particular attention is given to distinctions between market-based methods, where actual markets in resource services or in related commodities are assumed, and non-market-based methods, which make use of administered valuations or of hypothetical markets. Issues associated with the use of non-market-based methods in centrally planned economies are briefly reviewed.

Introduction

In 1983, the US spent approximately $24,000,000,000 on a variety of programs designed to to improve and protect the quality of the nation's water resources [US Bureau of the Census, 1985, p. 207]. Water quality improvement enjoys considerable public and political support in the US, but needed funds are always difficult to obtain. It would be easier to justify expenditures on water quality improvement if the expected benefits could be measured in monetary terms.

A related situation is found in the case of improper disposal of hazardous wastes. Under current US law, states can institute legal proceedings against those who cause hazardous materials to be released into the environment, and can recover the monetary value of the damage caused (natural resource damages). To seek such a recovery, however, assumes the ability to express environmental damage, including water quality degradation, in monetary terms.

In fact, methods exist for expressing the benefit of water quality improvement (or, conversely, the damage from water quality degradation) in monetary terms. Some of these methods have been in use for thirty years or more, and have

L. M. Brush et al. (eds.), Taming the Yellow River: Silt and Floods, 21–30.

been extensively tested. Others are recently introduced, and less completely understood. Unfortunately, no single method has proved to be useful in all situations.

Many estimates have been made of the value of water quality improvement for specific places, and some have attempted national aggregation. Also, natural resource damages associated with hazardous waste discharge are now routinely measured in monetary units. This paper outlines the theoretical underpinnings of benefit (damage) measures associated with water quality changes, briefly presenting the principal methods and their relative strengths and weaknesses.

The Value of a Natural Resource

BASIC CONCEPTS

The following definitions are provided for terms used here:

Resource: Something which can be relied upon for supply or support.

Natural Resource: A part of the natural environment which functions as a resource; e.g., a water resource.

Natural Resource Service: Services derived from a natural resource which, when consumed directly or combined with other goods or services, result in benefits to individuals and to society. Examples of water resource services include agricultural water supply, harvestable fish populations, water suitable for recreational boating.

Natural Resource Injury: A change in the characteristics of a natural resource which adversely affects subsequent flows of services. The introduction of a contaminant into a river is a natural resource injury.

Natural Resource Damage: The loss in benefits to society which results from natural resource injury. This is measured as the sum of the values of all natural resource services before the injury, less the sum of the values of all resource services after the injury.

Measuring the benefits of water quality improvement is simply the converse of measuring water quality damage. First, it is necessary to identify the increased flows in natural resource services which will result from removing a natural resource injury (water pollution). The increased value which results from these increases in service flows is the benefit attributable to water quality improvement.

The damage (benefit) measure of interest is the value

placed on the natural resource injury by society as a whole. No direct observation of such a value is possible, however, so it is customary to use the sum of individual valuations as a surrogate for social value.

Natural resource damage appears in two ways: (1) those who would have used the services of the resource, but cannot because of injury (pollution), lose the benefits they would have received; and (2) others who were uncertain future users, or who would not have used the resource services in the future, may feel a loss because of the services are no longer available. The first mechanism gives rise to use value, while the second is expressed in several ways, collectively known as intrinsic value.

USE VALUE

Use values are straightforward in concept: those who cannot use a resource service because it has been injured are worse off because of the injury. The monetary equivalent of their lost enjoyment is a measure of the benefit lost or, conversely, the benefit that would be gained if the injury could be removed. Use values are measured for each injured resource service, summed over all resource users, and aggregated over time (reduced to present value).

INTRINSIC VALUE

Intrinsic value has several forms:

Option value: Where future demand for a resource service is uncertain, the injury to the service is at least short-run irreversible, and the prospective user is risk-averse, it is believed that individuals place a value on retaining the option of resource service use. When the option is removed, as a result of water quality degradation, the option value is lost.

Existence value: Persons who never expect to use the services of a natural resource often place a value on the preservation of that resource. The knowledge that water quality has been improved in a major river, for example, provides satisfaction to people who will never see the river or use its services.

Bequest value: Persons who never expect to use the services of a natural resource, and who are personally indifferent to its condition, may believe that it is wrong to impose degraded natural resources on future generations. Such a person may obtain satisfaction from the knowledge that resources are being preserved in usable condition for the future.

Measurement Methods

Measuring benefits or damages associated with environmental quality change requires measuring the values of all affected resource services. Available methods can be divided into two kinds: (1) market-based methods which are based, directly or indirectly, on market transactions involving the resource service itself or a closely related good; and (2) non-market methods which make no assumption about the existence of a market. Most methods must be applied individually to each resource service, and the results summed over all injured services, although some exceptions will be noted.

MARKET-BASED METHODS

Market Price Method. Where the resource service itself is traded in a market (sold at a non-zero price), and supplies are not rationed or allocated except by price, it is possible to measure the value of the service through analysis of market data. To do so requires knowledge of present and future demand functions of consumers, as well as the present and future supply functions for the resource service, both with and without injury.

This method draws on standard microeconomic techniques and provides a direct measurement of use value. It gives no indication of intrinsic value. The market value method is rarely, if ever, applicable to environmental quality issues since affected resource services are not often directly sold in markets. Even if they were, the necessary data (demand and supply functions) would not usually be available.

Appraisal Method. Even where resource services are not traded in markets, or where necessary demand and supply information are not available, it may be possible to secure expert opinion on the value of the resource service, with and without injury. The basis for such an appraisal, however, must be knowledge of comparable market transactions elsewhere. Where no comparable or similar markets exist, the appraisal method would not be likely to be used. For a discussion of the appraisal method and comparison to market value approaches, see, Tittman and Brownell [1984].

Similar to the market value approach, the appraisal method measures only use value, and for a specific resource service.

Factor Income Method. There are cases where a resource service is not itself traded in a market, but is used as an input in a production process to produce a marketable good. Changes in the quantity or quality of the natural resource service input may bring about measurable shifts in the economics of the production process. For example, poor quality water drawn from a nearby river may require a manufacturing plant to undertake special water treatment steps, or to modify its manufacturing process. The net cost to the plant of these steps is a measure of the benefit which water quality improvement would create.

The factor input method measures only use value, and only with respect to specific users (those who use the injured service in a production process). Further discussion of this approach, along with detailed analysis of other methods, can be found in Freeman [1979].

Travel Cost Method. The travel cost method is an example of a benefit measurement technique which relies on analysis of a related market, one for a complementary good. Persons who must travel in order to use the services of specific resources (such as national parks, lakes, unique landscapes, etc.) reveal something of their values by their travel behavior. By observing the amounts spent for travel, and analyzing the frequency of travel by persons in various travel cost categories, the value attributed to the resource can be inferred.

The travel cost method has been used for nearly thirty years in the US to measure values associated with national parks and other outdoor recreation sites. It was first suggested by Clawson [1959], and has been extensively refined [see, e.g., Clawson and Knetsch, 1966; Freeman, 1979; Smith, Desvousges, and McGivney, 1983]. Results are confined to use value, and to the values expressed by actual present day users (future users and nonusers are not considered). Travel cost valuations apply to all services provided at a given site, and are subject to bias when more than one alternative recreational site exists.

Hedonic Price Method. Another approach utilizing a complementary good, (residential housing in most cases) is the hedonic price method. In the US, families frequently seek to live near particularly attractive environmental features, such as a beautiful lake, and to avoid others, such as a polluted river. This behavior is reflected in the market for residential property, inflating prices in the presence of desirable resource services, and depressing prices where environmental quality is degraded. Appropriate econometric analysis of housing prices can provide estimates of the value placed on the resource services at a given site, as compared to other locations.

There is considerable experience with this method for both air and water resource services; it has also been used to determine the value of urban services and civic amenities. Freeman [1979] provides a comprehensive discussion of application to water quality; other significant contributions include Rosen [1974], Polinsky and Rubinfeld [1977], and Smith and Gilbert [1983]. Hedonic price measures apply only to use value, but have the advantage of aggregating all resource services in use in the study area.

Non-Market Methods

CONTINGENT VALUATION METHOD

In the absence of a market for the injured resource service, or for any related good, it is possible to obtain valuations from individuals by creating a contingent market. A survey is conducted in which resource service users are asked how much they would pay to improve the quality or quantity of a resource service, _if_ there were a market requiring them to do so (willingness to pay). Alternatively, they may be asked what compensation they would require to accept a resource service in a degraded condition (willingness to accept).

The contingent valuation method has been applied in the US with increasing frequency in recent years, and has received much attention in the literature (see, e.g., Bohm [1972]; Rowe, d'Arge, and Brookshire [1980]; Mitchell and Carson [1984]; and Cummings, Brookshire, and Schulze [1986]). Various classes of potential bias have been identified, such as hypothetical bias, starting point bias, strategic bias, framing bias, instrument bias, etc. It was originally thought that the method would be invalid where the notion of an actual market in the resource service was not plausible (hypothetical bias), although empirical studies have generally not supported this view.

Unlike other benefit measurement techniques, contingent valuation surveys can be designed to capture all types of benefit, both use and intrinsic values. Furthermore, contingent valuation measures incorporate all affected resource services.

ALTERNATIVE COST METHOD

In some situations, an injured resource service can be completely replaced by providing access to some other service. If ground water has been contaminated, for example, it may be possible to import water from another source to supply all future needs. Where the replacement service is an exact substitute for the injured service, and where no other use of the replacement service is foregone, the lost value attributable to the resource injury is the cost of providing the substitute service.

Although the applicability of this method is limited, it is capable of providing reliable estimates in appropriate applications. The alternative cost method measures use value; since the existence of a substitute service means that injury is not reversible, option value is zero although existence and bequest values may still be present. There is some danger of over-estimating damages, where the cost of replacing a resource service is not justified by its subsequent value to users. This suggests the use of another damage measurement technique in conjunction with the alternative cost calculation.

UNIT DAY VALUE METHOD

The unit day value method has possibly the longest history of any resource service valuation method used in the US. It has been customary to set unit day values for certain kinds of outdoor recreation: e.g., $25 per fisherman-day or $15 per camper day. When a resource service is injured, the injury is expressed as lost fisherman-days, lost camping days, etc. The number of lost days is multiplied by the unit day value to obtain the damage estimate.

This method does not directly assume any market for the resource service or related goods, although unit day values are frequently obtained from studies which utilize one of the market-based methods [Dwyer, Kelly, and Bowes, 1977]. The method is limited to consideration of use values, since only users are counted. Reliability is considered poor, since user day values are unlikely to remain constant over all locations and environmental conditions [Loomis and Sorg, 1982].

Discussion

MARKET-BASED METHODS IN A NON-MARKET ECONOMY

Where data are available, market-based methods are almost universally preferred in the US. Their major deficiency is the inability to address intrinsic values and the requirement (for some) that each affected resource service be analyzed separately. In a society like the US, where most resource allocation is accomplished through decentralized markets, data taken from such markets give the most reliable information on social values.

On the other hand, where most resource allocation occurs as the result of central economic planning, and markets appear only in an ancillary role, such preferences may not be appropriate. Market data will be available much less often and its interpretation, when available, is more difficult. Except for the occasional instance where a travel cost approach might be applicable, it can be assumed that market-based measurement techniques will not be useful in non-market economies.

The Future of Non-Market Methods

Of the three non-market methods now in use, two (the contingent valuation method and the alternative cost method) are generally accepted as reliable, when used in the proper circumstances. In fact, the alternative cost method is almost without critics, provided that the injured resource service can be replaced perfectly and that there is some assurance that the resource service is not worth less than its replacement cost.

The contingent valuation method, the most widely applicable and comprehensive of all methods, is also the most controversial. In spite of repeated comparison studies which demonstrate good agreement between results of contingent valuation studies and measurements made by other methods [e.g., Davis and Knetsch, 1966; Bishop and Heberlein, 1980; Brookshire, Thayer, Schulze, and d'Arge, 1982], the debates go on. Meanwhile, techniques improve as more and more environmental economists develop the skills necessary to conduct these studies.

The possible future of contingent valuation techniques in centrally planned economies is an intriguing research question. Experience so far shows that results are not greatly affected by the relative plausibility of the contingent market. However, both US and European respondents are familiar with broadly similar markets and may be able, through analogy, to make judgments about unfamiliar markets. It would be difficult to know how to interpret responses from persons who have difficulty comprehending the notion of private markets in natural resource services.

Summary and Conclusions

Measurements of benefits on water quality improvements, or corresponding measurements of damages due to water quality degradation, are routinely performed in the US. Various methods are used, including some based on market data (markets for the affected resource service, or for a related good) as well as non-market methods. When data are available, the market-based methods are frequently preferred. Each method has limitations regarding type of application and data required. It is believed that market-based methods could not be generally applied in a centrally planned economy.

However, one of the non-market methods shows some promise. The contingent valuation method is nearly universally applicable to all types of environmental quality issues; it measures both use and intrinsic values; and it can be designed to account for planned future as well as present uses. If it could be adapted to produce reliable measurements of environmental quality improvement benefits within a centrally planned economy, the resulting information could be invaluable in the allocation of scarce funds to water quality-related projects.

References

Bishop, R.C., and T.A. Heberlein, 1980, "Simulated Markets, Hypothetical Markets, and Travel Cost Analysis," Report No. 187, December.

Bohm, P., 1972, "Estimating Demand for Public Goods: An Experiment," **European Economic Review**, vol. 3, no. 2, pp. 111-130.

Brookshire, D.S., M.A. Thayer, W.D. Schulze, and R.C. d'Arge, 1982, "Valuing Public Goods: A Comparison of Survey and Hedonic Approaches," **American Economic Review**, vol. 72, no. 1 (March), pp. 165-177.

Clawson, M., 1959, "Methods of Measuring the Demand for and Value of Outdoor Recreation," reprint no. 10, Washington, DC, Resources for the Future, Inc.

Clawson, M., and J.L. Knetsch, 1966, **Economics of Outdoor Recreation**, Baltimore, MD, Johns Hopkins Press for Resources for the Future.

Cummings, R.G., D.S. Brookshire, and W.D. Schulze, 1986, **Valuing Environmental Goods: An Assessment of the Contingent Valuation Method**, Totowa, NJ, Rowman and Allanheld.

Dwyer, J.F., J.R. Kelly, and M.D. Bowes, 1977, "Improved Procedures for Valuation of the Contribution of Recreation to National Economic Development," Research Report No. 128, Water Resources Research Center, University of Illinois at Champaign-Urbana, IL.

Freeman, A.M., III, 1979, **The Benefits of Environmental Improvement: Theory and Practice**, Baltimore, MD, Johns Hopkins Press for Resources for the Future.

Knetsch, J.L., and R.K. Davis, 1966, "Comparison of Methods for Recreation Evaluation," in A.V. Kneese and S.C. Smith, eds., **Water Research**, Baltimore, MD, Johns Hopkins Press for Resources for the Future, pp. 125-142.

Loomis, J., and C. Sorg, 1982, "A Critical Summary of Empirical Estimates of the Values of Wildlife, Wilderness and General Recreation Related to National Forest Regions," unpublished paper, US Forest Service, Denver, CO.

Mitchell, R.C., and R.T. Carson, 1984, "Willingness to Pay for National Freshwater Quality Improvement," report to US Environmental Protection Agency, Washington, DC, Resources for the Future, Inc., October.

Polinsky, A.M., and D.L. Rubinfeld, 1977, "Property Values and the Benefits of Environmental Improvements: Theory and Measurements," in L. Wingo and A. Evans, eds., **Public Economics and the Quality of Life**, Baltimore, MD, Johns Hopkins Press, pp. 154-180.

Rosen, S., 1974, "Hedonic Prices and Implicit Markets: Product Differentiation in Perfect Competition," **Journal of Political Economy**, vol. 82 (January/February), pp. 34-55.

Rowe, R.D., R.C. d'Arge, and D.S. Brookshire, 1980, "An Experiment on the Economic Value of Visibility," **Journal of Environmental Economics and Management**, vol. 7, no. 1 (March), pp. 1-19.

Smith, V.K., W.H. Desvousges, and M.P. McGivney, 1983, "The Opportunity Cost of Travel Time in Recreation Demand Models," **Land Economics**, vol. 59, no. 3 (August), pp. 259-278.

Smith, V.K., and C.C.S. Gilbert, 1983, "The Valuation of Environmental Risks Using Hedonic Wage Models," in M. David and T. Smeeding, eds., **Horizontal Equity, Uncertainty, and Well Being**, NBER Conference on Income and Wealth Series, Chicago, IL, University of Chicago Press.

Tittman, P.B., and C.E. Brownell, 1984, **Estimating Fair Market Value for Public Rangelands in the Western United States Administered by; USDA-Forest Service and USDI-Bureau of Land Management**, Washington, DC, US Government Printing Office.

US Bureau of the Census, 1985, **Statistical Abstract of the United States: 1986**, (106th ed.), Washington, DC.

WASH LOAD, AUTOSUSPENSIONS AND HYPERCONCENTRATIONS

Lucien M. Brush
The Johns Hopkins University
Department of Geography and Environmental Engineering
Baltimore, Maryland 21218, U.S.A.

ABSTRACT. A review of washload, autosuspensions and hyperconcentrated flow is made to sort out the similarities and differences of these processes or concepts. Both theory and experimental information show that for uniform, steady open-channel flow autosuspension cannot and does not occur. Washload, as defined by Einstein, is a most useful and descriptive concept which helps to portray the actual transport processes which occur in natural channels. Hyperconcentrated flows are characteristic of a relatively small number of rivers in the world, but it too is a very excellent descriptive term and reflects a state of suspension which behaves somewhat differently than ordinary suspensions. Some authors have attempted to relate these three entities even though one doesn't exist and the other two are independent. Washload and hyperconcentrations are different and attempts to relate them should be abandoned.

Introduction

Studies of suspended sediment transport have generated three entities--autosuspension, hyperconcentration, and wash load--all of which have significant meanings to individual authors. Recently attempts have been made to establish some interrelationships among them. For example, Wang (1984) has used the autosuspension concept (although he calls it the "principle of effective power of sediment") to establish the distinction between bed-material load and wash load and applies this to hyperconcentrated flows in the Yellow River of China and the Rio Puerco River of New Mexico, USA. If autosuspension is not a valid concept, as many feel is the case, then the wash load distinction made by Wang is meaningless, and any application to hyperconcentrated flows is misleading. The purpose of this discussion is to examine in depth the significance of these terms and their proper place

L. M. Brush et al. (eds.), Taming the Yellow River: Silt and Floods, 31–44.

in the literature pertaining to suspended sediment transport.

Wash Load

As noted by Einstein (1950) and well described by Einstein and Chien (1953),

> "The 'wash load' refers to the part of the moving particles which is washed through the stream channel without any deposition. It is not found in any significant amount in the bed. The rate at which the 'wash load' is moving through a reach depends only on the rate with which these particles become available in the watershed, and not on the ability of the flow to transport them."

Einstein and Chien go on to point out that some wash load gets to the bed, and that these particles for a short time move in steps just as does the bed-material load. Einstein (1950) goes on to point out that within the suspended load there is both suspension of bed-material load as well as wash load and that the only way of trying to separate the two would be to collect detailed particle size distributions from the suspended load which could then be compared to the size distribution of the bed material. At no point is any hint ever given that the suspended load which may contain wash load follows any different concentration profile than that which is usually found. For convenience, Einstein suggests that the finest 10% of the material by weight found in the bed be considered as wash load in recognition of the fact that some wash load is trapped in the interstices of the larger particles.

It should be noted that material which might be wash load can be trapped in the bed quite easily under ripples or dunes as they advance over these fine particles deposited in the trough of bed forms, see Brush and Brush (1972).

In any event, the wash load concept is an important one and recognizes that many streams can at times carry more wash load than anything else. However, in terms of defining the sizes representing the wash load, only an arbitrary cutoff can be made. Einstein chose 10% but he might well have selected 5% or some other small value. The important point is that there is and cannot be any precise physical criterion for making this cutoff.

The whole concept of wash load is blurred if one chooses a criterion based on instantaneous suspension from the bed as was supported by Nordin (1985). Nordin suggests that wash load should be defined by the following conditons:

$$\frac{w_s}{U_*} \leq 0.8 \text{ for } \Theta \geq \Theta_c \qquad ...(1)$$

where w_s is the settling velocity of the sediment, U_* is the shear velocity, Θ is the Shields parameter, and Θ_c is the critical Shields parameter. This criterion is deficient in that it establishes a quantitative criterion for defining the maximum particle size in the wash load without paying attention to the field conditions and the bed material. The criterion suggested by Wang (1984) is that the sizes in the wash load are those for which

$$w_s \leq VS \qquad ...(2)$$

where V is the mean velocity, and S is the slope of the stream bed. This criterion is equally inappropriate primarily for the same reasons as noted before, but also for some additional technical reasons which will be mentioned later.

If one remembers that wash load is a valid concept not meant to be too precise, then there is no real need to attempt to establish strict criteria based on fluid mechanics. Rather, it seems sufficient to use Einstein's (1950) method of using smallest 10% of the bed material to define wash load.

Autosuspension

The claim that under certain circumstances sediment adds more energy to a stream than is required to suspend the same sediment is the backbone of an hypothesis called autosuspension. Apparently originated by Knapp (1938), rediscovered by Bagnold (1962), and nurtured by many others including Nordin (1963) and Wang (1984), the concept is basically as follows:

The net power per unit volume required to hold sediment in suspension is

$$(\rho_s - \rho) g C w_s$$

where ρ_s is density of the sediment, ρ is the fluid density, g is the acceleration due to gravity, and C is the sediment concentration. The net power transferred to the stream by the additional sediment is

$$(\rho_s - \rho) g C V \sin \alpha \approx (\rho_s - \rho) g C V S \qquad ...(3)$$

where V is the mean velocity and S is the slope. Therefore, the net power per unit volume contributed by the fluid for suspension is stated to be

$$(\rho_s - \rho)g \quad (w_s - VS)$$

According to Bagnold (1962) and Knapp (1938), if

$$w_s \leq V S \qquad \text{...(2)}$$

no net energy expenditure is required for suspension and the sediment concentration in the stream may increase indefinitely provided material in these sizes is available for transport. No recognition is given to the fact that there is also a net energy loss due to flow resistance caused by the presence of the sediment.

A number of researchers disagree with this notion of autosuspension either explicitly or implicitly. However Nordin (1985) supports the concept and ascribes much of the original notion (with a noted correction) to Gilbert (1914). While Gilbert (1914) recognizes the two terms noted in equation 2, he recognizes that there is also a "viscosity factor" which he was unable to fully decipher.

Rubey (1933) in his analysis while not writing directly to autosuspension formulates an equation which implicitly makes equation 2 invalid. His equation, while incorporating the basic terms leading to equation 2, contains an additional term that recognizes resistance to the fluid motion and one which also contains an additional component involving the sediment-fluid mixture. This is important for it suggests that the components leading to equation 2 are incomplete.

Even more important is the experiment by Southard and Mackintosh (1981), who showed conclusively that autosuspension does not hold for suspensions flowing in pipes. They also felt that their results are equally applicable to open channel flow. Parker (1982) in a discussion of this paper agrees in principle with their findings, and attempted to write his own energy balance to show this. In a new paper, but really a response to the discussion by Parker (1982) and by Pantin (1982), Paola and Southard (1983) present a very detailed energy analysis to justify their position and to show very convincingly that Knapp and Bagnold, and therefore Nordin, are incorrect.

While no one has performed a specific experiment to test the autosuspension hypothesis for open channel flow, there are data in the literature which are applicable. The classic work of Kalinske and Hsia (1945) may be used to test the hypothesis of autosuspension. In fact, if one wished to test autosuspension it would most likely be an experiment conducted like that of Kalinske and Hsia. The experiments were conducted in a recirculating flume. Eleven runs were made, nine of which had suspended loads. Ground silica having a median diameter of 0.011 mm (S.G. = 2.67) was used for the experiments. In addition to the mean flow data shown in Table I, measurements were also made of:

TABLE 1. RUN DATA FROM KALINSKE AND HSIA (1945) WITH DATA CONVERTED TO THE METRIC SYSTEM

Run	Slope	Depth cm	Discharge l/s	Velocity cm/s	Shear Velocity cm/s	K	Concen-tration % Wt.	Bed State
1	.0005	9.3	28.6	44.8	1.89	.41	0	steel
2	.0005	15.4	58.0	55.2	2.29	.43	0	steel
3	.00025	11.3	19.8	25.6	1.40	.40	.64	ripples
4	.00025	14.9	35.7	34.8	1.65	.34	1.29	ripples
5	.00025	20.1	55.2	39.9	1.86	.32	1.67	ripples
6	.0005	11.0	28.6	38.1	2.10	.40	1.95	rugged
7	.0005	15.9	51.3	46.9	2.44	.33	2.24	and
8	.0005	19.5	73.6	55.5	2.56	.40	2.27	irregular
9	.0010	10.7	37.1	51.2	2.99	.40	3.36	"
10	.0010	17.1	76.5	65.2	3.81	.34	6.81	"
11	.0013	15.9	90.6	83.2	3.90	.40	11.10	smooth

1. The velocity distributions for each run.
2. The concentration distributions for each run.
3. The particle size distributions for samples of the suspended load as well as for the bed material for each run.

For the purposes of examining autosuspension, the data have been arranged slightly differently in Table II. As can be seen for each run with sediment, VS is greater than w_s (based on d_{50} of the suspended load) indicating that in each run "autosuspension" should exist in the eyes of autosuspension proponents. This means that if material with a fall velocity equal to or less than VS, the concentration should continue to rise until the supply of these sizes is exhausted. On Figure 1, a plot of the total concentration versus shear velocity is shown with the percentage of material available for autosuspension remaining in the bed and in the suspended load noted for each point. Obviously there is abundant material left in the bed to continue to feed the "autosuspension" unless some other mechanism is preventing this from happening. Each run came to equilibrium indicating that, despite the availability of autosuspension sizes in the mobile bed, they remained there. Also the return flow system was designed so that no sediment could be stored there to inhibit autosuspension. Ripples were formed in three runs, irregular and rugged forms in five runs and a plane bed occurred for the run with the greatest flow. The moving bed forms continually exposed materials of various sizes in the bed and gave these sizes the opportunity of becoming suspended. As continually increasing suspension did not occur it must be concluded that autosuspension as defined by Knapp (1938) and Bagnold (1962) and others did not occur during these experiments. There is every evidence to support the conclusion that the same results would apply to natural streams and rivers. Furthermore, nothing unusual occurred during these runs. The velocity distributions were logarithmic, the Karman K ranged from .32 to .44, although unlike other experiments, no systematic change in K occurred with increased concentrations. The concentration distributions followed the Rouse-type relationship and because of the small sizes involved and the low Z values (where $Z = 2.5\ w_s/U_*$), the sediment was nearly uniformly distributed in the vertical.

If one were to use the definition of wash load proposed by Wang (1984), using the autosuspension criterion

$$w_s \leq VS \qquad \ldots(2)$$

at least 81 to 98 percent of the material in suspension, depending on the run, would be classified by Wang as wash load, which seems out of accord with the fact that all of the

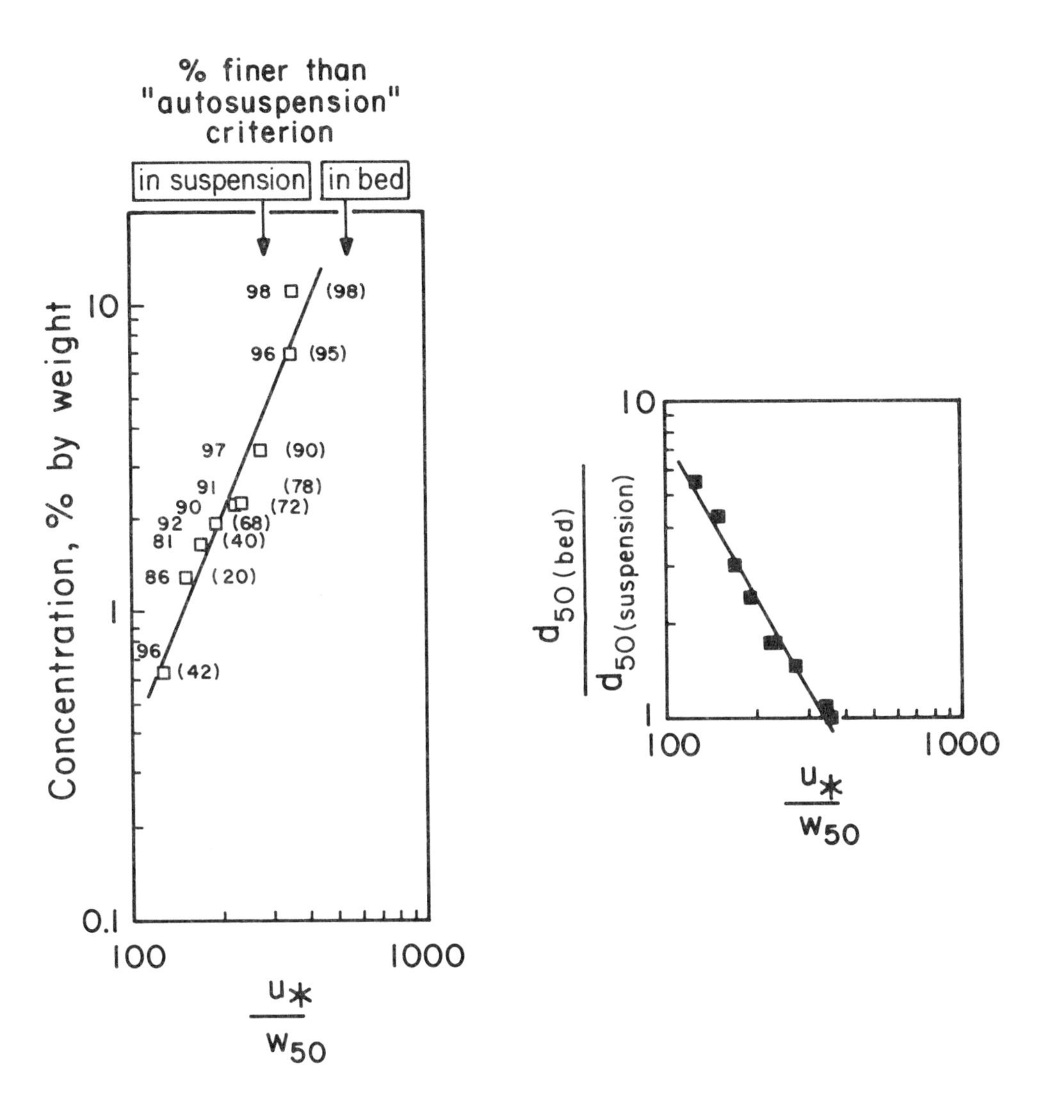

Data from Kalinske & Hsia (1945)

TABLE 2. DATA COMPUTED FROM RUN DATA OR READ FROM GRAPHS IN KALINSKE AND HSIA (1945)

Run	Bed Material, d_{50} mm	Suspended Load, d_{50} mm	Velocity slope product cm/s	Equivalent particle size[1] mm	% Mat. finer in suspension %	% Mat. finer in bed %	Shear velocity-fall velocity ratio
1	--	--	.0224	--	--	--	--
2	--	--	.0276	--	--	--	--
3	.010	.0017	.0064	.0084	96	42	127
4	.014	.0034	.0087	.0098	86	20	150
5	.0125	.0043	.0100	.0107	81	40	169
6	.0115	.0050	.0191	.0147	92	68	191
7	.010	.0061	.0235	.0162	90	72	222
8	.010	.0061	.0275	.0175	91	78	233
9	.010	.0069	.0512	.0240	97	90	272
10	.010	.0091	.0652	.0270	96	95	346
11	.012	.0120	.1085	.0350	98	98	355

[1] Equivalent size determined from settling velocity graph for temperature of 20°C. (U.S. D.A., S.C.S., California Institute of Technology)

material moving either as bed load or suspended load was derived from an identical source of bed material. The study by Kalinske and Hsia (1945) does not support the hypothesis of autosuspension and therefore the attempt by Wang (1984) to relate wash load to autosuspension also reinforces the conceptual framework so ably set up by Einstein (1950) with regard to the bed material load and wash load.

Without rewriting the energy balances as may be found in Paola and Southard (1983) or Hinze (1963), an attempt will be made to reexamine the energy relationships which occur in a stream with a suspension and to give even more evidence that autosuspension is an hypothesis which has no merit for suspensions in steady, uniform flow in open channels.

As noted by Hinze (1975, p. 68),

> "...an average steady state can only exist, if there is equilibrium between the energy supplied to the turbulent motion, the transport by the mean motion, and the diffusion plus dissipation of turbulent energy."

In arriving at the total mechanical energy which is the sum of mean mechanical energy and the turbulent mechanical energy, each of which has its own balance, a number of approaches may be taken. One was given by Hinze (1975) and another more recently by Paola and Southard (1983). The most important point is that one must be careful to place the correct terms in the appropriate energy balance. This has not always been the case as some authors have either omitted a term or mixed up terms for use in the mean and turbulent mechanical energy balances. For the total mechanical energy, Rubey (1933) states in words the terms which should occur. He says that,

> "...in any unit of time the loss of potential energy of water and debris in flowing down hill, E, plus the original kinetic energy K_1 = the energy consumed in friction, F, plus the energy consumed in supporting the debris, D, plus the subsequent kinetic energy, K_2."

Of course for uniform steady flows the kinetic energy terms cancel.

Paola and Southard (1983) look separately at the mean kinetic energy and the turbulent kinetic energy. For the mean kinetic energy, they say that the energy is added to the mean flow through gravity and is equal to the energy lost in working against the Reynolds shear stress (assuming work against the viscous shear stress is negligible). For the turbulent kinetic energy they show that the energy transferred from the mean motion to turbulence (as noted above) is mostly dissipated to heat and the small part that is left over is used to work against the concentration gradient. Their total energy relationship should be somewhat equivalent to that suggested by Rubey (1933), and to Parker (1982) in his response to Southard and Mackintosh (1981). In contrast, Knapp (1938), Bagnold (1962), and Wang (1984) omit recognition of any friction or turbulent dissipation in the problem. Therefore, they obtain equations which do not represent balances of either mean flow energy, total flow energy, or turbulent flow energy.

A few additional comments seem appropriate pertaining to the lack of autosuspensions in steady uniform flow in open channels. As noted previously, the basic criterion for autosuspension is

$$w_s \leq V\ S \qquad \text{...(2)}$$

However, in developing this criterion, Knapp (1938), Bagnold (1962) and Wang (1984) really start with a criterion such as

$$w_s \leq u\ S \qquad \text{...(4)}$$

where u is the local velocity in the downstream direction. This means that it is possible for one grain size, reflected by w_s, to be in "autosuspension" in the upper part of the vertical profile and not in the lower part. For example, all of Kalinske and Hsia's (1945) sediment-laden runs show this. In particular, for run 6, VS = 0.0191 cm/sec, which would be equivalent to a particle size of about 0.0147 mm. A velocity distribution for this run shows that a uS of this magnitude would occur at a relative depth of about 0.31. Presumably material of this size occuring above this depth would be in autosuspension and that below, not. Other runs and other sizes show the same trends, but the resulting concentration profiles are in accord with previous studies such as Vanoni (1946). Drawing from another source, Laursen (1958) ran experiments for total load and some of his data are pertinent. For example, run 106 has a uS product of 0.034 cm/sec at about mid depth. This is equivalent to a diameter of about 0.0195 mm, which from the particle size analysis of the sediment occurs at about d_{20}. In other words, 20% of the material present in the bed was this size or smaller. Again, the concentration distribution followed the usual curve with a Z = 0.098. This example was chosen because the criterion for autosuspension of this size occurred at mid depth. Therefore with turbulent mixing causing eddies to move both up and down past the midpoint, a buildup in concentration in the upper level would be expected as long as a source existed at the bed. Obviously this did not happen. The important point is that the eddies which are transferring sediment randomly throughout the water column do not know about the autosuspension criterion but merely move sediment down the concentration gradient (upward in space) as sediment settles downward back through the water column. This continues until a balance is struck, maintaining an average concentration gradient which is in accord with the turbulent diffusion strength and settling of the particles.

It is apparent through theory as well as experiment both locally and for the average stream flow parameters, autosuspension as described does not exist.

Hyperconcentrated Loads

Under the current definition (National Resource Council, 1982), hyperconcentrated flows are roughly subdivided into:

Water floods	$\leq$ 40% by weight
Mud floods	40 to 67% by weight
Mud flows	43 to 50% by volume
Landslides	$\geq$ 50% by volume

Under certain circumstances the middle Yellow River sometimes would be classified as a "mud flood." Even so, the division between river floods and mud floods is arbitrary and it would be helpful to have a definition based on some valid physical principle. Nevertheless, the usual definition of hyperconcentration is a concentration by weight of about 40%, see Beverage and Culbertson, 1964.

Some of the apparent properties of hyperconcentrated flows include logarithmic velocity distributions with the K's decreasing with concentration but then recovering to about .4 or larger, with concentrations of 800 to 900 Kg/m^3 or more, as noted by Qian, Zhang, Wang and Wan, 1985. In addition, it was noted that the concentration distributions obeyed the usual diffusion theory, although the Z's for the coarser particles were quite small. Long and Qian (1986) note similar relationships for Z and state that a certain amount of fine silt and clay is indispensible for hyperconcentrations. It is under these conditions that the fine particles are believed to form a floc structure which will reduce the settling velocities of the coarser particles, and thereby reduce the Z values. Others including Qi and Zhao (1985) discuss the non-Newtonian aspects of hyperconcentrated flows and use a Bingham-type model to explain its rheological behavior.

Different observers describe the qualitative nature of river flows with hyperconcentrations somewhat differently, but most agree that a turbulent structure remains in order to suspend the sediment even though at some places the water surface is more placid than is expected in a large turbulent river.

Only Wang (1984) suggests that his "potential energy velocity" which is really a surrogate for Knapp's "auto suspension" is operative. As noted previously, "autosuspension" is not a valid concept for open channels. While it may well be a fact that a great deal of the material passing through the middle reaches of the Yellow River is actually wash load, it is wash load in the Einstein sense and not generated through autosuspension.

Summary

Wash load, autosuspension and hyperconcentrations are entities which attempt to describe certain aspects of suspended load in open-channel flow. Wash load as defined by Einstein both quantitatively (the finest 10% of the bed material) and qualitatively in the sense that it recognizes that certain fine material is often carried through a reach without being controlled by the processes which govern the bed-material load even though the concentration distributions in the vertical obey the physical laws governing turbulent dispersion of small particles. Hyperconcentrations of sediment do occur in some rivers such as the Yellow, and the effects are dramatic. The rheological aspects of hyperconcentrated suspensions cause them to behave differently than ordinary suspensions, but many of the traces of ordinary suspensions remain. Velocity distributions remain logarithmic in nature although possessing different constants. Concentration profiles obey the turbulent dispersion laws although the forms are altered by changes in the fall velocities of particles carried in the suspension. Nevertheless, hyperconcentrations are a fact of life in certain rivers and cause unusual erosional and depositional processes to occur. Autosuspensions, on the other hand, in open channel flows are a myth. Both theory and experiments have conclusively shown that the supposed criterion for generating autosuspensions in open channel flows as well as in pipe flows is incorrect. This does not preclude that forms of autosuspension may occur in turbidity currents, avalanches, or certain types of landslides, but the processes governing the onset of such flows is not generally applicable to steady uniform flow in open channels. Researchers should be aware of the concepts of wash load and hyperconcentrated flows, but should be careful not to suggest that certain aspects of these flows are governed by the use of an hypothesis of autosuspension. The presence of wash load is a fact, hyperconcentrated flow is a fact, but autosuspensions in open channel flows are fictitious having arisen from a fallacious hypothesis.

References

Bagnold, R.A., 1962, Autosuspension of transported sediment; turbidity currents, Proc. Roy. Soc. A, V. 265, pp. 315-319.

Beverage, J.P., and Culbertson, J.K., 1964,

Hyperconcentrations of suspended sediment, Proc. A.S.C.E., Jour. Hydraulics Div., V. 90, HY6, pp. 117-128.

Brush, G.S. and Brush, L.M., Jr., 1972, Transport of pollen in a sediment-laden channel: a laboratory study, Amer. Jour. Sci., V. 272, pp. 359-381.

Einstein, H.A., 1950, The bed-load function for sediment transportation in open channel flows, Tech. Bull. No. 1026, U.S. Dept. Agriculture, 77 pp.

Einstein, H.A., and Chien, N., 1953, Transport of sediment mixtures with large ranges of grain sizes, MRD. Sediment Series No. 2, U. California, Berkeley, and Missouri River Div., U.S. Army Corps of Engineers, 44 pp.

Gilbert, G.K., 1914, The transportation of debris by running water, U.S. Geol. Surv. Prof. Paper No. 86, 263 pp.

Hinze, J.O., 1975, Turbulence, McGraw-Hill, 2nd ed., 790 pp.

Hinze, J.O., 1963, Momentum and mechanical-energy balance equations for a flowing homogeneous suspension with slip between the two phases, Applied Sci. Res., V. 11, pp. 33-46.

Kalinske, A.A., and Hsia, C.H., 1945, Study of transportation of fine sediments by flowing water, Univ. of Iowa Studies in Engineering, Bull. No. 29, 30 pp.

Knapp, R.T., 1938, Energy-balance in stream-flows carrying suspended load, Trans. Amer. Geophys. Union, pp. 501-505.

Laursen, E.M., 1958, The total sediment load of streams, Proc. Amer. Soc. Civil Engrs., V. 84, No. HY, pp. 1530-1 to 1530-36.

Long, Y. and Qian, N., 1986, Erosion and transportation of sediment in the Yellow River basin, Internat. Jour. Sediment Research, V. 1, No. 1, pp. 2-38.

National Resource Council, 1982, Selecting a methodology for delineating mudslide hazard areas for the National Flood Insurance Program, Committee on Methodologies for Predicting Mudflow Areas, National Academy Press, Wash., D.C.

Nordin, C.F., Jr., 1985, Discussion of "The principle and application of sediment effective power," Jour. of Hydraulic Engineering, Amer. Soc. Civil Engrs., V. 111, No. 6, pp. 1023-1024.

Pantin, H.M., 1982, Comments on: Experimental test of autosuspension, Earth Surface Processes and Landforms, V. 7, pp. 503-505.

Paola, C., and Southard, J.B., 1983, Autosuspension and the energetics of two-phase flows: reply to comments on "Experimental test of autosuspension," Earth Surface Processes and Landforms, V. 8, pp. 273-279.

Parker, G., 1982, Discussion of: Experimental Test of Autosuspension, Earth Surface Processes and Landforms, V. 7, pp. 507-510.

Qi, P., and Zhao, Z., 1985, The characteristics of sediment transport and problems of bed formation by flood with hyperconcentration of sediment in the Yellow River, Internat. Workshop on Flow at Hyperconcentrations of Sediment, Beijing, pp. III 3-1 to III 3-16.

Qian, N., Zhang, R., Wang, X., and Wan, Z., 1985, The hyperconcentrated flow in the main stem and tributaries of the Yellow River, Internat. Workshop on Flow at Hyperconcentrations of Sediment, Beijing, pp. III 4-1 to III 4-18.

Rubey, W.W., 1933, Equilibrium-Conditions in Debris-Laden Streams, Trans. Amer. Geophys. Union, pp. 497-505.

Southard, J.B. and Mackintosh, M.E., 1981, Experimental test of autosuspension, Earth Surface Processes and Landforms, V. 6, pp. 103-111.

Vanoni, V.A., 1946, Transportation of suspended sediment by water, Trans. Amer. Soc. of Civil Engrs., V. 111, pp. 67-102.

Wang, S., 1984, The principle and application of sediment effective power, Jour. of Hydraulic Engineering, Amer. Soc. Civil Engrs., V. 110, No. 2, pp. 97-107.

AN OUTLINE FOR PLANNING DEVELOPMENT OF RIVER BASINS: A REGIONAL APPROACH

Nathan Buras, Ph.D.
Professor and Head, Department of Hydrology and Water Resources
University of Arizona, Tucson, AZ 85721, U.S.A.

ABSTRACT. River basin development programs are complex frameworks, in which every project is required to be formulated as a sub-system of the larger plan. Some of the major issues encountered when planning a river basin scheme on a regional basis are the utilization of surface flows and groundwater resources (including the maintenance of a desired level of water quality), the determination of alternative cropping patterns in irrigated agriculture, evaluation of tradeoffs between competing uses of water, and sequencing of investments in both space and time. The proposed outline will probably be useful when considering the various projects in the Yellow River basin, so that they may fit an overall plan. The methodology was adapted from previous work on river basin development performed for the World Bank in South Asia.

Planning Principles

INTRODUCTION

Water resources development planning involves the identification and evaluation of alternatives designed to meet the short-term and the long-term needs of society in a region. If the emphasis is placed on the natural hydrologic phenomena which are to be utilized for the satisfaction of these needs, then the region is congruent with a river basin. In arid regions, groundwater basins may also define the boundaries of a region; in the humid tropics, the river basin seems to be an adequate geographic unit for planning purposes.

Planning the development of river basins on a regional basis has been attempted since early in this century: the negotiations between the States within the Colorado River basin which culminated in the Colorado River Compact of 1922 represent such an early attempt. Some fifteen years later,

L. M. Brush et al. (eds.), Taming the Yellow River: Silt and Floods, 45–72.

the Tennessee Valley Authority took one step further in planning the utilization and management of streamflows in the Tennessee River basin. Other parts of the world tried to emulate the TVA with various degrees of success, and from this mixed record, emerged the need for a methodology which would be adequate for the complexities of river basin planning. Perhaps the first such methodological approach was made under the auspices of the United Nations [ECAFE, 1972], with special reference to the humid tropics of South Asia and the Far East.

The decade of the seventies has seen a major effort in the introduction and adaptation of the systems approach into the water resources planning process which resulted, among others, in a proliferation of scientific and professional papers and reports, some of which were used in actual development projects. A number of systems engineering methods were applied in the planning of Rio Colorado in Argentina [Major and Lenton, 1979]. However, notwithstanding the high caliber of this river basin plan, the project was not implemented, probably due to the existing priorities for the public investment in that country.

More recently, a further step was taken in the search for improving regional development: river basin planning was viewed as a socio-ecological megasystem [Saha and Barrow, 1981]. According to this view, the river basin megasystem is composed of three main systems and several subsystems as shown in Figure 1. An important feature, shown in this diagram, is the interaction identified between the various subsystems.

Planning the development of river basins is an art based on scientific principles. Some fundamental concepts are derived from natural and social sciences such as hydrology, applied mathematics, economics, public administration, and others. The practice involves more than the mastery of these concepts--which is a necessary prerequisite--and requires also an understanding of the interactions between these factors as well as a sensitivity to the regional socio-economic environment.

DEVELOPMENT OBJECTIVES

One should distinguish between two categories of development objectives: (a) socio-economic goals; and (b) technological-engineering targets. One could view category (b) as a subset of group (a).

Socio-economic goals. A river basin development plan makes sense only insofar as it helps in attaining regional socio-economic goals, or is used as an instrument for implementing a regional socio-economic policy. Two major goals can be identified in this category:

(a) Maximization of economic efficiency, meaning that

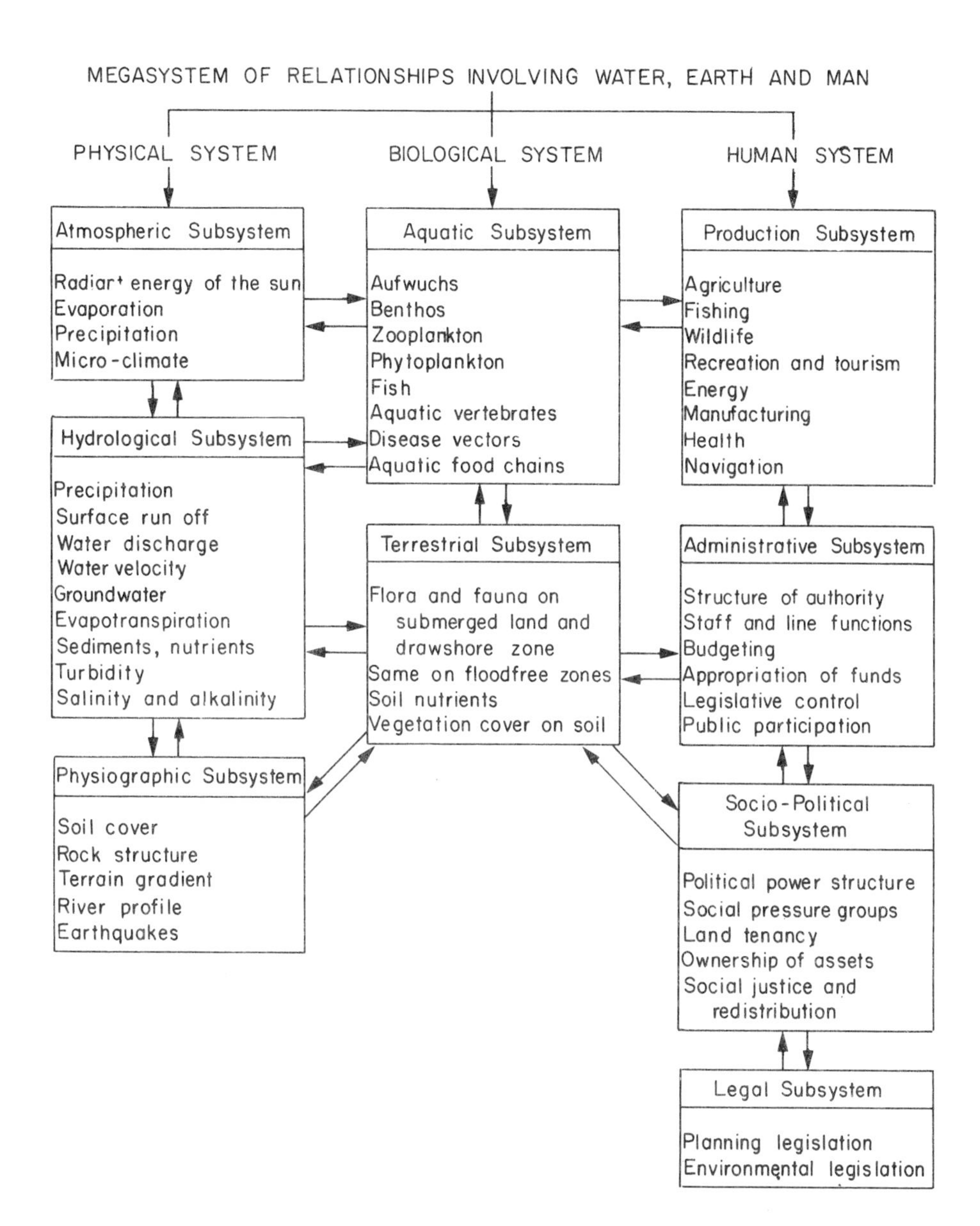

Figure 1

SOCIO-ECOLOGICAL MEGASYSTEM OF RIVER BASIN PLANNING

Source: Saha and Barrow, 1981

the river basin is developed so that overall benefits are increased without any one individual being worse off than before. Observe that this objective may favor one group in the region, so that although the lowest income social stratum does not lose anything, the disparity between incomes may increase social tensions.

(b) Redistribution of regional income. The benefits received by various groups in a region may be given different weights in the objective function, so that the economic status of specific social strata will be enhanced. Of course, one could use goal (b) in the objective function, subject to economic efficiency, or conversely, maximize economic efficiency subject to a specified raise in the income level of given water-user groups.

Technological-engineering targets.

(a) Control of moisture in the root zone of soils. In humid tropics it would be especially misleading to consider irrigation alone as the hydraulic infrastructure necessary for river basin development. Supplementary irrigation may be required in many areas of a river basin, but drainage--surface and sub-surface-- and mitigation of flooding and inundation damages are often more important.

For planning agricultural development, it is convenient to divide a river basin into three zones:

(i) The upper zone usually has light-textured and well-drained soils. Irrigation may be needed for special crops, such as paddy rice, if at all.

(ii) The middle zone is covered mostly by medium-textured and deep soils. Supplementary irrigation may be necessary to increase yields of crops, and much of it could be probably based on groundwater development (see, for example, Los Naranjos irrigation scheme in the Papaloapan basin, Mexico).

(iii) The lower zone usually has deep, fine-textured soils, difficult to manage, and which need mostly drainage. Low-lying lands have to be protected against flooding and inundations, protection which could be attained by regulating reservoirs in the upper part of the river basin, by levees in the protected area, or by a combination of both. One has to emphasize that no combination of reservoirs and levees will handle all floods: The flood containment measures have to be designed by balancing accepted risks and costs.

(b) Hydropower generation. The objective of hydropower development is to maximize net revenue produced by the generated electricity. However, in order to attain his broadly stated objective, a number of details must be clarified on the basis of comprehensive studies and analyses. For example:

(i) Projections of power demands to be satisfied by electricity generated by the potential hydroelectric plants should be made both within the river basin and for export requirements. Base load capacities and peaking needs should be identified.

(ii) Combined operation of reservoirs should be studied so as to determine installed capacities which would maximize power generation, given certain diurnal, seasonal, and yearly demand patterns. Sizes of dams, reservoirs and power plants should be established.

(iii) Consideration should be given to possible multiple uses of water. Even in the case of single-purpose projects, water which generates electricity may be used subsequently elsewhere downstream for a variety of purposes, particularly for irrigation. Hence, operation of a hydropower project will affect, beneficially or otherwise, other water users in the river basin. Trade-offs between power and irrigation should be identified.

(iv) Some of the power generated in the basin will have to be used by some of the major subsystems of the region itself, primarily by the groundwater subsystem and by transbasin diversions, if any. These aspects have to be evaluated before formulating the objectives of hydropower development.

(c) Water for energy. Future economic growth in a river basin will probably induce further development of the energy sector. The impact of the energy sector on the planning of a river basin development could be identified at least in two directions.

(i) The construction of new thermal power plants--coal-fired or nuclear--and the replacement of old plants will require substantial amounts of cooling water. Areal distribution, as well as time patterns, of these demands will have to be established.

(ii) If considerable coal or other energy resources exist, probably they will be used at an increased rate for power generation. The coal will be transported over long distances, both within the river basin and also exported. One mode of transportation is by barge on the river itself. An alternative mode is via coal-slurry pipelines. Both water-related alternatives will have to be considered in detail before formulating a river basin development project.

THE SYSTEMS APPROACH

Systems analysis. The planning of a river basin development project, the design of its components, the construction of the major structures, the operation of its subsystems and the monitoring of its performance needs methods and techniques which will ensure the attainment of maximum economic and social objectives at the least cost. This methodology is termed "systems analysis" and it involves the integration of different perspectives (engineering, economic, agricultural, social, ecological, etc.) into one strategy for the solution of complex issues.

Systems analysis is a methodological approach designed to assist decision-makers choose the "best" (defined in some sense) alternative course of action. Whether the issues are related to design of structures, to operation of subsystems, or to implementation of projects, alternatives have to be identified and evaluated in order to reach rational policy decisions. The systems approach includes some or all of the following steps.

(a) Identification of a problem or of a group of interrelated problems which need clarification leading to alternative solutions.

(b) Definition of the problem boundaries so that the problem is analytically tractable, yet the simplification does not transform it into a trivial issue.

(c) Specification of goals, objectives and targets to be attained by the solution of the problem. In more complex settings, there may be a hierarchy of objectives, and/or objectives which are noncommensurate (multi-objective decision analysis).

(d) Decomposition of the problem into sub-problems and the analysis of components and their interrelations, including the uncertainties involved. This step usually entails the use of mathematical formulations of various complexity.

(e) Synthesis of alternative solutions, their evaluation and the selection of the optimal solution. This step also involves the formulation of mathematical models, especially those belonging to decision sciences (operation research, mathematical programming).

(f) Implementation of accepted solution(s), including the timely scheduling of construction.

(g) Monitoring of system performance, adjusting operating policies, and/or modifying system components.

Specifically, a planning problem such as a river basin development requires analysis, clarification and resolution of the following issues, among others.

(a) Dams: What should be the capacity of the various reservoirs? Are all reservoirs needed? Should additional storage facilities be constructed? What should be the installed capacity of hydro-power plants?

(b) Water resources: Should some of the project areas be supplied by groundwater rather than by surface flows? How should the considerable groundwater resources be integrated with the surface system?

(c) Water allocation: How much water should be allocated for irrigation and for other uses in different areas, and at what times during the year?

(d) Operation: How should the system of reservoirs be operated to maximize irrigation and power net benefits, given the stochastic nature of streamflows?

(e) Performance: What standards and criteria should be used for the evaluation of the system performance?

(f) Reliability: How should individual canal systems, if any, be operated to ensure a timely and reliable water supply with minimum water losses?

(g) Scheduling: When should the different project components be built to ensure that benefits will begin accruing as early as possible? What impact does the timing of different projects have on their design?

The analysis of these issues involves mathematical models of several kinds [Loucks, 1981]. Thus issues (a), (b) and (c) require screening (optimization) models; issue (d), (e) and (f) are analyzed primarily through simulation; and issue (g) is formulated as a sequencing (again, optimization) model. The systems approach to planning is shown in Figure 2.

Sizing of major structures. A surface reservoir created by a dam should be viewed in a manner similar to a production process: a set of inputs is used by a system to generate one or more outputs. In the case of reservoirs, the inputs (land, cement, labor, steel, etc.) can be commonly expressed in monetary terms, while the output is expressed as regulated flow (Maf/yr), with a given probability. Thus the reservoir equivalent of the industrial production function is the yield-cost relationship. This relationship is specific for each dam site. However, in a multi-reservoir water resources system, the hydrological interdependence of the various structures has to be considered in order to evaluate accurately the benefits generated at the various dams, on the basis of the yield-cost functions. This analysis is carried out by means of screening models.

A screening model, in its simplest form, uses relatively simplified descriptions of project components (such as the yield-cost functions for dams) and average seasonal streamflows. The purpose of the model is to evaluate the economic viability of various project configurations and produce initial estimates of reservoir capacities, power plant installed capacities, sizes of main canals and operating

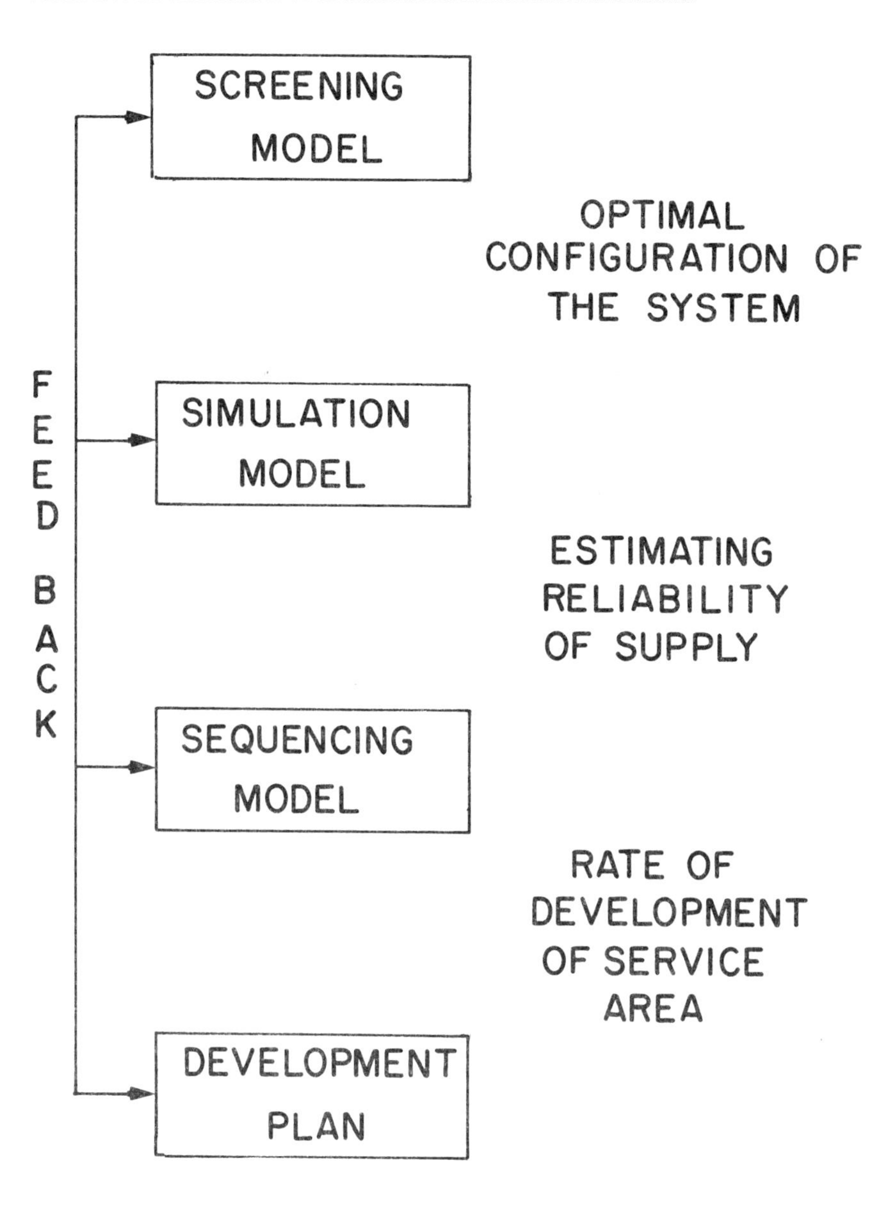

Figure 2

THE SYSTEMS APPROACH TO PLANNING

policies. The project configuration should represent quantitatively the hydrological interdependence of reservoirs (mentioned in the previous paragraph) in the sense that upstream dams usually increase the regulated flow produced by downstream reservoirs, thus augmenting water availability for irrigation, for power generation and for other uses. Consequently, the analysis relevant to the sizing of major structures should use models which optimize simultaneously all reservoirs.

The data required for such an analysis include (a) hydrological data at each proposed dam site; (b) cost-capacity relationships for every reservoir; (c) estimates of power output at different reservoir levels for every projected hydropower installation; (d) initial assessments of irrigable lands; (e) estimated value of irrigation water and power output at different times of the year; (f) groundwater resources and their potential integration with the surface system.

A necessary prerequisite to the sizing of major structures in a river basin is a hydrological study leading to estimates of streamflows at each proposed reservoir site. This study should begin as soon as possible in the planning process and should be based on current studies, if such studies exist.

The full screening model for sizing the major structures of a river basin requires a great deal more information. The data required are sketched in Figure 3. A partial list of studies for the aquisition of this information appears in the Appendix.

<u>Operation of a multi-reservoir system</u>. The ability of a river basin authority to utilize its waters to the best advantage depends not only on the optimal capacity of the major reservoirs but also on the operating policies for these structures.

Simulation models are required for deriving operating policies of multi-reservoir systems, for evaluating their performance, and for estimating the reliability of water supply for irrigation, for hydropower generation, and for other uses. They differ from screening models not only by the fact that the latter are optimizing formulations and the former are not, but also by the type of hydrological information used as input. More specifically, the stochastic aspect of hydrology is handled differently in screening and simulation models.

The statistical variability of streamflows is treated in screening models by means of parameters such as mean, variance, and serial correlation (i.e., the probabilistic relationship between streamflows in serial, or adjacent, seasons). The performance of water resources systems depends, however, not only on these parameters, but also on sequences of abundant (or poor) years. Simulation models use as one of

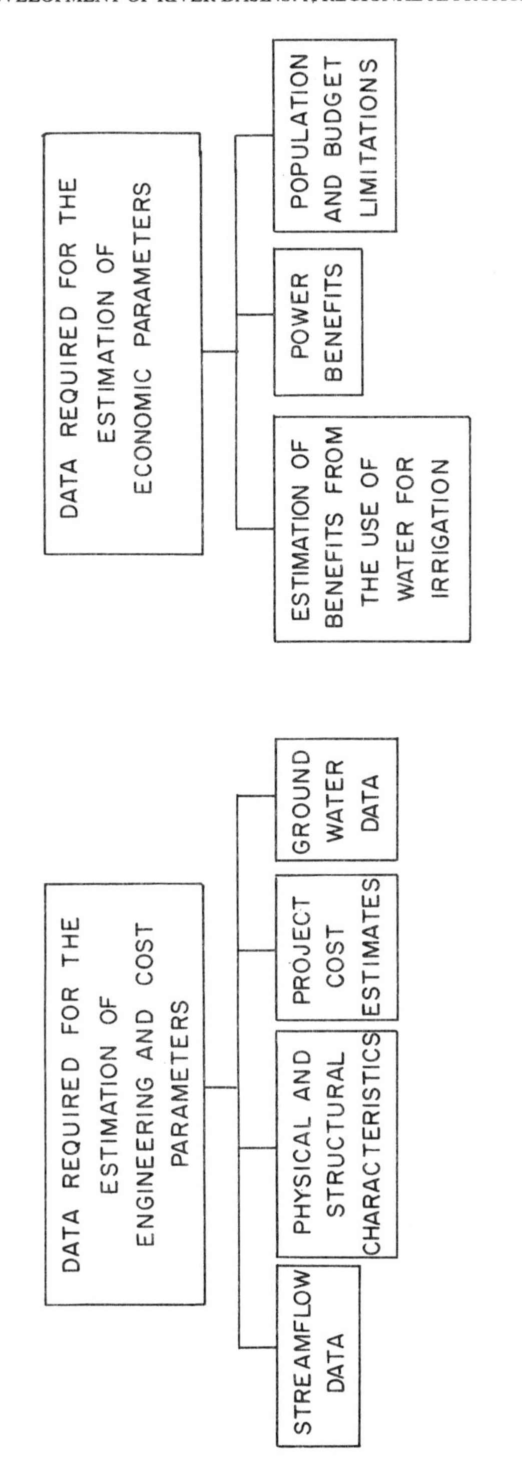

Figure 3

DATA REQUIRED BY THE SYSTEMS APPROACH TO THE PLANNING PROCESS

the inputs this type of hydrological information, i.e., seasonal flows which vary from year to year. Since the future sequences of abundant and poor years will be certainly different from those in the past, it is necessary to generate series of numbers which will represent equally likely hydrological sequences. These series are called synthetic (or operational) hydrology; their derivation, based on mathematical-statistical models representing the streamflow phenomena, should begin as soon as the planning activities are initiated.

The formulation of the simulation model should start when some of the information required by the basin-wide screening model has been collected. Meanwhile, the synthetic (operational) hydrology study should also be in progress. It is estimated, therefore, that production runs of the simulation model can be obtained as early as eighteen months after the start of the hydrological studies.

The simulation runs will yield estimates of how often a multi-reservoir system will not meet specified demands, what would be the magnitude of shortages, and how long would they last under given operating policies. Similarly, estimates will be obtained of expected surpluses and their magnitudes. These results will be used to evaluate parameters in the screening model and adjust them, if necessary. It is envisaged that two or three iterations of screening-simulation would be sufficient in order to reach conclusions on which the design of some key components can be based.

Timing and sequencing of investments. Generally, projects that provide the highest social and economic benefits should be built before others. However, due to hydrological interdependence, the benefits generated by one project will depend on which of the other projects have already been completed or will be completed before the project under consideration. Thus, the timing of projects within a river basin development is important.

Since timing and design are interrrelated in multi-structure projects, the scheduling of investments needs to be determined at the same time as the design parameters (e.g., reservoir capacities, irrigation deliveries and power installations). An optimal investment program depends on an optimal schedule for constructing the various projects of a river basin development plan, and on optimal rates of development of the service areas. Scheduling of construction and development of service areas can be optimized by means of sequencing models.

Sequencing models include some of the most important implementation constraints:

(i) Budgetary constraints expressed as availability of funds for development in the irrigation and power budgets of the regional authority;

(ii) Manpower limitations which may restrict engineering, economic and other studies necessary for design;

(iii) Technical and administrative factors influencing the rate of construction such as land acquisistion, capability of contractors, availability of skilled, semi-skilled and unskilled labor;

(iv) The rate at which farmers in prospective service areas are prepared technically, economically and managerially to shift from rainfed to irrigated agriculture.

These models may also include restrictions that certain components of the system be constructed before others, and/or conditions that two or more components should be completed within a specified period of time.

As already mentioned, timing and design are interrelated. Thus different configurations of screening models may yield different optimal schedules of construction, and sequencing model analysis may raise issues relevant to the screening and simulation models. For example, when analyzing periods of construction, opportunities for improving irrigation deliveries and/or power generation may be identified for stages prior to full development. Thus the systems approach has a feedback component (as shown in Figure 2) which makes it an iterative process, narrowing down options to those which best satisfy project objectives.

Planning Issues

A BASIN-WIDE DEVELOPMENT PLAN

The hydrological interdependence of the major streamflow regulating structures envisaged for a river basin was already mentioned and the importance of the systems approach in their planning was emphasized. However, the structural components of a water resources system are merely instruments for the implementation of development policies. It is important, therefore, to view the development of a river at the basin-wide level, rather than project by project, so that an optimal development plan maximizing net social and economic benefits for the river basin as a whole could be produced.

The implementation of the various projects in a river basin could take several decades, and the development of the service areas will probably span an even longer period of time. During this considerable number of years, social, economic and political conditions will evolve so that devlopment aims and goals will change. An "once and for all" planning exercise, even if it will need the next few years to be completed, will hardly be relevant in the context of future conditions and objectives. It follows that planning should be a continuous process, albeit varying in intensity, involving data collection, identification of new issues as they may arise, and studies for the analysis and resolution of these issues, all these activities closely linked with the design and construction of the various development projects within the river basin.

It becomes clear now that planning is not only an activity continous in time, but also has an impact on every project, subsystem and component comprising a river basin development plan. The planning process can then be viewed as a hierarchical structure exhibiting different degrees of detail at different levels, as described briefly below.

The basin-wide development plan will define overall development goals and will evaluate alternatives for their attainment. This plan has to be reviewed and, if necessary, revised periodically. The plan would evaluate past experience, would summarize on-going planning studies and would outline future development goals and options. It will also specify targets and set time schedules for all planning, design, construction and monitoring activities in the short and medium range. It might be useful to perform the periodical revisions of the plan at intervals of about five years.

Framework plans for major projects. These projects will include (i) all multi-purpose structures; (ii) hydroelectric projects with an installed capacity above a specified minimum; (iii) transbasin diversions; (iv) irrigation projects servicing cultivated areas in excess of a minimum acreage. Initial drafts of Framework Plans would present development options, assessed in terms of benefits and costs, to be used in the preparation of the first version of the Basin-Wide Development Plan. The Framework Plan would include detailed programs for further project preparation and construction.

Feasibility studies for each major component and/or block of development of major projects and for smaller projects, covering technical and economic aspects of the scheme. Feasibility studies are particularly useful when preparation and implementation of one component can proceed independently of other components of a major project.

Design reports for major structures, such as hydropower installations.

Studies needed for the prepartion of the initial version of the Basin-Wide Development Plan are listed in the Appendix.

GROUNDWATER AND CONJUNCTIVE USE

Preliminary surveys should estimate the exploitable groundwater in the river basin. The groundwater potential appears to be greatest in recent alluvial deposits wherever they exist. The development of surface irrigation projects may increase the groundwater recharge considerably. Given that water losses in major surface schemes--even when fully lined--are much higher than in groundwater projects, the objective of optimal utilization of the water resources in a river basin indicates that a larger area could be irrigated with groundwater than with surface flows.

Groundwater development usually has a significant energy requirement and high capital costs. A rough estimate of energy requirements at a power plant for high yielding deep tubewells with a lift in excess of 20 m, with an efficiency of the pump-motor aggregate of about 50%, and with power transmission losses of some 30% would be around 0.25 kwh/m^3 of water delivered to an irrigated crop. One m^3 of water released for power generates 0.0023 kwh per meter of head. This means that in order to pump one m^3 of groundwater, it is necessary to release for power one m^3 of water over a drop of 0.25/0.0023 = 109 m. However, this is not the break-even point for substituting surface water for groundwater because of the different efficiencies of supplying water by wells and by surface systems; the efficiency of unlined surface systems is much lower than the distribution systems of groundwater. Therefore, every m^3 of groundwater is substituted by more than one m^3 of surface water which, if diverted away from hydropower generating turbines, can result in substantial economic losses.

The capital cost of groundwater development appears to be lower than that of surface systems. Given the serious drainage problems that may arise rapidly in parts of the service areas, groundwater development in a river basin should usually be given higher priority. It is anticipated that significant economic benefits could be derived from an optimum use of surface and groundwater, especially when these resources are developed conjunctively. Planning the conjunctive use of surface and groundwater would involve a more accurate assessment of the resource base, as indicated broadly in Figure 4.

USE OF IRRIGATION WATER

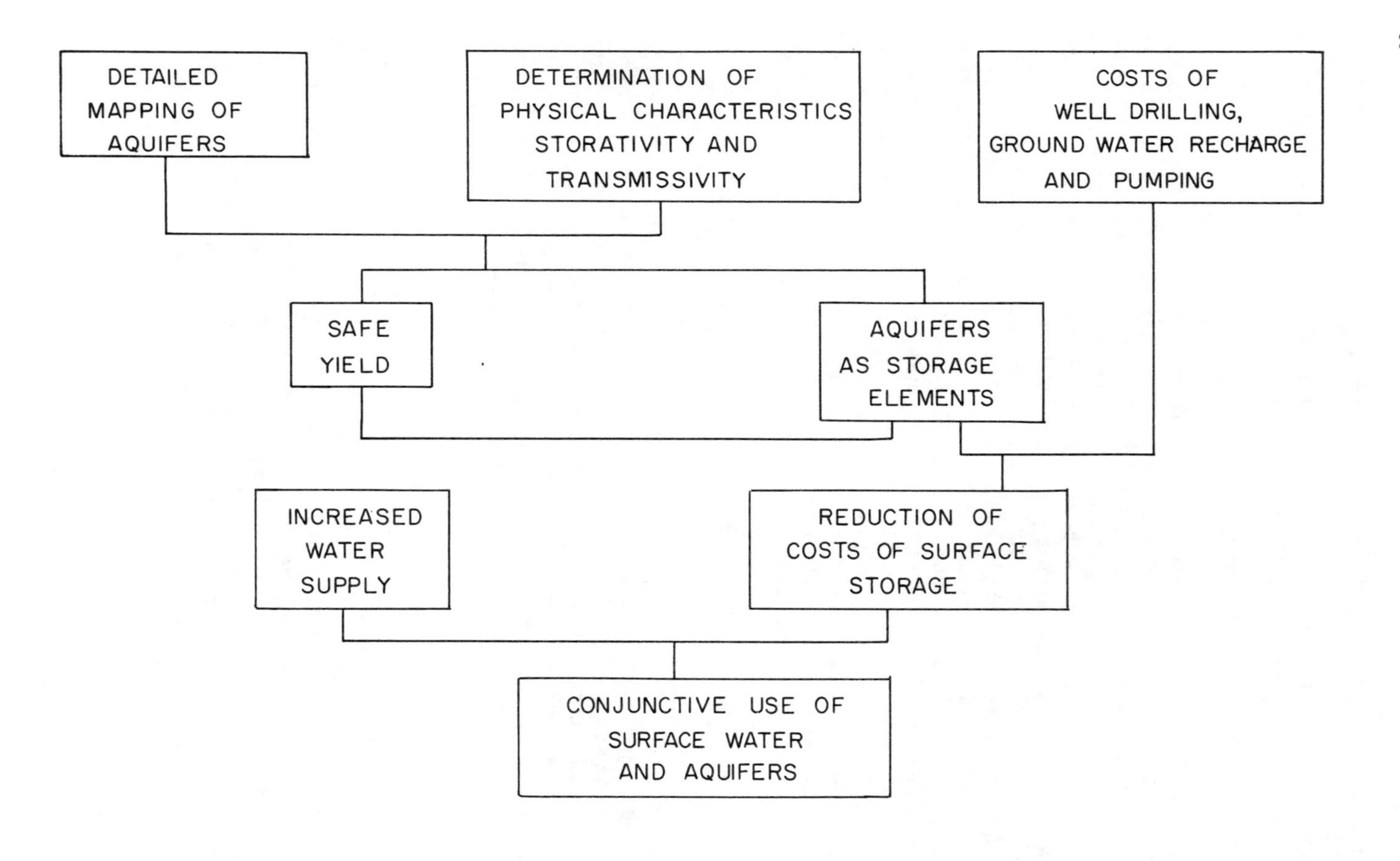

Figure 4

GROUND WATER DATA

In a complex river basin development project, irrigation water may have different values depending on its location within the basin and on the season. Consider, for example, the topographically lowest irrigation project in the river basin. One cubic meter of water released from an upper basin project at full development can generate considerable electricity while dropping over a cascade of reservoirs. Depending on the efficiency with which this cubic meter is used subsequently for irrigation, it is estimated that its power and irrigation benefits could be substantial. This indicates that water has a fairly high opportunity cost in the upper part of the basin, i.e., its diversion outside the basin will generate substantial production losses inside the basin.

The seasonal effect is also significant. Surplus water is likely to be available, in many years, during the rainy months, so that the opportunity cost of water may be lower than some other estimates. Transbasin diversions, if any, would involve high investment (and possibly energy) costs. Therefore, the social and economic implications of such diversions need careful evaluation.

Planning of irrigation projects must meet performance criteria which would maximize net social and economic benefits; some of these criteria are as follows:

(1) Surface irrigation projects must allocate water to users by volume and not by flow or by area irrigated.

(2) The supply must be reliable both in quantity and in timing, and it must be as flexible as possible to match actual crop water requirements.

(3) The water distribution from the last irrigation district outlet to the farm gates must be "manageable" by the farms themselves.

The performance criteria can be attained in a number of different ways, and underlying all of them is the necessity to increase efficiency and to introduce operational flexibility into the planning and design of irrigation projects. The necessary flexibility for efficient irrigation can be achieved by creating storage facilities at intermediate levels between the major dams and the farm gate, and by planning in greater detail the infrastructure connecting a water source with an irrigated field which may be tens of kilometers distant. This infrastructure should be divided into three subsystems:

(1) A <u>conveyance</u> subsystem for the efficient delivery of water from a source to a balancing (intermediate) reservoir.

(2) A <u>distribution</u> subsystem which draws its water from

the intermediate reservoir and conveys it to farm gates in accordance with farm demand patterns.

(3) A field irrigation subsystem on the farm itself which delivers water in accordance with actual crop water demand.

Operational management objectives of each of the subsystems would differ and their analysis should be integrated into irrigation project design at the planning stage.

TRADE-OFFS BETWEEN POWER AND IRRIGATION

Hydropower generation and irrigation supply may be complementary uses of water or competing activities. The complementarity is obvious when water is diverted for irrigation after passing through a power plant; otherwise, a competing situation may be created. Under competition, priorities have to be established on the basis of trade-off analysis, or in some other way. In general, irrigation takes precedence over hydropower generation. Nevertheless, it is essential to clarify the trade-offs between power and irrigation when planning a project as complex as river basin development.

The relative worth of irrigation and power production varies significantly on a daily basis and also seasonally. Ample hydropower capacity is usually available during the rainy season, giving an opportunity to take thermal plants out of service for major maintenance works. During the night hours of that season, hydropower probably has its lowest value; this can be estimated primarily by the saving in coal needed to operate thermal power stations. The daylight hours in the dry season are probably the peak demand period; then the total generating capacity of an electrical network may become a serious constraint and the additional kwh generated by hydropower has a high value.

The value of water used for irrigation exhibits a similar pattern showing differences between the rainy and dry seasons. Rough estimates indicate that one cubic meter of water used by irrigated crops is worth about 50% more in the dry season than during the rainy period.

In order to evaluate the trade-offs between power and irrigation, the value of water used for either purpose should be expressed in monetary units per m^3. It is especially important to make this evaluation when the main stream in a river basin has a number of favorable storage sites which, when developed, will produce a cascade of reservoirs and dams. Then the worth of water for hydropower at the upper dams must take into account the entire potential drop clear to the lowest point of the river basin, yielding comparatively high values. Of course, for the dams further downstream, the value of water used for hydropower will decrease.

Some general qualitative conclusions, subject to verification and refinement, can now be reached:

1. In the upper part of the basin, water generally produces more economic benefits if used for power than if diverted for irrigation. (The converse appears to be true in the lower basin.) However, due to differences between the seasons, irrigation releases should be made in the season when their benefits will be maximized.

2. Irrigation water in the dry season appears to be more valuable than supplementary irrigation in the rainy season. Thus it is definitely worthwhile to place the planning emphasis so as to supply irrigation water primarily in the dry season.

3. Storage benefits appear to be higher in the upper part of the basin. This does not mean that reservoirs in the middle and lower basin should not be built: it simply re-emphasizes the need for the basin-wide planning of a given region and for more detailed socio-economic studies on which the planning should be based.

OTHER ISSUES

Planning the development of a river basin involves considerably more than evaluating the irrigation and power objectives and the trade-offs between them. Other objectives should be considered, even if during the initial phases of planning their importance may seems negligible: at some point in the future, they may surface as key issues in the futher development of the river basin. Among the additional issues which should be considered are regional economic growth (however defined), income distribution (basin-wide and between sub-basins), demographic problems (overall population growth and rural-urban migration), food self-sufficiency and agricultural surpluses, balance of trade with other regions within the country and with foreign countries, energy dependence (or self-sufficiency), environmental quality and public health. Each of these issues, to a greater or lesser extent, is affected by the reliability of water quantity and quality in time and space.

A Time Frame

PLANNING DECISIONS TO BE TAKEN WITHIN A YEAR

In order to ensure the soundness and effectiveness of decisions pertinent to the planning of a river basin development project, it is essential to recognize and to accept the fact that one deals with a project of considerable complexity and its planning, design, construction, operation and monitoring cannot be based on concepts and methods developed in the last century nor on outdated information.

Once this fact is realized, the following decisions should be taken and implemented within the first twelve months of the planning process:

1. The development of a region should be based on a basin-wide planning effort and not on a component-by-component approach.

2. This complex planning task should be performed by the best available engineers, economists, agronomists, and other scientists expert in the disciplines involved in planning river basin development.

3. Scientifically sound evaluation of available surface and groundwater resources (see Appendix):

(a) Detailed hydrological studies within the basin.

(b) Statistical-mathematical analysis of streamflows leading to the generation of synthetic hydrological series.

(c) Reconnaissance survey of groundwater resources, including measurements of tubewell discharges and their power consumption.

4. Evaluation of land resources and their suitability for irrigation:

(a) Land-use capability mapping of the proposed service area.

(b) Investigation of the drainage (and potential salinity) problems in the river basin, including the establishment of a network of observation wells.

5. Determination of irrigated agriculture potential: cropping patterns; yield estimates with irrigation and without; peak water demands and their time distribution; marketing surveys; estimation of socio-economic impacts of irrigation.

6. Initiate reservoir sizing studies by constructing

a multi-reservoir mathematical model for the main stem of the river basin.

7. Determine standards and criteria for system performance so as to ensure a supply of water for irrigation, power generation and other uses at a high level of reliability both in quality and in time.

DECISIONS FOR THE FOLLOWING TWO YEARS

During the first year of the planning process, most decisions should be either in matters of principle (e.g., the basin-wide planning approach) or in regard to the data acquisition phase of the planning process (e.g., determination of the irrigated agriculture potential): they form the necessary basis for a development policy. In the ensuing two years, the acquired basic information should be used for policy decisions.

Some of the decisions to be taken and implemented are the following:

1. Groundwater development: determine areas to be irrigated by groundwater, and areas where both surface and groundwater supplies will meet irrigation requirements; specify groundwater management (pumpage and recharge) criteria.

2. Initiate power studies for the purpose of determining optimal sizes and operating policies of hydropower stations when integrated within an existing electricity network, and for estimating power-irrigation trade-offs.

3. Specify criteria for the operation of the river basin as an integrated system and develop one or more simulation models for the evaluation of alternative operating policies.

4. Initiate irrigation studies to gather the information necessary for framework planning of each of the individual projects:

(a) Irrigation efficiency at the field level under different irrigation methods and regimes.

(b) Study advanced irrigation technologies for eventual adaptation to the local conditions.

(c) Evaluate irrigation return flows for more accurate regional water balances.

5. Study proposed transbasin diversions, if any,

including their potential socio-economic impact both within the river basin and outside it, and determine which diversion should be implemented.

6. Develop cost functions for every component of irrigation, power and other sub-systems of the river system, including alternative designs of the same component.

7. Formulate framework plans for each major project in the river basin, highlighting the development options in terms of cost, benefits, uncertainties involved, and the social equity aspect.

8. Project short-, medium-, and long-term uses of water for irrigation, hydropower, municipal demands, industry, and energy (cooling thermal power plants); evaluate their impact on planning decisions and on water and environmental quality.

THE MORE DISTANT TIME HORIZON

Projections in time beyond three years are at best tentative, especially under conditions of rapid change. Furthermore, decisions which the regional authorities would have to take in the future will depend, to a great extent, on the policy followed during the initial years of the planning process and on the degree of its implementation. Nevertheless, a number of issues which will require action can be perceived at this stage of the planning process. Some of them are:

1. A first version of the river basin development plan should be ready for formulation and approval by regional and other governmental authorities.

2. Detailed design of individual projects within the river basin should be initiated.

3. The matter of social benefits and social equity will have to be studied in great detail, in order to provide a basis for the next version of the river basin development plan. Impact of irrigation on farmers' incomes will have to be evaluated, as well as the effect of irrigation projects on intra- and inter-regional income distribution patterns.

4. Determine the complementary investments required to insure the effectiveness of the river basin development plan and to protect the very large investment involved in its implementation: land shaping, drainage, roads, crop storage facilities,

sorting and packing of produce, primary processing installations (e.g., dairies). Improve agricultural vocational training, research and extension services. Initiate the establishment of farmers' organizations and institutions necessary for efficient self-management of irrigation projects.

5. Develop an initial optimal schedule for project implementation and construction. Identify major constraints and bottlenecks (manpower, budgets, materials, etc.) and determine alternative ways of dealing with them.

6. Determine rate of development of service area, including on-farm activities, compatible with the progress of major construction.

Summary

An outline of activities--studies, evaluations, decisions--required for the comprehensive planning of a regional approach to river basin development is presented. The water resources of the basin available to the region should first be estimated. Considering the relatively large amounts of water involved--surface flows and groundwater aquifers--and the considerable impact which the river basin development may have on the regional economy, the maximum economic and social benefits will be attained only if the whole basin will be developed as one comprehensive system rather than constructing irrigation and power projects in isolation one from another.

Development objectives are seldom stated in a clear and precise fashion. Irrigation systems, for example, are designed mostly in isolation. If a river basin development project is to serve a progressive, forward looking economy in the 21st century, development objectives must be stated in as exact terms as possible. Irrigation, for instance, has to be specified in terms of timeliness and reliability of deliveries at different points in the service area, so as to attain specified economic and social objectives. Power generation should be planned for maximum net economic benefits and for the maximization of social welfare. Other objectives of water resources development, such as those related to industrial water supply or to the energy sector, should be also stated in terms of net economic and social benefits.

Planning the development of a large river basin of considerable complexity requires a clear departure from the procedures traditionally used in smaller and/or less intricate projects and the adoption of a modern planning methodology based on advanced engineering-economic-social concepts, using tools of mathematical analysis, and supported by computer-solved models: systems analysis. The systems analysis, through its analytical approach, is capable not only to determine optimal sizes of major structures (dams, power plants, main canals, etc.), but also to indicate the studies necessary for the acquisition of data relevant to the planning activity. In addition to the sizing of major components, systems analysis, through its simulation models, can test and evaluate alternative operating policies for the system as a whole, estimating the magnitude and frequency of shortfalls as well as those of surpluses. Finally, by means of sequencing models, the systems approach can generate optimal schedules for project implementation, given budgetary, manpower, technical, administrative and other constraints.

The effective application of the systems approach requires that planning the development of a major river in a region should be at the basin-wide level, rather than considering individual projects separately. The basin-wide development plan should also be a sustained activity allowing for periodic revision of the plan in response to changing conditions. The planning process should have a hierarchical structure which includes the following levels (with increasing detail): the basin-wide development plan; framework plans for major projects; feasibility studies for major components which can be implemented independently of other components of the same project; design reports for major structures. An important issue to be considered in detail during planning is the integration of surface and groundwater resources within the basin. Another crucial issue to be analyzed and resolved by the planning process is the supply and use of irrigation water in a timely and reliable fashion so as to maximize economic and social benefits. Of course, irrigation analysis should consider alternative uses of water (primarily for hydropower) and develop trade-off criteria. In addition, other issues, such as economic growth and income distribution, should be considered by the planning process.

The planning of a river basin development project involves a series of decisions to be taken by the appropriate governmental authority at its highest level. During the first year, the decisions should be in matters of basic concepts (e.g., basin-wide systems analysis) and in regard to the acquisition of fundamental information (e.g., assessment of available surface and groundwater resources). In the following years, policy decisions (including specific design criteria for major subsystems, such as groundwater development) should be based on the fundamental concepts and basic information developed during the first year of planning. Their implementation will yield alternative approaches to the major development objectives, which when evaluated will lead to the initial version of the river basin development plan.

It would be useful for the planning group to prepare as soon as feasible a PERT-like network for the various studies involved in the planning activity so as to identify a critical path. A similar network related to the implementation of projects should be prepared at a later date.

A specific application of the methodology presented herein is being carried out in a river basin development project in India [Buras, 1981].

References

Buras, N., 1981, The Narmada River Basin Planning in Madhya Pradesh, Report submitted to the World Bank.

ECAFE, 1972, Water Resources Project Planning, Water Resources Series No. 41, United Nations, New York.

Loucks, D.P., J.R. Stedinger and D.A. Haith, 1981, Water Resource Systems Planning and Analysis, Prentice-Hall, Englewood Cliffs, N.J.

Major, D.C. and R.L. Lenton, 1979, Applied Water Resource Systems Planning, Prentice-Hall, Englewood Cliffs, N.J.

Saha, S.K. and C.J. Barrow, 1981, River Basin Planning: Theory and Practice, Wiley, New York.

Appendix: List of Studies for a River Basin Development Project

BASIC STUDIES

Parametric hydrology. Establish a study program based on the Stanford Watershed Model (OPSET) for the estimation of maximum flows at major reservoirs; calibrate and validate the model in order to extend the streamflow data on the basis of the existing precipitation measurements; determine the statistical cross-correlation between sub-basins defined in the river basin (this is an important input to the following study).

Synthetic (operational) hydrology. Use existing streamflow data as well as estimates generated by the above study to generate a number of equally likely sequences of hydrologic events at selected points in the river basin. Length of each sequence--about 50 years. The results of this study are a necessary input to simulation of system operation and response.

Groundwater mapping. Perform a reconnaissance survey of groundwater of the service area. Identify areas where detailed mapping should be done and carry out such mapping.

Determine transmissivity of aquifers (so as to estimate potential water yield) and their storativity (for possible use as storage elements within the river system). Assess alternative technologies for artificial recharge of aquifers and evaluate proportion of groundwater recovery. Conclude with a preliminary identification of mixed surface water/groundwater projects.

Land use capability mapping. Classify the service area with respect to suitability of irrigation. Estimate cost for installing effective irrigation systems in each of the various classes of irrigatable lands. Estimate crop yield differences in each of the different land use classes, which will affect ultimate project benefits.

Agricultural survey. Determine cropping patterns in various (zones, subzones, regions and) subregions of the proposed service area; determine past trends and changes. Estimate yields with and without irrigation, so as to reach an approximate value of irrigation water. Determine peak demands for irrigation water under different cropping patterns and their impact on maximum capacities of main canals. Survey marketing practices of agricultural products and determine alternative improvements for increasing farmers' revenues.

PLANNING STUDIES

Reservoir sizing. Mathematical modeling of the major dams as a multi-reservoir system. Determine alternative operating rules of hydropower generation in relation with the demand patterns of the regional power grid. Estimate trade-offs between irrigation and power benefits. Determine the impact of power generated at major dams. Integrate all the proposed projects within the river basin into a unified regional system, so as to screen out sub-optimal projects and/or sizes of proposed reservoirs.

Operation (simulation) studies. Use synthetic (operation) hydrological records developed by the basic reservoir sizing study to determine long-term socio-economic costs and benefits. Estimate frequency and magnitude of shortages, as well as those of surpluses, and refine operating policies developed in the above study.

Integration of groundwater. Evaluate (in terms of quantity and cost) groundwater aquifers as additional sources of water on a regional basis. Study the possibility of substituting groundwater for surface supplies for irrigation, and the subsequent use of substituted surface flows for hydropower generation. Evaluate the use of aquifers as regional and local storage elements, and determine the value of groundwater. Identify and estimate the feasibility of local and subregional projects based entirely on groundwater, on both surface and groundwater, or on surface water alone.

Cost estimates. The purpose of cost estimates is primarily to develop capital cost functions in terms of monetary units versus irrigated hectares for each project in the river basin. The irrigation cost functions should include also the drainage component, whenever applicable. Similar functions need to be developed for hydropower projects or for power components of multi-purpose projects. Cost functions should also be developed for alternative designs of the same system components. The cost functions are necessary for the proper evaluation of economic and social benefits of the projects considered in the planning process.

Water use. Project short-, medium-, and long-term uses (withdrawal demands as well as consumptive uses) in irrigation, domestic use, industry, mining, energy (e.g., cooling of thermal power plants), and in-stream use (hydropower). Determine magnitude and seasonal variability of return flows.

IRRIGATION AND POWER STUDIES

Irrigation efficiency. Perform field measurements of seepage losses from any existing canals of any size and capacity, in different soils, with and without lining, and under different operating policies. Determine irrigation efficiency at the field level for different irrigation methods and regimes. Evaluate impact of alternative irrigation efficiencies (overall) on capacities of canals (main, branches, majors, minors, sub-minors, etc.) and of possible en route and regional (subregional, local) reservoirs.

Irrigation technology. Study the possible impact of sprinkling irrigation development in the region: its possible rate of development; possible active incentives; its shortcomings. Evaluate possible impact on planning, design, and sequencing of projects in the region. Estimate the potential influence on the overall plan for the river basin development.

Irrigation return flows. The return flows are a function of: (i) the cropping pattern; (ii) the amount of water applied; (iii) the irrigation technology adopted; (iv) the soils; and (v) the drainage and groundwater conditions. Without a detailed analysis of these factors, estimates are unreliable. The return flows will not only have a significant impact on the total availability of water, but also on the sizing of reservoirs, power stations, etc. An analysis of the expected return flows would also include an assessment of the effects on water quality of such flows.

Power studies. Estimate the value of hydropower when generated for base load and when satisfying peak demand. Determine marginal cost of producing power at different times. Estimate opportunity cost of power in terms of lost production (in the various sectors of the economy, including agriculture) due to failure to meet power demands at peak periods. Evaluate the integration of the hydropower within the alternative operating rules developed by the reservoir siting study. Assess the impact of this study on reservoir sizing and operation.

SOCIO-ECONOMIC STUDIES

Complementary investments. Determine the additional investments, on a regional basis, necessary to insure the effectiveness of the river basin development scheme: land shaping, drainage, roads. Estimate the rural infrastructure required to protect the very large investment proposed for the major structures: crop storage facilities, sorting and packing of produce, on-farm (regional) primary processing (e.g. milk pasteurization). Assess the needs for agricultural research and extension. Evaluate a probable evolution of farmers' organizations and institutions and their possible impact on the rate and timing of service area development activities. Determine the impact of such complementary investments and activities on the overall regional development plan.

Social Benefits. Determine the impact of irrigation in various zones, subzones, regions, and subregions of the proposed service area on farmers' incomes. Evaluate the effect of irrigation projects on income distribution patterns.

PROJECT SEQUENCING STUDY

Implementation of project. Determine optimal patterns of construction sequencing and identify major constraints and bottlenecks (manpower, budgets, farmers' readiness to utilize beneficially the constructed irrigation system, etc.). Determine alternatives to deal with constraints and to overcome bottlenecks. Estimate rate of development of service area compatible with progress of major construction. Determine on-farm development activities complementary to the service area development projects: planning, design, and implementation.

NEW RESEARCH DIRECTIONS IN AGRICULTURAL HYDROLOGY AND EROSION

Klaus W. Flach
Special Assistant for Science and Technology
Soil Conservation Service, U.S. Dept. of Agriculture
Washington, D.C. 20012, USA
and
John M. Laflen
Research Leader, National Soil Erosion Laboratory
U.S. Dept. of Agriculture
West Lafayette, IN, USA

ABSTRACT. This paper addresses two rather contrasting areas of research that have developed rapidly during the last 10 years and that promise to have major impact on soil erosion control and agricultural hydrology.

One area addresses conservation tillage, which includes various tillage systems that minimize soil disturbance during tillage and maintain a cover of vegetation or crop residues on the soil. Conservation tillage has proven itself an effective practice for controlling erosion. It has in many places replaced conventional erosion control practices, such as terraces, or has made conventional erosion control practices more effective. In recent studies, conservation tillage has been shown to improve infiltration and decrease runoff and may, therefore, show promise for improving hydrologic regimes of streams.

The other area of research has to do with a new model for erosion prediction. This new model, which has been given the preliminary acronym WEPP (water erosion prediction project), is being developed to replace the Universal Soil Loss Equation, USLE, the main tool for conservation planning during the last thirty years. WEPP is only one example of a family of new hydrologic and erosion models that concentrate on hydrologic and erosion processes and are driven by detailed, usually daily, meteorologic data. We believe that these new models will greatly improve our capability to predict runoff and sediment delivery. We also believe these models will enable us to develop more effective soil conservation systems.

L. M. Brush et al. (eds.), Taming the Yellow River: Silt and Floods, 73–84.

Conservation Tillage

Conservation tillage, reducing soil erosion by maintaining as much plant residue cover on the soil as possible, has been an exciting new development in soil conservation in the United States during the last 20 years. Conservation tillage is now generally accepted either as a replacement for conventional erosion control practices, such as terraces, or as an invaluable supplement to such practices.

There are many forms of conservation tillage. Generally accepted definitions of some of the most important forms are as follows:

No till. -- The soil is left undisturbed prior to planting. Planting is done in a narrow seedbed, 3 to 8 cm wide that is generated during the planting operation.

Ridge-till. The soil is left undisturbed prior to planting. During cultivation of the preceding crop, approximately one-third of the soil's surface is formed into ridges that are 8 to 15 cm high and the crop is planted in these ridges.

Strip-till. The soil is left undisturbed prior to planting. About one-third of the soil surface is tilled and the crop planted into the tilled areas.

Mulch-till. The entire soil surface is tilled prior to planting with non-inverting tillage tools such as sweeps, disks, or field cultivators. At least 30 percent of the soil remains covered by crop residues at all times.

For statistical purposes, conservation tillage is defined as any form of tillage that leaves at least 30 percent of the land covered by crop residues at all times. All figures on the areas under conservation tillage in this paper are based on this definition.

EFFECTIVENESS OF CONSERVATION TILLAGE IN EROSION CONTROL

Conservation tillage was primarily intended as a technique for controlling sheet and rill erosion, and for this it has been very effective.

Sheet and Rill Erosion. The following data are selected values from erosion plots in various parts of the United States. Tables 1 and 2 show data from rainfall simulator plots for rotations of corn and soybeans for loess soils in Iowa. The simulated rainfall intensity for the erosion measurements was 6.35 cm/hr.

TABLE 1. Soil loss from various tillage systems with continuous row cropping (Laflen and Colvin, 1981). 1/

SOIL TYPE	TILLAGE METHOD	RESIDUE COVER (%)	SOIL LOSS (kg/ha)
	Conventional	5	4080
Clarion	Reduced	15	2080
	No-Till	42	930
	Conventional	0	53,690
Monona	Reduced	11	38,030
	No-Till	29	15,140

1/ Rainfall simulator plot data. Values are averages of four rotations of corn and/or soybeans.

Table 3 shows similar data for wheat on loess soils in the eastern part of the State of Washington. These are average values for wheat following another cereal or spring peas over a period of 5 years.

TABLE 3. Runoff and soil loss from winter wheat (Don McCool, personal communications).

Previous Crop	Tillage	Residue Cover (%)	Runoff (cm)	Soil Loss (Mg/ha)
Fallow	Conventional	10	7.4	19.0
Cereal	Reduced	35	2.3	0.9
Cereal	No Till	96	2.0	0.2
Spring Peas	Reduced	30	3.0	1.6
Spring Peas	No Till	51	3.0	1.3

The negative correlation between percentage of ground cover and erosion shown in Tables 1, 2, and 3 suggests that the effectiveness of conservation tillage can be attributed to the effect of crop residues on raindrop impact and consequent reduction of splash erosion; and to the reduction of water in rills and consequent reduction of shear.

GULLY EROSION

As conservation tillage affects runoff, one should expect that

TABLE 2. Soil loss from various tillage systems (Laflen et al., 1978). 1/

Tillage Method	Ida Soil Residue Cover (%)	Ida Soil Soil Loss kg/ha	Tama-Kenyon Soil Residue Cover (%) Tama-Kenyon	Tama-Kenyon Soil Soil Loss kg/ha	Tama Soil Residue Cover (%)	Tama Soil Soil Loss kg/ha
Conventional	9	29,000	4	12,500	4	12,500
Till-Plant	17	30,000	2	21,000	20	5,000
Chisel	23	17,500	10	12,500	21	4,000
Disk	45	13,000	12	11,500	21	3,500
Ridge	46	10,500	24	9,000	31	2,000
Coulter	58	5,000	27	6,000	63	1,000

1/ Rainfall simulator plot data. Simulated rainfall of 6.35 cm/hr with water added at upper end of plot. Soil loss values are estimated from figure 5 of the cited paper.

it also reduces erosion from concentrated flow as in ephemeral gullies and gullies. Ephemeral gullies, those that are small enough to be obliterated by annual tillage operations, are beginning to be recognized as important contributors to erosion. At this time, we have only a very few data on the effect of conservation tillage on the formation of such gullies.

One such set of data comes from a watershed on loess at Treynor in the western part of the State of Iowa. Total soil loss, including sheet and rill and concentrated flow erosion, for the part of a watershed under conservation tillage was 2.0 Mg/ha as compared to 13.0 Mg/ha for conventional tillage. The conservational tillage consisted of a ridge-till system with the ridges laid out on the contour, the crops were corn (zea mays) and soybeans (glycine max), and the experiment was run over a period of 12 years. The conventionally tilled field was laid out on the contour. Hence, erosion from it was not as high as it could have been under very poor management. Total runoff was 25 mm/yr for the ridge-tilled part of the watershed as contrasted to 113 mm/yr for the conventionally tilled watershed. The contribution of concentrated flow to the total erosion was studied in a separate one-year experiment. During this year, the ridge-till system reduced erosion from the concentrated flow channels by 90 percent, from 17 Mg/ha to 1.8 Mg/ha.

The results clearly show that conservation tillage on steep land can be extremely effective in reducing soil erosion by sheet and rill and concentrated flow if the ridges are parallel to the contour, are sufficiently high to contain all the water during major storms, and if any runoff is disposed of through grassed waterways. Nevertheless, serious erosion may take place if the ridges overtop.

The data (Table 2) also suggest that conservation tillage increases infiltration and, accepting concentrated flow channels as precursors, may at least help to reduce the formation of deep gullies. However, at this time, we have no definitive data on the influence of conservation tillage on the formation of deep gullies. Better infiltration will, of course, also increase the amount of water available to crops, ground water recharge, and base flow.

CONSERVATION TILLAGE AND INFILTRATION

Data on the effect of conservation tillage on infiltration are available from a few sites in the United States. An extreme example (Edwards et al., 1988) was reported from central Ohio where runoff decreased to essentially zero after four to six years of conservation tillage of maize (Table 4, watershed 191).

The adjacent, conventionally managed watershed (Table 4, watershed 123), also in maize, had extremely high rates of runoff in the four years for which records are available.

TABLE 4. Total rainfall and runoff for watersheds in Ohio (Edwards et al., 1988)

Year	Total Precipitation	Intense Rainfall 1/	Runoff Watershed 191	Runoff Watershed 2/ 123
	mm	mm	mm	mm
1979	1124	92	3.81	140.0
1980	1175	80	4.90	313.0
1981	1057	88	0.14	142.0
1982	889	137	0.00	113.0
1983	1027	174	0.00	---
1984	909	198	2.31	---
1985	929	139	0.01	---
avg.	1016	130	1.60	172.0

1/ Rain falling at intensity greater than 15 mm/h.
2/ 4-y avg. Not measured after 1982.

The gradually increasing effectiveness of conservation tillage over time as shown in Table 4 seems to be typical of most conservation tillage systems and has been attributed to the formation of large pores in the surface soil that conduct water readily as long as they are open to the soil surface. These pores are formed by decaying roots and by earthworms. Earthworms are probably the more important agent. In colder areas, where earthworms are less prevalent, the effect of conservation tillage on infiltration seems to be much smaller. In an experiment at Morris, Minnesota, where the mean annual temperature is 5C lower than that in Ohio, annual runoff was only reduced from 116 mm for conventional tillage to 94 mm for no till.

The experiments in Ohio and Minnesota differed in design and may not be strictly comparable, but the results are consistent with qualitative observation in other locations.

SUITABILITY OF CONSERVATION TILLAGE FOR VARIOUS FARMING SYSTEMS

Conservation tillage is rapidly being accepted by farmers in the United States. A 1986 survey (CTIC, 1986) showed that about 40 percent of the maize acreage, 33 percent of the small grains, and 34 percent of soybeans were under some form of conservation tillage. Actually, one of the main impediments to the acceptance of conservation tillage has been the pride farmers take in the appearance of neat, cleanly tilled fields and reluctance to accept the unkempt appearance of no-till fields. After the initial few years, yields of crops in conservation tillage tend to be about the same as under conventional tillage; but in climates where crops suffer from moisture stress, crops often benefit from the increased infiltration of water under conservation tillage, and yields are higher. In general, conservation tillage is more readily accepted in moderate climates (mean annual temperature 8 - 13C) than in cooler climates where reduced soil temperatures under conservation tillage in the spring delay germination and depress yields. In the same areas, and for the same reasons, conservation tillage has had more difficulties on poorly drained than on well drained soils. The acceptance of conservation tillage has been slower in warmer climates because of difficulties with pest control.

Equipment for conservation tillage for most crops is available. Since the power requirements for no-till are significantly lower than for conventional tillage, savings in machinery costs and for fuel can be significant. These savings are partly obviated by increased expenditures for pesticides and fertilization. As the soil organic matter level increases during the first few years of conservation tillage, somewhat higher N fertilization is required until a new equilibrium level is reached.

Conservation tillage is still a relatively new technology. We believe that over time more farmers will shift to no-till and that improved techniques and equipment will make conservation tillage more effective. For example, controlled traffic, using the same wheel tracks for all field operations, has been shown to mitigate problems of soil compaction due to wheel traffic during planting and harvesting operations which has been a problem with no-till in some areas. The conversion of farm equipment to uniform wheel spacings is expensive and poses certain technological problems that have not yet been completely resolved.

Conservation Tillage: Summary and Conclusions

We have accepted conservation tillage as an effective and inexpensive practice to combat sheet and rill erosion. We believe that it shows promise for the control of concentrated flow channels which not only contribute significantly to the total soil loss but also serve as critical pathways for the movement of sediments to streams. Consequently, conservation tillage may prove itself as an effective way of reducing the sediment load of streams. Under certain conditions, conservation tillage may increase infiltration and consequently increase base flow and reduce peak flow of streams.

Modeling of Erosion and Runoff

The ability to predict the impacts of various soil management and conservation alternatives on erosion and runoff is extremely important for the practical application of conservation practices. These predictions have to be valid under the great variety of soil and climatic conditions encountered by the conservation planner.

In the United States, the main tool for conservation planning of individual fields has been the Universal Soil Loss Equation (USLE) (Wischmeier and Smith, 1978). This equation, which contains soil erodibility, rainfall characteristics, crop characteristics, conservation practices, and slope gradient and length as determinative components, was developed in the 1940's and 1950's based on the stochastic and statistical treatment of the results of research on numerous erosion plots in various parts of the country. These plots had similar dimensions and slope gradients. The experiments were carried on for many years under various soil and climatic conditions and allowed the measurement of total erosion under various kinds of crop management and erosion control practices. Since only the total sediment leaving the plots was measured, the experiments integrated splash erosion and sheet and rill erosion. Also, the plots were straight and did not consider the effects of slope shape on erosion.

The USLE has served as an extremely useful tool for conservation planning. However, renewed concerns about the effects of erosion on soil and water quality in the 1980's has led to State and National legislation that required farmers to meet certain minimum standards in erosion control. The equitable implementation of this legislation required more accurate erosion estimates than the USLE provided. Also, new water quality legislation mandated that we obtain more accurate estimates of the amount and nature of sediments that contribute to the pollution of streams and lakes.

In the meantime, continuing research has indicated that factors not considered in the USLE, such as the shape of slopes, were important variables in erosion and that the effects of slope length on erosion really could not be estimated without knowing the amount of runoff and the relationship between runoff and rill or concentrated flow erosion. At the same time, the widespread availability of microcomputers--the Soil Conservation Service of the United States Department of Agriculture expects to have microcomputers in each of its field offices by 1990--made it possible to use rather sophisticated mathematical models in the everyday field operations of a conservation agency. Consequently, and also because the USLE used a mature technology that did not lend itself to further development, we decided to develop an entirely new erosion prediction model.

THE WATER EROSION PREDICTION PROJECT (WEPP)

Water Erosion Prediction Project (WEPP) is our name for the coordinated research activities for the development of the model that will replace the USLE. In the following discussion, the term WEPP will be used for the project as well as its product, the WEPP model. The project is a major effort involving four United States government agencies and scientists at more than 12
locations throughout the United States. Interestingly, the coordination of research by so many scientists is being achieved largely through the use of an electronic telecommunication system that allows the instant communication of messages or computer programs between microcomputers on the individual scientists' desks. We believe that this technology makes it possible for large teams of scientists at many locations to cooperate effectively on the solution of problems that affect large geographic areas.

At this time, intensive field research using a modified rainfall simulator that allows estimates of soil detachment by raindrops and by shear are being conducted in all parts of the United States. Also, careful measurements of the number and volume of rills are made through micro-photogrammetric techniques that were specifically developed for such purposes. A preliminary version of the model is operational and is being used to evaluate the results of the field research. The model uses major components of tested hydrologic models such as CREAMS (Foster 1981). The following discussion is based on reasonably conservative predictions of the model's performance.

The WEPP model will use our fundamental knowledge about erosion processes in order to predict erosion under a variety of conditions. It will assist a new generation of technicians to select agricultural practices and design conservation systems on our Nation's lands.

WEPP will use the concepts of rill and interrill soil erosion and sediment transport in prediction. Interrill erosion will be expressed as a function of the square of rainfall intensity,

$$Di = KiI^2$$

where Di is the detachment and transport of soil due to raindrops. This soil is usually transported very short distances to a small stream of flowing water--a rill. Ki is the interrill soil erodibility, and I is rainfall intensity.

Rill erosion is caused by the shearing force of flowing water when this force exceeds a critical value. This may be expressed as

$$Dr = Kr(T-Tc)$$

where Dr is rill detachment, Kr is the rill erodibility, T is hydraulic shear, and Tc is critical hydraulic shear.

Because of the fundamental nature of the equations, i.e. the separation of the rill and interrill processes of detachment, the technology can be used for almost all situations of soil erosion by water. For example, a modification of the interrill equation would suffice for many sprinkler irrigation problems, or the use of the rill portion alone would allow estimates of furrow irrigation by the same technology.

Snow-melt erosion could also be computed by ignoring the interrill portions of the equations.

Because it is a fundamental process technology, WEPP lends itself to the prediction of erosion for individual storms or for short time frames. Values of soil erodibility and critical shear can be selected for the time period of interest, rather than long-term annual averages. Since the technology relies upon computation of the forces involved in detachment, hydrologic components are needed to compute the runoff rates and volumes.

The model continuously estimates changes in the soil's infiltration capacity. It uses a plant growth and evapotranspiration model to estimate the degree of saturation of the soil profile, and we hope to be able to consider changesin infiltration rates due to tillage and surface soil crusting. In general, the hydrologic components of the model will use technology developed for existing small watershed models such as CREAMS (Foster, 1982). In turn, we expect that experimental work done for WEPP will improve the infiltration and runoff and erosion components of CREAMS and similar models.

WEPP can be run with actual historical daily weather data using established relationships between daily rainfall and peak rainfall intensities or with data from a daily weather generator, a statistical estimate of daily weather events based on long-term records. We expect that actual data will be used mostly for the testing of the model and data from the weather generator for its operational use.

The fundamental nature of the technology also makes the model more transportable to other regions than is the USLE. The same basic erosion processes occur the world over, but parameter values may vary widely. WEPP can be used in locations where it has not been used before by estimating local parameter values. Also, we expect that WEPP, like other models based on fundamental processes, can be continuously updated as new scientific knowledge is generated or as new demands for the use of the model become evident. This will be a major advantage over stochastic models, such as the USLE, that have very limited capability for being updated.

WEPP will also compute sediment transport. Hence it will be able to compute erosion and sedimentation within a field and the amount of water and the amount and composition of sediments leaving the field. It will be able to compute soil erosion in concentrated flow channels, and consequently to give at least an estimate of the likely impact of various land management alternatives on gully formation. We believe that elements of WEPP can be incorporated into watershed models and contribute significantly to the improvement of existing models.

The WEPP will be implemented in several versions. One version, the profile model, will consider a two-dimensional profile of the field including changes in slope gradient and slope shape. This version will run on a small personal computer or a lap-size computer which the conservation technician can take to the field. The second, or watershed version, will include three dimensional aspects of a small watershed within a field as they can be estimated by a technician without detailed measurements other than slope gradient. It will include erosion and deposition in concentrated flow channels. The third, or grid version, will depend on surveyed grid points within the field and will provide similar, but more quantitative, results than the watershed version. It will require a moderate size microcomputer in a field office setting.

Modeling: Summary and Conclusions

We believe that the new model and the research that is being carried on as part of the development of the WEPP model will lead to a better understanding of water movement at the soil surface and in shallow ground water systems. By using digitized maps of topography, soils, and geology and geographic information, we hope to incorporate this knowledge into watershed models of various sizes. We expect that the model will lead to major changes in soil conservation technology. It will allow us to depart from the present emphasis on long-term average erosion and to focus on critical erosion events and the statistical probability of such events. Also, it will allow us to plan more specifically for the control of certain aspects of soil erosion, such as the formation of temporary (ephemeral) and deep gullies. It will allow us to make better estimates of the total amount and composition of sediments leaving the field. Finally, we believe that the model will improve our ability to identify critical areas and critical storm events that deliver the bulk of sediments to bodies of water and to plan conservation measures accordingly.

Literature Cited

CTIC, 1986. National Survey, Conservation Tillage Practices. Conservation Technology Information Center, West Lafayette, IN

Edwards, W.M., L.D. Norton and C.E. Redmond, 1988 Soil Sci. Soc. Amer. Journ.52:483-487 Characterizing macropores that affect infiltration into non-tilled soil.

Foster, G. R., L.J. Lane, J.M. Laflen, and R.A. Young, 1981. Estimating erosion and sediment yield on field-sized areas. Trans. Amer. Soc. Agric. Eng. 24 (5): 1253-1262

Foster, G.R., 1982. Modeling the erosion process. In C.T. Hann (ed). Hydrologic modeling of small watersheds, ASEA Monograph Number 5, Amer. Soc. Agric. Eng., St. Joseph, Michigan.

Laflen, J.M. et al., 1978, Soil and water loss from conservation tillage systems. Trans. Amer. Soc. Agric. Eng. 21 (5):881-885.

Laflen, J.M. and T.S. Colvin, 1981. Effect of crop residue on soil loss from continuous row cropping. Trans. Amer. Soc. Agric. Eng. 24 (3):605-609.

Wischmeier, W.H. and D.D. Smith, 1978. Predicting rainfall erosion losses. A guide to conservation planning. United States Department of Agriculture, Agriculture Handbook No. 537, Washington, D.C. 20402

EXPERT SYSTEMS AND RIVER BASIN MANAGEMENT

Mark H. Houck
Professor of Civil Engineering
School of Civil Engineering
Purdue University
West Lafayette, Indiana 47907 U.S.A.

ABSTRACT. Expert systems are an emerging technology that may have significant implications for river basin management. Expert systems incorporate in a computerized environment the heuristics, judgments, experience and knowledge of one or more experts in a particular domain. The expert system is then available for consultation by the non-expert to gain advice and guidance on new problems in the domain. There has been a radical improvement in expert systems technology in the past decade, resulting in order of magnitude improvements in the quality of applications and in the development time for applications. There have been successful applications of expert systems that relate to river basin management problems and there are additional applications currently under development. The potential for successful application of expert systems and the potential for significant improvements in river basin management are great.

Introduction

One of the emerging technologies that will have significant effects on the way we manage our water resources is expert systems. Expert systems represent a new way of viewing problems. It is not a panacea but it is a new tool that can be used in conjunction with others to improve decision making substantially. Expert systems can be embedded in other techniques, such as optimization and simulation models, or these models plus other procedures such as graphics, data base management systems, geographic information systems and other information processing procedures can be embedded in the expert system.

Although expert systems technology has been developed only recently, there are already a significant number of applications in water resources. These applications span a

L. M. Brush et al. (eds.), Taming the Yellow River: Silt and Floods, 85–97.

wide range including the calibration and use of hydrologic and hydraulic models, water supply and waste water management, and hazardous waste management.

The purpose of this paper is to provide an introduction to expert systems and their application in river basin management. To begin, expert systems will be defined and reasons for their construction and application will be considered. The architecture of expert systems and an example of the kind of reasoning or knowledge processing that is the heart of an expert system will be presented. Problems with the development of expert systems and several examples of expert system applications that have been developed or are currently under development will be described.

Expert Systems

"An expert system is a knowledge-based reasoning system that captures and replicates the problem-solving ability of human experts" (Boose, 1986). The definition of an expert system is very general and it could be argued that a great deal of work in the past could be, at least partially, considered to be development of expert systems. Typically, expert systems as they are known today have greater structure than is indicated by Boose's definition. This structure will be described later in more detail.

Why would we want to build an expert system? There are many reasons that an automated reasoning system might be appropriate. Among these are:

* Experts retire, taking their knowledge with them.
* There may be better uses of an expert's time than answering users' questions.
* Expertise may be scarce.
* Expertise may be expensive to deliver.
* Expertise may not be available when it is needed.
* Experts are not always consistent.

In any particular application these reasons or others may be important in deciding to construct an expert system.

The basic components of expert systems have been defined and described by many authors (e.g. Boose, 1986; Hayes-Roth et al., 1983; Ginn and Houck, 1986; Harmon and King, 1985). A typical expert system can be described as having six basic components: a <u>knowledge acquisition facility,</u> a <u>knowledge base,</u> an <u>inference engine</u> or <u>reasoning mechanism,</u> a <u>context,</u> an <u>explanation facility,</u> and a <u>user interface</u> (Maher, 1987). The relationship of these basic components to one another is shown in Figure 1. The <u>knowledge base</u> is the repository for information that is static and domain-wide. The domain of an expert system is the area within which the expert system would

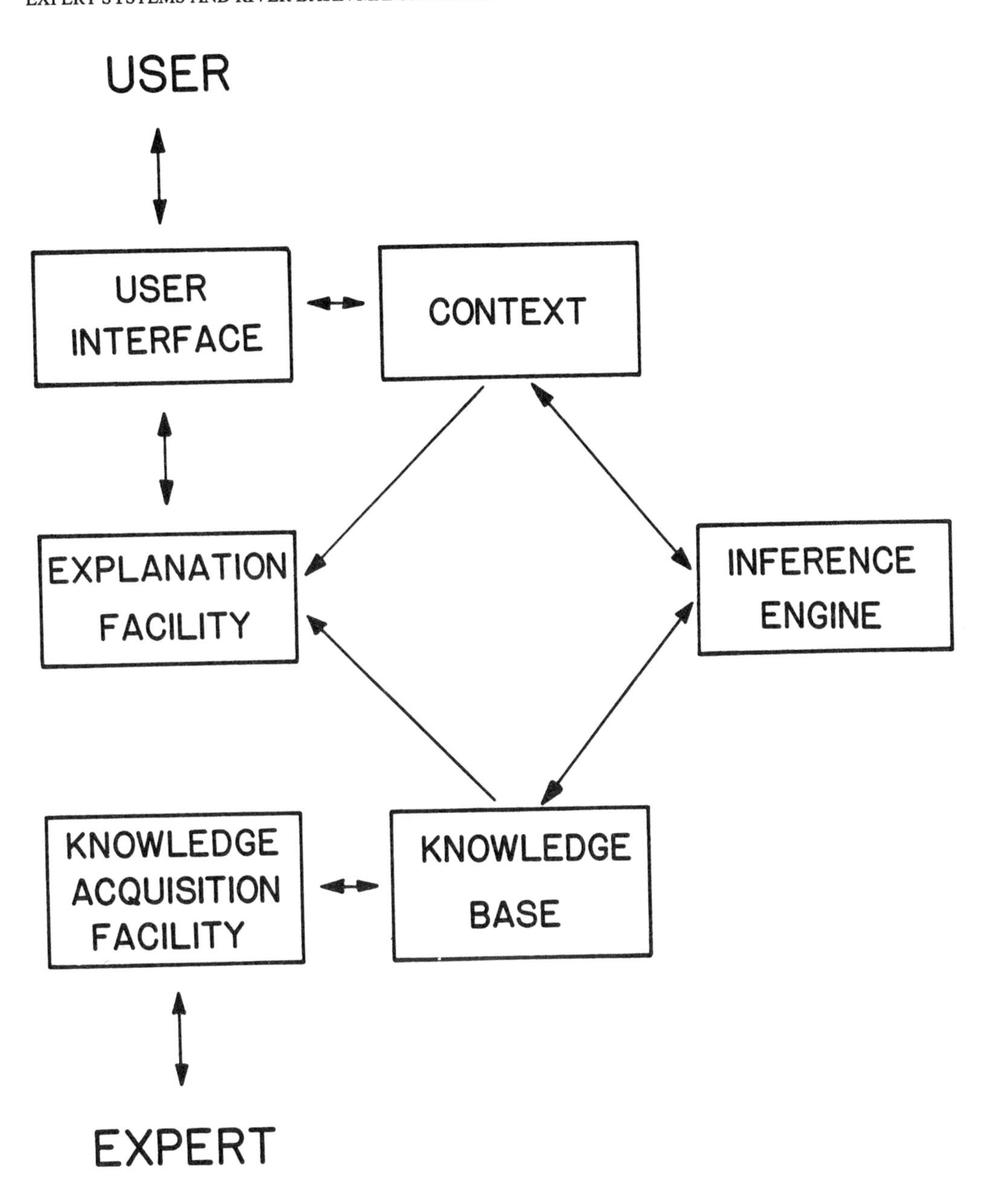

FIGURE 1 Architecture of an Expert System
(reprinted with permission of ASCE)

be called upon to solve problems. The knowledge base would contain not only data that apply to the domain and would not be changing from one problem to the next, but it also may contain empirical rules, theoretical rules or laws, heuristics, and models that may be employed as part of the solution.

The knowledge acquisition facility is the interface between the human expert and the knowledge base. The purpose of the knowledge acquisition facility is to acquire or extract information about the decision making process employed by the expert in the domain of interest. Knowledge acquisition is considered by many to be the largest bottleneck in expert systems development.

There are many ways that have been proposed and have been used to extract knowledge from a human expert. There is even a title for a person who facilitates the knowledge acquisition. That person is known as a knowledge engineer. Among the techniques employed by a knowledge engineer to extract knowledge from the expert or to get the expert to articulate how he or she resolves problems are (Boose, 1986):

* On-site observation - the knowledge engineer observes an expert solving current problems.
* Problem discussion - the knowledge engineer and the expert discuss the data, knowledge, and procedures required or used to solve specific problems.
* Problem description - the expert attempts to define categories into which problems in the domain may be assigned.
* Problem analysis - the knowledge engineer provides a problem from the domain to the expert who then analyzes the problem and describes the procedures and knowledge he or she would use to resolve the problem.

The techniques described above all focus on beginning the process of extracting knowledge from the expert. Each of these must be followed by the analysis and synthesis of the information and knowledge into the knowledge base. Once this is accomplished, then refinement of the knowledge base can continue with one of these techniques:

* Systems refinement - permit the expert to assess the knowledge base that has been developed by testing the expert system with new or different problems.
* System examination - permit the expert to review the knowledge base directly to assess its validity and completeness.
* System validation - permit other experts to criticize the expert system either by directly reviewing its knowledge base or by testing it with other problems familiar to them.

These are only a sample of the possible techniques to extract information and knowledge from a human expert. In any particular case, one or more of these would be appropriate for use in the construction of the expert system.

Unfortunately, all of these have problems associated with them. "Most knowledge engineers do not have sufficient training to interview experts efficiently and effectively" (Boose, 1986). To be an effective knowledge engineer you must be capable of understanding the computerized end of the expert system as well as effectively deposing, interrogating, and interacting with human experts on a complex problem to extract their problem-solving methods, approaches and heuristics. In addition, "domain experts have difficulty expressing their problem-solving knowledge explicitly. They tend to create plausible lines of reasoning which may have little to do with how problems are actually solved" (Johnson, 1983). "Explanations may be in terms of idealized verbal descriptions learned in school, while much expertise seems to be compiled in heuristics accumulated over the years .... Additionally, experts may be wrong when the describe their procedures to knowledge engineers" (Boose, 1986). Although these are significant problems, progress has been made in the last five years developing successful methods for knowledge acquisition.

Another basic component of the expert system is the inference engine or reasoning mechanism. It contains all the procedures for manipulating, searching, and exercising the knowledge base. There are many ways to structure the inference engine to use the knowledge base, and expert systems may be distinguished by the inferencing mechanism used.

When the expert system is to be consulted by a user, there must be an interface between the user and the expert system. This user interface is one of the basic components of the architecture of the expert system and provides all the usual support for the user to input information and acquire information in return from the expert system. This support will depend on the level of sophistication of the user and the computing environment.

One of the key functions of the user interface is to permit the user to describe to the expert system a new problem to be solved. This requires input of data about the problem and these data are stored in another basic component of the expert systems called the context. As the expert system works on the problem to find a solution, it will add information to the context. Eventually, when the expert system has completed its analysis of the problem and arrived at the recommended solution, the complete information about the solution will be stored in the context.

One of the distinguishing characteristics of an expert systems is that the user may ask the questions: how? and why?

The result is a listing of the series of steps taken by the expert system to arrive at an answer to the problem. This may include a listing of the rules from the knowledge base that were employed as well as the order in which they were applied to the problem and any data that were input as well as any information that was deduced by the expert system along the way to finding the answer to the problem. All of these functions are performed by the last basic component of an expert system, the explanation device.

A small example of how an expert system might work may be helpful. Suppose that the knowledge base contained only two rules of the type "if predicate then consequent":

Rule 1: if A then B
Rule 2: if B then C

Assume that A, B and C are some conditions that are either true of false. The expert system begins by the user asking the question: is C true?

This particular expert system uses the inferencing mechanism known as backward chaining. Therefore, the inference engine when asked if C is true, will attempt to find all rules in the knowledge base that conclude C is true. The knowledge base is examined and there is only one rule that concludes with C is true, and that is rule 2. The question is then posed to the user: is B true? If the user can provide the information about whether B is true, then the entire process can stop. In this case, the user has responded with: don't know. Therefore, the expert system will again search the knowledge base to find rules that will help it answer the question: is B true? It finds that rule 1 concludes that B is true if A is true. Therefore, the user is again queried for whether or not A is true. In this case the user has responded: yes. And the expert system has concluded: then C is true! At this point the user may respond with the question: why? An expert system would then respond with the line of reasoning it used to arrive at the conclusion that C is true. This is just one small example illustrating how an expert system might function. There are many variants that make the use of expert systems more appropriate for different environments and different problem domains.

Expert Systems Applications

The secret to successful development of an expert system may rest in three characteristics of the expert system (Rehak, 1983). These three characteristics are obvious yet essential. The first is that the expert system must be useful: it must address a problem of importance. It must also perform well: it must permit the user to address the problem better than using other techniques. And lastly, the expert system needs to be transparent: it must be easy to use and able to provide adequate explanations for all the procedures invoked to address the problem entered by the user.

One of the earlier expert systems developed was DENDRAL (Buchanan and Feigenbaum, 1978), an interpreter of the mass-spectrogram. Its immediate successor was MYCIN (Shortliffe, 1978), which is an expert system that diagnoses bacterial infections in the blood and prescribes suitable drug therapy. MYCIN's knowledge base consists of hundreds of rules such as:

IF (1) the infection is primary-bacteremia, and
(2) the site of the culture is a sterile site, and
(3) the suspected portal of entry of the organism is the gastro-intestinal tract
THEN there is suggestive evidence (0.7) that the identity of the organism is bacteroides.

Note that the rules in this case express a degree of belief; a factor of 0.7 is associated with the inference. Certainty factors that represent a degree of belief in a particular rule are commonly included as an integral part of an expert system. The user can also interrogate the system to find out the reasoning process for the diagnosis and the certainty factor associated with that diagnosis.

Another early expert system is PROSPECTOR, a geologic expert system developed by SRI (Duda et al., 1979) which is used for mineral exploration. This development was supported by the U.S. National Science Foundation and by the U.S. Geological Survey. PROSPECTOR contains submodels for several types of deposits: porphyry copper, porphyry molybdenum, sandstone uranium, nickel sulfide, carbonate lead/zinc. In a consultation with PROSPECTOR, the user enters basic field observations. The program matches these against its submodels, requests additional information as needed and finally draws the conclusions and summarizes the findings.

There are many other expert system applications that have been successfully developed in the past decade that cover fields as diverse as medicine, chemistry, manufacturing and engineering (e.g. Baffaut et al., 1987; Barnwell et al., 1986; Cuena, 1983; Destrigneville et al., 1986; Engman et al., 1986; Feigenbaum, 1981; Gaschnig et al., 1981; Harmon and King, 1985; Harris et al., 1987; Houck, 1985; Hushon, 1986; Kostem and Maher, 1986; Levitt and Kunz, 1987; Maher, 1987; Mullarkey, 1985; Reboh et al., 1982; Ritchie et al., 1986; and Wilson et al., 1986). In the water resources field, there are a number of successful applications of expert systems that have already been described in the literature (Maher, 1987) as well an even greater number that are currently under development and soon will be appearing in the literature

(Ortolano and Steinemann, 1987). These applications are listed in Tables 1 and 2. Table 1 includes those water resources applications that have already appeared in the literature (Maher, 1987) and Table 2 contains a listing of those applications that are still under development and have not been described and discussed in the literature yet (Ortolano and Steinemann, 1987). These applications are divided into three sections. The first concerns applications in hazardous waste management. The second concerns water supply and waste water management expert systems. And the third concerns applications of expert systems for calibration and use of models.

Many of the applications in Tables 1 and 2 are important for river basin management. For example, during the analysis of river basin problems and the development of designs and policies for river basin management, models to consider these parts of the water resource might be constructed: meteorology and climate, rainfall and runoff, watershed hydrology, urban hydrology, erosion and sediment production, sediment transport, river hydraulics, water quality, etc. Many of these are complex and require expert judgment for calibration and use. Expert systems have been shown to be effective in model calibration and the interpretation of model results. There is a great potential in this type of application for expert systems to improve the quality of decision making.

Another area of potential application for expert systems in river basin management is in the real-time operation of control facilities such as reservoirs. There have already been efforts to develop expert systems to consider multiple-purpose reservoir operations, special operations of a reservoir system during droughts, and special operations of reservoir systems during floods. These efforts have met with initial success and are continuing.

TABLE 1. DEVELOPED EXPERT SYSTEM APPLICATIONS

Hazardous Waste Management

* Advise on the emergency response to chemical spills
* Generate work plans for abandoned waste site rehabilitation
* Rank sites for safety and environmental damage
* Evaluate hazardous waste sites
* Assess the efficiency of hazardous waste site liners
* Diagnose failures of waste incineration facilities

Water Supply and Wastewater Management

* Evaluate and issue permits for sewage disposal sites
* Operate activated sludge systems in response to toxic inputs

* Estimate the probable maximum flood

Calibration and Use of Models

* Calibrate a water quality model (QUAL2E)
* Calibrate a snowmelt-runoff model

TABLE 2. NEW EXPERT SYSTEM APPLICATIONS

Hazardous Waste Material

* Act as response tool for containment of accidental spills
* Guide reviewers of hazardous waste facility closure plans under the Resource Conservation and Recovery Act (RCRA)
* Assist EPA reviewers of hazardous waste facility permit applications
* Identify and screen remedial technologies feasible for a given site
* Identify theneed for external leak detection monitoring of underground storage tanks
* Assist in siting hazardous waste disposal facilities near wetlands environments
* Analyze unknown contents of waste containers
* Interpret goundwater sample data from a mass spectrum analysis
* Evaluate impacts of potentially hazardous dredged material disposal
* Provide remedial action advice for controlling leaks from underground storage tanks
* Select and price remedial technologies at Superfund sites to forecast budget needs

Water Supply and Wastewater Management

* Assist trickling filter plant operators by diagnosing failures and suggesting remedies
* Improve operation of aerobic digester processes and diagnose problems due to poor stability in small treatment plants
* Schedule water treatment plant pumping operations, considering projected water demand and energy costs
* Assist water treatment plant operators in assessing water quality and quantity; suggest required maintenance
* Analyze survey data collected from a scan of the water distribution system; determine points of leakage
* Assist activated sludge plant operators in diagnosing problems and improving plant performance
* Operate reservoirs to maximize hydropower and mitigate adverse environmental impacts

* Operate water supply systems during drought conditions

Calibration and Use of Models

* Select the best mathematical model to predict water quality where wastewater discharge mixes with receiving water
* Use groundwater flow models and calibrate model parameters using field data
* Tutor laboratory students and act as a "front end" to large computer programs
* Assist non-experts in use and calibration to EPA's Storm Water Management Model

---

The potential for improvement for decisions making in this area is also great.

There are also numerous applications of expert systems that are already under development to control hazardous and obnoxious wastes. These are important for a river basin because hazardous waste can contaminate the water resource, either the goundwater or surface waters, and affect the management of the entire basin. In addition, this category of hazardous and obnoxious wastes may also include such things as dredge material disposal that may be obnoxious and undesirable, but not necessarily hazardous. Once again, in this area of application of expert systems technology, the potential for significant improvement exists.

Summary

Expert systems are an emerging technology that may have significant implications for river basin management. Expert systems incorporate in a computerized environment the heuristics, judgments, rules of thumb, experience and knowledge of one or more experts in a particular problem domain. The expert system is then available for consultation by the non-expert to gain advice and guidance on new problems in that domain. There has been a radical improvement in expert systems technology in the past decade, resulting in order of magnitude improvements in the quality of applications and in the development time for applications.

There have been successful applications of expert systems that relate to river basin management problems. There are additional applications currently under development. And the potential for successful application of expert systems and the potential for significant improvement in river basin management are great.

References

1. Baffaut, C., S. Benabdallah, D. Wood, J. Delleur, M.H. Houck, J.R. Wright, "Development of an Expert System for the Analysis of Urban Drainage Using SWMM (Storm Water Management Model)", *Technical Report 180*, Purdue University Water Resources Research Center, West Lafayette, Indiana, June, 1987.

2. Barnwell, T.O., Jr., L.C. Brown, and W. Marek, *Development of a Prototype Expert Advisor for Enhanced Stream Water Quality Model QUAL2E*, Interim Report, U.S. Environmental Protection Agency, Environmental Research Laboratory, Athens, Georgia, September, 1986.

3. Boose, J.H., *Expertise Transfer for Expert System Design*, Elsevier, New York, New York, 1986.

4. Buchanan, B., and E. Feigenbaum, "DENDRAL and META-DENDRAL: Their Applications Dimension", *Artificial Intelligence*, Vol. **11**, 1978.

5. Cuena, J., "The Use of Simulation Models and Human Advice to Build an Expert System for the Defense and Control of River Floods," *Proceedings of the Eighth International Joint Conference on Artificial Intelligence*, pp. 246-249, August, 1983.

6. Destrigneville, B., P. LeCloitre, G. Lecoeur, D. Trayaud, and R. Macgilchrist, "Sewerage Rehabilitation Planning Expert System, *Water and Information Systems Colloquim*, Ecole Nationale des Ponts et Chaussees, Paris, pp. 189-195, May, 1986.

7. Duda, R., Gaschnig, J. and Hart, P., "Model Design in the Prospector Consultant System for Mineral Exploration", in *Expert Systems in the Micro-Electronic Age*, D. Michie (ed.), Edinburgh University Press, pp. 153-167, 1979.

8. Engman, E.T., Rango, A., and Martinec, J., "An Expert System for Snowmelt Runoff Modeling and Forecasting," *Proceedings, Water Forum '86*, American Society of Civil Engineers, Vol. **1**, pp. 174-180, August, 1986.

9. Feigenbaum, E.A., "Expert Systems in the 1980's" in *Machine Intelligence, Infotech State of the Art Report*, A. Bond (ed.), series 9, no. 3, Pergamon Infrotech Limited, 1981.

10. Gaschnig, J., R. Reboh, and J. Reiter, *Development of a Knowledge-Based Expert System for Water Resource Problems*, SRI International, Menlo Park, California, August, 1981.

11. Ginn, R., and M.H. Houck, "Calibration of a Reservoir Operations Model and an Expert Systems Framework for Reservoir Operations," Technical Report CE-HSE-86-11, School of Civil Engineering, Purdue University, West Lafayette, Indiana, September, 1986.

12. Harmon, P., and D. King, Expert Systems, John Wiley & Sons, Inc., New York, New York, 1985.

13. Harris, R., L. Cohn, and W. Bowlby, "Designing Noise Barriers Using the Expert System CHINA, Journal of Transportation Engineering, Vol. **113**, No. 2, 1987.

14. Hayes-Roth, R., D.A. Waterman, and D.B. Lenat (eds.), Building Expert Systems, Addison-Wesley, Reading, Massachusetts, 1983.

15. Houck, M.H., "Designing an Expert System for Real-Time Reservoir System Operation", Civil Engineering Systems, Vol. **2**, pp. 30-37, March, 1985.

16. Hushon, J.M., "Response to Chemical Emergencies, Environmental Science and Technology, Vol. **20**, No. 2, pp. 118-121, 1986.

17. Kostem, C. and M.L. Maher (eds.), Expert Systems in Civil Engineering, American Society of Civil Engineers, April, 1986.

18. Levitt, R.E., and J.C. Kunz, "Using Artificial Intelligence Techniques to Support Project Management", Journal of Artificial Intelligence in Engineering Design, Analysis and Manufacturing, Vol. **1**, No. 1, 1987.

19. Maher, M.L. Expert Systems for Civil Engineers: Technology and Application, American Society of Civil Engineers, New York, New York, 1987.

20. Mullarkey, P.W., "CONE - An Expert System for Interpretation of Geotechnical Characterization Data from Cone Penetrometers", Doctoral Dissertation, Carnegie-Mellon University, Pittsburg, Pennsylvania, 1985.

21. Ortolano, L., and A.C. Steinemann, "New Expert Systems in Environmental Engineering", Journal of Computing in Civil Engineering, forthcoming.

22. Reboh, R., J. Reiter, and J. Gaschnig, Development of a Knowledge-based Interface to a Hydrological Simulation Program, SRI International, Menlo Park, California, May, 1982.

23. Rehak, D.R., "Expert Systems in Water Resource Management", Proceedings of a Conference on Emergency Computer Techniques in Stormwater and Flood Management, William James (ed.), American Society of Civil Engineers, New York, 1983.

24. Ritchie, S., C. Yeh, J. Mahoney, and N. Jackson, "Development of an Expert System for Pavement Rehabilitation Decision Making", Transportation Research Record, No. 1077, pp. 96-103, National Academy of Science, Washington, D.C., 1986.

25. Shortliffe, E.H., Computer-Based Medical Consultants: MYCIN, American Elsevier, New York, 1978.

26. Wilson, J.L., G.K. Mikroudis, and H.Y. Fang, "GEOTOX: A Knowledge-Based System for Hazardous Site Evaluation", Applications of Artificial Intelligence in Engineering Problems, D. Sriram and R. Adey (eds.), Springer-Verlag, April, 1986.

SEDIMENT YIELD VARIATIONS AS A FUNCTION OF INCISED CHANNEL EVOLUTION

S.A. Schumm and Allen C. Gellis
Department of Earth Resources
Colorado State University and
Water Engineering and Technology, Inc.
Fort Collins, Colorado 80521, U.S.A.

ABSTRACT. Part of the high sediment load in the Yellow River is the result of gully erosion in the Loess Plateau, and part of the high sediment load in the Colorado River is the result of incised-channel (arroyo) development in the 19th century. Therefore, processes contributing sediment loads to both rivers may be similar. Field investigations in southwestern (Arizona, New Mexico) and southeastern (Mississippi, Tennessee) United States show that incised channels follow an evolutionary development that strongly influences sediment production through time. When the channels incise, sediment loads are initially high, but they decrease as the channels evolve to a new condition of relative stability. Experimental studies of incised channels and rejuvenated drainage basins also show this type of sediment yield change with time. Therefore, it is reasonable to expect a natural decrease of sediment loads through time in rivers that derive much of their sediment load from incised channels. Land managers and river engineers should be aware of this natural sequence of events and plan for it.

Introduction

The Yellow River transports a very high sediment load that is derived in part from gully erosion in the Loess Plateau. The Colorado River also transports a high sediment load that is derived in part from arroyo incision that occurred in the alluvial valleys of the southwest in the late 19th century. Therefore, a large part of the sediment loads of both rivers may be derived from tributary-channel incision. The development of a model of incised channel (gullies, arroyos, channelized streams) evolution through time can provide a basis for predicting future sediment yields from tributary

L. M. Brush et al. (eds.), Taming the Yellow River: Silt and Floods, 99–109.

basins and future sediment loads in the main channels. Therefore, the objective is to use the model of incised-channel evolution to predict future trends in the sediment loads of major rivers such as the Yellow and Colorado Rivers. Both field and experimental studies will be used to demonstrate the basis for such a prediction.

Field Investigations

Numerous field studies of incised channels reveal that they evolve from an initial condition of channel instability and high sediment production to a new condition of relative channel stability and lower sediment production (Schumm, et al., 1984, 1987a). The initial channel deepening and subsequent bank erosion greatly increases sediment contributions to downstream channels. Hotopha Creek in Mississippi provides an example of the morphologic and hydraulic changes of such an incised channel (Harvey, et al., 1987). A five-stage incised-channel evolution model shows channel cross section changes through time (Fig. 1), and Table 1 provides data on average channel characteristics at each stage of this development. It should be noted that the model is based on location-for time substitution that uses present changes along the channel to show channel evolution at one cross section through time (Schumm, et al., 1984). Therefore, the Stage 1 cross section (Fig. 1) is the farthest upstream, and the Stage 5 cross section is the farthest downstream. In Hotopha Creek the channel was deepened and straightened (channelized) in 1961 in order to reduce the frequency of overbank flooding. The result of channelization was incision that proceeded in or at the initial stage of incision. The steepened gradient and the confinement of flood waters in the incised channel caused further incision (Stage 2), bank collapse, and channel widening (Stage 3). The result of widening and continued sediment contributions from upstream reaches caused the channel to aggrade (Stage 3), which increased bank stability (Stage 4).

Formation of an inner channel and vegetative colonization of the recently deposited sediment completed the channel adjustment (Stage 5) in a period of about 30 years in this area of high annual precipitation (1270 mm/yr). Table I shows that sediment transport and sediment delivery to downstream channels is a maximum during Stages 2, 3 and 4. Studies of similar channels in Tennessee confirm this change of sediment load with time (Andrew Simon, 1988, written communication). For example, sediment loads in the South Fork Obion River increased from 50 tons per day before channelization (1958-65) to 624 tons per day immediately following incision (Stages 2 and 3) (1966-70) and decreased to 226 tons per day as the channel adjusted (1977-85) to Stage 5.

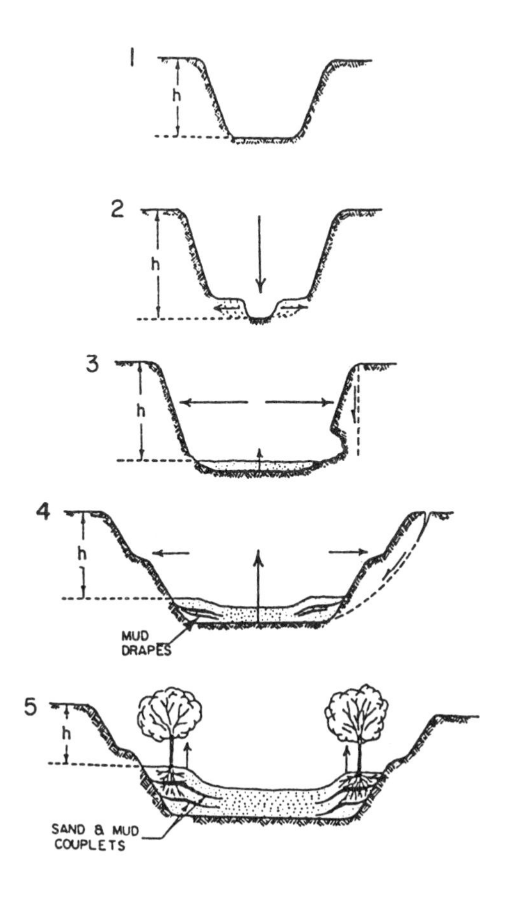

Figure 1. Location for time substitution shows evolution of a channelized stream at a cross section by comparing cross sections from upstream (1) to downstream (5). Channelized stream (1) initially deepens (2), widens (3) and then stabilizes (4, 5) as deposition decreases exposed bank height (h) and width-depth ratio increases (after Schumm, et al., 1984).

TABLE I. Changing Character of Hotopha Creek Channel Following Channelization (from Harvey, et al., 1987). Numbers in brackets are one standard deviation.

Mean Channel Character	Stage of Evolution (Figure 1)				
	1	2	3	4	5
Top width(m)	20(1.3)	19(4.3)	35(18.4)	35(5.1)	48(12.3)
Bank height(m)	4(0.6)	7(0.2)	7(0.7)	7(0.7)	3(0.7)
Width/depth ratio	5(0.1)	3(0.6)	5(2.7)	5(1.0)	16(4.6)
Gradient	.0047	.0035	.0036	.0015	.0013
Sediment load m^3/sec at effective discharge ($34m^3$/sec)	.017	.037	.031	.076	.017

The Rio Puerco, an arroyo that is tributary to the Rio Grande, has a history of sediment production and channel evolution similar to the channelized streams. Bryan and Post (1927) calculated that 33 million tons per year of sediment left the valley between 1887 and 1928 after channel incision. However, during the period 1948-1968 when the lower part of the Rio Puerco had evolved to Stage 5 (Fig. 1), only 6 million tons per year of sediment left the valley.

Experimental Studies

In order to obtain informtion on channel and drainage network evolution a number of experimental studies of stream channel, drainage basin, and alluvial fan response to rejuvenation have been performed (Schumm, et al., 1987b). Two of these experiments relate to the problem of changing sediment yield and sediment loads through time.

Begin, et al. (1980, 1981), created an incised channel by lowering baselevel at the outlet of a large flume. The immediate result was a dramatic increase in the sediment leaving the flume. This high sediment yield reflected channel bed degradation due to knickpoint recession and bank failure. However, during the 700 minute course of this experiment, sediment yield decreased logarithmically from a maximum during the first few minutes of channel incision as follows:

$$Q_s = 135t^{-0.48} \tag{1}$$

where Q_s is sediment discharge (g/sec) and t is time since lowering of base-level. We believe that logarithmic decrease of sediment yield replicates the change of sediment loads in the incised channels studied in the field.

Another experimental study (Parker, 1977; Schumm, et al., 1987b) involved the rejuvenation of a drainage network in a 10 by 15 m rainfall-erosion facility. When base level was lowered, the existing drainage network incised. As a result, there was considerable variability of sediment yield from the drainage basin, as the system hunted for a new condition of equilibrium. Nevertheless, following an initial maximum a logarithmic decrease of sediment yield occurred as follows:

$$Q_s = 850\ V^{-0.86} \tag{2}$$

where Q_s is sediment discharge (g/sec), and V is the volume of rainfall (m^3) applied to the surface of the drainage basin, which is an index of the passage of time. Even without base level lowering there was a decrease of sediment discharge with time, as the drainage network grew and stabilized as follows:

$$Q_s = 78\ V^{-0.15} \tag{3}$$

The experimental studies confirm that, following base level lowering, there will be very high initial sediment production that is caused by rapid channel incision, but a natural decrease will occur as the system adjusts to a new condition of stability.

Discussion

As suggested earlier, the documented changes of sediment production in incised channels leads us to believe that the high sediment loads of the Yellow and Colorado River may decrease naturally through time, as the incised channels adjust and stabilize.

Recent studies of the upper Colorado River basin (Fig. 2) appear to provide an example of this postulated change. In the late 19th century and early 20th century many valleys in the Colorado River basin were incised as arroyos developed. These arroyos delivered vast amounts of sediment to the Colorado River, and sediment loads in the Colorado River were high.

More than 18 million people and 1.7 million acres of agricultural land require water from the Colorado River (U.S. Dept. of Interior, 1987). The variety of demands on this water places great emphasis not only on the quantity of water, but also on its quality. The situation is complicated nationally because seven states are involved: Wyoming, Colorado, Utah, New Mexico, Arizona, Nevada and California (Mann, et al., 1974), and it is further complicated internationally by Mexico's use of this water (Holburt, 1977).

Since sediment sampling in the Colorado River at Grand Canyon started in the mid-1920s, the sediment load has decreased and it has been suggested that the change was relatively abrupt during the period 1940-45. In the Grand Canyon the decrease during this period was from 50 to 100 million tons per year (Fig. 3). This change of sediment load was first reported by Thomas, et al. (1963), but except for mention by Hadley (1974) and Graf (1985), it has escaped the serious attention of hydrologists and geomorphologists. Thomas, et al. (1963) attributed the significant decreases of sediment load to drought in the major sediment-producing areas of the drainage basin. The change has also been attributed to changes of sampling procedures by the U.S. Geological Survey (Schumm, et al., 1987a). Others have claimed that the decrease is the result of over 20 years of conservation and flood control efforts in the Colorado River drainage basin. Hadley (1974) for example considered the decrease in sediment load in the Colorado River and in major tributaries such as the San Juan and Green Rivers to be due to a major reduction in livestock numbers and to the erosion control efforts during

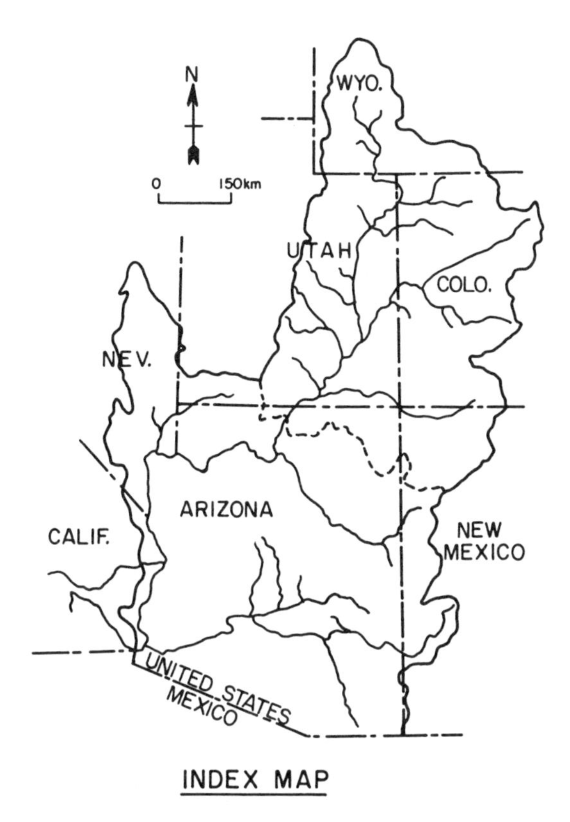

Figure 2. Index map, Colorado River basin. Dashed line indicates boundary of upper Colorado River basin.

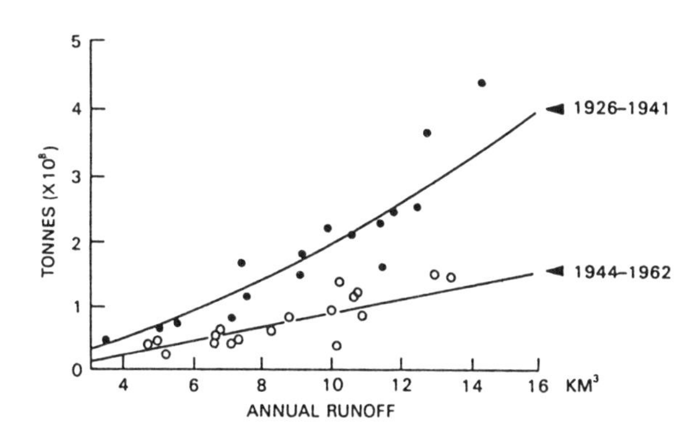

Figure 3. Relation between annual runoff and suspended sediment load of Colorado River at Grand Canyon, 1926 to 1962 (from Graf, et al., 1987).

the previous decades.

The field and experimental studies suggest another reason for the decrease of sediment load. As noted above, during the latter part of the 19th century many arroyos formed, and channels incised throughout the Colorado Plateau, and large quantities of sediment were delivered to the Colorado River from incised tributary streams. The experimental and field studies indicate that, following a major period of sediment production, there will be a decrease as the readily available sediment supply is exhausted, as the incised channels widen and become less efficient transporters of sediment and as new floodplains develop in the valleys. Studies in Escalante River, Kanab Cr. and Little Colorado River (Hereford, 1984, 1985; Graf, et al., 1987) reveal that sediment storage is taking place in many valleys of the southwestern United States. However, this explanation of decreased Colorado River sediment loads requires that there be a progressive decrease rather than an abrupt decrease (Fig. 3). Indeed, when sediment load data are plotted as a time series, a different interpretation of the sediment load change emerges (Fig. 4). It is probable that the decrease of sediment load as indicated by the experimental results is progressive, but the record at Grand Canyon is complicated by major flood events and drought years. The curve of Fig. 4 corresponds well with the results of experimental studies of channel and drainage basin rejuvenation (Eqs. 1 and 2).

Conclusions and Applications

The field and experimental studies have demonstrated that time is an important variable, as incised channels evolve (Fig. 1) and therefore it should be an important variable in any scheme to manage incised-channel erosion and sediment loads. Fig. 5 is a conceptual diagram that shows the change in sediment yield and active or incised channel drainage density (length of incised channels per unit area) with time. In a drainage basin that has been rejuvenated and in which the drainage network is incising, sediment production will increase as the length of incising channels increases (Fig. 5, Times 1 to 4). However at Time 4, maximum channel headward growth has occurred and the channels begin to stabilize between Times 4 and 7. By understanding this cycle of channel incision and gullying from initial stability (Time 1) to renewed stability (Time 8), it is possible to select the times in the cycle when land management and incised-channel control practices will be most effective. For example, gullies just forming (Times 1 or 2) and gullies almost stabilized (Times 6, 7 or 8) will be the most easily controlled. The efforts at Times 1 and 2 will be most effective in preventing erosion and high sediment production whereas efforts at Times 6 and 8 will have little

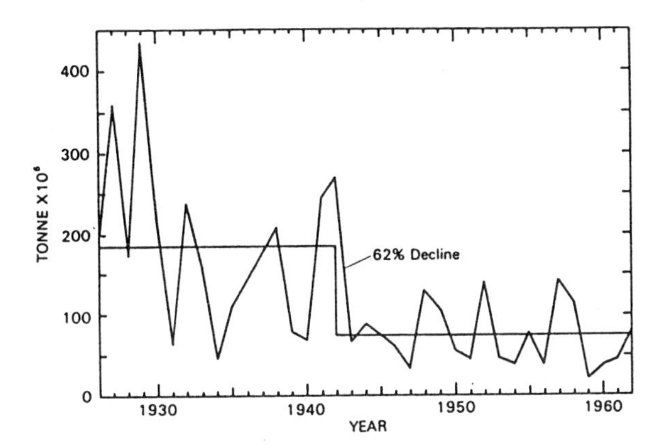

Figure 4. Annual sediment and water discharge, Colorado River at Grand Canyon (from Graf, et al., 1987). Abrupt decline in 1942 is suggested by horizontal lines, which are the median sediment loads for the periods 1926-1941 and 1942-1962.

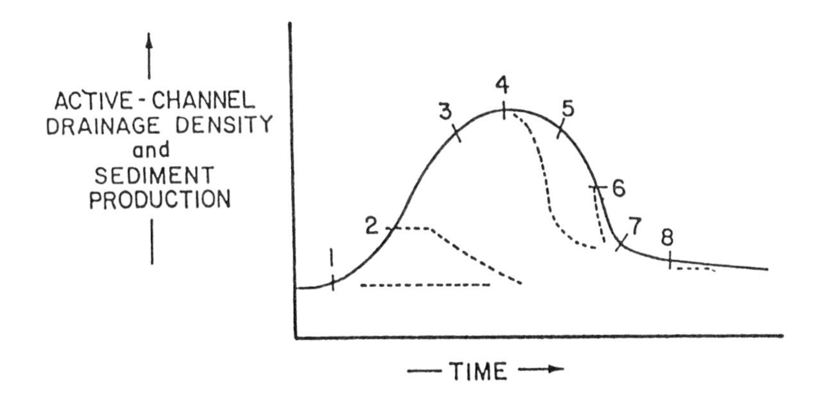

Figure 5. Hypothetical model of change of active-channel (gully) drainage density and sediment production at different times following channel incision.

effect, as the channels are stabilizing naturally. At Time 4 control will be difficult and expensive. Obviously, consideration of this complex evolving system for only short time spans can yield erroneous conclusions.

The field and experimental studies indicate that with time there should be a natural decrease of sediment production from incised channels and, therefore, a decrease of sediment loads in the main channel to which they are tributary. We suggest that there may be an analogy between the Colorado and Yellow Rivers, if indeed much of their sediment loads are derived from incised-channel erosion. Therefore, additional studies are needed in the Loess Plateau in order to determine if incised channels are major producers of sediment and to determine if such channels follow an evolutionary development as outlined in Fig. 1.

References

Begin, Z.B., Meyer, D.F., and Schumm, S.A., 1980, Sediment production of alluvial channels in response to base level lowering: Amer. Soc. Agric. Engr., Trans., v. 23, p. 1183-1188.

Begin, Z.B., Meyer, D.F., and Schumm, S.A., 1981, Development of longitudinal profiles of alluvial channels in response to base level lowering: Earth Surf. Proc. and Land Forms, v. 6, p. 49-68.

Bryan, Kirk, and Post, G.M., 1927, Erosion control of silt on the Rio Puerco, New Mexico: unpublished report, 99 p.

Graf, W.L., 1985, The Colorado River; Instability and basin management: Washington, D.C., Assoc. American Geographers, 88 p.

Graf, W.L., 1987, Late Holocene sediment storage in canyons of the Colorado plateau: Geol. Soc. America, Bull., v. 99, p. 261-271.

Graf, W.L., Hereford, Richard, Laity, Julie, and Young, R.A.: Colorado Plateau: in Graf, W.L. (ed.) Geomorphic systems of North America, Geol. Soc. America, Centennial Spec., v. 2, p. 259-302.

Hadley, R.F., 1972, Sediment yield and land use in the southwest United States: Internat. Assoc. Sci. Hydrology, Pub. 113, p. 96-98.

Harvey, M.D., Watson, C.C., and Peterson, M.R., 1987, Recommended improvements for stabilization of Hotopha Creek

Mississippi: unpublished report, Water Engineering and Technology, Inc., Fort Collins, Colorado, 97 p.

Hereford, Richard, 1984, Climate and ephemeral stream processes; 20th century geomorphology and alluvial stratigraphy of Little Colorado River, Arizona: Geol. Soc. America, Bull., v. 95, p. 654-668.

Hereford, Richard, 1985, Modern alluvial history of the Paria River drainage basin, southern Utah: Quaternary Research, v. 25, p. 293-311.

Holburt, M.B., 1977, The 1973 agreement of Colorado River salinity between the United States and Mexico: Proc. of Nat'l. Conf. Irrigation Return Flow Quality Management, Fort Collins, Colorado, p. 325-333.

Mann, D., Westerford, G., and Nichols, P., 1974, Legal-political history of water resource development in the Upper Colorado River basin: Lake Powell Research Project Bull. 4, Inst. Geophysics and Planetary Physics, Univ. of California, Los Angeles, California, 62 p.

Parker, R.S., 1976, Experimental study of drainage basin evolution and its hydrologic implications: unpublished Ph.D. dissertation, Colorado State University, 331 p.

Patton, P.C., and Schumm, S.A., 1981, Ephemeral stream processes: Implications for studies of Quaternary valley fills; Quaternary Research, v. 15, p. 24-43.

Schumm, S.A., 1977, The fluvial system: Wiley, New York, 338 p.

Schumm, S.A., Harvey, M.D., and Watson, C.C., 1984, Incised channels: Morphology, dynamics, and control: Water Res. Publ., Littleton, Colorado, 200 p.

Schumm, S.A., Mosley, M.P., and Weaver, W.E., 1987, Experimental fluvial geomorphology: Wiley, New York, 413 p.

Schumm, S.A., Gellis, A.C., Watson, C.C., 1987a, Variability of sediment and salt loads in the upper Colorado River basin: Significance for conservation and management; unpublished report, Water Engineering and Technology, Inc., Fort Collins, Colorado, 101 p.

Thomas, H.E., et al., 1963, Effects of drought in the Colorado River basin: U.S. Geol. Survey Prof. Paper, 372-F, 50 p.

U.S. Department of Interior, 1987, Quality of water Colorado River basin: Prog. Rept. No. 13

SYNTHESIS OF DATA ON BED CONFIGURATIONS IN ALLUVIAL CHANNELS, AND THE EFFECT OF WATER TEMPERATURE AND SUSPENDED-LOAD CONCENTRATION

John B. Southard
Dept. of Earth, Atmospheric, and Planetary Sciences
Massachusetts Institute of Technology
Cambridge, MA 02139, USA

Introduction

Turbulent flows in alluvial channels mold beds of gravel, sand, and coarse silt into a wide variety of configurations. These configurations are in-channel phenomena (the microforms and mesoforms of Jackson, 1975, and Crowley, 1983). The relationships between these in-channel features, on the one hand, and elements of channel geometry itself (macroforms), like bars of various kinds, on the other hand, are not always clear; this paper will be restricted to in-channel features. Bed configurations in rivers, and the oceans as well, have had a long history of study by hydraulic engineers, marine geologists, sedimentologists, and geomorphologists.
This paper concentrates on rational empirical analysis of bed configurations in rivers. It is in three parts: the first part deals with the existence and general characteristics of bed configurations, with emphasis on useful empirical predictors, and the second and third parts discuss the effects of water temperature and suspended-load concentration on the bed configuration. The paper is a review of previous experiments and observations on these topics, along with a summary of some new data synthesis on the occurrence and stability of the various bed configurations in flumes (Southard and Boguchwal, in review a), as well as analysis of water-temperature effects (Southard and Boguchwal, in review, b). Useful summaries of the occurrence, characteristics, and dynamics of bed configurations may be found in Vanoni (1975), Yalin (1977), Allen (1982), and Middleton and Southard (1984).

It is useful to distinguish between the bed configuration (the overall bed geometry that exists at a given time in response to the flow), a bed form (any individual element of this overall geometry, like a ripple or a dune), and the bed state (the aggregate or ensemble of like bed configurations

L. M. Brush et al. (eds.), Taming the Yellow River: Silt and Floods, 111–138.

that can be produced by a given mean flow over a given sediment bed). One can also use the term bed phase for recognizably or qualitatively different kinds of bed configurations that are produced over some range of flow and sediment conditions.

Geologists and hydraulic engineers alike have marveled at bed forms. Apart from their intrinsic fascination as a physical phenomenon, bed forms are of practical importance: they constitute obstacles to navigation and a threat to submarine structures; when present in a channel they are the most important determinant of flow resistance; and they have an important and complicating effect on bed-material sediment transport rate. Moreover, the existence of rugged bed configurations leads to the development of some of the most common stratification features in sedimentary deposits, and these structures among the most useful tools available in the interpretation of ancient sedimentary environments.

Nature offers a rich variety of bed configurations produced by unidirectional sediment-transporting flow. Over a wide range of conditions the bed forms that constitute these configurations are ripple-shaped forms that are generally oriented transverse to the flow and have gently sloping upstream surfaces and steeply sloping downstream surfaces. These bed forms move downstream by erosion on the upstream sides and deposition on the downstream sides. In contrast, antidunes are smooth and symmetrical undulations, also transverse to flow, in phase with water-surface waves. In certain ranges of flow a planar transport surface is the stable configuration. Flow-parallel bed forms, while important, are beyond the scope of this paper.

The dynamics of flow-transverse bed configurations have been elusive; although much progress has been made (e.g., Engelund, 1970; Smith, 1970; Engelund and Fredsøe, 1974; Richards, 1980) since the pioneering work of Exner (1925) and Kennedy (1963), prediction of the scale, geometry, and even range of existence of these features has been largely empirical. The underlying reasons for the existence of bed forms are not obvious. At first thought it might seem that the natural mode of transport would be over a planar bed. Only in laminar flow, however, is this invariably so. In turbulent flow, rugged bed configurations are the dominant transport-surface geometry over a wide range of conditions, although plane beds are important also. The dynamics of bed configurations is one of the most difficult problems in turbulent flow. The essential element of complexity that makes theory so difficult is this: as the flow molds the bed by erosion and deposition, the bed geometry thus generated changes the structure of the flow itself and therefore the nature of the sediment transport in fundamental ways, so there is a strong interaction or feedback between the bed and the flow.

Even aside from theory, the status of observational

understanding of bed configurations leaves much to be desired. It is easy to observe bed configurations in flumes, and there have been many extensive series of experiments, perhaps the most noteworthy being those by Gilbert (1914) and by Simons and coworkers (Simons and Richardson, 1962, 1963; Guy et al., 1966). Even in the laboratory, however, there is room for improvement, because it is difficult to observe the bed when the transport rate is high, and the usually small width-to-depth ratios tend to inhibit full three-dimensional development of the bed configuration, which is of great importance for flow resistance as well as for sedimentary structures. And even the largest flumes are restricted to depths at the lower end of the range of conditions in rivers. Observations of the bed configuration in rivers have been not only limited by technical difficulties but also complicated by unsteadiness of the flow, which causes the bed configuration to lag behind the flow (Allen, 1973, 1974, 1976a, 1976b, 1976c; Gee, 1975; Allen and Friend, 1976; Fredsøe, 1979). This effect of bed-form lag or disequilibrium makes it much more difficult to decipher the relationships among bed phases.

Survey of the Hydraulic Relations of Equilibrium Bed Configurations

INTRODUCTION

This section reviews the status of knowledge of the relations among the various kinds of bed configurations in equilibrium with steady and uniform flows. For a more detailed treatment, see Southard and Boguchwal (in review, a). A great many investigators have attempted to develop predictors, or existence diagrams, for bed configurations from flume experiments or field studies. Notable examples are those of Simons and Richardson (1966), Bogardi (1961), Southard (1971; Middleton and Southard, 1984), Engelund and Fredsøe (1974), and Vanoni (1974); see also the detailed summary by Allen (1982, p. 336-343). Most of these diagrams have been two-dimensional, presumably because of what has been perceived as the difficulty of developing a useful and practical diagram in three or more dimensions. I hope to show here that a satisfactorily complete existence diagram must have at least three dimensions but that such a diagram can be useful and practical for both engineers and scientists interested in fluvial bed configurations.

DIMENSIONAL ANALYSIS

Variables characterizing the bed configuration are those that describe the flow, the sediment, and the fluid. Fluid density ρ and viscosity μ are needed because of their effect on fluid forces on grains and on the bed forms themselves. Sediment size D is needed because of its effect on particle weight and therefore on entrainment and settling of grains, and on skin friction. Sediment density ρ_s is needed separately from its role in the submerged specific weight of the sediment γ' because of its role in the inertial forces produced by accelerations of sediment particles. The acceleration of gravity g is important in the behavior of surface gravity waves in certain ranges of flow, but it need not be included separately as long as γ' is included in the list. The three variables ρ, ρ_s, and γ' therefore must all appear separately.

Choice of variables to characterize the flow is more controversial. If channel width is ignored, two variables are needed. One of these must express the scale of the flow normal to the bed, and is most naturally the mean flow depth d. The other must be a measure of flow strength, and could be bed shear stress τ_o, mean velocity U (or surface velocity), or stream power P. The first might seem best, in that the shear stress is what moves the sediment. But because of the great decrease in form resistance in the transition from dunes to plane bed (see below), there is a range of τ_o for which three values of U are possible (Brooks, 1958; Kennedy and Brooks, 1963); this results in serious ambiguity in use of τ_o (or P) to characterize the bed configuration. Although less directly related to the movement of the sediment, U does not suffer from this ambiguity and will be used below, together with d, to characterize the flow. (Even better would be use of the skin-friction component of τ_o, if a simple and accurate way of partitioning the drag were available.) The goal here is to show unambiguously the relationships among bed phases. If a one-to-one correspondence between bed configurations and combinations of variables is achieved, then one can be confident that the graph predicts the bed configuration that would result from a given combination of flow, sediment, and fluid; in this respect U is better than τ_o as a flow-strength variable.

Dimensional analysis provides a set of four dimensionless variables equivalent to the seven variables (U, d, D, ρ_s, ρ, μ, γ') discussed above. One such set, which has the advantage of being the most directly related to the physics of the flow, contains the mean-flow Reynolds number $\rho Ud/\mu$, a Froude number $\rho U^2/\gamma' D$ based on γ', the relative roughness d/D, and the density ratio ρ_s/ρ. An equivalent set (Southard, 1971; Southard and Boguchwal, in review a), which separates the variables of greatest interest--flow depth, flow velocity, and sediment size--is $(\rho\gamma'/\mu^2)^{1/3}d$, $(\rho^2/\mu\gamma')^{1/3}U$, $(\rho\gamma'/\mu^2)^{1/3}D$, and ρ_s/ρ. The first three can be viewed as dimensionless depth, d^o, dimensionless velocity U^o, and dimensionless sediment size D^o. A similar set of four variables was used by Vanoni (1974) in one of the few attempts in the literature to develop a three-dimensional predictor for bed configurations. For quartz-density sand in water, stability or existence fields for bed phases can be represented in a three-dimensional graph filled with field volumes for the various bed phases. This will here be called the dimensionless depth-velocity-size diagram, or d^o-U^o-D^o diagram.

DEPTH-VELOCITY-SIZE DIAGRAM

The writer (Southard and Boguchwal, in review a) has undertaken, by use of data reported in the literature, to construct the best present approximation to the d^o-U^o-D^o diagram. Data were taken from 35 flume studies that report water temperature and equilibrium bed configuration as well as d, U, and D. An important requirement was an adequate description of the bed configuration by words, pictures, or profiles, so that bed-phase assignments could be made consistently. Bed configurations were classified into six bed phases: no movement on plane bed, ripples, dunes, lower-regime plane bed, upper-regime plane bed, and antidunes. This is generally in accordance with classifications proposed by Simons and Richardson (1963), American Society of Civil Engineers (1966), Japanese Society of Civil Engineers (1974), and Ashley (in review). It is widely accepted that there is a fundamental break in scale and dynamics between ripples, on the one hand, and the various large-scale flow-transverse bed forms, variously termed dunes, megaripples, or sand waves, on the other hand. The writer suspects, but cannot prove, that these large-scale flow-transverse forms form a continuum in scale and dynamics.

Two kinds of section through the d^o-U^o-D^o diagram are useful for visualizing relationships among bed phases: depth-velocity sections for a given sediment size, and velocity-size sections for a given flow depth. Figure 1 shows three d^o-U^o sections along with their schematic versions, for sediment sizes of 0.10-0.14 mm, 0.50-0.64 mm, and 1.30-1.80 mm, and Figure 2 shows a U^o-D^o section for a flow depth of 0.16-0.25 m along with its schematic version. All of these plots were standardized using a reference water temperature: the axes are labeled not with values of d^o, U^o, and D^o, which convey no concrete picture of flow conditions, but with the equivalent values of d, U, and d at an arbitrary reference water temperature of 10°C. The plots thus have the advantage of being in dimensionless form while still expressing concrete values of d, U, and D. To find the 10° values, first d^o, U^o, and D^o were computed from the original data and then d, U, and D were computed from these using values of ρ and μ corresponding to 10°C water.

The d^o-U^o diagram for 0.10-0.14 mm (Figure 1A) shows fields for only three phases: ripples, upper-regime plane bed, and

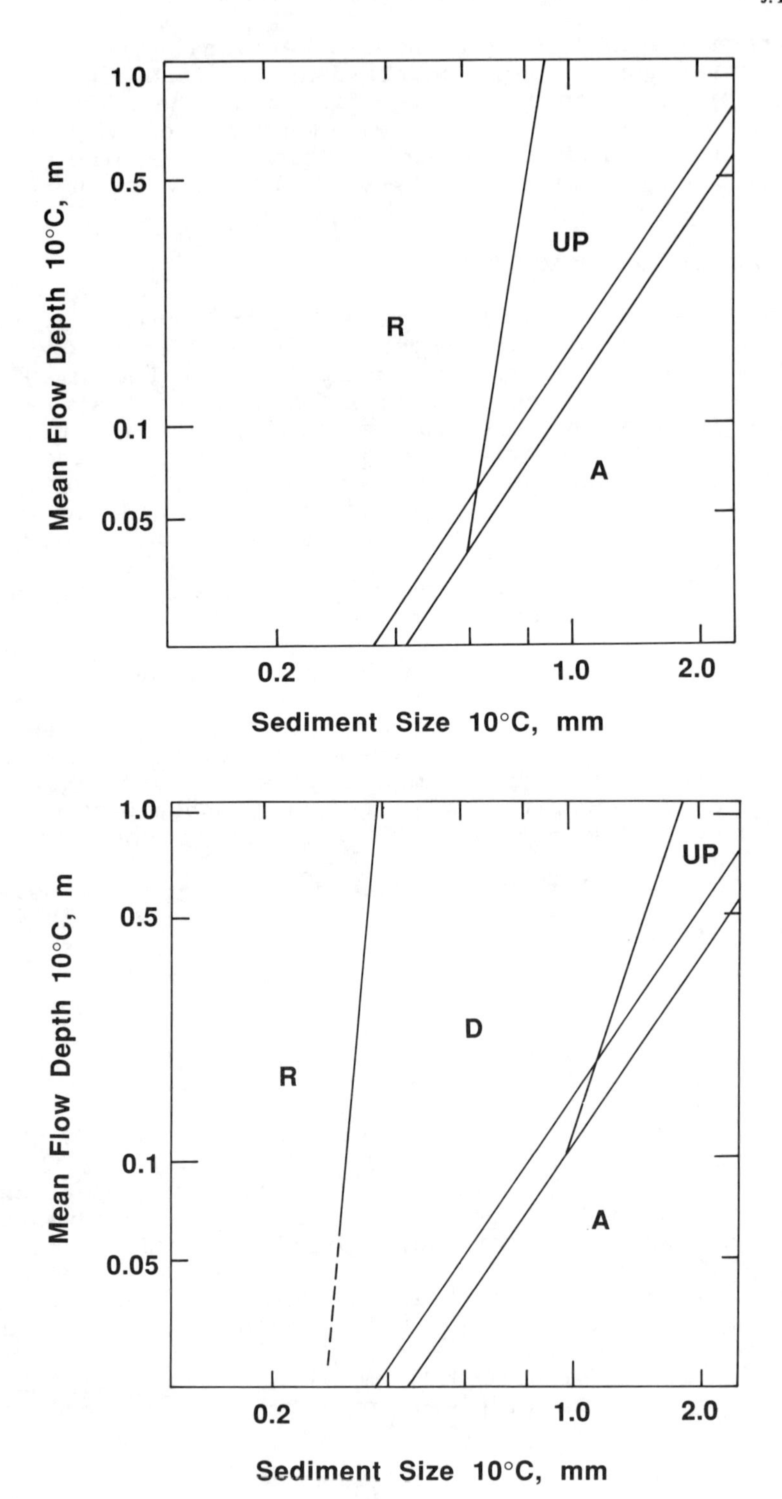
1.0
0.5
0.1
0.05
Mean Flow Depth 10°C, m
UP
R
A
0.2
1.0
2.0
Sediment Size 10°C, mm
1.0
0.5
0.1
0.05
Mean Flow Depth 10°C, m
UP
D
R
A
0.2
1.0
2.0
Sediment Size 10°C, mm

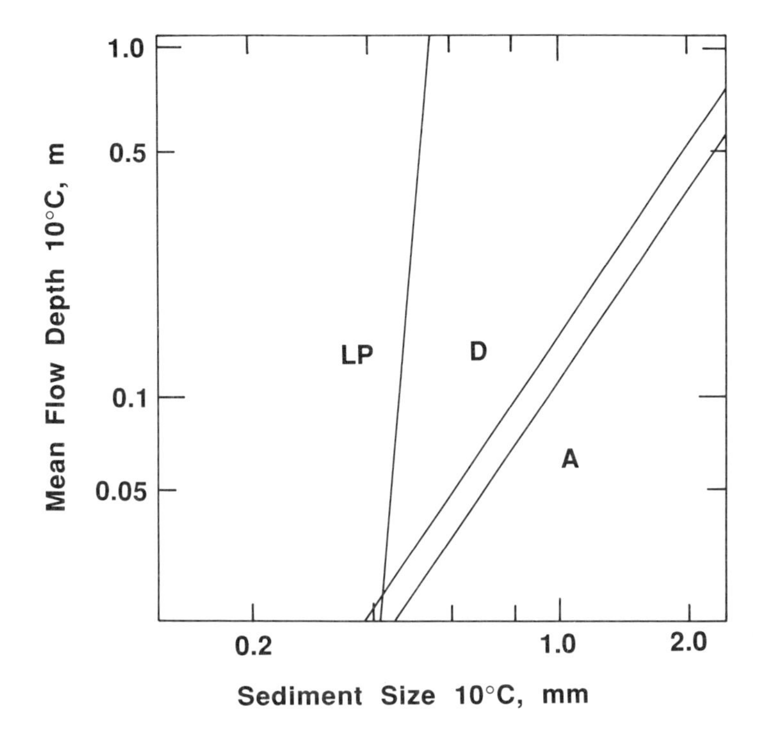

Figure 1. Three depth-velocity sections through the d^o-U^o-D^o diagram for existence fields of equilibrium bed phases in steady uniform flow of water over quartz-density sediment ($\rho_s/\rho = 2.65$). Flow depth and flow velocity are standardized to 10^oC water temperature; see text. (A) 0.10-0.14 mm sediment; (B) 0.50-0.64 mm sediment; (C) 1.30-1.80 mm sediment.

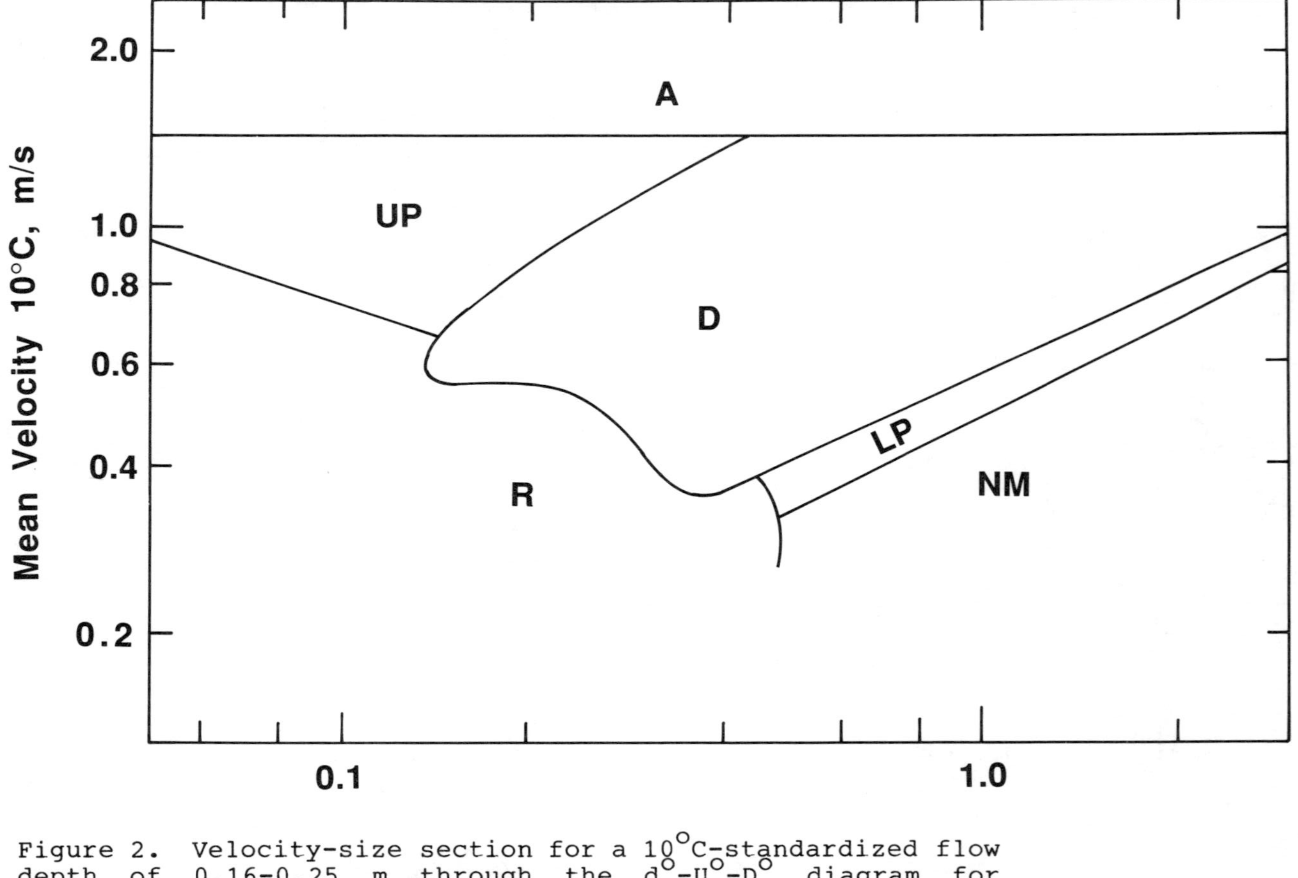

Figure 2. Velocity-size section for a 10°C-standardized flow depth of 0.16-0.25 m through the d^{o}-U^{o}-D^{o} diagram for existence fields of equilibrium bed phases developed in steady uniform flow of water over quartz-density sediment ($\rho_s/\rho = 2.65$).

antidunes. The boundary for the abrupt change from ripples to upper plane bed is steeper than that for the gradual change from upper plane bed to antidunes. The latter boundary is shown extending downward to truncate the former. As velocity is increased over ripples at shallow depths, surface waves develop to be in phase with the ripples and gradually become dominant, and the bed forms move upstream as antidunes. On the left are two boundaries between movement and no movement; one is the Shields curve for incipient movement on a plane bed (transformed into this graph), and the other, farther left (not shown), describes the minimum velocity needed to maintain pre-existing ripples at equilibrium.

The d^o-U^o diagram for 0.50-0.64 mm sand (Figure 1B) shows an additional field for dunes, with a lower-velocity boundary that slopes more steeply than the higher-velocity boundary. Again the antidune field truncates of the dune field, and again there are two kinds of boundary between movement and no movement, not shown.

In the d^o-U^o diagram for 1.30-1.80 mm sand (Figure 1C), a lower-regime plane bed replaces ripples at low flow velocities. A field for upper-regime plane bed would be present beyond the upper right corner of the graph if data for sufficiently high velocities were available. Upper-regime plane beds succeed antidunes with increasing velocity and decreasing depth in the lower right; apparently the bed becomes planar once again as the Froude number becomes sufficiently greater than one. The boundary between no movement and lower plane bed corresponds to the Shields curve.

The U^o-D^o section (Figure 2) is better for visualizing relationships among bed phases as a function of sediment size. Ripples are stable for sediment sizes finer than about 0.6 mm. The range of flow velocities for which ripples are stable narrows with increasing sediment size; the ripple field terminates against fields for plane beds with or without movement. Relationships in this region are difficult to study because in these sand sizes it takes a long time for the bed to adjust to equilibrium. In medium sands ripples give way abruptly to dunes with increasing flow velocity, but in finer sediment, ripples give way (also abruptly) to upper plane bed. Although not well constrained, the boundary between ripples and upper plane bed clearly rises to higher velocities with decreasing sediment size.

Dunes are the stable phase over a wide range of velocities in sediments from medium sand to indefinitely coarse gravel. Flow velocities for dunes increases with increasing sediment size. In the finer sizes, dunes pass fairly slowly into upper plane bed with increasing flow velocity. The boundary between dunes and upper plane bed slopes upward to higher flow velocities with increasing sediment size until truncated by the antidune boundary. For sediments coarser than about 0.6 mm there is a narrow field below the dune field for lower-regime plane bed; the lower boundary of this field is represented by the Shields curve.

The shape of the boundary between ripples and dunes in Figure 2 is noteworthy: there is a triple point among ripples, dunes, and upper plane bed at a sediment size of about 0.2 mm. The dune field forms a salient pointing toward finer sizes, and the boundary between ripples and upper plane bed intersects this salient well above its almost horizontal lower bounding curve. We speculate that the boundary between ripples and upper plane bed passes unaffected beneath the dune field at this triple point and emerges again at coarser sediment sizes and lower velocities as the boundary between ripples and lower plane bed. The stability fields for ripples and dunes thus seem to be controlled by dynamically separate effects, and the particular position of the boundary between the two is the outcome of a contest for dominance between two competing effects. This is consistent with the existence of a geometrically confused transitional bed geometry over a narrow range of velocities between ripples and dunes (Boguchwal, 1977; Boguchwal and Southard, in review a).

PREDICTORS INVOLVING BOUNDARY SHEAR STRESS

A great number of two-dimensional bed-form predictors have been developed; few have seen wide use. Those with dimensionless versions of boundary shear stress τ_o or stream power $\tau_o U$ along one axis and dimensionless sediment size along the other axis have been most widely used (Simons and Richardson, 1966; see review by Allen, 1982). When τ_o is used instead of U, the effect of flow depth is much less, so a single suitably nondimensionalized two-dimensional graph of bed shear stress against sediment size should therefore be expected to represent bed states reasonably well.

Figure 3 shows bed-phase stability fields in a plot of dimensionless boundary shear stress $T^o = \tau_o(\rho/\mu^2\gamma'^2)^{1/3}$ (a dimensionless form of τ_o that does not contain D) against D^o from flume experiments; for a more complete version showing the data, see Southard and Boguchwal (in review, a). All τ_o values were corrected for sidewall effects. Scatter or overlapping of points for different bed phases is much greater than in Figures 1 and 2, mainly because form resistance is large over rugged flow-transverse bed forms but disappears in the transition from ripples to upper plane bed or from dunes to upper plane bed, so τ_o actually decreases with increasing U in these transitions before increasing again. This means that there is a broad band approximately parallel to the D^o axis in the T^o-D^o graph in which the fields for ripples and dunes, on the one hand, overlap or fold onto the fields for upper plane bed and antidunes. Otherwise, relationships among the bed phases are qualitatively similar to those in Figures 1 and 2.

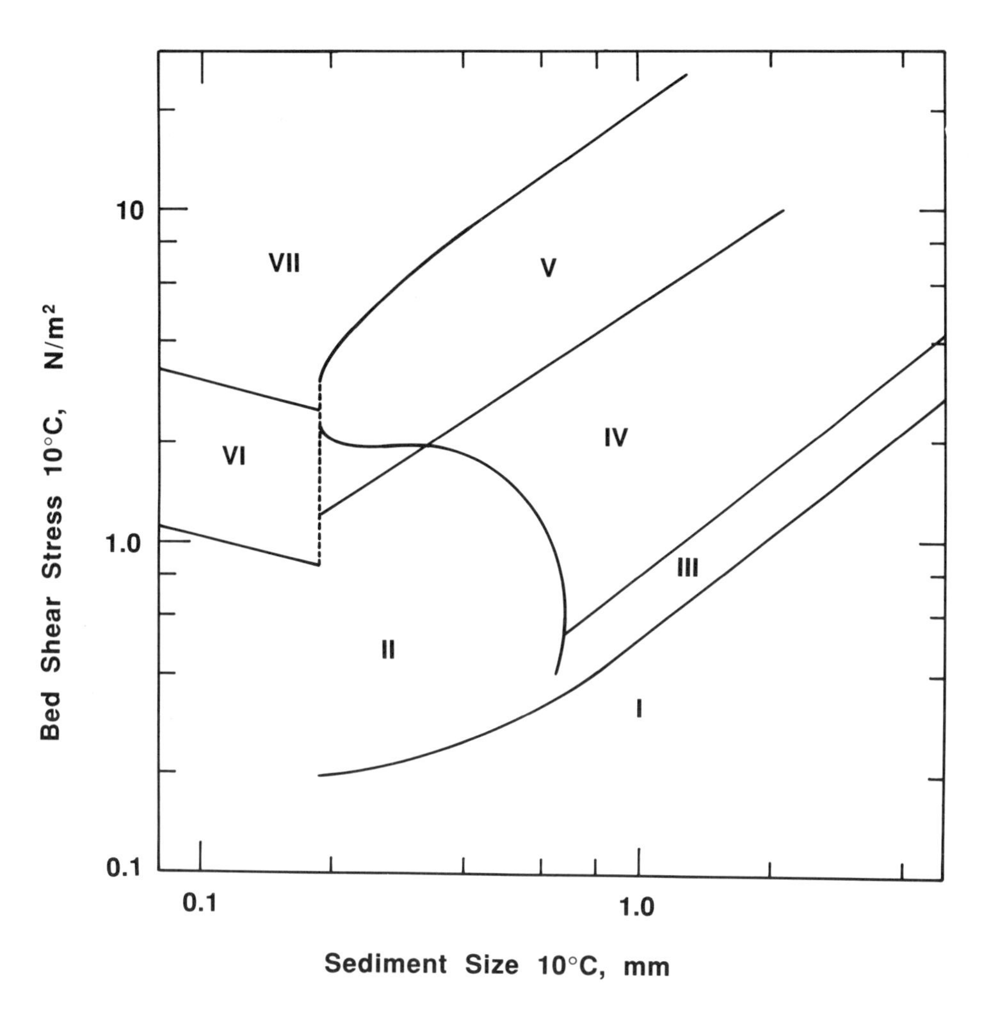

Figure 3. Existence fields for bed phases in steady unidirectional water flow over quartz-density sediment ($\rho_s/\rho = 2.65$), in a graph of dimensionless bed shear stress T^o vs. dimensionless sediment size D^o.

EXTENSION TO GREATER FLOW DEPTHS

Flume data on bed configurations are limited to flow depths mostly less than one meter. Although there have been numerous studies on the bed configuration in rivers with a wide range of discharges and bed-material sediment sizes, firm data on the details of the bed configuration as a function of flow and sediment in rivers are not abundant. One should expect that the depth-velocity-size diagram described above could be extrapolated to the greater flow depths characteristic of rivers; the results of this extrapolation would be that all boundaries would shift to higher flow velocities, the shape of the fields would change somewhat, although not drastically, and the boundary for antidunes would rise much more rapidly than the other boundaries.

Rubin and McCulloch (1980) have used their own data and those of others from both rivers and shallow marine currents to extend the depth-velocity-size diagram to much greater flow depths. Their graph extends from the range that is well constrained by flume data (cf. Figures 1 and 2 of the present paper) into conditions of much deeper flow depths. The relationships are similar to those in Figure 2 except that, as expected, the dune field is fuller and its boundaries lie at higher U. Rubin and McCulloch schematically contoured the dune field for dune height H, with the following results: (i) for given d and U, H increases with D; (ii) for given D and U, H increases with d; (iii) for given d and D, H first increases and then decreases with increasing U.

Water-Temperature Effects

GENERAL

Water temperature in sediment-transporting natural flows ranges from 0°C in the coldest flows to well over 30°C in rivers and warm shallow seas. The substantial and even dramatic effects of temperature changes on sediment discharge and bed configuration have been documented in rivers (Hubbell and Ali, 1961; Colby and Scott, 1965; Burke, 1966; Fenwick, 1969; Shen, 1977; Shen et al., 1978) and in flumes (Ho, 1939; Neubauer, 1948; Straub et al., 1958; Guy et al., 1966; Franco, 1969; Taylor and Vanoni, 1972a, 1972b; Abou-Seida and Arafa, 1977; Hong et al., 1984). These studies have demonstrated complicated and conflicting effects of temperature on sediment transport, with the magnitude and sense of change depending on the particular sediment size and flow conditions involved.

Understanding of the mechanics of temperature-related changes in sediment transport is still rudimentary. The intent here is not to attack the dynamical causes of these changes (which presumably involve the effects of changing water viscosity on such things as sediment fall velocity, near-bed turbulence structure, and fluid forces on bed grains) but to introduce a unified framework for analyzing temperature effects. This framework, involving the dimensionless depth-velocity-size diagram discussed in the preceding section, serves to generalize the problem by placing it in the context of the question why different dimensionless bed configurations differ from each other in the first place. At the same time, to the extent that the diagram is known for (or can be extrapolated to) the flow depths of natural environments, this approach provides a useful predictive tool for the effect of temperature on bed configuration even when the fundamental dynamics behind the changes are still not well understood. The following sections are based on an analysis by Southard and Boguchwal (in review, b).

ANALYSIS OF TEMPERATURE EFFECTS

As noted earlier in this paper, every point in the d^o-U^o-D^o diagram represents a dimensionless bed state. Corresponding to each dimensionless bed state are an infinite number of actual bed states arising from combinations of values of the seven variables d, U, D, μ, ρ, ρ_s, and γ' that give the values of d^o, U^o, and D^o for that dimensionless bed state. These bed states are all equivalent to one another, in the sense that all of the characteristics of the bed configuration, when expressed in dimensionless form, are the same from state to state. These equivalent bed states must differ in the values of at least two of the seven variables; two actual bed states that differ in only one of the seven variables, say μ by virtue of a difference in water temperature, cannot be equivalent. And in general, even if two or more of the seven variables are different, the bed states will not be equivalent, because they will not represent the same dimensionless bed state. This concept of equivalent and nonequivalent bed states is a useful framework for examining changes in bed configuration with change in temperature.

As a hypothetical example (Figure 4), suppose that the bed state in a flow of water at a certain mean flow depth and mean flow velocity and a water temperature of 5°C is ripples. The point corresponding to this bed state is at some point A in the U^o-D^o section for a flow with this d^o. If the water temperature is changed to 25°C but d, U, and D are kept the same, as is often approximately the case in rivers and can easily be prearranged in flume work, the dimensionless bed state is different, and the point corresponding to this bed state, Point B, lies in a different U^o-D^o section for a different value of d^o. Because μ enters into d^o, U^o, and D^o to a negative power, the line connecting the two points A and

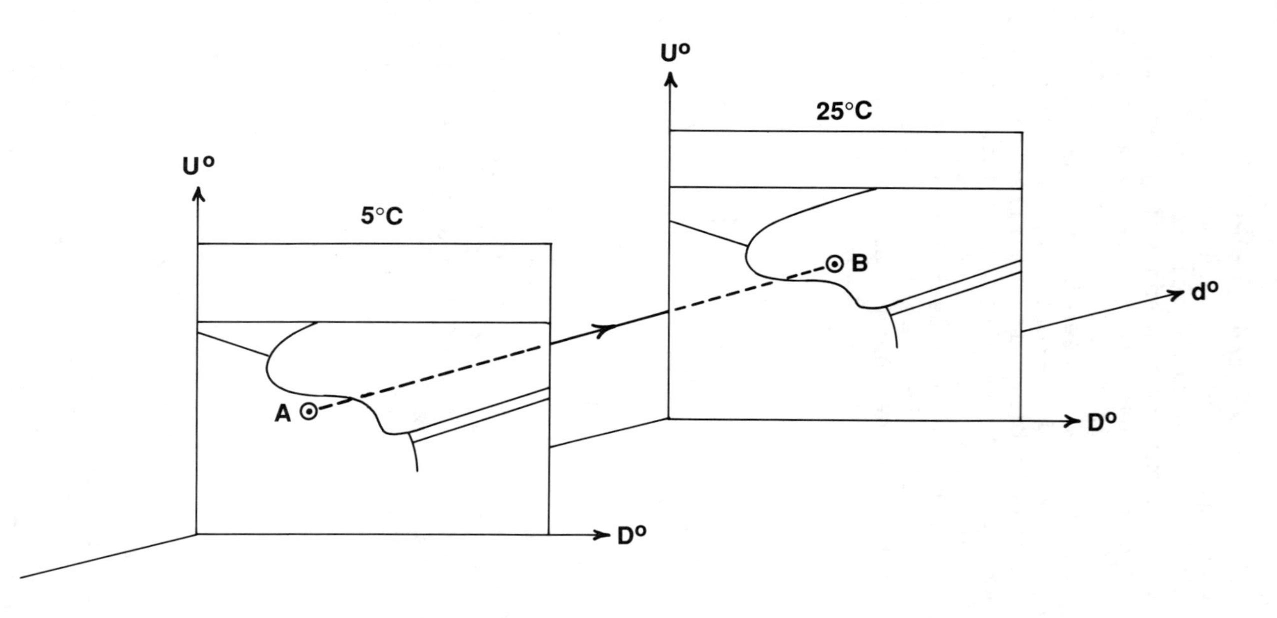

Figure 4. Hypothetical example of the effect of a change in water temperature in a flow at given mean flow depth, mean flow velocity, and median sediment size on the bed configuration. The dimensionless bed state changes from point A to Point B in the d^o-U^o-D^o diagram, leading to a change from ripples to dunes. See text for details.

B extends diagonally outward through the dimensionless depth-velocity-size diagram. The change in water temperature is thus likely to change the bed state from ripples, which are the stable bed phase in the part of the diagram occupied by the cold-water bed state, to dunes, which are the stable bed phase in the part of the diagram occupied by the hot-water bed state.

Prior knowledge of the arrangement of bed-phase stability fields in the d^o-U^o-D^o diagram thus allows prediction of the new bed phase. Moreover, once the diagram is contoured for other dependent sediment-transport variables, such things as bed-form size and geometry, bed-form speed, sediment discharge, and friction factor could also be predicted.

Whatever it is about the interaction in fluid flow and sediment movement that leads to the development of ripples in dimensionless bed state A and dunes in dimensionless bed state B in the example above--and the mechanics of the transition between ripples and dunes is still a matter of controversy--the difference in conditions could have been brought about not just by a change in water temperature but by changes in sediment size and flow conditions as well. The mechanics of temperature effects are thus seen to be logically and usefully approachable not just by arguments in specific situations, as has been the case up to now, but by a comprehensive approach to sediment-transport mechanics as well. Viewed in this way, the problem of temperature effects on bed configuration becomes an inseparable part of the overall problem of the mechanics of bed configurations; what is basic is the difference in dimensionless bed states, not the difference in temperature in itself.

AN EXAMPLE FROM A FLUME

To demonstrate the utility of the foregoing approach, I here examine two previous studies on the effect of water temperature on bed configuration in which a change of bed phase was documented, one (Taylor, 1974) in a flume and the other (Shen et al., 1977), treated in the following section, in a river. In each case, use of the d^o-U^o-D^o diagram accounts well for the change.

Taylor (1974) made seven pairs of runs with 0.23 mm sand in a large flume to study the effect of temperature on bed configuration when flow depth and flow velocity are held constant. In one of these pairs, with flow depth about 0.1 m and flow velocity about 0.5 m/s (his Runs F-33 and F-34), the bed configuration for 22.9°C was ripples and that for 38.3°C was dunes. Table 1 gives the original data together with d^o, U^o, D^o, and the 10°C-equivalent values d_{10}, U_{10}, and D_{10} computed therefrom.

When the two sets of computed dimensionless or temperature-standardized flow velocities and sediment sizes are plotted in Figure 5 in U^o-D^o sections similar to those of Figure 1 for the two values of d^o, it is seen that the low-temperature run lies in the ripple field and the high-temperature run lies in the dune field. The effectively greater sediment size and flow velocity combine to push conditions across the boundary between ripples and dunes even though that boundary is itself shifted upward by the temperature increase.

AN EXAMPLE FROM A RIVER

Shen et al. (1978) examined field data on sediment, flow, and bed configuration taken in a controlled reach of the Missouri River near Omaha, Nebraska between 1966 and 1975. In the period 1966-1969 most of the data are for discharges that varied only between 31,400 cusecs (900 cumecs) and 35,200 cusecs (1000 cumecs); both flow depth and flow velocity varied within fairly narrow limits, but water temperature ranged from 36°F (2.2°C) to 73°F (22.8°C). During most of the observation intervals, dunes with heights of about a meter and spacings of about a hundred meters covered most of the bed, but at times of lowest water temperature the percentage of the bed covered with dunes decreased sharply and the bed configuration is described by the authors as plane bed.

We chose the three observations at highest water temperature and the three observations at lowest water temperature for analysis. As above, dimensionless and 10°-standardized flow velocity, flow depth, and sediment size were computed from these data (Table 1), and the dimensionless size and velocity were plotted in Figure 6 in U^o-D^o sections similar to those of Figure 1 for the corresponding values of d^o. (Since our data synthesis reported above did not extend to flow depths much greater than 1 m, we had to extrapolate the U^o-D^o plots to a d_{10} of about 2 m to obtain Figure 6.) The high-temperature flows fall well within the field for dunes, but the low-temperature flows, with their smaller effective sediment sizes and flow velocities, cross over leftward into the field for upper plane bed.

CONCLUSION

Temperature effects on bed configurations in flumes can thus be accounted for and predicted by use of the existing body of data on bed states obtained from studies made over a wide range of depths, velocities, and sediment sizes without the need for a large range of water temperature. To the extent that the d^o-U^o-D^o diagram can be extended to greater flow depths either by extrapolation or by inclusion of field data, the effect of temperature on bed configurations in rivers can be predicted in the same way.

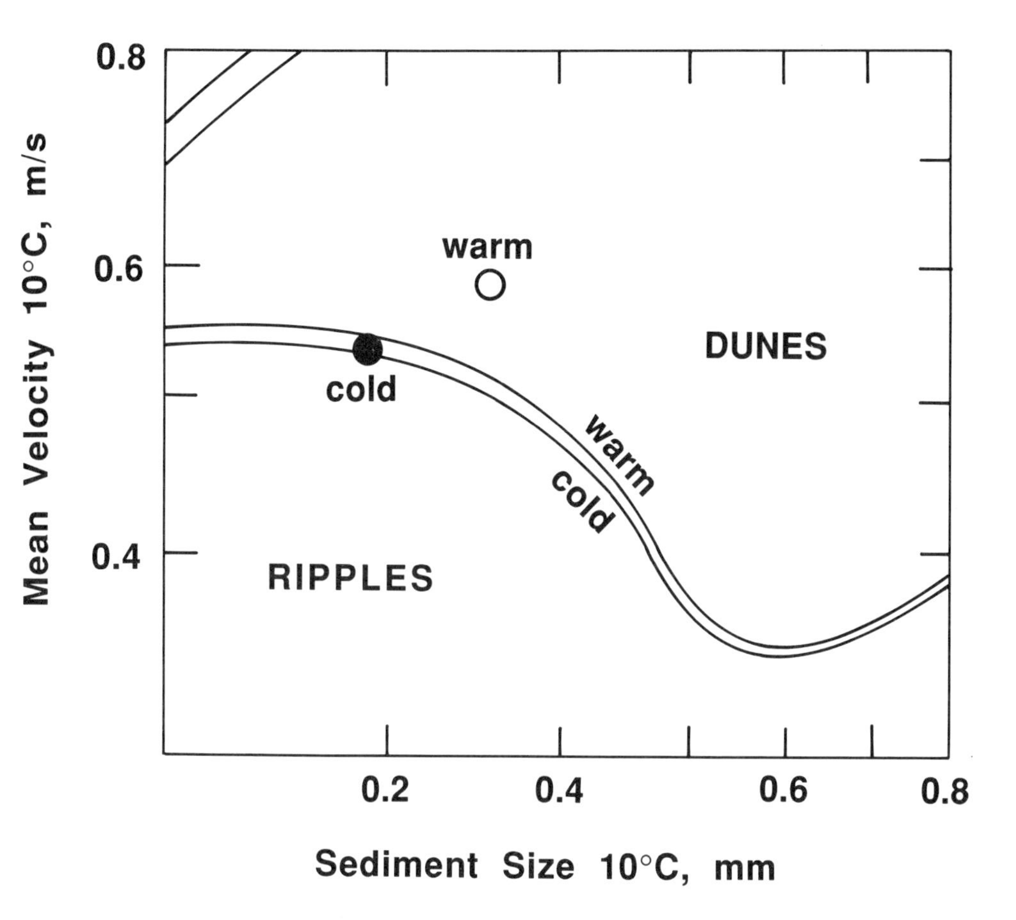

Figure 5. Flume example of a change from ripples to dunes with increasing water temperature, other variables being approximately constant. Dimensionless velocity-size graphs are interpolated from a set from Southard and Boguchwal (in review, a), exemplified by the graph in Figure 2, for dimensionless flow depths corresponding to Run F-33 (cold water; solid circle) and Run F-34 (warm water; open circle) of Taylor (1974). See Table 1.

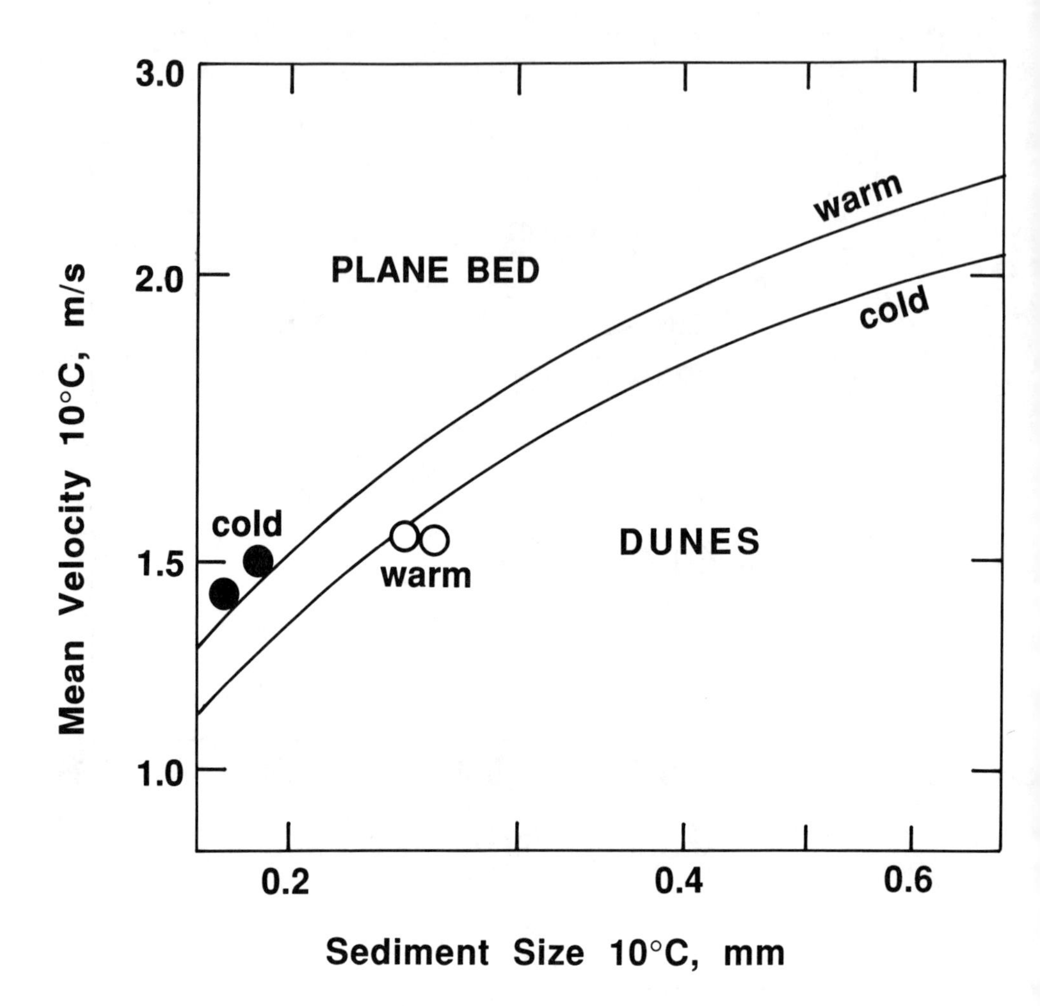

Figure 6. Field example of a change from plane bed to dunes with increasing water temperature, other variables being approximately constant. Dimensionless velocity-size graphs are extrapolated from a set from Southard and Boguchwal (in review, a), exemplified by the graph in Figure 2, for dimensionless flow depths corresponding to three of the warmest-water flows (open circles) and three of the coldest-water flows (solid circles) reported by Shen et al. (1978) from the Missouri River near Omaha, Nebraska. See Table 1.

Effects of Suspended-Load Concentration

INTRODUCTION

There have been surprisingly few studies of the effect of high wash-load concentrations on bed configurations in alluvial-channel flows: the writer is aware only of studies by Simons et al. (1963), Kikkawa and Fukuoka (1969), and Wan (1985). Observations of the bed configuration in rivers with unusually high suspended-sediment concentrations, sometimes reaching hyperconcentrations of tens of percent by volume, are scarce as well, presumably owing to the great difficulty in measuring anything about the bed under such conditions. Yet such suspended-load concentrations are common in rivers, and at least in the lower range of concentrations the bed configuration must often be broadly similar to that under normally low concentrations. One might speculate that, as the concentration increases, the bed configuration at first changes quantitatively and even qualitatively but can still be considered in the same context as the clearer "normal" flows, but then, as concentrations rise into the hyperconcentrated region and even approach those of debris flows, either the bed no longer exhibits nonplanar configurations or the nonplanar configurations bear no obvious relationship to those within our experience.

Wash load is usually distinguished from suspended bed-material load. Quite apart from the possible presence of wash load, the concentration of suspended bed-material load increases with increasing flow strength and with decreasing sediment size. Strong flows over beds of coarse silt to very fine sand, as in the Yellow River basin in China, often carry in suspension enormous concentrations of sediment whose sizes are abundantly represented in the bed and are therefore technically not wash load, even though the concentrations may be much higher than the "equilibrium" concentration that would be attained in a sediment-recirculating flume and are therefore in the nature of wash load.

The effect of suspended load on the bed configuration might therefore naturally be examined in two distinct (but intergrading) cases: (i) flows carrying abundant fine bed-material load over fine beds; and (ii) flows carrying abundant wash load over coarse beds. From the standpoint of the bed configuration alone, the first case is less interesting than the second, because in sediments finer than about 0.15 mm the only stable bed configurations are ripples and plane bed, with increasing flow strength; the only effect an increase in suspended-bed-material concentration can have is to (i) shift the position of the boundary between ripples to plane bed, (ii) alter the shape and/or scale of the ripples, or (iii) prevent the existence of ripples entirely.

These two cases will be examined in turn below.

HIGH BED-MATERIAL CONCENTRATIONS OVER FINE BEDS

There have been several systematic flume studies of transport and bed configurations in silts (Kalinske and Hsia, 1945; Rees, 1966; Jopling and Forbes, 1979), but the only study known to the writer that involved excess concentrations of suspended sediment is some unpublished work by Sprunt (1972).

HIGH WASH-LOAD CONCENTRATIONS OVER COARSER BEDS

We begin by developing a crude tentative model for the effect of fine suspended sediment on some aspects of the bed-material transport, and then examining the experimental evidence on effects of wash load on bed configuration in terms of that model.

It seems reasonable to suppose that at fairly low concentrations the presence of fine sediment in the flow has three main effects on flow and sediment transport, having to do with (i) an increase in the effective viscosity of the fluid, (ii) an increase in the effective density of the fluid, and (iii) an increase in the effective submerged specific weight of the bed-material particles:

(i) The effective viscosity of the fluid. First, provided that the particle size is smaller than the minimum eddy size, the presence of the particles increases the viscous dissipation of turbulent kinetic energy at the lowest end of the spectrum, so the turbulence feels an increase in the fluid viscosity but does not sense the particles directly. Indirect effects on transport of bed material come about by changes in near-bed flow structure, instantaneous fluid forces on particles on the bed and within the flow, and larger-scale flow structure over bed forms. Second, the same reasoning applies also to the effective viscosity of the sediment-laden fluid as it is locally sheared by the passage of a large bed-material particle nearby. If there are no significant interparticle forces within the suspension, then in both cases the fluid should remain approximately Newtonian.

(ii) The effective density of the fluid. Provided that the suspended particles are much smaller than the bed-material particles, these suspended particles follow closely the accelerated motions of fluid located near a bed-material particle in motion relative to that fluid. The bed-material particle thus senses the presence of the suspended particles as a greater density of the fluid it accelerates.

(iii) The effective submerged specific weight. Since the relative motion between the fluid and the suspended particles is small, the fluid itself supports the suspended particles. The effect of the suspended particles is therefore to add weight to the fluid overlying any given bed-material particle, thus increasing the hydrostatic pressure exerted by the fluid on the surface of that particle. The effect is as if the bed-material particle were submerged in a homogeneous fluid with greater specific weight. The submerged specific weight of the bed-material particle is thus smaller, and its fall velocity is smaller.

If it is assume tentatively that (i) the complicated effect of suspended sediment on the effective viscosity of the fluid can be described simply by the increase in suspension viscosity measured in a standard viscometer, (ii) the effect of relative density does not shift the d^o-U^o-D^o diagram greatly, and (iii) the effect on submerged specific weight of the bed-material load can be ignored, then one can analyze the effect of fine sediment on the bed configuration in the same framework as the effect of water viscosity (discussed earlier in this paper) in the following way.

Since μ enters into d^o and D^o to the -2/3 power and into U^o to the -1/3 power, changes in μ shift the bed state along straight lines in the d^o-U^o-D^o diagram (with logarithmic axes) which extend diagonally from large values of d^o, U^o, and D^o to small values. At certain points these lines cross boundaries between two adjacent bed-phase existence fields. (It was by tracing out such changes along viscosity-change lines that the two examples in the previous section on temperature effects were analyzed.) This approach could be successful, however, only to the extent that a nonnegligible yield strength due to interparticle forces in the fine suspension does not develop. In the following paragraphs the meager experimental evidence is examined on these assumptions.

Both Simons et al. (1963) and Kikkawa and Fukuoka (1969) made series of flume runs using the same bed sediment (0.45 mm and 0.15 mm, respectively), and approximately the same depths and velocities of flow but widely varying wash-load concentrations. Wash-load concentrations ranged up to 10% by weight in the former study and up to about 6% by weight in the latter. One important difference between the two studies is that Simons et al. used bentonite clay as the wash-load sediment, whereas Kikkawa and Fukuoka used silica silt. In both studies two tendencies were noted as the wash load concentration increased: dunes were converted to plane bed, presumably reflecting a shift in the the transition from dunes to plane bed to a lower flow velocity; and plane bed was converted to antidunes, presumably reflecting a shift in the transition from plane bed to antidunes to a lower flow velocity. Both of these kinds of change are in qualitative accord with the idea that bed states shift along viscosity-change lines in the d^o-U^o-D^o diagram as described above; in fact, the change from dunes to plane bed is the same as noted in the work of Shen et al. in the Missouri river caused by temperature change. A further step, not yet undertaken, would be to plot the bed states represented by the runs made by Simons et al. on the d^o-U^o-D^o diagram to see whether the changes they documented are in quantitative agreement with the changes predicted based on their measured suspension viscosities. Unfortunately this cannot be done for the runs made by Kikkawa and Fukuoka because they did not report suspension viscosities.

The study by Wan (1985) affords a tantalizing glimpse into the effects of flow of suspensions with much higher wash-load concentrations over coarse bed material. In Wan's study too there was a tendency for transition from dunes to plane bed with increasing suspension viscosity. At even higher concentrations, however, dunes made no appearance--a phenomenon for which the foregoing analysis based on the d^o-U^o-D^o diagram for quartz-density sediment in water has no explanation. One problem in interpreting these results, however, is that the bed material was only slightly denser than water, 1.29 g/cm^3, so as the effective density of the suspension increased, the effective density ratio must have become close to unity, meaning that the d^o-U^o-D^o diagram for quartz-density sediment in water is wholly inapplicable. It is possible that the disappearance of dunes is related not to non-Newtonian behavior of the suspension but to the role of a greatly different effective density ratio. The answer must await further experimental work on bed-load transport of natural materials in high-concentration suspensions.

References

Abou-Seida, M.M., and Arafa, F., 1977, Effect of temperature on channel resistance: American Society of Civil Engineers, Proceedings, Journal of the Hydraulics Division, 103, p. 251-263.

Allen, J.R.L., 1973, Phase differences between bed configuration and flow in natural environments and their geological significance. Sedimentology, 20, p. 323-329.

Allen, J.R.L., 1974, Reaction, relaxation and lag in natural sedimentary systems: general principles, examples and lessons. Earth Science Reviews, 10, p. 263-342.

Allen, J.R.L., 1976a, Bed forms and unsteady processes: some concepts of classification and response illustrated by common one-way types. Earth Surface Processes, 1, 361-374.

Allen, J.R.L., 1976b, Computational models for dune time-lag: general ideas, difficulties, and early results. Sedimentary Geology, 15, p. 1-53.

Allen, J.R.L., 1976c, Time-lag of dunes in unsteady flows: an analysis of Nasner's data from the river Weser, Germany. Sedimentary Geology, 15, 309-321.

Allen, J.R.L., 1982, Sedimentary Structures; Their Character and Physical Basis. Elsevier, Developments in Sedimentology 30, 2 Vols.

Allen, J.R.L., and Friend, P.F., 1976, Relaxation time of dunes in decelerating flows. Geological Society of London, Journal, 132, p. 17-26.

American Society of Civil Engineers, 1966, Nomenclature for bed forms in alluvial channels. American Society of Civil Engineers, Committee on Sedimentation, Task Force on Bed Forms in Alluvial Channels. American Society of Civil Engineers, Proceedings, Journal of the Hydraulics Division, 92 (No. HY3), p. 51-64.

Ashley, G.M., in review, Classification of large-scale aqueous bed forms: a new look at an old problem. Journal of Sedimentary Petrology.

Bogardi, J., 1961, Some aspects of the application of the theory of sediment transportation to engineering problems. Journal of Geophysical Research, 66, p. 3337-3346.

Boguchwal, L.A., 1977, Dynamic scale modeling of bed configurations. Massachusetts Institute of Technology, Dept. Earth and Planetary Sciences, Ph.D. Thesis, 149 p.

Boguchwal, L.A., and Southard, J.B., in review, Bed configurations in steady unidirectional water flows. Part 1. Scale model study using fine sands. Journal of Sedimentary Petrology.

Brooks, N.H., 1958, Mechanics of streams with movable beds of fine sand. American Society of Civil Engineers, Transactions, 123, Part I, 526-549.

Burke, P.P., 1965, Effect of water temperature on discharge and bed configuration, Mississippi River at Red River Landing, Louisiana. U.S. Army Engineering District, New Orleans, Louisiana.

Colby, B.R., and Scott, C.H., 1965, Effects of water temperature on the discharge of bed material. U.S. Geological Survey Professional Paper 462-G, 25 p.

Crowley, K.D., 1983, Large-scale bed configurations (macroforms), Platte River Basin, Colorado and Nebraska: primary structures and formative processes. Geological

Society of America, Bulletin, 94, p. 117-133.

Engelund, F., 1970, Instability of erodible beds. Journal of Fluid Mechanics, 42, p. 225-244.

Engelund, F., and Fredsøe, J., 1974, Transition from dunes to plane bed in alluvial channels. Technical University of Denmark, Institute of Hydrodynamics and Hydraulic Engineering, Series Paper 4.

Exner, F.M., 1925, Über die Wechselwirkung zwischen Wasser und Geschiebe in Flussen. Vienna, Akademie der Wissenschaften, Sitzungsberichte, Mathematische-Naturwissenschaftliche Klasse, Abteilung IIa, 134 (No. 3-4), p. 166-204.

Fenwick, G.B., 1969, Water-temperature effects on stage-discharge relations in large alluvial rivers. U.S. Army Corps of Engineers, Committee on Channel Stabilization, Technical Report 6, Vicksburg, Mississippi.

Franco, J.J., 1968, Effects of water temperature on bed-load movement. American Society of Civil Engineers, Proceedings, Journal of the Waterways and Harbors Division, 94, p. 343-3532.

Fredsøe, J., 1979, Unsteady flow in straight alluvial streams: modification of individual dunes. Journal of Fluid Mechanics, 91, p. 497-512.

Gee, D.M., 1975, Bedform response to unsteady flows. American Society of Civil Engineers, Proceedings, Journal of the Hydraulics Division, 101, p. 437-449.

Gilbert, G.K., 1914, The transportation of debris by running water. U.S. Geological Survey, Professional Paper 86, 263 p.

Guy, H.P., Simons, D.B., and Richardson, E.V., 1966, Summary of alluvial channel data from flume experiments, 1956-61: U.S. Geological Survey, Professional Paper 462-I, 96 p.

Ho, P.Y., 1939, Abhängigkeit der Geschiebewegung von der Kornform und der Temperatur. Preussisch Versuchanstalt für Wasser-, Erd- und Schiffbau, Mitteilungen, No. 37, Berlin.

Hong, R.J., Karim, M.F., and Kennedy, J.F., 1984, Low-temperature effects on flow in sand-bed streams. Journal of Hydraulic Engineering, 110, 109-125.

Hubbell, D.W., and Al-Shaikh Ali, K., 1961, Qualitative effects of temperature on flow phenomena in alluvial channels. U.S. Geological Survey, Professional Paper 424-D, p. D21-D23.

Jackson, R.G. II, 1975, Hierarchical attributes and a unifying model of bed forms composed of cohesionless material and produced by shearing flow. Geological Society of America, Bulletin, 86, p. 1523-1533.

Japanese Society of Civil Engineers, 1974, The bed configuration and roughness of alluvial streams. Japanese Society of Civil Engineers, Task Committee on the Bed Configuration and Hydraulic Resistance of Alluvial Streams. Japanese Society of Civil Engineers, 5, p. 107-119.

Jopling, A.V., and Forbes, D.L., 1979, Flume study of silt transportation and deposition. Geografiska Annaler, 61A, p. 67-85.

Kalinske, A.A., and Hsia, C.H., 1945, Study of transportation of fine sediments by flowing water. University of Iowa, Studies in Engineering, Bulletin 29, 30 p.

Kennedy, J.F., 1963, The mechanics of dunes and antidunes in erodible-bed channels. Journal of Fluid Mechanics, 16, p. 521-544.

Kennedy, J.F., and Brooks, N.H., 1963, Laboratory study of an alluvial stream at constant discharge. U.S. Department of Agriculture, Agricultural Research Service, Miscellaneous Publication 970, Proceedings of the Federal Inter-Agency Sedimentation Conference, p. 320-330.

Kikkawa, H., and Fukuoka, S, 1969, The characteristics of flow with wash load. International Association for Hydraulic Research, 13th Congress, Proceedings, 2, 233-240.

Middleton, G.V., and Southard, J.B., 1984, Mechanics of Sediment Movement. Society of Economic Paleontologists and Mineralogists, Short Course 3, 2nd edition, 401 p.

Neubauer, L.W., 1948, Some effects of viscosity variation upon the movement of bed-sediment in an open channel. University of Minnesota, Department of Civil Engineering, Ph.D. Thesis, 179 p.

Rees, A.I., 1966, Some flume experiments with a fine silt. Sedimentology, 6, p. 209-240.

Richards, K.J., 1980, The formation of ripples and dunes on an erodible bed. Journal of Fluid Mechanics, 99, p. 597-618.

Rubin, D.M., and McCulloch, D.S., 1980, Single and superimposed bedforms: a synthesis of San Francisco Bay and flume observations. Sedimentary Geology, 26, p. 207-231.

Shen, H.W., 1977, Analysis of temperature effects on stage-discharge relationships in a Missouri River reach near Omaha. U.S. Army Corps of Engineers, Missouri River Division, M.R.D. Sediment Series No. 15, Omaha, Nebraska.

Shen, H.W., Mellema, W.J., and Harrison, A.S., 1978, Temperature and Missouri River stages near Omaha. American Society of Civil Engineers, Proceedings, Journal of the Hydraulics Division, 104, p. 1-20.

Simons, D.B., and Richardson, E.V., 1962, Resistance to flow in alluvial channels. American Society of Civil Engineers, Transactions, 127, Part I, 927-952.

Simons, D.B., and Richardson, E.V., 1963, Forms of bed roughness in alluvial channels. American Society of Civil Engineers, Transactions, 128, Part I, p. 284-302.

Simons, D.B., and Richardson, E.V., 1966, Resistance to flow in alluvial channels. U.S. Geological Survey, Professional Paper 422-J, 61 p.

Simons, D.B., Richardson, E.V., and Haushild, W.L., 1963, Some effects of fine sediment on flow phenomena. U.S. Geological Survey, Water-Supply Paper 1498-G, 47 p.

Smith, J.D., 1970, Stability of a sand bed subjected to a shear flow of low Froude number. Journal of Geophysical Research, 75, p. 5928-5939

Southard, J.B., 1971, Representation of bed configurations in depth-velocity-size diagrams. Journal of Sedimentary Petrology, 41, p. 903-915

Southard, J.B., and Boguchwal, L.A., in review a, Bed configurations in steady unidirectional water flows. Part 2. Synthesis of flume data. Journal of Sedimentary Petrology.

Southard, J.B., and Boguchwal, L.A., in review b, Bed configurations in steady unidirectional water flows. Part 3. Temperature effects. Journal of Sedimentary Petrology.

Sprunt, H.H., Jr., 1972, Bed aggradation experiments in a small recirculating flume employing 40 micron silt with special reference to turbidity current depositional structures. Massachusetts Institute of Technology, Dept. of Earth and Planetary Sciences, unpublished M.S. thesis, 160 p.

Straub, L.G., Anderson, A.G., and Flammer, G.H., 1958,

Experiments on the influence of temperature on the sediment load. U.S. Army Corps of Engineers, Missouri River Division, M.R.D. Sediment Series No. 10, Omaha, Nebraska.

Taylor, B.D., 1974, Temperature effects in flow over nonplanar beds. American Society of Civil Engineers, Proceedings, Journal of the Hydraulics Division, 100, p. 1785-1807.

Taylor, B.D., and Vanoni, V.A., 1972a, Temperature effects in low-transport, flat-bed flows. American Society of Civil Engineers, Proceedings, Journal of the Hydraulics Division, 98, p. 1427-1445.

Taylor, B.D., and Vanoni, V.A., 1972b, Temperature effects in high-transport, flat-bed flows. American Society of Civil Engineers, Proceedings, Journal of the Hydraulics Division, 98, p. 2191-2206.

Vanoni, V.A., 1974, Factors determining bed forms of alluvial streams. American Society of Civil Engineers, Proceedings, Journal of the Hydraulics Division, 100, p. 363-377.

Vanoni, V.A., editor, 1975, Sedimentation Engineering. American Society of Civil Engineers, Manuals and Reports on Engineering Practice, No. 54, 745 p.

Wan, Z., 1985, Bed material movement in hyperconcentrated flow. Journal of Hydraulic Engineering, 111, p. 987-1002.

Yalin, M.S., 1977, Mechanics of Sediment Transport, 2nd edition, Pergamon Press, 298 p.

List of Symbols

d mean depth of flow
d^o dimensionless flow depth $(\rho\gamma'/\mu^2)^{1/3}d$
d_{10} 10^oC-equivalent flow depth
D mean or median sediment diameter
D^o dimensionless sediment diameter $(\rho\gamma'/\mu^2)^{1/3}D$
D_{10} 10^oC-equivalent sediment diameter
g acceleration of gravity
H dune height
P flow power $\tau_o U$
T^o dimensionless bed shear stress $(\rho/\gamma'^2\mu^2)^{1/3}\tau_o$
U mean velocity of flow
U^o dimensionless flow velocity $(\rho^2/\mu\gamma')^{1/3}U$
U_{10} 10^oC-equivalent flow velocity
γ' submerged specific weight of sediment

μ fluid viscosity
ρ fluid density
ρ_s sediment density
τ_o bed shear stress

EROSION CONTROL AND RESERVOIR DEPOSITION

M. Gordon Wolman
Johns Hopkins University
Department of Geography and Environmental Engineering
Baltimore, Maryland 21218, U.S.A.

ABSTRACT. The path of sediment from hillslope to ultimate deposition involves many storage sites on hillslopes, streambeds, and in valley alluvium. The spatial and temporal distributions of such storage sites, some of which may involve decades to centuries, are poorly understood. Predictions of the impact of conservation measures on river systems thus involve significant uncertainty. The Universal Soil Loss Equation contains parameters difficult to evaluate and subject to considerable error at drainage basin scales. Isotope data indicate sediment delivery over the U.S.A. of only 4-5% of particles contaminated from air over a long period of time. Ten-fifteen percent delivery at drainage areas greater than 100 mi^2 is not uncommon. Land use changes including revegetation demonstrably reduce sediment yields, as much as ten-fold in places, but large reductions are generally confined to small drainage areas of uniform land use. At roughly 1000 mi^2, with mixed land uses, sediment yields converge to a regional average in temperate U.S.A.

Reservoir storage represents an efficient sediment control technique in some terms. However, reservoir lifetimes where capacity/watershed ratios are small (≤ 20) may be decades. Only large downstream reservoirs have lifetimes of centuries. Long-term sediment control clearly demands land management, even though socially and economically more difficult than reservoir construction. Precise identification of sources and connective links of sediment delivery may lessen the total area requiring careful land management.

Introduction

Much is known about the movement of sediment and associated pollutants from the watershed land surface into watercourses and thence downstream in the channel system. Since the 1930s a great deal of work has been done throughout the world on

L. M. Brush et al. (eds.), Taming the Yellow River: Silt and Floods, 139–161.

ways in which sediment can be controlled on the land, or managed in watercourses, to reduce the downstream effects of sediment transport and deposition. Both theoretical and empirical studies, the latter at a variety of scales, have contributed to this knowledge.

While much is known about these processes, several features remain quite uncertain. These include problems relating to the complex transport processes themselves, the episodic or step-like character of erosion, transport and deposition, the time scales involved in successive storage and transport steps, and the way in which sequential processes on the land surface and in channel systems are integrated over space and time.

This paper reviews very briefly some fundamental aspects of the cycle of sediment on the landscape and in rivers to illustrate both what is well known and some gaps in the understanding of scale and time in basin processes. These in turn provide a basis for discussing the possible long-term consequences of approaches to sediment control using land conservation and reservoirs. Integrating reasonable time scales into planning confronts the fact that different sediment control activities may have different consequences depending upon both location and the time horizon one has in mind. The presumed "conflict" between the use of land conservation techniques or reservoirs to control sediment should not be an issue as the two are complementary. However, in various spatial distributions and over different and long periods of time, reservoir construction for sediment control may not be a substitute for, nor an efficient complement to, conservation practices on the land.

The Cycle of Erosion and Sedimentation

Sediment detached from the land surface by running water moves progressively from the land surface itself into bottom lands, channels, and ultimately downstream to temporary storage sites or to final sinks. On occasion, the path of movement may be direct with detachment, transport from land, and flow in the channel continuous and uninterrupted (Figure 1). Over time, however, the movement of most particles is episodic with the distance of movement a function of the storm event, discharge, topography, channel configuration, and vegetation.

Rough estimates indicate that the duration of sediment storage in different stages of the transport sequence is highly variable. Most important, some storage intervals such as storage of sediment on alluvial fans in upland areas and in terraces and floodplains along rivers, may involve "geological time," certainly centuries or millennia and even longer. Storage times enter into the distinction between erosion and sediment yield, and again where long and variable lag times

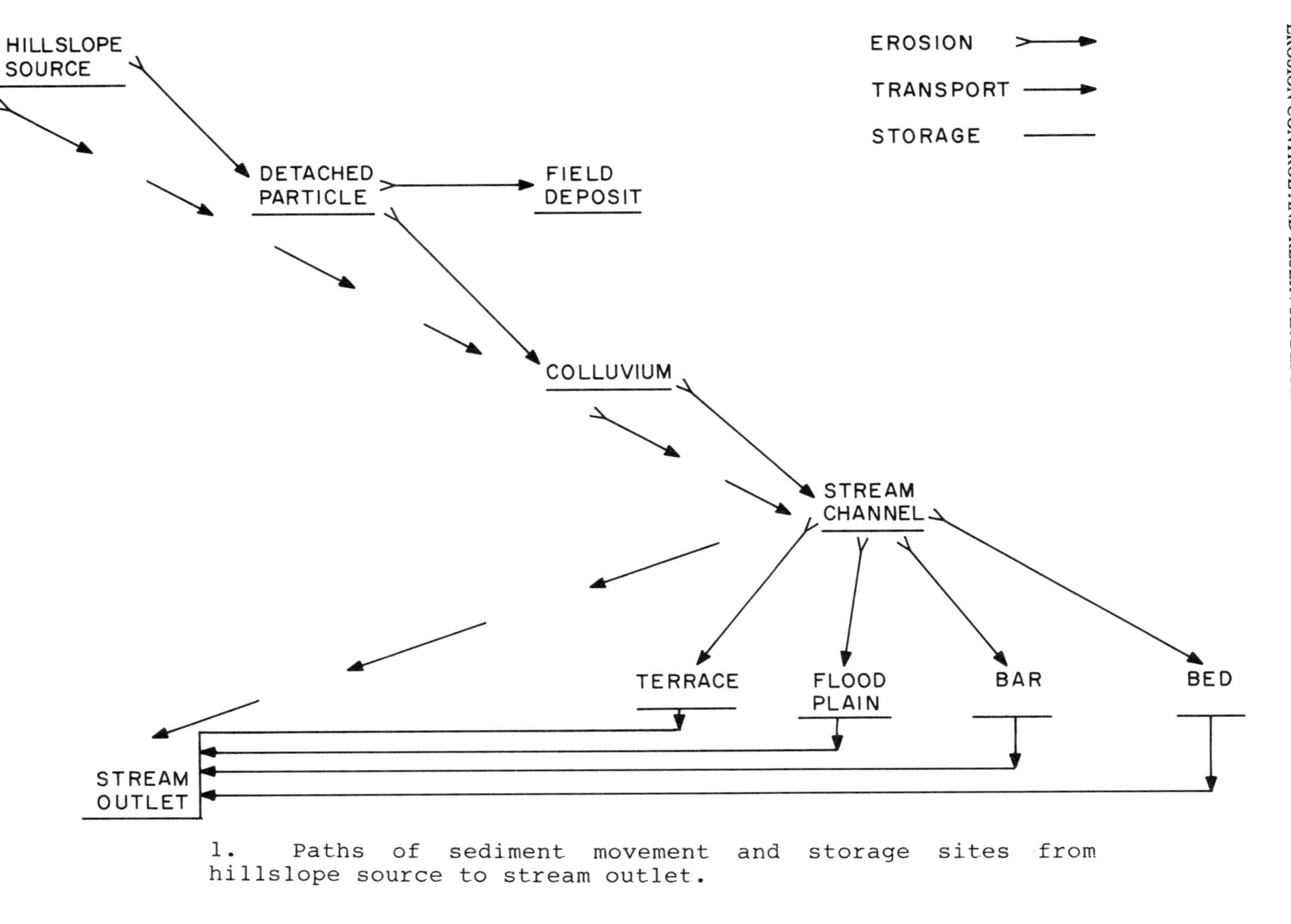

1. Paths of sediment movement and storage sites from hillslope source to stream outlet.

are involved, estimates of the impact of various control measures on the landscape may be markedly influenced by assumptions about the time scales involved.

Erosion of the Land Surface

Erosion of the land surface is a function of a number of variables including geology, soils, climate, vegetation and topography. These factors have been partially captured in the well-known Universal Soil Loss Equation (USLE), in which Soil Loss (t/ha/yr) = RKLSCP; where R is a rainfall factor, K erodibility, L slope length, S degree of slope, C a cropping and P an erosion control factor (W & S, 1978) (Wischmeier and Smith, 1978). This empirical equation was derived from statistical analysis of data from thousands of experimental plots in the U.S. It has been recalculated for different parameters in many parts of the world using appropriate regional measures of such factors such as rainfall intensity and soil characteristics (Hussein, 1986).

The equation is intended to describe erosion by sheet flow and rills on the surface. Calculation of sediment movement from gullies is presumably additive. However, because it is difficult to conceive of large areas in which sheet flow alone occurs, without significant rills and miniature drainage nets, it is likely that at the watershed scale erosion estimates include both sheet and rill erosion and conceivably some erosion from ephemeral gullies as well (Natl. Acad. of Sciences, 1986).

The USLE is also used to calculate erosion from larger areas and some verification of the results is available for small drainage basins where erosion estimates coincide with measurements of sediment yield. Application to the watershed is not without difficulties, however, particularly in the evaluation of the slope and hillslope length and in the integration of these values over drainage basin areas (Wilson, 1986).

Because erosion is variable in space even on a single field, and because deposition as well as erosion takes place on the land itself, calculated erosion from the landscape usually exceeds measured yields of sediment in the stream systems. The difference between calculated erosion, which may or may not include the contribution from gullies, has been expressed as the sediment delivery ratio.

Pollutants including nutrients often move on the watershed adsorbed to sedimentary particles. Although this paper does not include consideration of water quality, it is clear that an understanding of sediment processes is also essential to an understanding of the behavior of water quality and damages resulting from the effects of sediment and pollutants on water quality. Investigation of these problems has become of great importance in current policy debates about water quality in the United States.

Customarily, the sediment delivery ratio declines with increasing drainage area (Figure 2). This well-established relationship applies to diverse systems in different regions although no single curve covers the universe. In some areas such as the loess region in the Yellow River drainage, the sediment delivery ratio for small areas is 1 (Mou Jinze, 1986). This is also the case in other areas where sediment is primarily derived from gully systems or where steep hillslopes intersect stream channels directly.

The consequences of a declining sediment delivery ratio may be significant in considering sediment controls in stream channels or on the watershed itself. The sediment delivery ratio becomes very small in many regions even at drainage areas of less than one hundred square miles. Moreover, at the continental scale Foster and Hakonson (1983) have shown that only 4% of the sediment eroded from the continental landscape of the U.S. since the fallout of plutonium derived from testing of nuclear weapons in the 1950s and 60s will appear as yield in the river systems for a considerable period of time. Decreasing yield with increasing drainage area has been attributed in large measure to declines in erosion and yield associated with declines in the steepness of topography as drainage area increases.

Because large drainage areas are likely to have inhomogeneous climate and geology, a number of other factors may influence the relationship of yield to area. Thus, observations indicate that the yield from eroding areas in the watershed may be exceedingly high from very limited areas. The loess in China is certainly such a region. An extreme example in the U.S., Fivemile Creek contributed 56% of the sediment, but only 7% of the water to the Bighorn River in Wyoming, USA. In contrast, several major tributaries and the main stem upstream contributed 50 to 70% of the water, but only 10 to 14% of the sediment (Table I). Fivemile Creek, an ephemeral channel, became the drainage outlet for an irrigation project. The additional water to the system caused major erosion of the channel banks and bed supplying sediment downstream. A nearby tributary, Muddy Creek, while not gullied, also contributed much sediment and little water. On a much larger scale, the spatial disassociation of discharge and sediment yield is seen on many large rivers. For example, the Colorado, initially fed from snowmelt and rainfall in mountain regions eventually traverses arid and semi-arid lands downstream subject to high erosion rates. While examples from semi-arid regions are striking, the same limited and local character of sediment sources is reflected in a study of the Anacostia River near Washington, D.C., where Scatena (1987, p. 128) notes, "that 77 to 85% of the sediment comes from the 5% of the landscape in quarries, stream channels, and

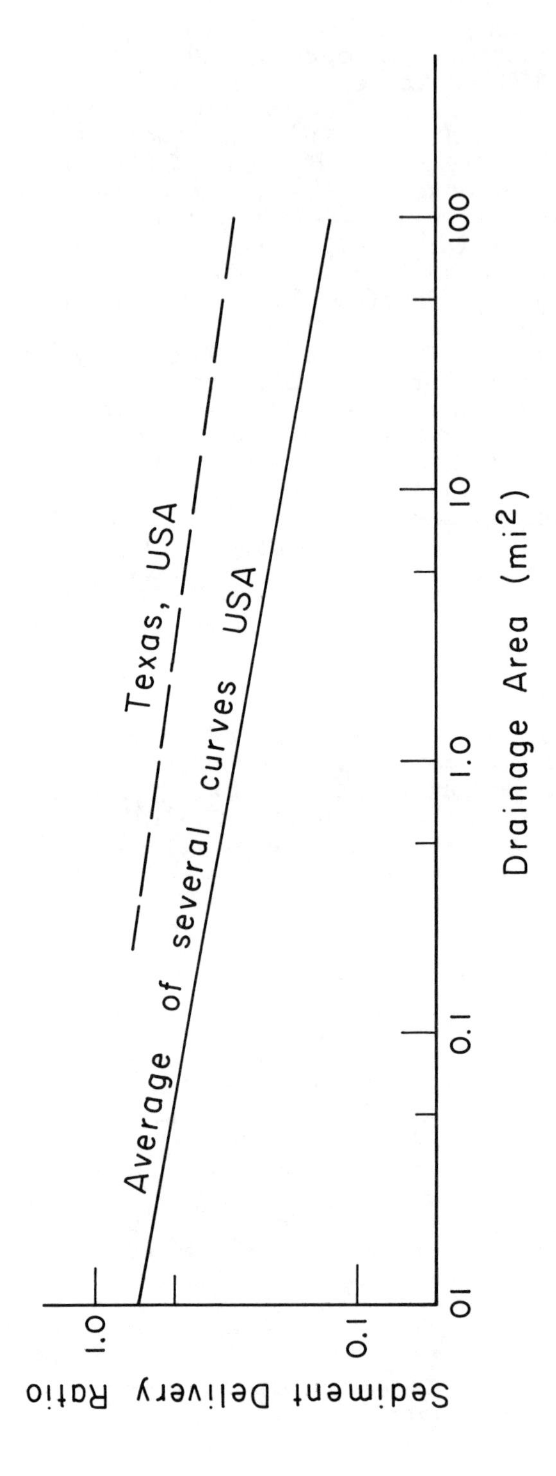

2. Relaltionship between size of drainage basin and sediment delivery ratio.

construction sites." The same study indicated that even on a single farm only a few fields actually contributed sediment to the channel system.

TABLE I. Average sediment discharge and streamflow in percentage of sediment discharge and streamflow of the Bighorn River at Thermopolis, Wyo., for selected upstream locations showing assymetry of sources of water and sediment.

Tributary	Drainage Area, mi^2	Percentage Flow	Percentage Sediment
Wind R.	178	67	14
Wind R.	2320	51	10
Popo Agie R.	2010	35	6
Fivemile Creek	397	7	56
Muddy Creek	340	1.4	16
Bighorn River	8080	100	100

(from Colby et al., 1956)

Identification of the loci of erosion, coupled with additional evidence of the likelihood that eroded sediment from specific sites will actually be transported to the stream networks, becomes critical in a cost-effective approach to control of sediment or non-point sources of pollutants. Not infrequently, such local sources are masked by computations of average sediment yields over larger areas. While the resultant yields may be the same, their sources and hence potential for control may be very different.

The Effect of Vegetation and Conservation Practices

It is a tenet of conservation that land use practices and changes in cover can significantly alter the yield of sediment. This is implicit in the USLE and supported by a number of independent observations. For example, throughout the United States broad land use classes yield distinctively different quantities of sediment (Table II).. Sequential changes in sediment yield with land use change are particularly well illustrated on the Gunpowder Falls. These land use changes mimic curves relating sediment yield to mean annual runoff, seasonality of runoff, and intensity of precipitation at regional scales (Langbein and Schumm, 1958, Dendy and Bolton, 1976, Wilson, 1972). In contrast to clearly identified differences in sediment yields from smaller

TABLE II. SEDIMENT YIELD FROM DRAINAGE BASINS UNDER DIVERSE LAND USE

River and Location	Drainage Area, mi^2	Sediment Yield tons/sq.mi./yr.	Land Use
Broad Ford Run, MD	7.4	11	Forested: entire area
Helton Branch, KY	0.85	15	Same
Fishing Creek, MD	7.3	5	Same
Gunpowder Falls, MD	303	808	Rural-Agricultural 1914-1943, farmland ~ 300,000 AC.
Same		233	Rural-Agricultural 1943-1961, farmland, ~ 200,000 AC.
Seneca Creek, MD	101	320	Same
Building Site, Baltimore, MD	0.0025	140,000	Construction: entire area exposed
Little Falls Branch, MD	4.1	2,320	Construction: small part of area exposed
Stony Run, MD	2.47	54	Urban: entire area

drainage areas where a particular land use covers much of the basin, over larger areas, 800-1,000 mi^2, in the Mid-Atlantic region of the U.S.A. sediment yields converge to an average for the region (Scatena, p. 25). This is in direct contrast to persistent, 6-fold differences in sediment concentrations from areas where less than 1/3, 1/3 to 2/3, and greater than 2/3 of the land is in cropland in the Midwest (Brune, 1948).

From the standpoint of management, perhaps of greatest significance is the evidence of large changes in sediment yields associated with major changes in land use over historical periods. A classic early study from the Tennessee Valley Authority (TVA) in White Hollow Creek showed a reduction of sediment yield from 473 t/mi^2 to 36 t/mi^2 in a period of 21 years. This heroic change in sediment yield resulted from complete reforestation of the 2.74 mi^2 watershed. On a broader historical base, Trimble (1974) reviewed the changes in sediment yield resulting from changes in land use over a major area of the Southern Piedmont of the U.S.A., suggesting a 10-fold increase in sediment yield in places as forest land was cleared for agricultural cropland, and a subsequent decrease in yield in the modern era. While no change in runoff occurred as a result of reforestation of White Hollow Creek in the TVA, reforestation reduced runoff in the Piedmont region, as it did in New York State. In areas deficient in moisture, runoff losses may conflict with the objective of reduction in sediment yield.

While the impact of conservation practices is clearly seen on some small watersheds, documentation of reductions in sediment yield from conservation practices on the landscape for other than the smallest watersheds is more difficult to identify. However, the classic literature of the 1930s and 1940s of the Soil Conservation Service of the U.S. Department of Agriculture included some outstanding examples (Table III). By virtue of its size, the most impressive is the 38% reduction on an area of 1,666 mi^2.

The best evidence for the impact of land use change on sediment yield is, of course, either at very small scales from plots, fields, or small drainage areas or where major changes have been made in land use practices and land cover. The latter are rarer, although Hadley (1977) documents extensive increases in plant cover and associated declines in runoff and aggradation of sediment due to climatic variations and overgrazing in the semi-arid western U.S.A. Over small areas, controlled grazing experiments document runoff from grazed lands roughly 130-150% and sediment yield 50% greater than from ungrazed lands in a region of 220 mm of annual precipitation (Hadley, 1977, p. 547).

Complete coverage by conservation practices or reforestation is not common in much of the world, even in some areas where conservation efforts have been pursued for decades or longer in the modern era. The economic, social, political,

TABLE III.

Selected examples of successful effects of conservation measures and watershed treatment on sediment yield (from Glymph, 1951).

River or Reservoir	Location	Drainage Area, mi^2	Reduction in Sediment, %	Surveys, years
Basque	Texas	1,666	38	1930-36-47
High Point	North Carolina	62.3	24-35	1934-1938
Issaquenna	South Carolina	14	53	1938-41-49
Jones Crk*	Iowa	2.26	98	1940-1949

*includes land treatment, gully control, reservoir.

and occasionally climatic reasons for the failure or absence of conservation practices is not central to this discussion. There are, however, some notable success stories. Thus, Bowonder et al. (1985) report a 30% reduction in sediment yield on a drainage area of 28,120 km^2 in India, and successful sediment control on another area of 2,222 km^2. Both involved integrated management of conservation efforts over large areas and application of a wide variety of conservation practices.

Despite some striking successes, the evidence suggests that while conservation measures can be very effective in reducing sediment yields locally, to achieve effects over large areas appears to require major changes in land use practices at selected locations or extensive coverage, or both.

Time Lags: Storage of Sediment and Basin Integration

Sediment is stored for varying periods of time at diverse sites from hillslope to channel bed to valley floor. Studies in the Piedmont of Maryland, USA, indicate that roughly 60% of the sediment eroded from the land accumulates as colluvium, 25% in the valley floodplain, while only 15% traverses the distance from source to outlet in a 50 mi^2 drainage (Costa, 1975). While the data in Tables IV and V are for a variety of conditions, some based simply on broad estimates, a few have been investigated in detail. In a small (1.5 mi^2) drainage area sediment in transit on the bed of the channel in a 2 mile reach was equivalent to 3 years annual yield (Wolman, 1987). As the data in Tables IV and V indicate, storage times may be very long, amounting to centuries, millennia, or longer. On Coon Creek, Trimble (1983) showed that while conservation practices were applied extensively to the uplands, sediment yield downstream at a drainage area of 138 mi^2 remained nearly constant between the period of very high erosion (1853 to 1938), and continuing through the period of conservation from 1938 to 1975. Sediments derived from the rapidly eroding uplands were initially stored in the bottom lands of Coon Creek and continuity in sediment yield was maintained by sediment derived from erosion of the alluvial floodplain and terraces. Similar processes were observed at Crab Orchard Lake in Illinois (Stall et al, 1954).

TABLE IV. Hydrologic storage sites and approximate detention times.

Location	Water	Sediment*
Land surface detention	minutes	unknown
Colluvium	?	10^2 years
Groundwater	10^1-10^3 years	
River pools	days-months	
Floodplain	days	10^1-10^2 years
Terraces		10^2-10^3 years
Lakes and reservoirs	months-years	10^1-10^2 years
Ocean and estuaries	10^2 years	10^6 years

*partially soluble constituents partitioned

At a larger scale, it is interesting to note that erosion rates in the U.S.A. are estimated to have increased 2 to 10-fold as a result of the replacement of forest and prairie by agriculture over large areas. Yet, the Mississippi River draining roughly 1/3 of the continent reveals no changes in the quantity of sediment transported annually to the Gulf of Mexico. Only since the completion of a sequence of large dams on the Missouri River has the yield of sediment in the Mississippi downstream declined significantly, by about 33% (Williams and Wolman, 1984).

Storage of sediment is significant in analyzing the likely time history and distribution of sediment derived from upland sources. The sediment delivery ratio may include not only errors in the estimation of erosion, but lags in the delivery of sediment to downstream points. Moreover, because storage times may be long, the impact of control measures on the landscape may be difficult to discern, and hence the effectiveness of upland control measures difficult to measure, at important downstream locations.

Reservoirs as Sediment Traps and Valuable Resources

In many instances, a major objective of either land treatment or sediment control by structural works is protection of storage volumes in reservoirs. At the same time, reservoirs elsewhere upstream on a drainage basin are seen as appropriate structural traps to control sediment. Multi-purpose reservoirs may serve a variety of purposes including sediment storage. Because the reservoirs can be effective sediment traps the percentage of sediment controlled by land treatment over a large area is likely to be considerably smaller than

TABLE V. Sediment Storage Times in Selected Rivers in the United States

River	Approximate % of Transport	Storage Time	Location of Storage	Termino-logy*	Reference
Mississippi	20-50	season	channel bed	E	Meade, in prep.
Potomac (MD)	3-5	season(?)	bed and adjacent	E	Miller, in prep.
Baisman Run (MD)	500	5 years	bed and bars	E	Wolman, in prep.
Redwood Creek (CA)	---	9-26	channel & floodplain	E	Madej, 1984
Redwood Creek (CA)	---	700-7200	terrace (?)	T	Madej, 1984
Sacramento (CA)	1000	50-100 yrs	valley bottom	A	Gilbert, 1917

* E = equilibrium, T = terrace, A = aggradation

the control provided by reservoirs and check dams. On a tributary of the Yellow River sediment yield was reduced by 8.7% by reservoirs, 48.7% by silt traps and 4% by conservation practices on the land (Zhang, 1986, Gong, S., 1987). (Of course, the objectives of land treatment may differ from those associated with control of sediment alone. Ideally, these objectives include the conservation of the soil resource itself and the retention of nutrients or pollutants on the landscape.) Hadley (1977) postulated that in a semi-arid region small reservoirs, by controlling discharge, reduced flows resulting in channel aggradation. If downstream sediment control is the principle objective, then reservoirs appear to be quite effective. As the classic curves of Brune (1953) show, the trap efficiency (effectiveness) of the reservoir increases rapidly as the ratio of capacity to annual inflow (or contributing drainage area) increases. At the same time, sediment yields per unit area decline downstream while the capacity to watershed area, based on empirical data, remained relatively constant for reservoirs on areas larger than about 10 mi^2 (Table VI). Thus, if the purpose is simply to control sediment to protect a downstream point, the "most effective" control is likely to be a reservoir at the most downstream dam site available in a drainage basin.

TABLE VI. Distribution of capacity-watershed ratios for a number of reservoirs in the United States.

		Capacity to Watershed Area (af/mi^2)					
Drainage Area (mi^2)	No.	Range	Distribution 0.1-1	1-10	10-100	100-1000	mean (apprx)
0.1-1	11	0.7-130	1	2	6	2	20
1-10	18	0.4-1000	1	4	6	8	60
10-50	9	0.2-84	2	0	2	7	150
50-100	4	18-420	0	0	1	3	150
100-1000	14	8-790	0	1	6	7	110
1000-10,000	15	15-310	0	0	9	6	62

(Data from Gottschalk and Brune, 1950, Glymph, 1951, Renfro and Moore, 1951, Stall et al., 1952 and 1954, Hains et al., 1952, Herb, 1980.)

The time history of reservoir sedimentation is of major interest in developing plans for sediment control over a large area. Trap efficiency declines roughly logarithmically with time as capacity is reduced by sediment accumulation. Similarly, as Figure 3 shows, the annual rate of storage loss through sedimentation is inversely proportional to the ratio of reservoir capacity to contributing drainage area. In the

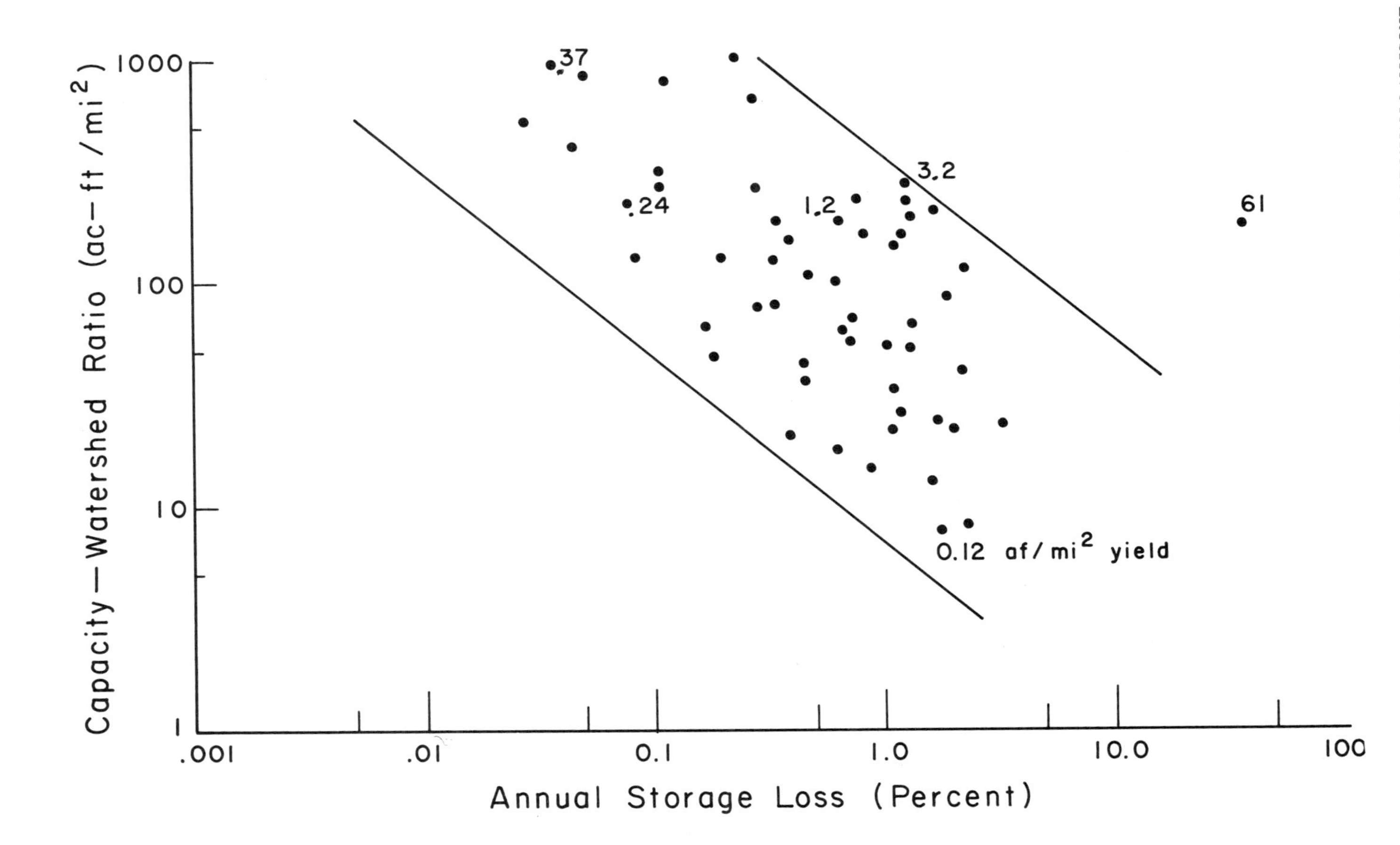

3. Relationship between size of drainage basin and sediment delivery ratio.

U.S.A. a number of reservoirs in the mid-continent region as well as in the East fall within the envelopes shown in Figure 3. (A reservoir in the Missouri Basin loess hills with very high sediment yield falls well beyond the upper envelope.) Although trap efficiency declines with time, the lifetimes of many small reservoirs are likely to be measured in decades (Eakin and Brown, 1939). Very large reservoirs, of course, have estimated life expectancies of hundreds of years in regions of moderately high sediment yields, although some in regions of very high sediment yield such as Sanmenxia, or Mangla dam, may also fill within decades.

Location and Scale in the Effectiveness of Reservoir Sediment Control

The results of a very simple model calculation are presented in Figures 4 and 5 to suggest the lifespan and trapping of sediment by reservoirs in a hypothetical river basin. The assumed sediment yield, bifurcation ratio, runoff and erosion rate are given on Figure 4, which shows the common decline in sediment yield with increasing drainage area. Values of the capacity/watershed-area ratio are based upon empirical results from many sites in the United States (Table VI). Trap efficiency of reservoirs is assumed to decline logarithmically with loss in capacity over time. In the model, forty-two reservoirs are located on each of the 2nd order tributaries of a fifth order basin, and reservoir size is varied from capacity-watershed ratios of 60 to 1, encompassing the empirical data. The assumed sediment yield of 0.40 af/mi^2 corresponds to cropland values observed in the midwestern Great Plains and portions of the southwestern United States (Brune, 1948).

The larger reservoirs, as expected, require 200 to 300 years to fill to 95% of capacity where located on 2nd order tributaries, and as much as 500 years when located on 3rd order tributaries. The decline in reservoir life with both increasing sediment yield and declining capacity/watershed ratio is readily apparent in Figure 5 for the model case of 42 reservoirs on each 2nd order stream. Life expectancy declines as capacity declines and is less than 150 years for capacity-watershed ratios of 20 or less. The latter value may not be uncommon for reservoirs on drainage areas smaller than 100 mi^2. Smaller reservoirs fill within decades and the loss of trap efficiency limits their effectiveness before capacity does. At the opposite extreme, three large reservoirs on 4th order basins are very effective sediment traps and have very long lifetimes (Table VII) although the rate of increase in longevity declines with increasing capacity-watershed ratio (Figure 5).

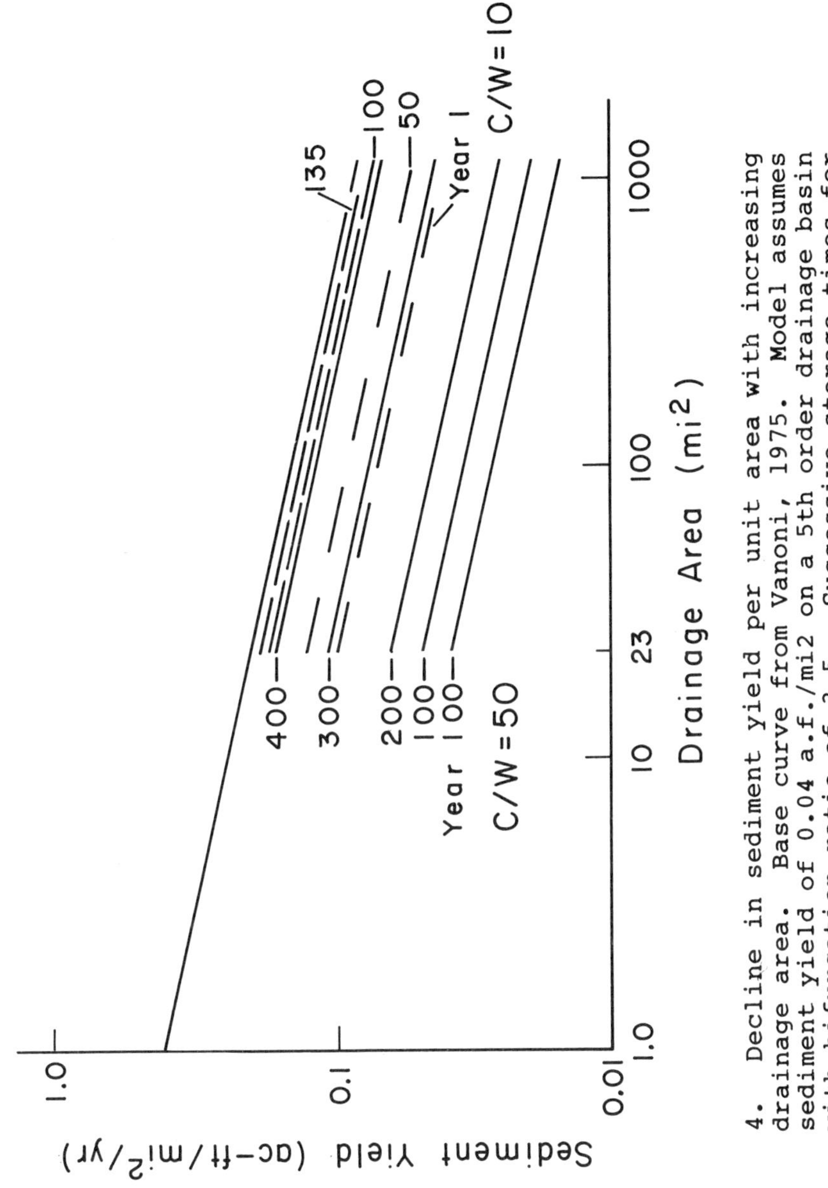

4. Decline in sediment yield per unit area with increasing drainage area. Base curve from Vanoni, 1975. Model assumes sediment yield of 0.04 a.f./mi2 on a 5th order drainage basin with bifurcation ratio of 3.5. Successive storage times for capacity/watershed ratio (c/w) = 50 (solid lines) and c/w = 10 (dashed line) for 12 reservoirs on 3rd order basins.

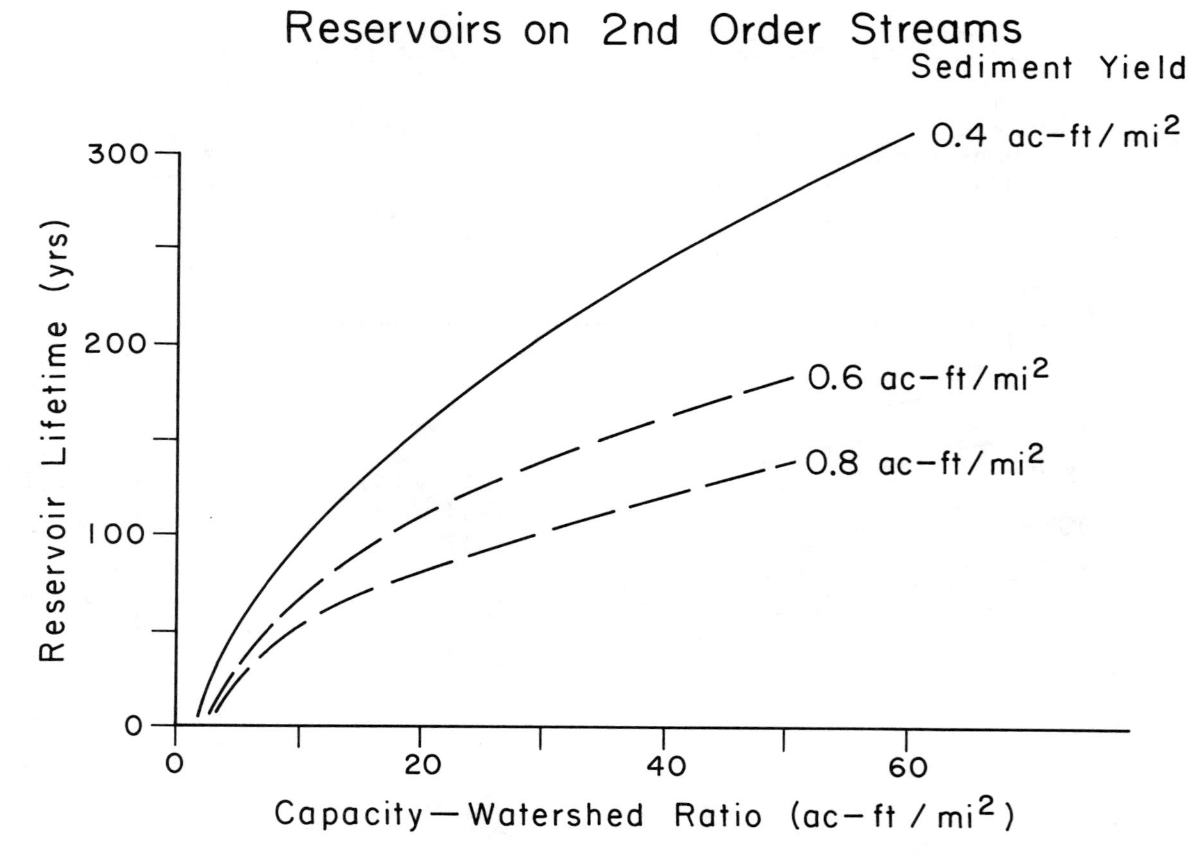

5. Reduction in reservoir lifetime with declining capacity/watershed ratio and increasing sediment yield. Forty-two reservoirs, one on each 2nd order tributary of a 5th order drainage basin.

TABLE VII. Reservoir siting model for three 4th order basins. c/w equals ratio of reservoir capacity to area of watershed.

C/W ratio ratio ac.ft/mi^2	Initial sed. yld. ac.ft/mi^2	Initial trap eff.	Years of Operation and comments
100	.40	84%	>500 yr.
50	.40	81%	>500 yr.
40	.40	78%	>500 yr.
30	.40	74%	425 yr. (95% full)
20	.40	68%	320 yr. (eff. 10%)*
10	.40	51%	196 yr. (eff. 10%)
5	.40	34%	109 yr. (eff. 10%)
4	.40	29%	86 yr. (eff. 10%)
3	.40	22%	57 yr. (eff. 10%)
2.5	.40	17%	38 yr. (eff. 10%)
2	.40	12%	13 yr. (eff. 10%)
1	.40	< 10%	--- (eff. 10%)

*less than 10% of capacity remaining.

"Ultimately" all of the reservoirs in upstream areas will fill in relatively short periods of time. Moreover, their efficiency in a large basin-wide sense cannot be very high simply because the amount of sediment from upstream areas delivered downstream declines in many river systems quite rapidly. Thus the distributed small reservoirs are efficient traps for a short period of time for a local region, but increasingly ineffective as time and drainage area increase.

The effectiveness of reservoirs as sediment traps in upstream regions is also partially undercut by the fact that erosion downstream from the dams in alluvial systems can also become significant in providing a renewed source of sediment. In addition, whether the sediment load is high or not, damages resulting from degradation downstream from reservoirs and dams may in themselves be significant.

Long Term Reduction of Sediment and Pollutant Yields

If the above reasoning is correct, then even though conservation measures on the land often do not appear to be as effective in controlling sediment as reservoirs distributed on the stream system, ultimately sources on the landscape need to be controlled in any event. One could argue, as implied earlier, that this approach is both difficult and thus far not shown to be universally effective. The difficulty in part lies with the fact that many individuals must be dealt with in providing guidance and incentives to individual farmers to practice conservation measures. Building a dam is not without its political and social difficulties, but once accomplished it is an engineering operation. Conservation on the landscape requires continuity of practice among thousands of individuals.

Nevertheless, bearing in mind two characteristics of erosion and sediment yields from the landscape, it may well be possible to improve the outcomes of sediment control practices on the landscape. The two principles involved are isolation and connectivity. Much evidence supports the fact that areas subject to intense erosion often provide the bulk of the yield of sediment even from large areas. For these areas subject to high accelerated erosion, the second attribute needed to make them contributors to both erosion and yield is "connectivity." The latter, the inverse of a declining sediment delivery ratio, consists of a set of direct connections between the source of erosion and downstream points of transport and deposition.

One might suggest an approach to an effective planned sediment control program as one in which sources of intense erosion and connecting links of sediment delivery were mapped. Remote sensing techniques may well facilitate this operation (Pelletier, 1985). This combination would identify the locations for the most intense efforts at sediment control.

These procedures are indeed applied now. At the same time, in some regions considerably less selectivity is exercised in applying conservation techniques. In many regions of the U.S.A., concern for control of non-point sources of pollution is leading to more detailed analysis of the location of such sources and attempts are being made to fashion programs for sediment and pollution control that are both effective and efficient. Great strides have been made on the Yellow River in identifying and controlling sediment sources (Gong, S., 1987). The author looks forward to learning the extent to which additional work may be warranted in integrating reservoir management and conservation practices to achieve optimum control of sediments and pollutants over a long period of time.

References

Bowonder, B., Ramana, K.V., and Rao, T.H., 1985, Management of watersheds and water resources planning, Water Internatl. V. 10:121-131.

Brune, G.M., 1948, Rates of sediment production in midwestern U.S., U.S. Dept. of Ag. SCS TP-65, 40 pp.

Brune, G.M., 1953, Trap efficiency of reservoirs, Trans. Am. Geophys. Un., V. 34:407-418.

Colby, B.R., Hembree, C.H. and Rainwater, F.H., 1956, Sedimentation and chemical quality of surface waters in the Wind River basin, Wyo., U.S. Geol. Surv. Water-Supply Paper 1373, 336 pp.

Costa, J.E., 1975, Effects of agriculture on erosion and sedimentation in the Piedmont Province, Md., Geol. Soc. Am. Bull., V. 86:1281-1286.

Dendy, F.E., and Bolton, G.C., 1976, Sediment yield-runoff-drainage area relationships in the United States, J. Soil and Water Cons., V. 31: 264-266.

Eakin, H.M. and Brown, C.B., 1939, Silting of reservoirs, U.S. Dept. of Ag. Tech. Bull. 524, 168 pp.

Foster, G.R. and Hakonson, T.E., 1983, Erosional losses of fallout plutonium, Los Alamos Natl. Lab. unpubl. 47 pp.

Glymph, L.M., Jr., 1951, Relation of sedimentation to accelerated erosion in the Missouri River basin, U.S. Dept. Ag. SCS-TP 102, 23 pp.

Gong, G. and Xu, J., 1987, Environmental effects of human activities on rivers in the Huanghe-Huaihe-Haihe Plain, China, Geograf. Ann., V. 69A:181-188.

Gong, S., 1987, The role of reservoirs and silt-trap dams in reducing sediment delivery into the Yellow River, Geograf. Ann., V. 69A:173-179.

Gottschalk, L.C. and Brune, G.M., 1950, Sediment design criteria for the Missouri Basin Loess Hills, U.S. Dept. Ag. SCS TP-97, 21 pp.

Hadley, R.F., 1977, Evaluation of land-use and land-treatment practices in semi-arid western United States, Phil. Trans. Ray. Soc. Land. Bu. 278:543-554.

Hains, C.F., Van Sickle, D.M. and Peterson, H.V., 1952, Sedimentation rates in small reservoirs in the Little Colorado river basin, U.S. Geol. Surv. Water-Supply Paper 110-D, 153pp.

Herb, W.J., 1980, Sediment trap efficiency of a multiple-purpose impoundment, North Branch Rock Creek basin, Montgomery County, MD, 1968-76, U.S. Geol. Surv. Water-Supply Paper 2071, 41 pp.

Hussein, M.H., 1986, Rainfall erosivity in Iraq, J. Soil and Water Cons., V. 4:336-338.

Langbein, W.B. and Schumm, S.A., 1958, Yield of sediment in relation to mean annual precipitation, Am. Geophys. Un. Trans. V. 39:1076-1084.

Madej, M.A., 1984, Recent changes in channel-stored sediment in Redwood Creek, Cal.: Redwood Natl. Park, Tech. Rept. No. 11, 54 pp.

Mou Jinze, 1986, Sediment sources and yield from small drainage basin, in Hadley, R.F., ed. Drainage Basin Sediment Delivery, IAHS Publ. No. 159, pp. 19-29.

Natl. Acad. of Sci., 1986, Soil Conservation: assessing the national resources inventory, Natl. Acad. Sci. Press, Washington, D.C., V. 1:114 pp., V. 2:314 pp.

Pelletier, R.E., 1985, Evaluating nonpoint pollution using remotely sensed data in soil erosion models, J. Soil and Water Cons., V. 40:332-335.

Renfro, G.W. and Moore, C.M., 1958, Sedimentation studies in the western Gulf states, J. Hydr. Div., V. 84, No. HY5, pp. 1-16.

Scatena, F.N., 1987, Sediment budgets and delivery in a suburban watershed, Johns Hopkins Univ. dissertation, unpubl., 200 pp.

Stall, J.B., Fehrenbacher, Bartelli, L.J., Walker, G.O., Sauer, E.L., and Melsted, S.W., 1954, Water and Land Resources of the Crab Orchard Lake Basin, Ill. State Water Survey Bull. No. 42, 53 pp.

Stall, J.B., Klingsbiel, A.A., Melsted, S.W., and Sauer, E.L., 1952, The silting of Lake Calhoun, Ill. State Water Survey Rept. of Investigation No. 15, 26 pp.

Tennessee Valley Authority, 1961, Forest cover improvement influences upon hydrologic characteristics of White Hollow watershed 1935-1958, Rept. No. 0-5163A, 104 pp.

Trimble, S.W., 1974, Man-induced soil erosion on the Southern Piedmont 1700-1970, Soil Conserva. Soc. Am., 177 pp.

Trimble, S.W., 1983, A sediment budget for Coon Creek basin in the Driftless area, Wisconsin 1853-1977, Am. J. Sci., V. 283:454-474.

Vanoni, V., 1975, ed., Sedimentation Engineering, Am. Soc. Civ. Eng., M&R No. 54, 463 pp.

Williams, G.P. and Wolman, M.G., 1984, Downstream effects of dams on alluvial rivers, U.S. Geol. Surv. Prof. Paper 1286, 83 pp.

Wilson, J.P., 1986, Estimating the topography factor in the universal soil loss equation for watersheds, J. Soil and Water Cons., V. 41:179-184.

Wilson, L., 1972, Seasonal sediment yield patterns of U.S. rivers, Water Resources Res. V. 8:1470-79.

Wischmeier, W.H. and Smith, D.D., 1978, Predicting rainfall erosion losses--a guide to conservation planning, Agr. Handbk. No. 537, U.S. Dept. of Ag., Wash., D.C.

Wolman, M.G., 1987, Sediment movement and knickpoint behavior in a small Piedmont drainage basin, Geograf. Ann., V. 69A:5-14.

Zhang, Shenghi, 1984, Effects of comprehensive soil conservation measures on the reduction of sediment yield in the Wuding River Valley, J. of Sed. Res., No. 3:1-11 (in Chinese).

FLOOD FORECASTING AND ITS RELIABILITY

Ben Chie YEN*
University of Illinois at Urbana-Champaign
Department of Civil Engineering
205 North Mathews Avenue
Urbana, Illinois 61801, USA

ABSTRACT. Methodologies for forecasting flood hydrographs are discussed for (a) catchment floods from rainfall and (b) river floods with flow from upstream. Catchment flood forecasting methods include the unit hydrograph techniques, conceptual models, computer-based physical process models, and stochastic simulation models. River flood forecasting methods include physically-based routing models, deterministic lumped system models, station correlation methods, and stochastic process methods. Flood forecasting is inevitably under uncertainties from various sources. Therefore, it is desirable to have not only the forecast of a flood but also some measure of the reliability of the forecast. Techniques to evaluate the reliability for flood forecasting of a catchment and of a river are suggested.

Introduction

Methods to forecast flood events can be classified in different ways:

(a) According to the result produced: Those produce discharge or stage hydrographs and those produce only the peak discharge of the flood.

(b) According to the input water data for forecasting: Those for catchment flood forecasting for which the major water input is rainfall, and those for river flood forecasting where the major input is the flow from upstream channel and tributaries.

---

* Now at the University of Virginia, Center for Advanced Studies and Department of Civil Engineering, Charlottesville, Virginia 22901, USA.

L. M. Brush et al. (eds.), Taming the Yellow River: Silt and Floods, 163–196.

(c) According to systems concept in terms of spatial details: The lumped system methods which treat the catchment as a black box not considering the internal transformation process, and the distributed system methods which consider the distribution and transformation process inside the system.

(d) According to the use of probability: Deterministic methods and stochastic methods.

Before the commencement of use of digital computers for flood forecasting in the 1960's, most of the forecasting methods were lumped system formulas or graphic methods determining only the flood peak discharge with simple computations. Typical examples are the rational method for catchment flood peak calculation and flood frequency analysis for channel flood peak estimation. A few lumped system methods involving relatively unsophisticated calculations were developed for hydrograph determination for catchment floods (e.g., unit hydrograph method) and for channel floods (e.g., Muskingum method and two or more station discharge correlations).

With the rapid development of computer technology and numerical techniques since 1960's, the trend of flood forecast is shifting to the use of computer-based simulation models, mostly estimating the flood hydrographs rather than merely peak discharges. Since hydrographs contain the information on peak discharges, this paper will concentrate on flood hydrograph forecasting on a real-time single-event basis, whereas methods dealing only with peak discharges will not be discussed.

Catchment Flood Forecasting Methods

In catchment flood forecasting, rainfall is the major input which produces the flood at the point of interest. Catchment flood hydrograph forecasting methods can be classified into four groups: unit hydrograph models, conceptual catchment models, computer-based physical or semi-physical simulation models, and stochastic simulation models.

UNIT HYDROGRAPH METHODS

Since Sherman (1932) proposed the concept of unit hydrograph, the method has been a major tool for catchment flood hydrograph forecasting. In applying the unit hydrograph method to forecast the flood produced by a rainstorm, the following rather elementary and standard procedures are involved:

1) Determine the rainfall excess by subtracting the

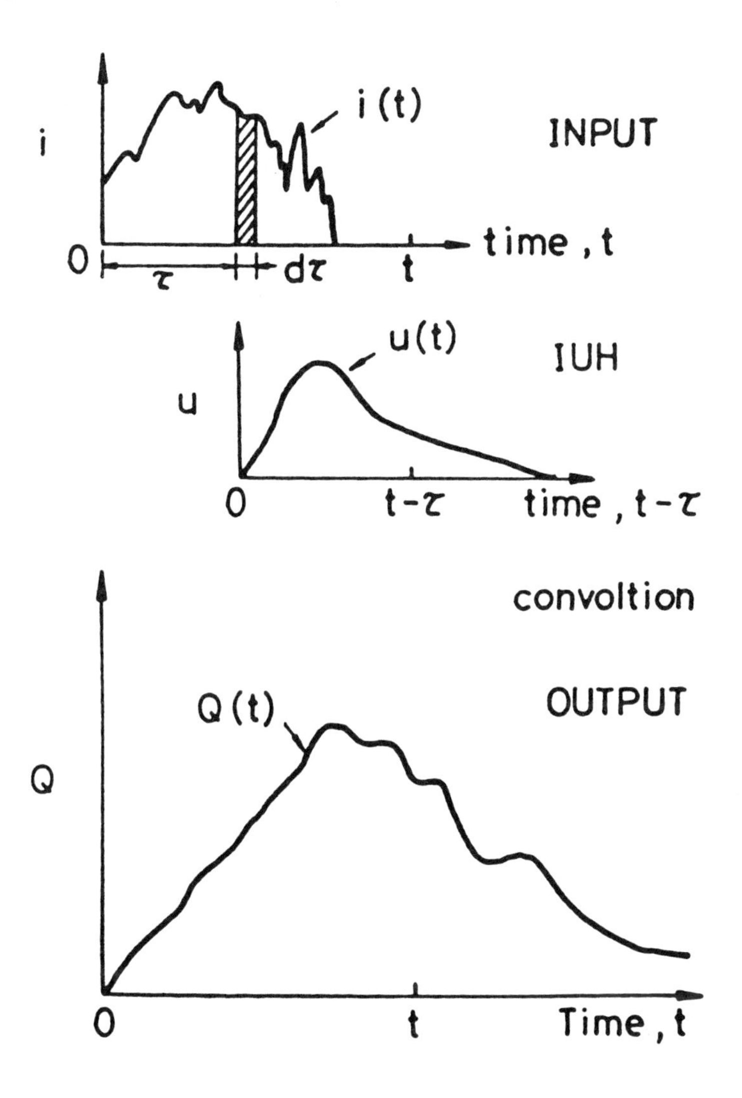

Fig. 1. Application of Unit Hydrograph through Convolution Integration

abstractions from the total rainfall.

2) Find the duration of the rainfall excess. If the rainfall excess is highly nonuniform over time, subdivide the rain excess into a number of successive components of reasonably uniform intensity and determine their respective durations.

3) Obtain the unit hydrographs of the proper durations.

4) Multiply the unit hydrograph ordinates (discharge values) by the depth of the rain excess to yield the direct runoff hydrograph. If there are more than one rain excess components, add the respective direct runoff hydrographs linearly with proper time consideration.

5) Add base flow (if significant) to the direct runoff hydrograph to yield the total runoff hydrograph.

Thus, the unit hydrograph method is a deterministic linear lumped system method. The unit hydrograph is a "fingerprint" of the catchment reflecting in a spatially cumulative way the characteristics of the catchment in producing runoff from rainfall excess.

Many methods have been suggested to derive the unit hydrograph for a catchment. The standard procedure is to deduce the unit hydrograph from observed runoff and rainfall data. For the case of no data, several synthetic unit hydrograph techniques have been proposed, including Synder's method, Gray's method, Doodge's method, and SCS method. The last method is described in Technical Paper 149 of the U.S. Soil Conservation Service. Description of the other methods can be found, e.g., in Gray (1973) and Chow (1964).

Recently geomorphology-based instantaneous unit hydrograph (GIUH) method has been proposed. A version is described briefly as follows. The channels in the catchment are organized in the order according to the Horton-Strahler law of stream orders (Chow, 1964, Section 4-II). A variable X (e.g., channel length, slope or flow time) for a channel is denoted as X_{ji2} where the subscript 2 indicates channel, i represents the stream order of the channel, and j identifies the channel from those of the same order. The catchment surface that flows into the channel is identified as X_{ji1} where the subscript 1 denotes surface.

The rainfall excess in the catchment will follow different flow paths from the point of origin to the catchment outlet according to the location of origin. The flow path may be expressed as

$$X_{ji1} \rightarrow X_{ji2} \rightarrow X_{j'(i+1)2} \rightarrow \ldots . \rightarrow X_{jk2} \rightarrow \ldots . \rightarrow X_{M2}$$

$$i = 1, 2, \ldots . M$$

$$k = i,\ i+1,\ \ldots . M$$

where M is the order of the catchment. The representative time of travel through an element of a path is defined as the service time T_{ji1} or T_{ji2}. Therefore, the total service time along a possible flow path r is

$$\tau_r = T_{ji1} + T_{ji2} + T_{j'(i+1)2} + \ldots + T_{j''k2} + \ldots + T_{M2} \tag{1}$$

The service time of a path is not constant. Its probability distribution function is $f_T(t)$ with a mean μ. Hence,

$$f_{\tau r}(t) = f_{Ti1}(t) * f_{Ti2}(t) * \ldots * f_{Tk2}(t) * \ldots * f_{TM2}(t) \tag{2}$$

in which * denotes convolution. The probability of the rainfall excess drained from the catchment is

$$P(T \leq t) = \sum_{r \epsilon R} P(\tau_r \leq t) \cdot P(r) \tag{3}$$

in which T is the time of a rain excess to runoff from the catchment, R represents all the possible paths, and P(r) is the transitional probability of the rainfall excess along the r path. Thus, the instantaneous unit hydrograph is

$$u(t) = \sum_{r \epsilon R} \left[[f_{Ti1}(t) * f_{Ti2}(t) * \ldots * f_{TM2}(t)]\ P(r) \right] \tag{4}$$

In applying the instantaneous unit hydrograph (IUH) to rainfall excess to produce direct runoff hydrograph, the convolution integration technique is used as shown graphically in Fig. 1,

$$Q(t) = \int_{o}^{t \leq t_d} i(\tau) u(t-\tau)\ d\tau \tag{5}$$

in which Q(t) is the catchment runoff rate at time t, t_d is the duration of the rainfall excess, and $i(\tau)$ is the rain excess intensity at time τ.

CONCEPTUAL HYDROLOGIC SYSTEM MODELS

In the conceptual hydrologic system models, the catchment transformation behavior from rainfall to runoff is assumed to be analogous to a simple hydraulic structure such as reservoirs or tanks. Thus, the models are basically lumped system deterministic models. The conceptual models utilize the continuity equation of the input (I), output (Q), and storage

(S):

$$I - Q = \frac{dS}{dt} \tag{6}$$

together with the storage equation

$$S = S(I, Q) \tag{7}$$

In a conceptual model the storage equation (Eq. 7) is not determined by hydraulic principles. In general, S can be a function of I and Q and their differential as well as integral functions, that is

$$S = \sum_i a_i \frac{d^i I}{dt^i} + \sum_j b_j + \frac{d^j Q}{dt^j} + \sum_k c_k \int \int_k \cdots\cdots \int \tag{8}$$

$$I(\tau)\, K_k\, (t - \tau)\, d\tau_1 \cdot d\tau_2 \cdots d\tau_k$$

$$+ \sum_p d_p \int \int_p \cdots \int \quad Q\,(\tau)\, H_p \quad (t - \tau\,)$$

$$d\tau_1\, d\tau_2 \cdots d\tau_p$$

where K_k and H_p are kernel functions, and the coefficients a_i, b_j, c_k and d_p can be functions of I and Q in addition to t. Neglecting the integral terms and assuming a_i and b_j independent of t, Eq. 8 can be simplified as

$$\frac{dS}{dt} = \sum_{i=0}^{m} a_i \frac{d^{i+1} I}{dt^{i+1}} + \sum_{j=0}^{n} b_j \frac{d^{j+1} Q}{dt^{j+1}} \tag{9}$$

Substituting this equation into Eq. 6 yields

$$Q = \left(- \frac{a_m D^{m+1} + a_{m-1} D^m + \ldots + a_1 D^2 + a_o D - 1}{b_n D^{n+1} + b_{a-1} D^n + \ldots + b_1 D^2 + b_o D + 1} \right) I = \phi I \tag{10}$$

where the operator D is defined as

$$D^a = \frac{d^a}{dt^a}$$

Equation 10 is referred by Chow (1982) as the general hydrologic system model. Its linearity depends on whether the operator Φ is linear or not.

In Eqs. 6 to 10, I can be any kind of input into the catchment system, not restricted to the rate of rainfall excess. It can be the rate of total rainfall without considering explicitly the abstractions, and correspondingly, the output is the total runoff rate without deducting the base flow. However, in practice because of different hydraulic characteristics and significant time scales of the surface and subsurface flow processes, the conceptual models are more reliable when dealing with rainfall excess as the input and surface runoff as the output.

In applying the conceptual model to a given watershed, the values of the coefficients a_i and b_j are first calibrated using observed data of rainfall excess rate and direct runoff rate. Presumably the method could be extended to catchments without data by prior determination of the variations of a_i and b_j with catchment properties, using the data of other catchments.

Conceptual catchment models can be further classified according to the assumptions made to represent Eq. 7. Most of the proposed models belong to the linear reservoir-linear channel type, in which the catchment is assumed to behave like a linear reservoir ($S = K_1Q + K_2$) or a linear channel ($Q(t) = I(t - \tau)$) or a combination of linear reservoirs and channels in series or in parallel. Best known models in this category are those proposed by Nash (1957) and Dooge (1959). Notable early development of this type of models is shown in Fig. 2.

Other conceptual models include the tank model (which is rather popular in Japan), weir model, orifice model, and Muskingum type catchment model (Yen, 1984). In recent years a number of nonlinear conceptual models have been proposed. For example, the Australian RORB model (Laurenson and Mein, 1985) which assumes $S = KQ^m$. Chiu and Huang (1970) considered the case K being also a function of time, thus making the model stochastic.

PHYSICAL PROCESS BASED COMPUTER-ASSISTED SIMULATION MODEL

A more intellectually satisfactory approach is to simulate the

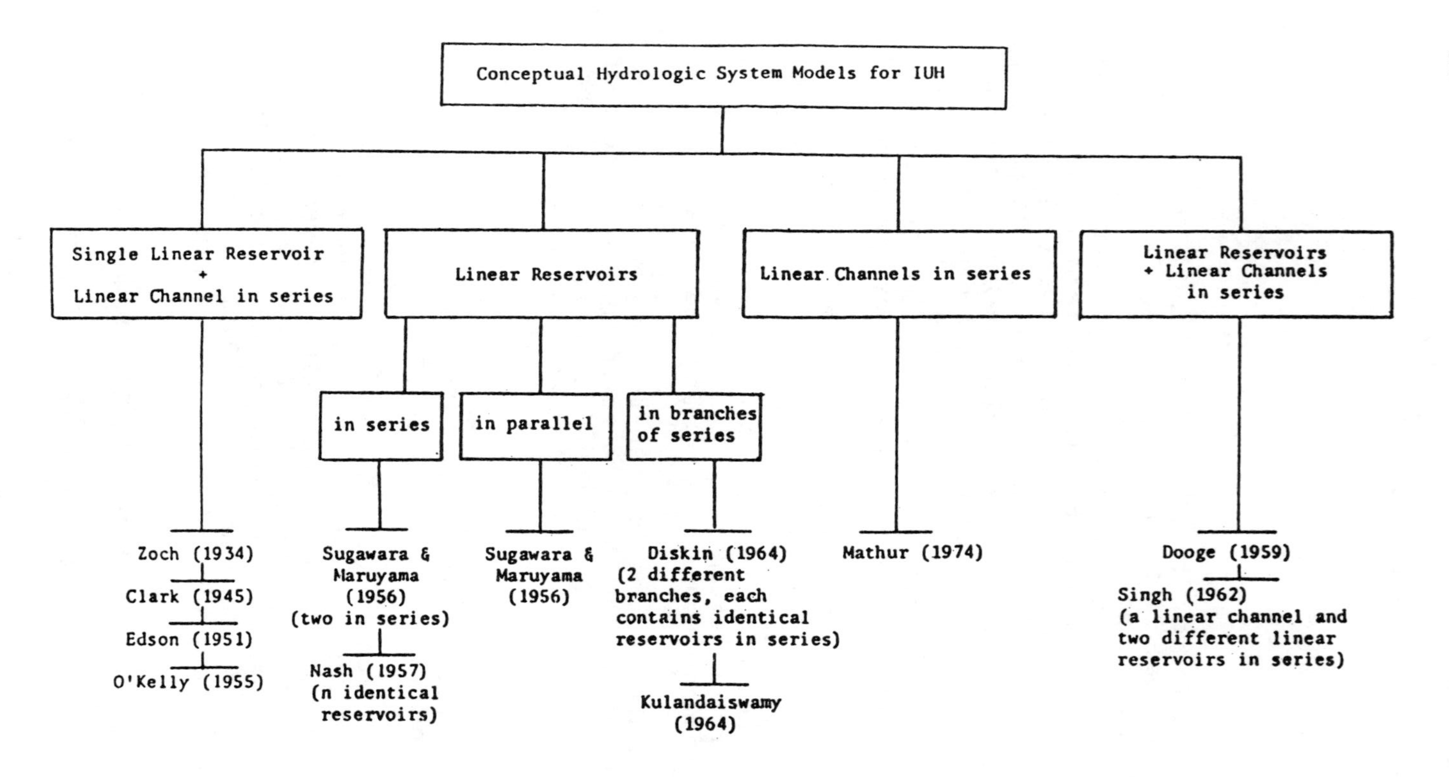

Fig. 2. Early Development of Linear Reservoir and Linear Channel Conceptual System Catchment Models (Yen, 1983)

catchment flood runoff process according to physical principles. This approach is gradually becoming practical with the advancement in computer technology. In this distributed system approach, detailed information on the drainage pattern within the catchment is needed. The catchment is decomposed into a network of components of channels and overland surfaces. The physical characteristics of each component are specified. Appropriate hydraulic equations are assigned to each component to simulate the hydrologic processes. Necessary values of coefficients and constants are specified or calibrated from observed data.

The basic hydraulic equations representing mathematically the flow in a channel as well as on an overland surface are the Saint-Venant equations. The continuity equation can be written as

$$\frac{\partial A}{\partial t} + \frac{\partial Q}{\partial x} = \int q_1 \, d\sigma \tag{11}$$

in which

x = flow longitudinal direction measured horizontally (Fig. 3)
A = flow cross-sectional area normal to x
Q = discharge through the cross section A
t = time
σ = perimeter bounding the cross section A
q_1 = lateral flow rate per unit length of perimeter σ along unit length of x, being negative for outflow

The momentum equation can be written in either discharge or velocity form (see Equations 12a and 12b on following page), in which

Y = depth of flow of the cross section, measured vertically
S_o = channel or overland slope, equal to $\tan\theta$
θ = angle between the horizontal plane and the channel bed or overland surface
S_f = friction slope
V = Q/A, cross-sectional average velocity along x direction
g = gravitational acceleration
U_x = x-component velocity of q_1

$$\beta = \frac{A}{Q^2} \int_A \bar{u}^2 \, dA$$

$\bar{u}$ = x-component of local point flow velocity where the bar indicates averaging over turbulencee

Since 1970 a number of computer-based catchment runoff

$$\frac{1}{gA}\frac{\partial Q}{\partial t} + \frac{1}{gA}\frac{\partial}{\partial x}\left[\frac{\beta Q}{A}\right]^2 + \frac{1}{gA}\int_\sigma U_x q_1 d\sigma \qquad + \frac{\partial Y}{\partial x} - S_o + S_f = 0 \tag{12a}$$

dynamic wave

quasi-steady dynamic wave

Non-inertia

kinematic wave

$$\frac{1}{g}\frac{\partial V}{\partial t} + (2\beta-1)\frac{V}{g}\frac{\partial V}{\partial x} + (\beta-1)\frac{V^2}{gA}\frac{\partial A}{\partial x} + \frac{1}{gA}\int_\sigma (U_x - V) q_1 d\sigma + \frac{\partial Y}{\partial x} - S_o + S_f = 0 \tag{12b}$$

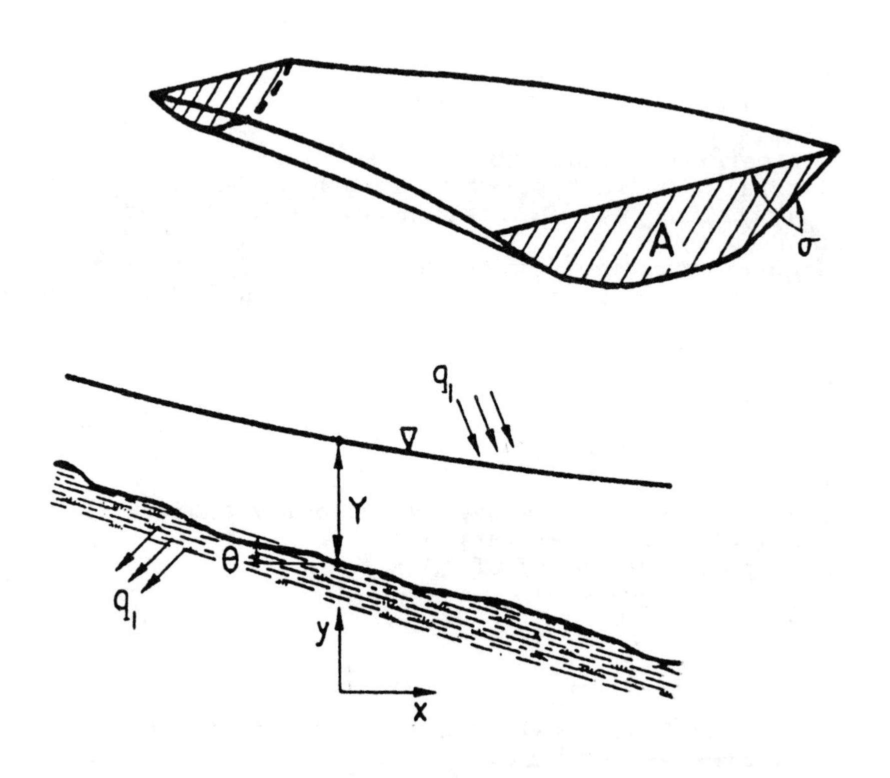

Fig. 3. Definition Sketch of Open-Channel Flow

models have been developed applying the complete Saint-Venant equations or their various simplified forms (see Eq. 12) to the components of a catchment. A popular version is the nonlinear kinematic wave approximation which retains the friction and channel slope terms but drops all the inertia and pressure terms in the momentum equation (Eq. 12) and solving this simplified equation together with the continuity differential equation (Eq. 11). If the continuity equation is written in a linear finite difference form and solved explicitly, sequentially (not simultaneously) and independently of the simplified momentum equation (which is solved in the form of the Manning, Darcy-Weisbach or similar equations), the version is a linear kinematic wave approximation. The nonlinear kinematic wave version is customarily called simply the kinematic wave approximation, whereas the linear kinematic wave version is usually referred to according to the equation used to substitute the momentum equation. One disadvantage of the kinematic wave approximation (linear or nonlinear) is its inability to account for the backwater effect from downstream.

Table 1 summarizes the major hydrologic features of five U.S. Government supported computer-based rainfall-runoff models applicable to rural catchments. All five models accept rainfall input allowing discretized temporal as well as spatial variations. User's guides are available for these models. The highest hydraulic level used in these models for channel or overland flow routing is kinematic wave which unfortunately, cannot account for the backwater effect from downstream. All five models allow or require calibration using observed data to determine the values of coefficients or constants for a given catchment. These models have been subjected to very little verification or no verification at all. Proper and adequate model verification remains a weak link in the different modes of hydrologic simulation model operation shown in Table 2.

STOCHASTIC SIMULATION MODELS

Contrary to the physical-based distributed system deterministic catchment simulation models,another group of computer-based catchment runoff simulation models is the probability-based lumped system models considering the flood hydrograph as a stochastic time process. The catchment runoff rate at a given moment is determined from the time series of the past runoff and rainfall of the event. Salas and Yevjevich (1972) proposed a general stochastic process expression as follows

$$Q(\tau) = \underbrace{Tm_\tau + Ts_\tau}_{\text{Trend Component}} \left(\mu_\tau + \sigma_\tau \left[\sum_{j=1}^{m} \alpha_{j,\tau-j}\,\epsilon_{\tau-j} + \left(1 \sum_{i=1}^{m} \sum_{j=1}^{m} \alpha_{i,\tau-i}\,\alpha_{j,\tau-j}\,\rho_{|i-j|,\tau-k}\right)^{1/2} \eta_\tau \right] \right) \quad (13)$$

(Trend Component) → Tm_τ, Ts_τ

(Periodic Component) → μ_τ, σ_τ

(Second-order stationary) → $\alpha_{j,\tau-j}$, $\alpha_{i,\tau-i}$, $\alpha_{j,\tau-j}$, $\rho_{|i-j|,\tau-k}$

(Independent stochastic component) → η_τ

Table 1. Selected U.S. Federal Supported Rural Catchment Runoff Simulation Models

Model	Abstractions		Runoff		Base Flow	Selected References
	Initial losses	Infiltration	Overland	Channel		
HEC-1	Yes	Horton's or SCS-Curve Number	Clark, Snyder or input unit hydrograph or unit hydrograph	Muskingum or linear reser- voir storage equation with variable coefficient	According to exponential recession equation	HEC (1981a,b)
SCS-TR-20	Depression (retention) storage	SCS-Curve Number	SCS nondimensional unit hydrograph	Convex method	Input constant	SCS (1973)
USGS DR3M	Depression (retention) storage	Green and Ampt	Kinematic wave	Kinematic wave	Input or constant	Dawdy et al. (1972) Alley and Smith (1982)
NWS Sacramento		Horton's		Kinematic wave or storage coefficient		Burnash et al. (1973)
HSPF (EPA)	Yes	Coupled with subsurface flow	Storage routing with constant depth using Manning's formula	Kinematic wave	Coupled with subsurface flow	Hydrocomp (1976) Johanson et al. (1980)

Table 2. Operation Modes of Simulation Models (Yen, 1986)

Mode	Input	Transformation		Output
		Parameters	Coefficient Values	
Prediction	known	known	known	?
Calibration	known	known	?	known
Verification	known	known	known	?(=known?)
Detection	?	known	known	known
Parameter identification	known	?	known(?)	known
Sensitivity	known	known	known	?

Table 3. Probabilistic Characteristics of Model Output

Input	Model	Output
Deterministic	Deterministic	Deterministic
Stochastic	Deterministic	Stochastic
Deterministic	Stochastic	Stochastic
Stochastic	Stochastic	Stochastic

in which

τ = sequence of intervals within the runoff duration, as fractions of the duration

Tm_{τ} = trend in the mean

Ts_{τ} = trend in the standard deviation

μ_{τ} = periodicity in the mean of detrended series

σ_{τ} = periodicity in the standard deviation of detrended series

$\epsilon_{\tau-j}$ = stochastic component having periodic dependence coefficients and having trends and periodicities in the mean and standard deviation removed

η_{τ} = independent, second-order stationary, stochastic component

$\rho_{|i-j|,\tau-k}$ = periodic autoregressive dependence coefficients of ϵ_{τ}

$\alpha_{i,\tau-i}$ = periodic autocorrelation coefficients of ϵ_{τ}

In a sense this equation is an alternative expression of Eqs. 6 and 8 with the coefficients being functions of time.

A number of stochastic process simulation techniques are adaptable and have been suggested for hydrologic and flood forecasting problems (see, e.g., Yevjevich, 1972; Salas et al., 1980; Bras and Rodriguez-Iturbe, 1984). The best known is the Markov model which can be written in a simple form as

$$Q(t) = mQ(t-1) + \epsilon(t) \qquad (14)$$

where m is a coefficient and ϵ is a standardized random variable with zero mean. A more general expression of Eq. 14 is the linear autoregressive (AR) model,

$$Q(t) = \epsilon(t) + m_1 Q(t-1) + m_2 Q(t-2) + .. + m_n Q(t-n) \qquad (15)$$

Often the moving averages (MA) are used in the AR formulation. Such ARMA models are popular in hydrology. Certain least-squares or square-root models have also been proposed. A promising one is the relatively simple linear least-squares estimation model called Kalman Filter. Hino (1973) applied Kalman Filter to simulate catchment runoff. Other examples can be found in Chiu (1978). One advantage of the Kalman Filter is that the model structure allows explicit consideration of the physical process in a lumped system manner.

River Flood Forecasting Methods

In river flood hydrograph forecasting the physical environment is a reach of channel. The major input is the inflow at the upstream end of the reach. The downstream end of the reach is the location of interest where the flood hydrograph is to be predicted. Water input from tributaries and rainfall may also be considered if necessary. The forecasting methods can be classified into four groups, namely, deterministic lumped system methods, physically based routing methods, station correlation methods, and stochastic process methods.

PHYSICALLY BASED FLOOD ROUTING MODELS

Recently the most popular river flood hydrograph forecasting method is to route the flood through the channel reach using unsteady flow equations. This approach becomes practical with the development of digital computer technology and numerical techniques in recent years. An example is the U.S. National Weather Service's Dynamic Wave Operational (DWOPER) model (Fread, 1978).

The basic hydraulic equations used in routing are the Saint-Venant equations of one-dimensional flow (Eqs. 11 and 12) given previously or the noninertial or kinematic wave simplifications. The friction slope is estimated from the Manning or Darcy-Weisbach formulas

$$S_f = \frac{n^2 R^{-4/3}}{K_n^2} V|V| = \frac{f}{8gR} V|V| \tag{16}$$

in which

R = hydraulic radius

n = Manning's roughness factor

K_n = 1 for SI units and 1.486 for English units

f = Weisbach resistance coefficient

In a river reach the main channel and major tributaries are divided into subreaches. The continuity and momentum equations are applied to each of the subreaches individually. Since the downstream boundary condition for a subreach is coupled with the upstream boundary condition of the next subreach, the flow equations of all the reaches are solved as a group. However, for the kinematic wave approximation, since no downstream boundary condition is required, the subreaches can be solved one by one in a cascading manner from upstream to downstream. It should be noted that the kinematic wave approximation cannot account for the backwater effect from the downstream.

No analytical solution is known for Eqs. 11 and 12 except for greatly simplified special situations which do not exist in natural rivers. In general, these equations are solved numerically with appropriate initial and boundary conditions. Alternatively, these equations may be transformed into two pairs of ordinary partial differential equations called characteristic forms, the solution of the differential equations depends on the numerical scheme used.

DETERMINISTIC LUMPED SYSTEM ROUTING MODELS

In this approach the basic working formulas are the equations of continuity and storage (Eqs. 6 and 7) applied to the entire river reach. Similar to catchment flood forecasting models, numerous models can be developed for rivers depending on the functional relationship assumed for the storage equation (Eq. 7). The best known model is the constant coefficient Muskingum method. The values of the coefficients are determined from observed inflow and outflow discharges of past floods passing through the reach. Cunge (1969) showed that the Muskingum method can be viewed as a special case of nonlinear kinematic wave allowing flood attenuation and with the values of the coefficient variable and related to the physical river characteristics.

In the standard Muskingum and other lumped system models, only the observed inflow and outflow data of the reach are needed to establish the working formulas. No information on the physical properties of the reach is needed.

STATION CORRELATION MODELS

If sufficient flood inflow and outflow data are available for a reach, the relationship between the inflow and outflow can be established through a regression analysis and the result can be used for flood forecasting. This correlation of the flows at the two ends of a reach can be established for either the discharges or river stages. In practice measurements are usually made on the stages and then converted to discharges by using the rating curve (stage-discharge relation) of the station. Therefore, establishment of the stage correlation between the stations is usually preferred. A correlation of the stages between Hamburg and Cuxhaven 90 km downstream on River Elbe presented by Plate and Ihringer (1986) is reproduced in Fig. 4 as an illustration.

No information on the channel property is needed. Hence, this approach is a lumped system stochastic method. This approach works well if the lateral flow from tributaries and other sources or losses are relatively insignificant and the characteristics of the river in response to floods do not change with time, i.e., hydrologically speaking, a stationary system.

METHODS OF STOCHASTIC TIME PROCESS

This approach requires observed data at the location of interest to forecast the flood values at future times beyond the latest observation. The basic working equations for the different versions of time series predictions are essentially the same as those described in Section II(D) for catchment runoff forecasting.

Reliability of Flood Forecast

Forecasting, by its nature, is subject to uncertainties. In other words, the predicted flood stage or discharge actually has a probability distribution. The value given in a forecasting is not necessary the expected value or the mode of the distribution. Sometimes confidence interval or similar statistical measures are used to give an indication on the prediction reliability.

Uncertainties prevail in practically all aspects of flood forecasting. The input data accuracy, the methodology, the natural randomness of incoming rainfall or inflow, and in the case of the distributed system models, the geometry of the channel and overland system, and the spatial and temporal discretization for computations all contribute to the uncertainties in flood forecasting.

Thus, the output provided by a flood forecasting is stochastic. As shown in Table 3, a stochastic output need not be generated from a stochastic model. A deterministic model (as most of the event-based catchment or channel simulation models are) with a stochastic input would generate a stochastic output. Since in flood forecasting the input is actually inevitably stochastic (although it may not be specified), the case of deterministic output obtained from deterministic input and model is irrelevant.

Methods to evaluate forecasting reliability accounting for all the different influential factors evolve gradually since 1950 in various engineering disciplines. The risk associated with an engineering system is defined here as the probability that the system fails to perform its intended function. For flood forecasting, performance will be related to the

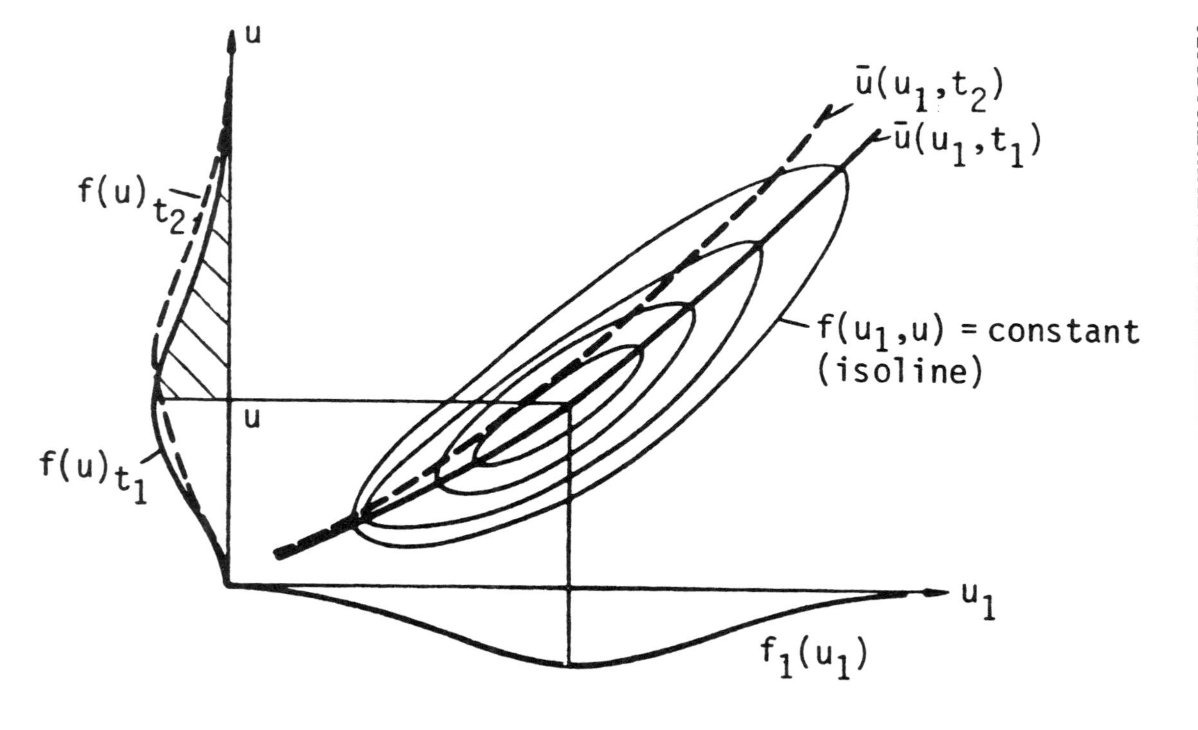

Fig. 4. Schematic of Probability Density Function of Stages at Hamburg (u) and Cuxhaven (u_1) on River Elbe (Plate and Ihringer, 1986)

prediction of the flood stage at a given location. The risk then is the probability of a specified stage, S, being exceeded by the variable river water level H, i.e.,

$$\text{Risk} = P(H > S) \tag{17}$$

$$\text{Reliability} = P(S < H) = 1 - \text{Risk} \tag{18}$$

In general, H and S are individually a function of a number of contributing components or parameters X_i, i=1,2,....n and Y_j, j=1,2,....m, respectively, i.e.,

$$H = H(X_1, X_2, ...X_n) \tag{19}$$

$$S = S(Y_1, Y_2, ...Y_m) \tag{20}$$

Each of these components is subject to its own uncertainties and it may perhaps be a function of further subcomponents. The uncertainties may be from one of the following four sources: (a) natural, (b) data, (c) model parameter, and (d) model structure. Natural uncertainties refer to the random temporal and areal fluctuations inherent in natural processes. Data uncertainties refer to: (i) the adequacy of the stream gage network, channel geometry data, etc.; (ii) measurement inaccuracy and errors; and (iii) data handling and transmission errors. Parameter uncertainties reflect errors in the determination of the proper parameter values to be used in modeling a given flood event. Model structure uncertainty refers to the ability of the model to reflect accurately the river system's true physical flow process.

By considering only those uncertainties pertaining to temporal randomness of natural events (source a above), a curve showing the "expected" discharge vs. exceedance probability can be obtained through a standard frequency analysis, shown as E(y) in Fig. 5. Because of the additional uncertainties (sources b, c and d above), the expected curve itself may vary. One approach involves simply taking a margin of discharge values (dashed line in Fig. 5) associated with the mean plus one standard deviation from the "expected" curve established above. A more rational approach would require the integration of all the uncertainties (including the model uncertainties) in the reliability analysis, thus obtaining the total risk-discharge curve R_F as shown in Fig. 5. Despite the apparent conservatism used in the first approach, significant underestimation on the unsafe side could result. These two curves, E[y] and R_F, reflect potentially different views to "forecasters" and "designers" or "operators". For a traditional forecaster, if he considers only the randomness of the natural process (such as in

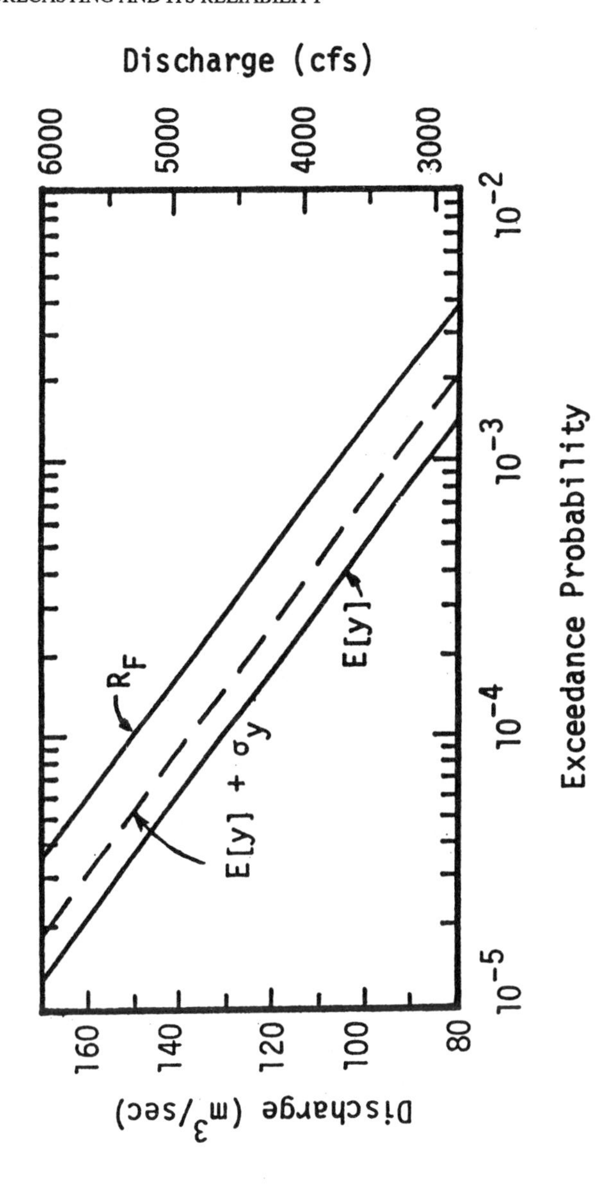

Fig. 5. Expected and total exceedance probabilities (Melching et al., 1986)

frequency analysis), he would probably predict on the basis of the expected value E(y). On the other hand, for a designer it is more reasonable to use R_F than E[y] (or any derived confidence interval) in risk cost evaluation and design.

Several methods that permit explicit calculation of flood risks accounting for uncertainties have been developed during the past two decades. These methods have become known in several engineering disciplines as probabilistic risk (or reliability) analysis, or simply risk analysis. They include the method of direct integration, MonteCarlo simulation, mean-value first-order second moment method, and advanced first-order second moment method. A comparative review of these risk analysis methods is given in Yen (1987). A summary table comparing these methods is reproduced here.

CATCHMENT FLOOD FORECASTING RELIABILITY

In a recent study, Melching et al. (1987) investigate explicitly the reliability of catchment flood forecasting by applying the first-order second-moment and Monte Carlo methods to a 1500 km^2 catchment at Pontiac in east central Illinois. The first-order analysis they used is an approximate analysis by truncating the second and higher order terms of the Taylor series expansion of the performance function. For instance, the first-order Taylor's expansion of the river water level H (Eq. 19) with respect to the point x_{ip}, i = 1,2,....n and of the specified stage S (Eq. 20) with respect to y_{jp}, j = 1,2,....m for uncorrelated variables yields

$$E(H) = G_1(x_{ip}) + \sum_{i=1}^{n} (\bar{x} - x_{ip}) \left[\frac{\partial G_1}{\partial X_i}\right] \tag{21}$$

$$Var(H) = \sum_{i=1}^{n} \left[\frac{\partial G_1}{\partial X_i}\right]_p^2 Var(X_i) \tag{22}$$

$$E(S) = G_2(y_{jp}) + \sum_{j=1}^{m} (\bar{y}_j - y_{jp}) \left[\frac{\partial G_2}{\partial Y_j}\right]_p \tag{23}$$

$$Var(S) = \sum_{j=1}^{m} \left[\frac{\partial G_2}{\partial Y_j}\right]_p^2 Var(Y_j) \tag{24}$$

Thus, Eqs. 22 and 24 allow a means to combine the uncertainties of the component variables. The risk (Eq. 17) can be evaluated as

$$P(H > S) = 1 - \phi[\beta] \tag{25}$$

Table 4. General Comparison of Risk Calculation Methods (Yen, 1987)

Method	Return period	Direct Integration	Monte Carlo	Reliablity Index	MFOSM	AFOSM
Capability to account for different factors	very limited	limited	yes	yes	yes	yes
Information needed on probability distribution of factors	indirectly	extensive	moderate	first two statistical moments	only the combined distribution, for factors the first two statistical moments suffice	only the combined distribution, for factors the first two statistical moments suffice
Complexity in application	simple	complicated	moderately complicated	moderate	moderate	moderate
Amount of computations	simple	moderate to extensive	extensive	moderate to simple	moderate to simple	moderate
Capability to estimate total risk	no	difficult	extensive computations	no	yes	yes
Result adaptable for risk cost analysis	partial	yes	yes	no	yes	yes

$$\beta = \frac{\ln(\bar{S}/\bar{H})}{\sqrt{\Omega_S^2 + \Omega_H^2}} \tag{26}$$

in which Ω denotes the coefficient of variation, and $\phi[\beta]$ denotes the cumulative standard normal distribution evaluated at a given value β. Note that Eqs. 21 to 26 require the use of only the mean and variance, i.e., the first two statistical moments of the random variables, in the probabilistic analysis.

In most engineering problems, particularly in the case of flood warning, the true probability distributions of the contributing variables are seldom known; the statistical information available may be limited only to the values of the mean and variance of the variables. Cornell (1972) showed that when properly applied the first-order second-moment approximation is adequate and suitable for reliability analysis and compatible with engineering conditions. The mean-value first-order second-moment (MFOSM) method expands the Taylor series at the mean values of the variables. However, in engineering projects, failures often occur at extreme values. Therefore, in the advanced first-order second-moment (AFOSM) method the Taylor series is expanded at a likely failure point on the failure surface in order to improve the accuracy of risk estimates in cases involving nonlinear function of variables. Detailed descriptions of the mean-value and advanced first-order methods for hydraulic and hydrologic problems can be found in Yen et al. (1986).

The reliability analysis can be applied to any catchment rainfall-runoff simulation models. The results of using HEC-1 are shown in Figs. 6 to 8. Thirty-two rainstorms were used for the HEC-1 model calibration to determine the values of the working constants. An example of fitted hydrograph as compared to the measured hydrograph is shown in Fig. 6. The exceedance probability of peak discharge from the advanced first-order reliability analysis for four rainstorms not used in the calibration is shown in Fig. 7. Also shown are the measured peak discharges for the four rainstorms. The probability distribution function of peak discharge for the May 4, 1965 rainstorm is plotted in Fig. 8. It is interesting to note that the observed peak is smaller than both the mean and mode of the distribution.

RIVER FLOOD FORECASTING RELIABILITY

A simple river flood forecasting reliability model is given here as a demonstration. A single failure mode of high water from upstream is considered. In this example the modeling and analysis of the major components of uncertainties associated

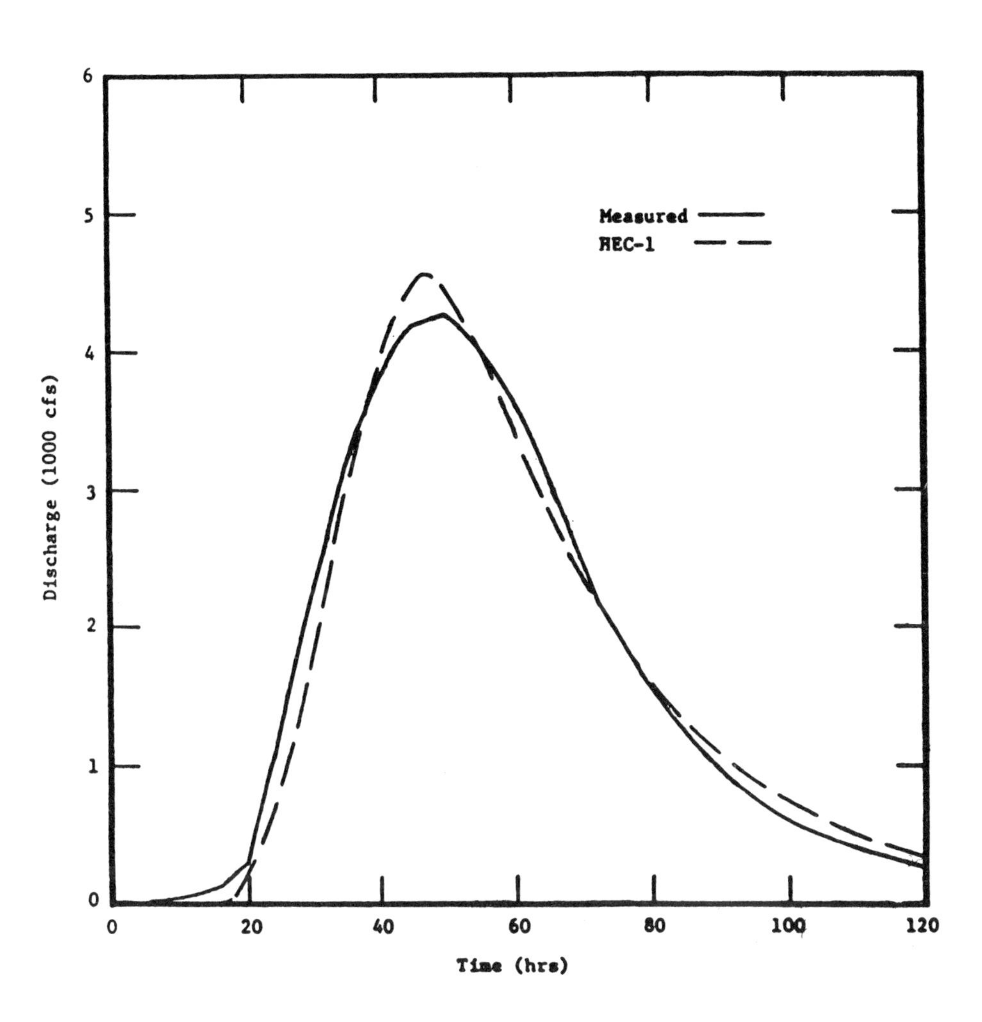

Fig. 6. Comparison of Measured and Best Fit Hydrographs for the April 12, 1983 Event

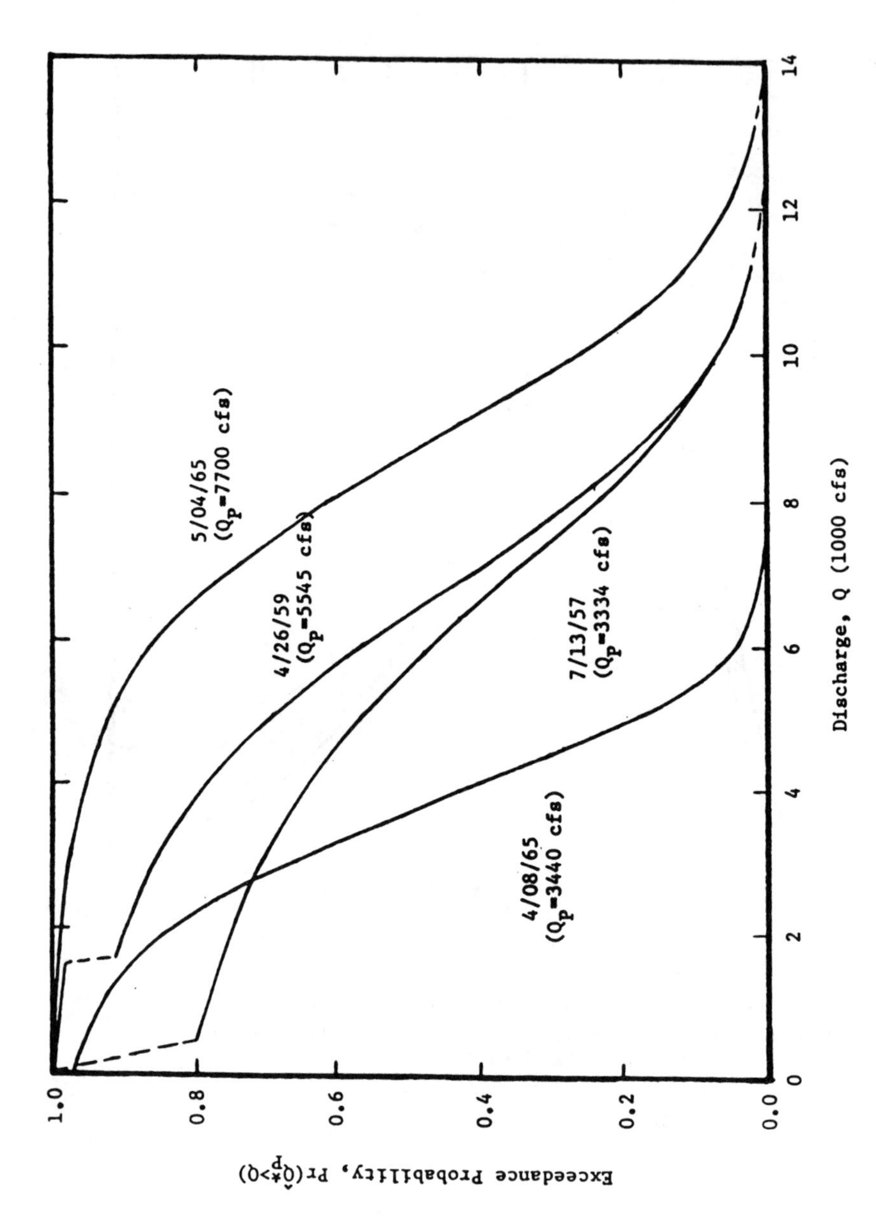

Fig. 7. Exceedance Probability as a Function of Discharge for HEC-1 Forecasts

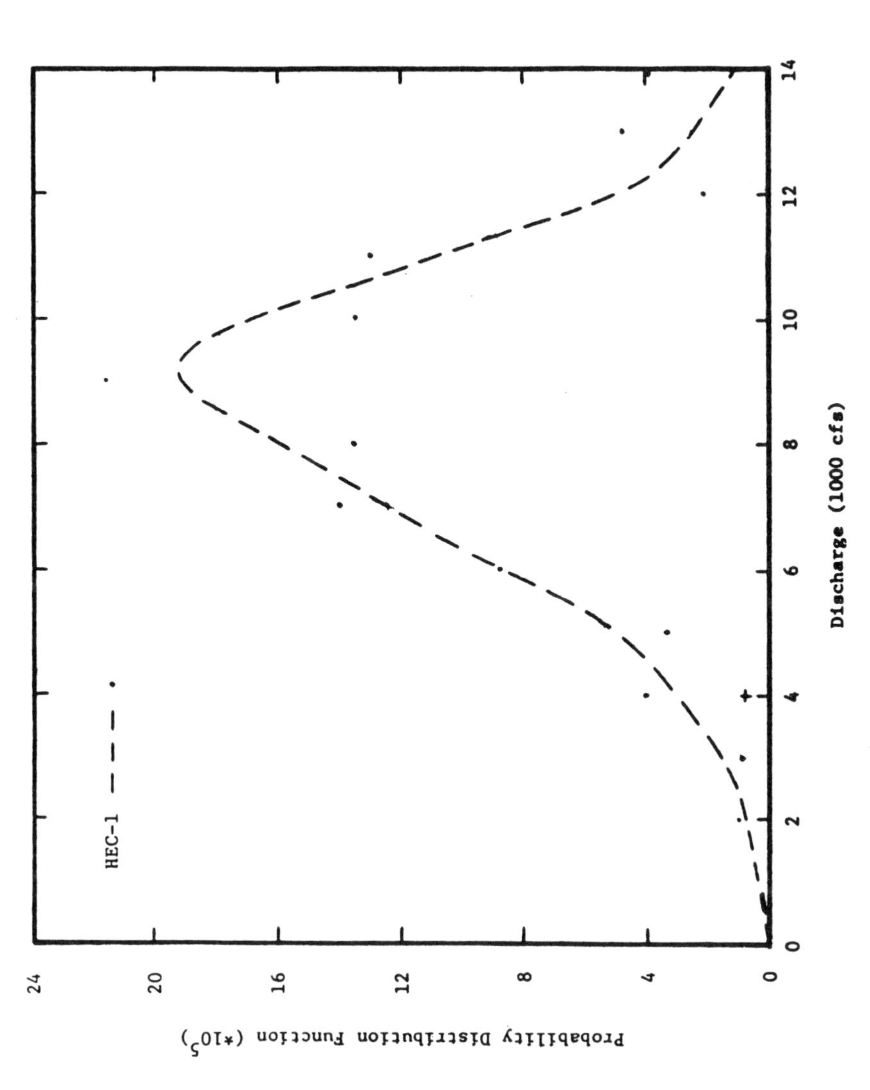

Fig. 8. Forecast Probability Distribution Function for the May 5, 1965 Event Estimated by the AFOSM Method

with flood forecasting are presented as follows:

(a) Rating Curve -- Since flood levels are measured as stages and most of the routing models accept the flow rate Q as input, conversion of stages to flow rate and vice versa through the rating curve is necessary in flood routing analysis. Suppose the rating curve is expressed as

$$Q = a + bH + cH^2 + \epsilon \tag{27}$$

where Q is the flow rate; H is the stage; a, b, c are coefficients determined from regression analysis; and ϵ is the scatter about the rating curve. Assume ϵ has mean zero and variance σ^2. For a known stage H, the expected flow rate is

$$E(Q|H) = a + bH + cH^2 \tag{28}$$

and the variance of flow rate is

$$Var(Q|H) = \sigma^2 \tag{29}$$

Furthermore, measurement error is usually incurred in recording the stage. Suppose this error is described by the variance σ^2_H, then incorporating this uncertainty through the first-order analysis, the variance of flow rate Q becomes (see Ang and Tang, 1975, p. 145)

$$Var(Q) = \sigma^2 + (b+2c\bar{H})^2 \sigma_H^2 \tag{30}$$

where $\bar{H}$ is the recorded stage.

At the downstream end, the reverse process occurs whereby the flood stage is to be determined from a given flow rate. In this case, separate regression analysis can be performed on the set of data on Q vs. H to obtain

$$H = g(Q) + \sigma' \tag{31}$$

where g(Q) is the regression curve and ϵ' (with variance σ'^2) is the corresponding scatter. It can be shown in this case that the variance of stage may be given by

$$Var(H) = \sigma'^2 + \left[\frac{\partial g(Q)}{\partial Q}\right]^2_{\bar{Q}} \sigma_Q^2 \tag{32}$$

where $[\ ]_{\bar{Q}}$ denotes that the expression within the bracket is evaluated at the mean flow rate $\bar{Q}$; and σ^2_Q is the variance describing the uncertainty of the flow rate in the downstream, which may be evaluated according to the procedure in (c) below.

(b) Estimation of Stages that are not Measured -- Not all upstream stages are recorded continuously. Hence the stage of some minor tribu-taries will be estimated. If data are available between such stages with those of adjacent stream where measurement is continuously monitored, correlation and multivariate normal analysis may be applied to obtain the mean and variance of such stages for a given value of the measured stages (H_m) in adjacent stream, namely $E(H|H_m)$ and $Var(H|H_m)$ may be determined. These may be combined with the uncertainty of the rating curve to obtain the overall uncertainty in the flow rate at the upstream end.

(c) Correction to the Downstream Flow Rate Determined from Routing Model -- For a given set of flow rates upstream, a routing model may be used to determine the flow rate downstream. This downstream flow rate Q_0 is only a predicted value. The actual value may differ significantly due to several factors, such as error in the simplified routing procedure, uncertainties in the transpiration and evaporation effects, seepage estimations, error in the area of stream bed assumed in the routing procedure, and others. As a simple illustration a multiplication model may be employed as follows:

$$Q^* = N_1 N_2 \ldots. N_m Q_0 \tag{33}$$

where Q^* is the actual flow rate; $N_i (i=1,2,\ldots.m)$ are the random corrective factors (with mean $\bar{N}_i$ and coefficient of variation Δ_i) accounting for the uncertainty in each factor described above. The information for evaluating $\bar{N}_i$ and Δ_i may come from any one or a combination of the following:

(i) subjective judgment
(ii) results of analysis from hydrologic models, especially the effect of range of parameters on the computed Q_0
(iii) results from simulated computer runs on the elaborate routing model under various conditions of each factor.

By assuming the N_i's being statistically independent, first-order analysis will give mean actual flow rate as

$$\bar{Q}^* = \bar{N}_1 \ldots. \bar{N}_m Q_0 \tag{34}$$

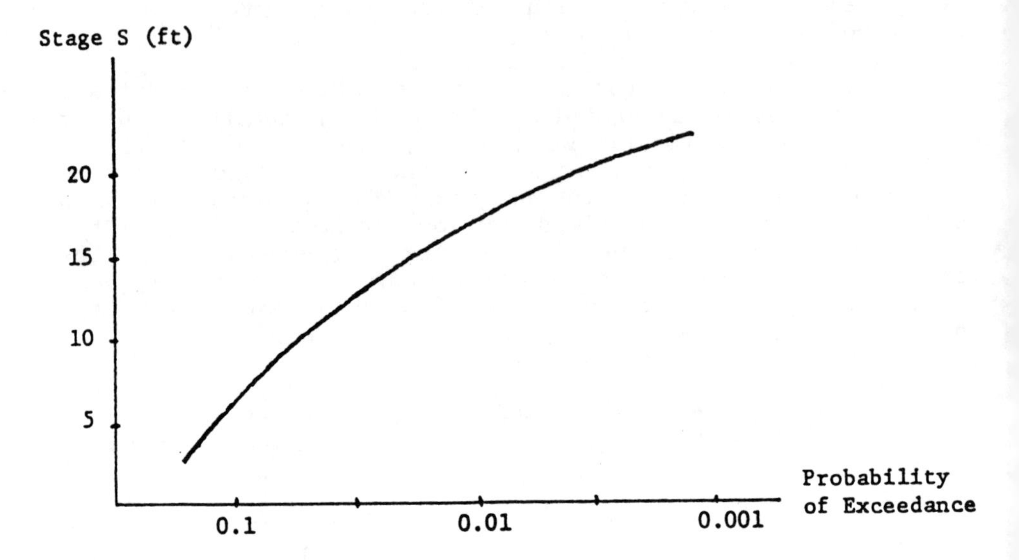

Fig. 9. Probability of Occurrence of Flood Stage Downstream

and its coefficient of variation (c.o.v.) as

$$\Omega_{Q*} = \sqrt{\Delta_1^2 + \; .... \; + \; \Delta_m^2} \tag{35}$$

Since the input flow rate to the routing model contains uncertainties, the value of Ω_{Q*} should be further modified to include this factor. It appears that the c.o.v. of Q_0 is equal to the c.o.v. of the input flow rates if a linear system between upstream and downstream flow rates is assumed.

Moreover, as more information become available, the statistics $\overline{N}_j$ and Δ_j evaluated can be updated, for example through the technique described in Tang and Ang (1973) or Cornell (1972). Hence, the reliability of flood warning will be always evaluated based on the up-to-date state of information.

(d) Establishment of Probability Distribution of Flood Stages Downstream -- Once the mean and variance of the downstream stage have been estimated according to procedure just described, a probability distribution may be prescribed. Lognormal distribution appears to be an appropriate choice since it is limited to positive flood stages and also the multiplication model in Eq. 33 would suggest lognormal model according to the Central Limit Theorem when the number of component variables becomes large. Hence the following curve (Fig. 9) may be plotted for a given set of selected measured stages upstream.

The above river flood forecasting reliability evaluation procedure just outlined is preliminary and merely provided as an example. Further study for improvements and alternatives is clearly desirable.

References

Alley, W.M., and Smith, P.E., "Distributed Routing Rainfall-Runoff Model -- Version II," Open-File Report 82-344, U.S. Geological Survey, GCHC, NSTL Station, Mississippi, 1982.

Ang, A.H.-S., and Tang, W.H., Probability Concepts in Engineering Planning and Design, Vol. I: Basic Principles, John Wiley & Sons, Inc., New York, 1975.

Bras, R., and Rodriguez-Iturbe, I., Random Functions and Hydrology, Addison-Wesley Publ. Co., Boston, 1984.

Burnash, R.J.C., Ferral, R.L., and McGuire, R.A., "A Generalized Streamflow Simulation System: Conceptual Modeling for Digital Computers," Report, U.S. National Weather Service/State of California Dept. of Water Resources,

Sacramento, California, 1973.

Chiu, C.L., ed., Applications of Kalman Filter to Hydrology, Hydraulics and Water Resources, University of Pittsburgh, 1978.

Chiu, C.L., and Huang, J.T., "Nonlinear Time-Varying Model in Rainfall-Runoff Relation," Water Resources Research, Vol. **6**, No. 5, pp. 1277-1286, Oct. 1970.

Chow, V.T., ed., Handbook of Applied Hydrology, McGraw-Hill Book Co., New York, 1964.

Chow, V.T., and Kulandaiswamy, V.C., "The IUH of General Hydrologic System Model," Journal of the Hydraulics Division, ASCE, Vol. **108**, No. HY7, pp. 830-844, 1982.

Cornell, C.A., "First-Order Analysis of Model and Parameter Uncer-tainty," Proceedings, International Symposium on Uncertainties in Hydrologic and Water Resources Systems, Vol. **2**, pp. 1245-1272, Tucson, Arizona, Dec. 1972.

Cunge, J.A., "On the Subject of a Flood Propagation Computation Method (Muskingum Method)," Journal of Hydraulic Research, IAHR, Vol. 7, No. 2, pp. 205-230, 1969.

Dawdy, D.R., Lichty, R.W., and Bergmann, J.M., "A Rainfall-Runoff Simulation Model for Estimation of Flood Peaks for Small Drainage Basins," Professional Paper 506-B, U.S. Geological Survey, 1972.

Dooge, J.C.I., "A General Theory of the Unit Hydrograph," Journal of Geophysical Research, Vol. **64**, No. 1, pp. 241-256, 1959.

Fread, D.L., "NWS Operational Dynamic Wave Model," in Verification of Mathematical and Physical Model in Hydraulic Engineering, (Proceedings, ASCE 26th Hydraulic Specialty Conference, College Park, Maryland), pp. 455-464, ASCE, New York, 1978.

Gray, D.M., ed., Handbook on the Principles of Hydrology, Water Information Center, Inc. Huntington, New York, 1973.

Hino, M., "On-Line Prediction of Hydrologic System," Proceedings, 15th Congress of the International Association for Hydraulic Research, Istanbul, Turkey, Vol. **4**, pp. 121-129, 1973.

Hydrocomp, Inc., "Hydrocomp Simulation Program: Operational Manual," 2nd ed., Hydrocomp Inc., Palo Alto, Calif., 1976.

Hydrologic Engineering Center, "Hydrologic Analysis of Ungaged Watersheds with HEC-1," Training Document TD-15, U.S. Army Corps of Engineers Hydrologic Engineering Center, Davis, Calif., 1981a.

Hydrologic Engineering Center, "Flood Hydrograph Package (HEC-1) User's Manual," Report CPD-1a, U.S. Army Corps of Engineers Hydrologic Engineering Center, Davis, Calif., 1981b.

Johanson, R.C., Imhoff, J.C., and Davis, H.H., Jr., "User's Manual for Hydrological Simulation Program - Fortran (HSPF)," Report EPA-600/9 -80-015, U.S. Environmental Protection Agency, Athens, GA, April 1980.

Laurenson, E.M., and Mein, R.G., "RORB-Version 3 Runoff Routing Program User Manual," Dept. of Civil Eng., Monash University, Clayton, Victoria, Australia, March 1985.

Melching, C.S., Wenzel, H.G., Jr., and Yen, B.C., "Application of System Reliability Analysis to Flood Forecasting," Int'l Symp. Flood Frequency and Risk Analysis, Baton Rouge, Louisiana, May 1986.

Melching, C.S., Yen, B.C., and Wenzel, H.G., Jr., "Incorporation of Uncertainties in Real-Time Catchment Flood Forecasting," Research Report 208, Water Resources Center, University of Illinois at Urbana-Champaign, 1987.

Nash, J.E., "The Form of Instantaneous Unit Hydrograph," IAHS Publication No. 45, Vol. **3**, pp. 114-121, 1957.

Plate, E.J., and Ihringer, J., "Failure Probability of Flood Levees on a Tidal River," in Stochastic and Risk Analysis in Hydraulic Engineering, ed. by B.C. Yen, pp. 48-58, Water Resources Publications, Littleton, Colorado, 1986.

Salas-La Cruz, J.D., and Yevjevich, V., "Stochastic Structure of Water Use Time Series," Hydrology Papers 52, Colorado State University, Fort Collins, Colorado, 1972.

Salas, J.D., Delleur, J.W., Yevjevich, V., and Lane, W.L., Applied Modeling of Hydrologic Time Series, Water Resources Publications, Littleton, Colorado, 1980.

Sherman, L.K., "Streamflow from Rainfall by the Unit-Graph Method," Engineering News-Record, Vol. **108**, pp. 501-505, April 7, 1932.

Soil Conservation Service, "Computer Program for Project

Formulation Hydrology," Technical Release No. 20, U.S. Dept. of Agriculture Soil Conservation Service, Washington, DC, June 1973.

Tang, W.H., and Ang, A.H.-S., "Modeling, Analysis and Updating of Uncertainties," presented at ASCE National Meeting on Structural Engineering, Conference Preprint No. 2016, San Francisco, April 1973.

Yen, B.C., Deterministic Surface Water Hydrology, Dienst Hydrologie, Vrije Universiteit Brussel, Belgium, 64 p., 1983.

Yen, B.C., "Small Watershed Flood Prediction," (in Chinese), Journal of Civil and Hydraulic Engineering (Taiwan), Vol. **10**, No. 4, pp. 79-99, Feb. 1984.

Yen, B.C., "Rainfall-Runoff Process on Urban Catchments and Its Modeling," in Urban Drainage Modeling, ed. by C. Maksimovic and M. Rodojkovic, pp. 3-26, Pergamon Press, Oxford, 1986.

Yen, B.C., "Reliability of Hydraulic Structures Possessing Random Loading and Resistance," in Engineering Reliability and Risk in Water Resources (Proceedings of the NATO Advanced Study Institute, Tucson, Arizona, May 1985), ed. by L. Duckstein and E.J. Plate, pp. 95-113, Martinus Nijhoff Publishers, Dordrecht, The Netherlands, 1987.

Yen, B.C., Cheng, S.T., and Melching, C.S., "First Order Reliability Analysis," in Stochastic and Risk Analysis in Hydraulic Engineering, ed. by B.C. Yen, pp. 1-36, Water Resources Publications, Littleton, Colorado, 1986.

Yevjevich, V., Stochastic Processes in Hydrology, Water Resources Publications, Littleton, Colorado, 1972.

Mr. Chairman, Ladies and Gentlemen:

We, American and Chinese scientists and engineers, have gathered here to discuss the flood prevention measures of the lower Yellow River. It is expected that some major problems for further collaborative research be recommended at the conclusion of the seminar. To this end, Professor M. Gordon Wolman, Professor Lucien M. Brush, Professor Ben Chie Yen, Dr. Long Yujian, Dr. Chen Xiande, Professor Liu Changming and I met in one evening and have pieced together suggestions from us or known to us to give birth to a provisional outline of four topics. This has been distributed to all participants in the Seminar to invite comments. It is subject to modification if more appropriate proposals come from the floor before the break. After the break, the discussion will continue in four separate groups, each centering around one of the four topics.

I would like now to speak a few words on my personal observations. The central theme of the present seminar is flood reduction. What matters most is too much sediment load and too little water in the lower reaches. Whether is it feasible to practically stop the build-up of the river bed by river training and reservoir management is a problem meriting careful investigation. Some experts are quite optimistic and predict that by taking advantage of the behavior of the non-Newtonian fluid formed of the superconcentrated flow in the lower Yellow, the sediment load may be transferred over long distances to depressions on the North China Plain and to the sea, while serious doubts are raised by others. Perhaps the majority in the scientific community are now taking a neutral stand. A definite answer to the question is certainly desirable. An alternative or complementary measure is to draw water from the upper or middle Yangtze. This stipulation, if realized, will no doubt be helpful in attempting to dispose of the excessive sediment load in the lower Yellow. Little study has been carried out on this and it seems to me it should be accorded research priority in conjunction with river training and reservoir management for the same purpose.

The reduction of sediment discharge from the middle course of the Yellow River is in a sense an alternative or complementary to the foregoing. In connection with this, I wish to stress three points. First, there is the need for us to know how much of the sediment discharge to the lower reaches stems from bank erosion, mass movement and related processes. Most of us have a deep impression of these features on the loess uplands. Perhaps it is not very difficult to gain

L. M. Brush et al. (eds.), Taming the Yellow River: Silt and Floods, 197–199.

an overall picture of their distribution. However, what is important to us is a quantitative assessment with reasonable accuracy of sediment yield and delivery in a period of five or ten years. In the absence of this information, we can hardly make a good judgment as to the magnitude of sediment discharge reduction by soil conservation.

Second, there is the need to conduct more vigorous research on soil erosion and soil conservation. We do have small-scale maps of soil erosion and numerous reports of successful soil conservation measures in many localities. I do not hesitate to say that much is left to be desired. The existing maps of soil erosion represent neither the annual mean over a certain number of years nor the data of one and the same year. Since soil erosion is so important in the case under consideration, more and better information is indispensable, especially in the some 100,000 square kilometers where it is the main source of sediment and the coarser fraction of sediment being delivered to the lower reaches. If a substantial portion of the beautiful descriptions of successes in soil conservation appearing in our newspapers in the last three decades tells the true story, our gathering here today would not be confronted with a problem of soil erosion. Reports come and go. The situation remains essentially unchanged. Some of the reported measures are really effective in reducing erosion. But their extension has been limited in the bygone years. Conservation measures are mostly area-specific. More and better work is required to evolve measures to suit the local conditions.

Third, serious consideration should be given to socio-economic factors. Unless constant attention is paid to the short- and long-term welfare of the local inhabitants, extension of soil conservation is doomed to failure. Soil conservation on the loess highlands should be directed in the same way as agroforestry in the tropical environment. In a unit of management, the pattern of land uses should lead to both soil amelioration and productivity enhancement. This is possible, but significant progress must be made through research and technical understanding. At the moment, opinions diverge widely as to the effectiveness of soil conservation in reducing sediment to the Yellow River. Sound predictions cannot be made with the existing knowledge. Much should be done before a conclusion can be reached. I do not think the task is as formidable as it appears. We may limit our efforts to the regions with a total area of about 100,000 square kilometers. Field observations and remote sensing image studies will reveal that only a part of this area is the scene of high sediment yield. If proper measures are taken to minimize the surface flow on the interfluves in this scene by increasing water infiltration into the soil, gully erosion will be weakened correspondingly.

Even if we are successful in dealing with the preceding problems, the threat of peak flood flows from the catchment between Sanmenxia and Huayuankou still exists. In 1761, this amounted to as much as 32,000 cubic meters per second. A flood originating from this region takes only ten to twelve hours to arrive at Huayuankou. Forecasting and warning are important. Its reduction is more so. The runoff from a small part of the area may be regulated by reservoirs. Wise land management may play a considerable role in the rest. The area is relatively little known to science. Research will bring forth meaningful results.

The growing importance of the present delta of the Yellow River as a result of the rapid development of the petroleum industry has rendered river conservancy in the area more and more complicated. While planned course shifting is desirable for river conservancy, this is in conflict with the interests of oil production and related infrastructure. It goes without saying that studies with tradeoffs between them in view are urgently needed.

The four areas of scientific enquiries are interrelated. It is my belief that any one of them would solve the problem we are facing, but the findings from them all will generate ample opportunities to reduce the flood hazard to a minimum.

Thanks.

Huang Bingwei

THE IMPORTANCE OF YELLOW RIVER SAFETY AND THE URGENCY OF RESEARCH FOR ITS CONTROL

Zuo Da-Kang
Institute of Geography
Chinese Academy of Sciences
Beijing, The People's Republic of China

With a total drainage area of 750,000 sq. km, the Yellow River valley has abundant resources and a large population. It is an important base for production of grain and cotton, energy, and heavy and chemical industries. The social and economic development in this area occupies a decisive position in the overall national economy.

For a long period of time, soil erosion in this area has been serious and flood damages in the lower reaches were frequent because of the vulnerability of the natural environment to anthropogenic activities. At present, there is an increasing demand for industrial, agricultural, and domestic water supply. The problems resulting from changes in the environment of the Yellow River valley involve the speed of economic development in this area as well as river safety.

The basic characteristics of the Yellow River are the small volume of water and the high concentration of sediment. The annual water discharge is 57.0 billion cu.m. Actual observations made at Huayuankou near Zhengzhou City show an average annual discharge volume of 47.0 billion cu.m. This figure is one-twentieth the amount for the Yangtze River, but its annual sediment discharge is 1.6 billion tons, or three times that of the Yangtze River. Ninety percent of the sediment comes from the heavily eroded Loess Plateau located in the middle reaches. The unbalanced proportion of water and sediment and the nature of high sediment concentration are the fundamental causes for the continuous aggradation of sediment in the lower part of river and the problems with flooding, dike breaches and avulsions.

The Threat of Flooding in the Lower Yellow River is Increasing

The most serious natural disasters in China came from the Yellow River. Ever since the begining of historical records, Yellow River floods have brought very grave damages to the

L. M. Brush et al. (eds.), Taming the Yellow River: Silt and Floods, 201–204.

Chinese people. From 602 B.C. to 1949 A.D., there were more than 1,500 floods, 7 big avulsions in the lower reaches and almost 2 dike breaches every 3 years. The floods can reach to Tianjin in the north and the Yangtze and Huaihe Rivers in the south. The affected areas were always large and the disasters were often serious.

In more than thirty years after the founding of People's Republic of China there have been no dike breaches during the summer flood peaks. Great achievements have been made in controlling the Yellow River. However, soil erosion in the Loess Plateau in the middle reaches has not been brought under permanent control and the essential nature of high sediment concentrations has not changed. The channel bed in the lower part is still aggrading at a rate of 10 cm per year. The threat of flooding becomes increasingly serious and there is a potential danger of dike breaches and avulsions.

The present channel of the lower river was formed in 1855 on base of the original Dagin River following a dike breach in Tongwaxiang. After more than 120 years of aggradation, the river has become a "suspended river," with the river dikes much higher than the land behind the levees. On average, the difference is 3-5 meters. It is 6-8 meters in the section between Huayuankou and Dongbatou. The maximum height difference is 11 meters. Since 1958 farmers have built a great number of dikes on the floodplain for agricultural production. These dikes confine the sediment in a narrow channel and have formed a "suspended river" within a "suspended river" between Fengqiou and Puyang. The formation of this double suspension has intensified the threat of dike breaches.

If a dike breach takes place, there will be devastating disasters. According to some studies, if it happens, either to the north or south, at a position west of Jinan City, the flooded area could be as large as 15,000-33,000 sq.km, and the population affected 7 to 18 million. Three important railways, Beijing-Guangzhou, Beijing-Shanghai, and Long-hai, would be cut off. As a result of sediment discharge, the problems of channel siltation and desertification of the land along the floodplain will be serious and the engineering works constructed over decades for harnessing the Huaihe River and Haihe River could be destroyed in one day. The economic loss would be tremendous.

The Yellow River delta is an intensely aggrading delta. Because of the intense aggradation of sediment, there have been frequent changes of the river channel near the mouth. In the past 90 years or so, avulsions near the mouth have taken place more than 10 times. On average changes occur about once every 8-10 years. There have been 3 avulsions in the past 30 years. Since a man-made channel shift to Qingshui Gully was completed in 1976, the present channel has been used for more than 10 years and is on the brink of another avulsion. The frequent swinging of the river course near the mouth imposes a constraint to the development of the oil field and economic construction in the delta.

In order to prevent the flooding danger in the lower Yellow River, there have been quite a few opinions proposed within scientific and technological circles. The prevention methods proposed include heightening the dike system, channel washing by means of artificial floods, diversion of water for sediment discharge, warping on the floodplain and washing off the sediments and deepening the channel, as well as construction of Xiao Langdi reservoir.

Controlling Soil Erosion in the Loess Plateau is Critical to Harnessing the Yellow River

The fundamental problem of the lower Yellow River is the excessive sedimentation. The main source of sediment is from the soil erosion in the Loess Plateau. Ninety percent of the 1.6 billion tons of sediment entering the lower Yellow River each year comes from this area. The area within the river basin is 430 thousand sq. km. The coarse sediment (with diameter greater than 0.05 mm) is most harmful to the lower river course and is mainly produced in 110 thousand sq. km located in the hill and gully regions in the delta occupied previously by the Shaanxi, Shanxi and Inner Mongolia Provinces as well as a region in the part of Gansu province. This area is responsible for 70% of the coarse sediment aggradation on the lower river bed.

Soil erosion has not only caused serious aggradation in the Yellow River but has also created great damage to the production and construction in the Loess Plateau. The Sanmenxia Reservoir had to be rebuilt in less than three years after its completion because of excessive sediment aggradation. Reservoirs constructed in Shanxi Province in the Loess Plateau are losing water storage capacity of 80 million cu.m per year. The damage from soil erosion is very alarming. The maximum erosion intensity can be as high as 10-30 thousand tons per sq. km per year.

From the above statement, it can be seen that the intense soil erosion in the Loess Plateau is the root of the trouble in the lower Yellow River, soil erosion causes the deterioration of the local ecological environment, and causes poverty and poor production. Therefore, controlling soil erosion is the fundamental approach for harnessing the Yellow River and land management in the Loess Plateau.

While severe soil erosion results from natural conditions and human activities, the latter is the main cause. According to some studies, annual sediment discharge of the Yellow River three thousand years ago was 1 billion tons. The increase in population, large-scale land reclamation, sloping land farming, excessive deforestation, over-grazing, and destruction of vegetation have greatly accelerated the soil erosion processes. The sediment production has been increasing steadily. It reached 1.6 billion tons per year in the 1950's and 2.2 billion tons per year in 1970's. In the past thirty years or more the population in the Loess Plateau has doubled, but the basic conditions for agricultural production have not changed significantly and farmers are still in a poor condition. There is a vicious cycle; the poorer the area, the more farming is done and the more farming is done the poorer it becomes. In addition to this, not enough attention has been paid to water and soil conservation in the construction of railways and highways, in proceeding hydraulic engineering projects and in development of mining industries. The problem of erosion has become more serious as a result of these activities. There exists an alarming phenomenon that positive land management goes along with negative destruction. In some river valleys such as the Malianhe River valley, the benefit of land management is exceeded by the negative affect of human activities.

In the studies of Yellow River control, we must consider the overall situation of the entire river valley. We must stress the importance of the overall considerations for the upper, middle and lower Yellow River. This includes area controls, the relationship between water and sediment discharges, the change in the ecological environment, and the harnessing and development of the river valley. To reduce sediment aggradation in the lower Yellow River comprehensive measures combined with a variety of useful methods must be taken.

TEMPORAL AND SPATIAL DISTRIBUTION OF HEAVY RAIN IN THE MIDDLE AND LOWER REACHES OF THE HUANGHE RIVER

Shen Jianzhu, Guo Yimei, Xu Shuying
Institute of Geography
Academia Sinica and State Planning Commission
The People's Republic of China

The Huanghe River (Yellow River) is the second largest river in China. It rises in the northern foot hills of Ba Yan Ke La mountain of the Qingzang Plateau, and flows through nine provinces - Qinghai, Sichuan, Gansu, Ningxia, Neimonggol, Shanxi, Shaaxi, Henan, Shandong and at last into the Bohai Sea (Chinese Acad. of Sci., 1981). It can be divided into three parts. The upper reach of the Huanghe River is in the Qingzang Plateau, over 3,000 meters above sea level. The middle reach is in the Loess Plateau, about 1,000-2,000 meters above sea level. The lower reach is in the plain of North China. In general, the channel of the Huanghe River is divided into the upper reach: River head - Liujia gorge, middle reach: Liujia gorge - Huayuankou, lower reach: Huayuankou - downstream. For practical purposes and convenience of investigation, the area covered in this paper extends to the north of the Yinshan and Yanshan Mountains, south to the Qingling Mountains - the Huaihe River, west to Liujia gorge, and east to the Bohai Sea.

The topography of the middle and lower valley of the Huanghe River is complicated. It is high in the west and low in the east. In the north, it is in the desert and loess plateau. To the south of the central part, hills are criss-crossed by river tributaries. The eastern part is a wide plain. The Huanghe River valley includes the Taihengshan, Funiushan, Luliangshan, Yinshan, Qingling, Liupanshan Mountains and among others the Weihe, Jinhe, Yihe, Luhe, Xinhe, and Fenhe River tributaries.

The Huanghe River valley is the most flood damaged area in Chinese history. The dikes of the Huanghe River broke 1,500 times in 2,000 years of recorded history. Flood damages in 1761 extended from the Huai River Valley in the south to the Fenhe, Xinhe and Haihe River in the north, and from a line from Guanzhong, Shaaxi province in the west to Huayuankou in the east. The flood bursting the dikes flooded 33 counties

L. M. Brush et al. (eds.), Taming the Yellow River: Silt and Floods, 205–222.

of Henan province, 12 in Shandong province, and 4 counties in Anhui province causing flood damage (Gao, 1983).

This paper analyzes the temporal and spatial distribution of the precipitation and heavy rains of the middle and lower reaches of the Huanghe River and demarcates the middle and lower reaches into some heavy rain regions according to rainfall characteristics.

Temporal and Spatial Distribution of Precipitation

Precipitation in the middle and lower reaches of the Huanghe River decreases from the southeast to the northwest. The rainfall of the lower reach is larger than that of the middle reach.

ANNUAL AND FLOOD-SEASON PRECIPITATION FEATURES

The annual rainfall of the middle and lower reaches of the Huanghe River is 200-900 mm. The rainfall of the lower reach is the largest, 600-700 mm on the average, and in its southern part is over 700 mm. In the southern part of the middle reaches, the Sanmen gorge - Huayuankou and Weihe River valley rainfall is 500-700 mm. The rainfall decreases from the southeast to the northwest part of the Great Bend of the Huanghe River falling from 500 mm to 200 mm (see Fig. 1). The mean annual number of rainy days of the middle and lower reaches of the Huanghe River is 70-120. Its distribution is basically analogous to the annual precipitation.

Fig. 1 The annual rainfall of the middle and lower reaches of the Huanghe River

---

The rainfall of the middle and lower reaches of the Huanghe River appears mainly in July-August comprising about 60-70% of the annual rainfall. The period of rainfall in the middle reach is more concentrated than that in the lower reach. The rainfall of the middle reach arrives mainly in the last ten days of July and the first ten days of August. Heavy rains often occur during this period.

The longest continuing rainfall ($\geq$ 0.1 mm) duration is 10-15 days in most parts of the middle and lower reaches except for the Toketo-Yuling river course. The longest continuing rainfall days appear mostly in August and September in the middle reach, but in the lower reach they mainly appear in July.

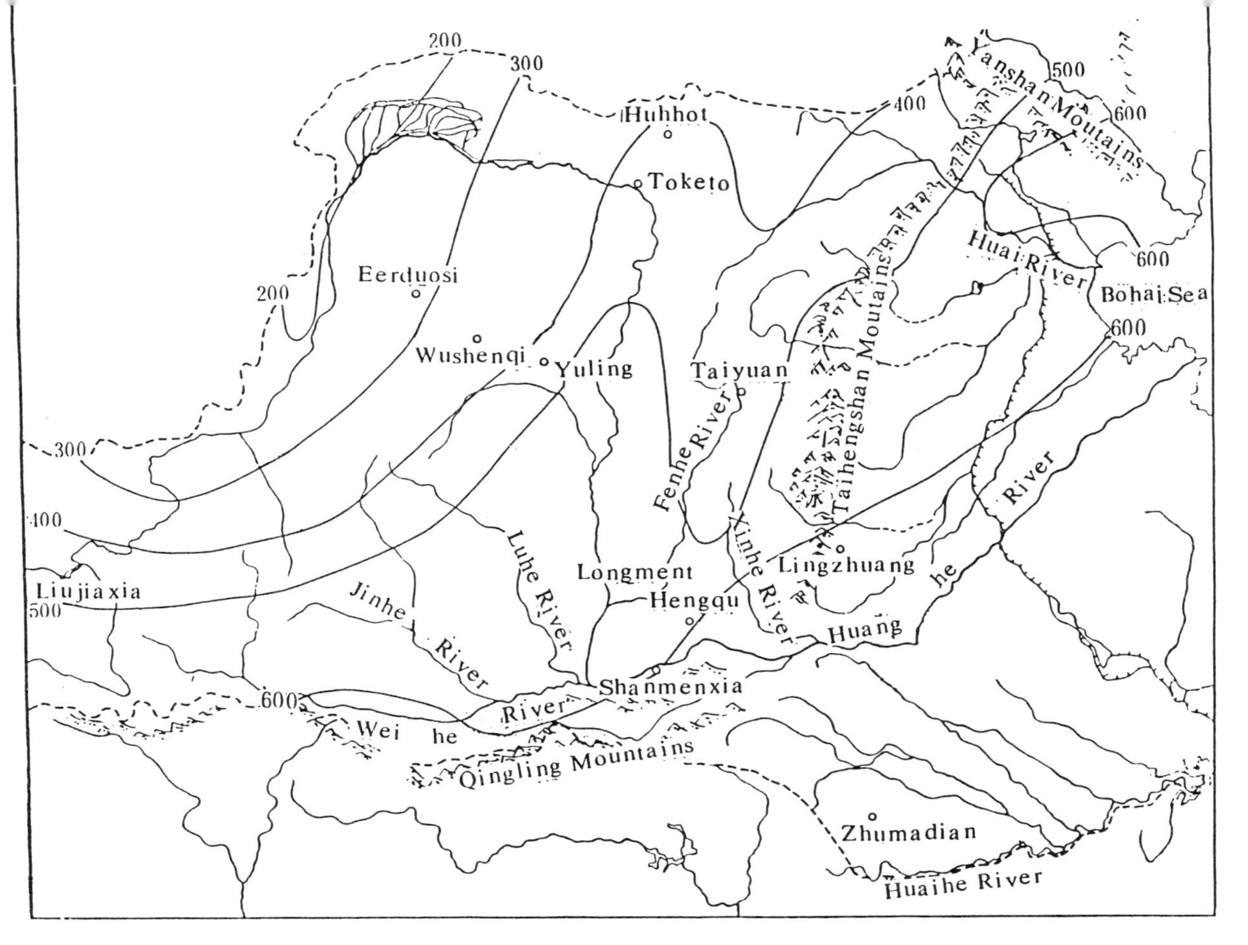

Figure 1. The annual rainfall of the middle and lower reaches of the Huanghe River

Fig. 2 The longest rainfall duration of the middle and lower reaches of the Yellow River (days)

Precipitation variability is larger in the middle and lower reaches. In the northern part of the Loess Plateau the annual relative variability of precipitation is over 30%, and gradually decreases toward the south. In the south part of the Weihe River Valley, it is 15% only, and increases to 25-30% in the Sanmenxia-Huayuankou area. This fact indicates that the Sanmenxia-Huayuankou river reach should receive more attention.

MODERATE AND HEAVY RAIN

The continuing moderate and heavy rain also easily causes flood damage in the middle and lower reaches. The moderate rain (≥ 10 mm) of the middle and lower reaches appears over 10-25 days on average, in about 10-20 days in the Toketo-Shanmen gorge section and in about 20 days in the Weihe River valley and Sanmenxia-Huayuankou (see Figure 3). Most moderate rain days are in summer. In the northwest part of the Loess Plateau, rain appears mostly in August, and in other parts of the middle and lower reaches of the Huanghe River, mostly in July. On average, it comes 4-6 times each July and August.

Fig.3 The moderate rain (≥ 10 mm) days of the middle and lower reaches of the Huanghe River

Although the distribution of heavy rain (≥ 25 mm) days in the middle and lower reaches is similar to that of the moderate rain days, the heavy rain days appear only about one-half as often as the moderate rain days.

The heavy rain days of the lower reach of the Huanghe River and Sanmenxia-Huayuankou are the longest in this river, about 6-9 days. They span about five days in the Weihe River Valley. In most parts of the Loess Plateau, heavy rains last 3-4 days; in a few places only 2 days. On the windward side of the mountain area, for example, on the southern slope of the Yanshan mountain the heavy rains last 7.5-8.5 days and 12.5 days in the Taishan mountain. The heavy rains mainly occur in July and August. In most parts of the middle and lower reaches, except for the Loess Plateau (see Fig. 4), they appear in July.

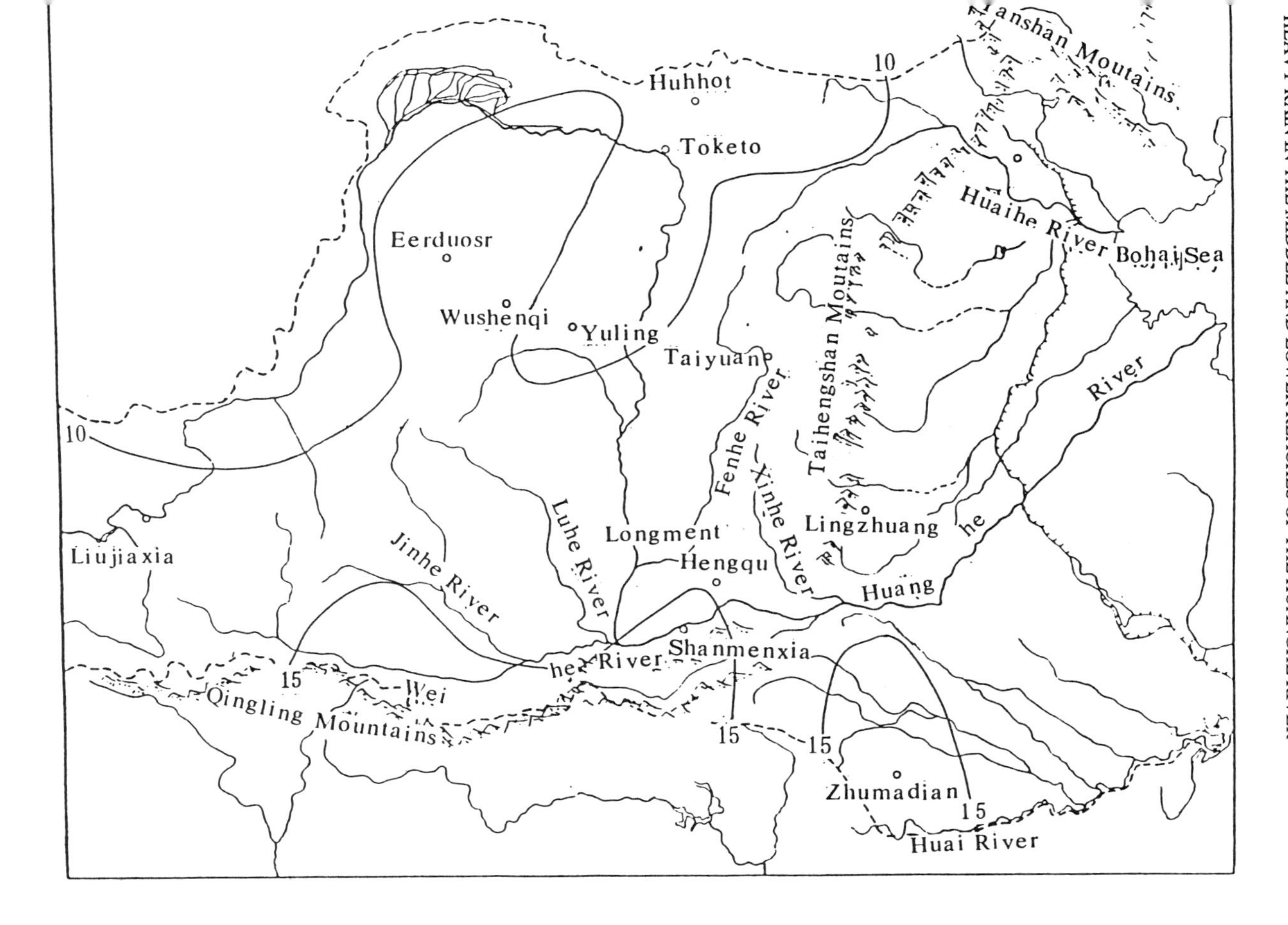

Figure 2. The longest rainfall duration of the middle and lower reaches of the Yellow River (days)

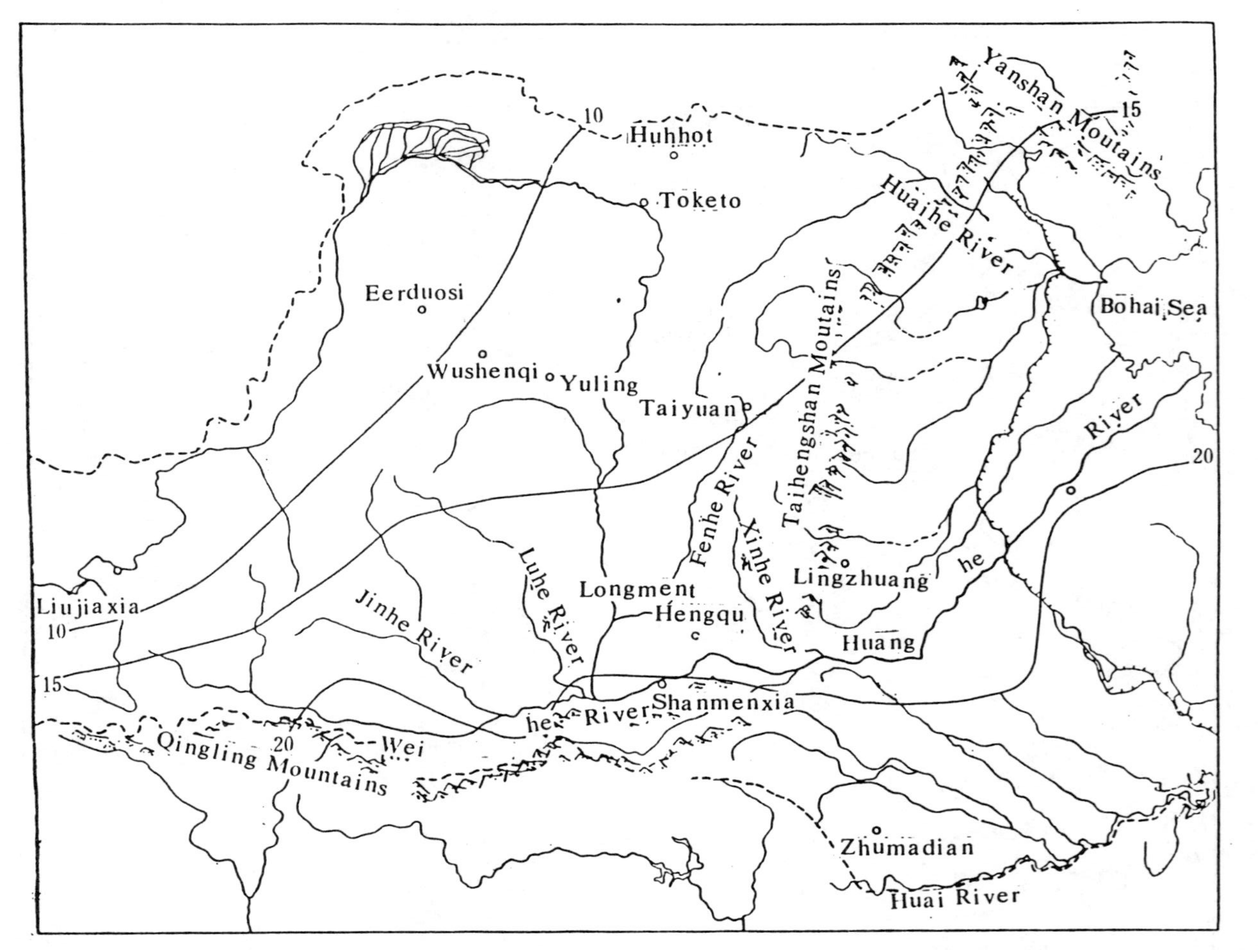

Figure 3. The moderate rain (≥ 10 mm) days of the middle and lower reaches of the

Fig. 4 The heavy rain (≥ 25 mm) days of the middle and lower reaches of the Huanghe River

Temporal and Spatial Distribution of the Rainstorms

This includes two kinds of rainstorms. The rainfall amount is ≥ 50 mm/day and ≥ 100 mm/day respectively. Rainstorms of the Huanghe River valley are of two types: (a) the rainfall field is small, rainfall intensity is large, and rainfall duration is short, and often appears in July and August in the Eerduosi area from Huhhot through Yuling to the northern part of Shanxi province or the boundary between Neimonggol and Shanxi provinces; and (b) the rain field is large, rainfall intensity is stronger, and rainfall duration is longer, about 4-5 days or more, and often appears in summer and autumn in the lower reaches of Jin and Wei rivers and Sanmenxia-Huayuankou. In general, the rainstorms of the middle and lower reaches of the Huanghe River do not occur at the same time.

RAINSTORM DAYS

Rainstorms (≥ 50 mm per day) decrease from the east to the west. They occur 1-3 times in the lower reach of the Huanghe River and Sanmenxia-Huayuankou. Downpours (≥ 50 mm per day) in the lower reach are more than those of the middle reach. Along the Huanghai Sea, sometimes rainstorms can occur 4 times, in the east part of the Taihengshan mountains on 1-2 days, and in the west part of the Taihengshan mountains it is less than one day.

Heavy rainstorms (≥ 100 mm per day) occur 0.5-1 day in the east part of the Taihengshan mountains, in the west part, less than 0.5 day.

The facts mentioned above show that the chance of a heavy rainstorm in the lower reach of the Huanghe River is more than that in the middle reach. The Sanmenxia-Huayuankou region is one of the heavy rainstorm areas in the Huanghe River Valley. Because the rainstorms appear more often and the river tributaries criss-cross the region, this region is a source of flood damage to the lower reach. Thus further discussion of the rainstorms of Sanmenxia-Huayuankou and the region mentioned above will be made.

Table 1 is the frequency of occurrence of rainstorms (area > ten thousands sq.km) from June-September of 1954-1977 in Sanmenxia-Huayuankou and the middle reach above Sanmenxia. It can be seen that the frequency of rainstorms at

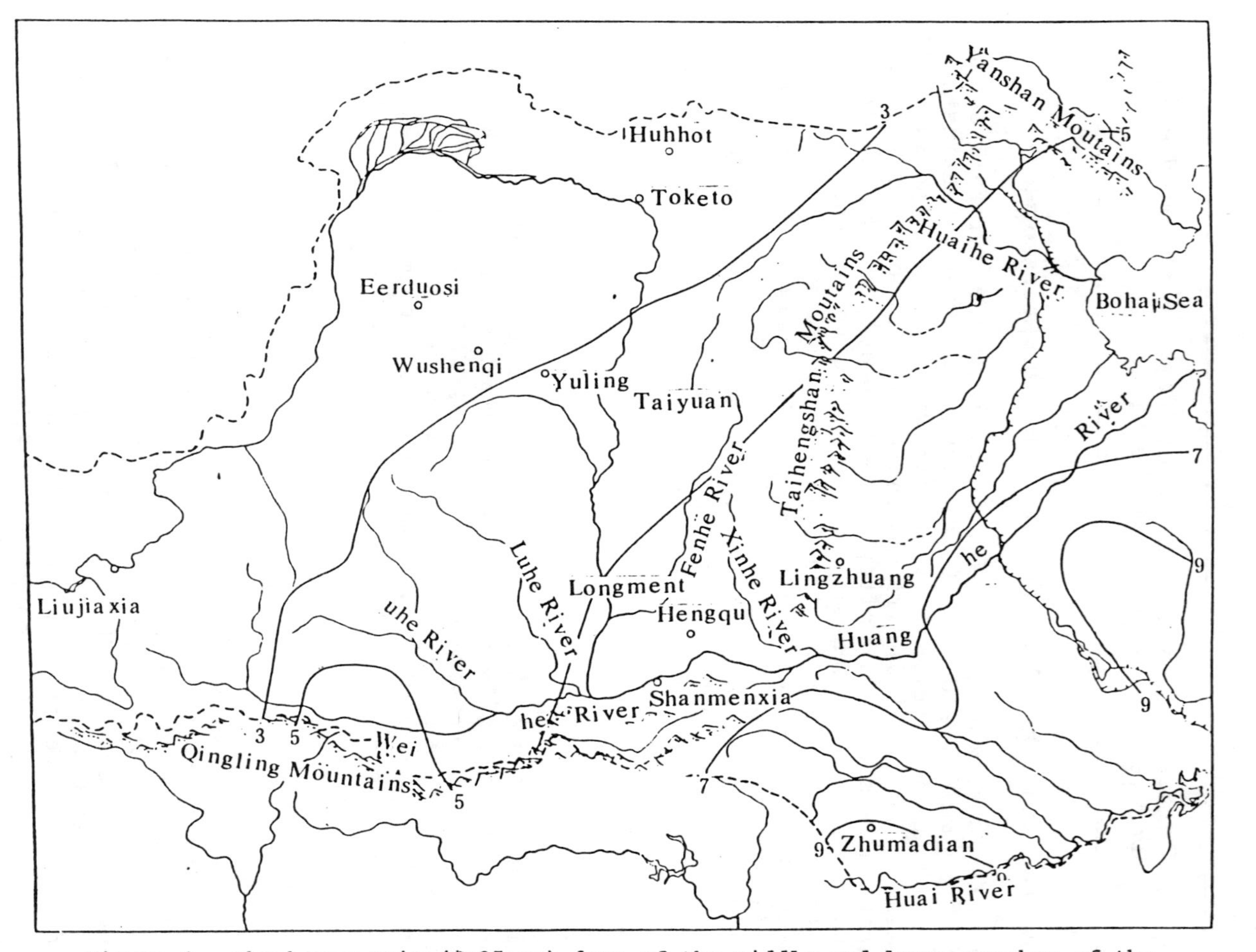

Figure 4. The heavy rain (≥25 mm) days of the middle and lower reaches of the Huanghe River.

Table 1. The frequency of rainstorms (area > ten thousands sq. km) in June-September

	June			July			August			September			total
	first	second	last	first	second	last	first	second	last	first	second	last	
Sanmenxia - Huayuankou	1	0	1	5	9	8	5	4	0	1	0	1	35
above Sanmenxia				4	7	4	6	6	3	2	0	1	33
total	1	0	1	9	16	12	11	10	3	3	0	2	68

Sanmenxia-Huayuankou is more than that above Sanmenxia. Rainstorms (area > ten thousands sq.km) can appear from June-September, but are usually concentrated in July-August. The monthly distribution of rainstorm frequency in June-September is as follows: It appears once every ten years or more in June in the middle reach of the Huanghe River, once every five years or more in September. But in July it can appear 1-2 times every year, in August one time every year. So the frequency of rainstorms with an area > ten thousands sq. km can appear 1-3 times every year in the middle reach of the Huanghe River. In addition, the Jinghe, Luhe, Weihe, Fenhe, and Xinhe River tributaries are in this area. Their runoff flows converge into the channel of the Huanghe River, and may cause flood damage.

RAINSTORM INTENSITY IN DIFFERENT PERIODS

Daily Maximum Precipitation. Daily maximum precipitation in the middle and lower reaches of the Huanghe River is 50-200 mm and decreases from the southeast to the northwest. The rainfall of the lower reach is 150-200 mm, and that of the middle reach, 50-150 mm (Fig.5). In the mountain area and windward side, along the sea of the east part, due to the rising topography, there are more rainstorm centers. For example, the daily maximum rainfall in the windward side of the Yanshan mountain and Taihengshan mountain is > 150 mm, and along the sea to the east is > 200 mm.

Fig. 5 Daily maximum precipitation in the middle and lower reaches of the Huanghe River

The daily maximum precipitation in the middle and lower reaches of the Huanghe River is observed by an outdoor survey after a rainstorm, the distribution of which is very much like that of the daily maximum precipitation according to the meteorological data. But the rainfall is much larger than the daily mean maximum rainfall. The daily maximum rainfall of the middle and lower reaches is about 100-1,000 mm. In the middle reach it is 100-300 mm and in the lower reach it is 300-500 mm. At some locations the rainfall can be over 1,000 mm. For example, 1,060 mm/day was once recorded at Zhumadian in August of 1975 in the Funiushan mountain front.

Heavy Downpours in Short Time Periods. Many heavy rainstorms occur in minutes or in a few hours. This kind of rainstorm is much more harmful to the Huanghe River. Table 2 shows the maximum rainfall for a short time. From Table 2, it can be seen that one maximum rainfall in Datong and Huangshui of

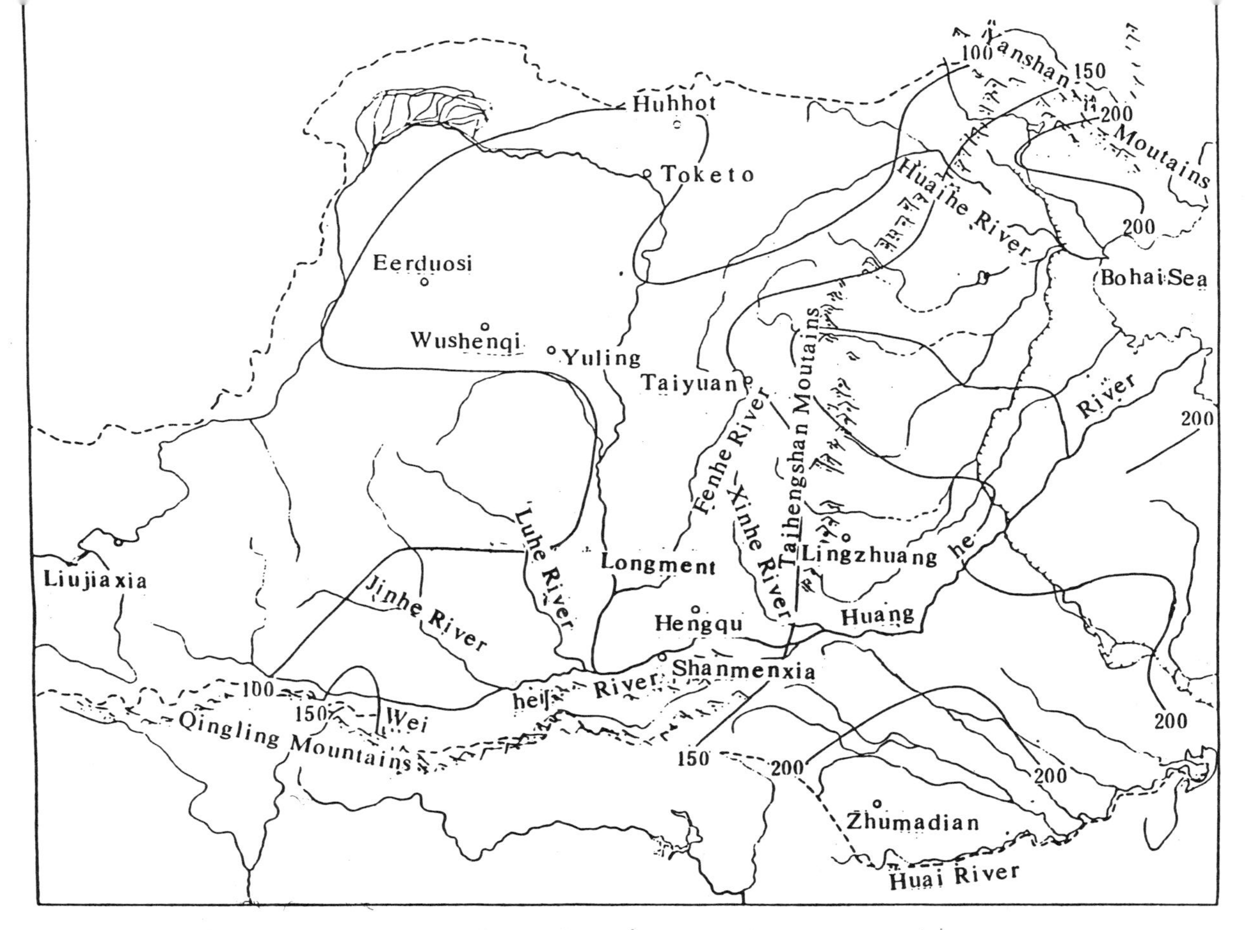

Figure 5. Daily maximum precipitation in the middle and lower reaches of the Huanghe River.

Table 2. Heavy Rainfalls in Short Time Periods (Zhan and Zuo, 1983).

Province	County	Year	Month	Date	Hours	Rainfall(mm)
Gansu	Gulin	1966	8	13	6	228*
	Hangchang	1976	7	15	6	330*
Qinghai	Datong	1976	6	19	0.5	240
	Huangshui	1977	8	1	1.5	204
Shaanxi	Shengmu	1971	7	25	12	409
	Anzai	1981	6	20	1	267
Neimong gol	Shangdu	1959	7	19	3-4	620*
	Xilinghot	1974	7	22	3	350*
	Wushenqi	1977	8	1	10	1440*
Shanxi	Taiyuan	1971	7	31	6	567*
	Fangzhi	1973	8	12	3	518*
	Huo County	1970	8	10	6	600*
Hebei	Wangquan	1962	7	23	1.5	400*
	Zhangyao	1963	8	4	3	218
	Kangbao	1972	6	25	2.5	550
	Shangyi	1973	6	25	1.5	430*
	Yangyuan	1973	6	28	3	600*
	Fengning	1974	7	30	1	315*
	Weichang	1974	8	5	2	300
Henan	Laojun	1975	8	7	1	190
	Laojun	1975	8	7	3	615
	Lingzhuang	1975	8	7	3	495
	Lingzhuang	1975	8	7	6	830

Qinghai province was 200-250 mm in 0.5-1.5 hours. In many places of Hebei province rains of 300-600 mm have fallen in 0.5-3 hours. Wushenqi is a place of scarce rainfall. In the Neimonggol Autonomous Region, a heavy rainstorm occurred in August of 1977 where the rainfall reached 1,440 mm in ten hours. Thus it can be seen, in many low rainfall sites, especially in the middle reach of the Huanghe River, rainfall intensity can be very large.

Division of Rainstorm Regions

The spatial distribution of rainstorms shows that there are great regional differences in the middle and lower reaches. So dividing the rainstorm area of the Huanghe River into different regions is meaningful in discussing the source of the floods of the lower reach and in determining the flood control measures.

According to the above precipitation analysis, annual rainstorm days, daily maximum rainfall, topography and the situation of the river basin, provide criteria to divide the Huanghe River basin into three regions (see Fig. 6):

Fig. 6 The rainstorm region of the middle and lower reaches of the Huanghe River

---

I, the region of the lower reach of the Huanghe River; II, Toketo-Huayuankou region of the middle reach; III, Liujiaxia-Toketo region of the upper reach.

According to the distribution of rainstorm centers in regions I, II, III and the features of the tributaries, the regions are divided again into 8 districts.

THE REGION OF THE LOWER REACH

The river flows gently in the channel of the lower reach of the Huanghe River. The silt is heavily deposited in the channel. It is a "Suspended River." In this river valley the frequency of rainstorms is large, rainfall intensity strong and rainfall duration is long. But there are no river tributaries converging with the channel of the lower reach. The rainstorms in the lower reach rarely cause flood damage. The floods of the lower reaches mainly come from runoff caused by rainstorms in the middle reach.

THE PLAIN DISTRICT OF THE TAIHANGSHAN MOUNTAIN FRONT

This district is located at the east foot of Taihangshan mountain. Due to the rising topography, rainstorms are larger

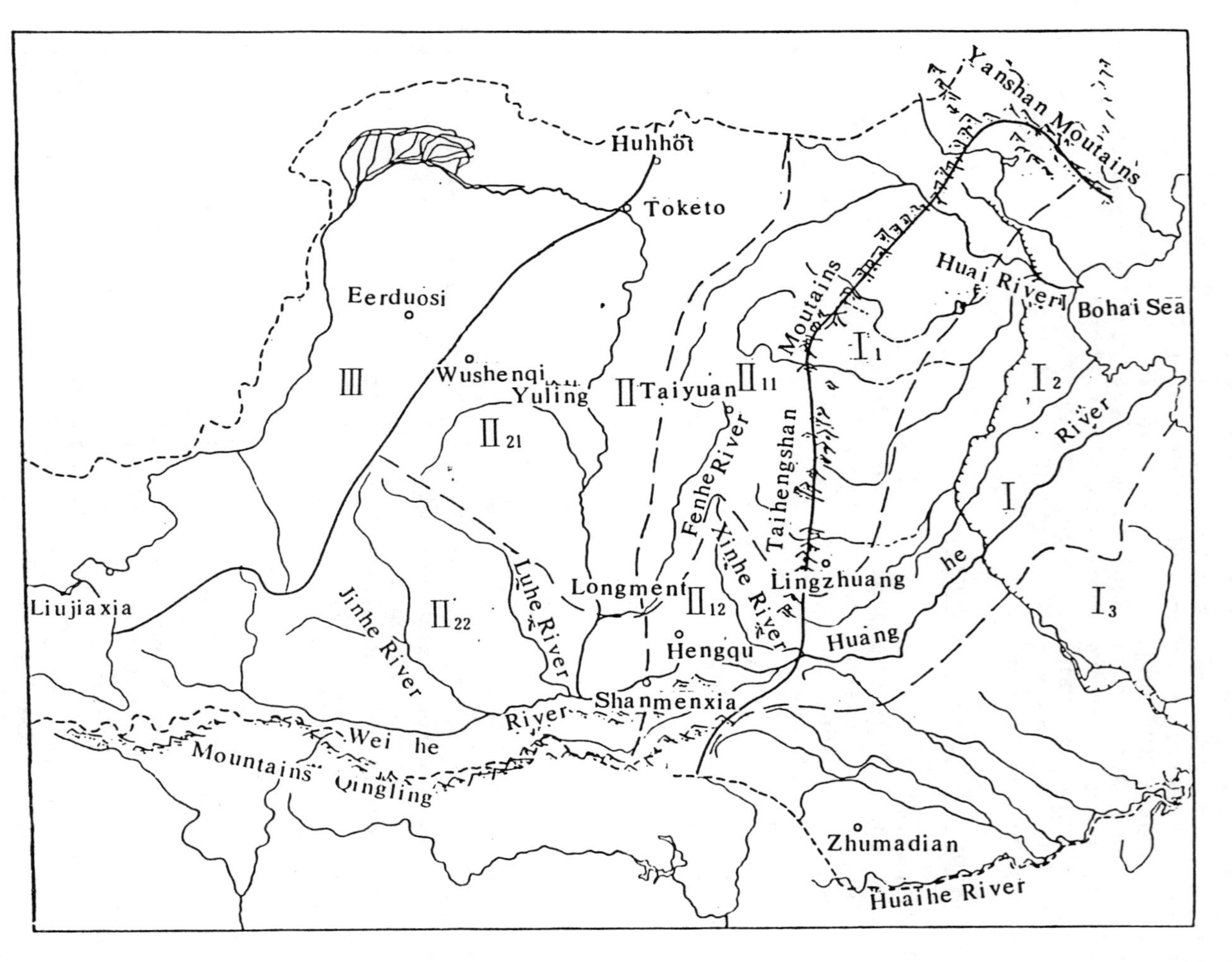

Figure 6. The rainstorm region of the middle and lower reaches of the

in this district. This is a center of rainstorms in the lower reach. It has more than three downpours each year. A heavy rainstorm occurred here in August of 1963. The rainfall duration was about seven days. The rainfall at Zhangyao during 24 hours was 950 mm. 1,458 mm fell during three days at that station. From the second to the eighth of August, the rainfall lasted for seven days, and amounted to 2,050 mm. The area where rainfall of over 1,000 mm occurred during the seven days covers 5,560 sq. km. This heavy rainstorm caused very great damage to the agricultural fields and the railway of the Haihe River basin. It was the largest heavy rainstorm ever recorded in North China.

THE PLAIN DISTRICT OF THE MAIN CHANNEL OF THE HUANGHE RIVER

The rainstorm intensity of the plain district of the main channel (III) is smaller than II, but due to low-lying terrain this district is easily flooded. The rainstorm days are usually 2 per year in this district. In Heilonggong, Hebei province it is less than 2 days, in the mountain area of south-central part of Shandong province and the area along the East Sea it is more than 3 days. In this district (III) the daily maximum rainfall is 150-200 mm. Because this channel of the Huanghe River in this district is a "Suspended River," no river tributaries converge with the channel. So rainstorm effects directly on floods are smaller. However a rainstorm of the middle reach poses a latent crisis for the channel of this district.

THE PLAIN DISTRICT OF THE NORTH OF THE HUAIHE RIVER

In the west part of the plain district north of the Huaihe River is Funiushan Mountain. In the east it faces the Huanghai Sea. The rainfall of this district (IV) is more than II and III. Rainstorms in the east and west of this district are more frequent than in the central part. In the west the area in front of Funiushan Mountain is the center of heavy rainstroms in the lower reaches. The days of rainstorms in the district IV are more than three every year. Rainfall at the center of a rainstorm during 24 hours amounted to 200 mm or more. The district had a heavy rainstorm in August of 1975. The maximum rainfall during 24 hours amounted to 1,060 mm at Lingzhuang, Henan province. The rainfall during six hours and one hour amounted to 830 and 189.5 mm, respectively. The area for which the rainfall was over 400 mm was over 19,410 square km. This heavy rainstorm damaged dams and reservoirs and the flood ran wild. In the east of IV, rainstorms due to the effects of typhoons occur for over 4 days every year, and the daily maximum precipitation is more than 200 mm.

THE REGION OF THE MIDDLE REACH OF THE HUANGHE RIVER

The surface slope in the middle reach of the Huanghe River is steep and it has many river tributaries. The soil texture of the Loess Plateau is very loose, which is a source for silt deposits in the channel of the middle reach. The Weihe, Jinhe, Yihe, Xinhe and Fenhe Rivers located in the south of this region are the sources of runoff which cause flood damage to the lower reach. The rainstorm intensity of this region (II) is smaller than that of region I, but region II is more important than region I in causing flood damages and sedimentation of the channel in the lower reach.

SANMENXIA-HUAYUANKOU COURSE

The district of Fenhe River basin. The Fenhe River basin district located in the west of the Taihangshan mountain and includes some parts of Shanxi province. The rainstorms of this district are relatively smaller than the other districts of region II. The number of rainstorm days is less than one each year. The daily maximum precipitation is 50-200 mm. Flooding caused by the rainstorms in this district is not very serious.

Sanmenxia-Huayuankou District. There are tributaries such as the Yihe, Luihe, etc. in the district. The landforms of the district are favorable for converging the water into the rivers. This district is also a district of the maximum of rainstorm intensity and causes flood damages in the middle reach. The daily maximum precipitation is 100-200 mm. A heavy rainstorm lasting 5 days occurred in July of 1958 in the Sanmenxia-Huayuankou and Yihe, Luihe, Xinhe River basins. The daily rainfall was 366 mm at Hengqu, Shanxi province in the rainstorm center. The maximum discharge at Huayuankou reached 22,300 m/sec. This was the maximum discharge for a flood since the liberation of China.

TOKETO-SANMENXIA COURSE

Toketo-Longmen District. This district is located in the Loess Plateau. There are many river tributaries in the district. It is also a rainstorm region in the middle reach. Rainstorm intensity is large, rainfall duration is not too long, and rainfall coverage is wide. Rainstorms mostly appear at the windward side of the Daqingshan mountain and at a border of the desert zone. The daily maximum precipitation is 50-150 mm. A heavy rainstorm occurred here in August, 1977. The rainfall at Wushenqi, Neimonggol Autonomous Region was 1,400 mm in nine hours. The area where the rainfall was 500 mm or more covered 74,000 sq.km, and the area of rainfall 100 mm or more covered 8,000 sq.km. This is an exceptionally large rainstorm for a desert area.

The District of the Jinhe, Weihe, Luihe River Basins. The topography of this basin allows the surface water to run into the Huanghe River channel easily. Although the rainfall from storms in this district is relatively small, the daily maximum precipitation is about 50-150 mm, the discharge is large due to the convergence of the waters from tributaries.

LIUJIAXIZ-TOKETO REGION OF THE UPPER REACH

The surface slope in this region is large, the characteristic rainfall is of a long duration and the rain field area is wide, but the rainfall intensity is small. The number of rainstorm days is less than one a year. In addition, forest, grassland, and marshlands in the region have a regulative effect on the flood level and make the surface waters of the region flow gently. Thus the discharge of the upper reach is only base flow for a flood in the lower reach.

Summary

The characteristics of the rainstorms mentioned above can be summed up as follows:

(1) Three rainstorm regions are identified in the Huanghe River basin:

a. The rainstorm region of the lower reach, below Huayuankou. In this region, rainstorm intensity is large, and the rainfall area is wide. In addition, the Huanghe River channel of the region is a "Suspended River." So rainstorms pose a latent crisis for flood damage on the Huanghe River.

b. Toketo-Huayuankou is the rainstorm region of the middle reach. The rainfall of this region is smaller than that of the rainstorm region of the lower reach. When a rainstorm occurs, waters from the tributary rivers in this region converge carrying a great amount of silt causing flood damage on the channel of the lower reaches. This rainstorm region is a source area for flood damages to the lower reach.

This region can be divided into four districts: Fenhe River basin, Sanmenxia-Huayuankou, Toketo-Longmeng, and the Jinhe, Weihe, Luihe River districts. The rainstorm intensity of the Sanmenxia-Huayuankou rainstorm district is largest. It is a source of floods on the lower reach. In the Toketo-Longmen rainstorm district, the rainstorm intensity is smaller and rainfall duration is also short, but this district is located in the Loess Plateau and the desert edge, so it transports the silt of the Loess Plateau to the channel of the Huanghe River.

c. The rainstorm region of the upper reach. The discharge formed from rainstorms in this region of the upper reach is only a base discharge in causing floods in the lower reaches.

(2) Daily maximum rainfall is 50-150 mm in the middle reach and 150-200 mm in the lower reach. According to the rain data and surveys after rainstorms, the daily maximum rainfall is about 100-300 mm in the middle reach, and 300-500 mm in the lower reach. At a few places in the middle and lower reaches, for example, Wushenqi, Neimonggol Autonomous Region, Lingzhuan, Henan province and Zhangyao, Hebei province, daily maximum rainfall can reach 1,000 mm.

(3) The rainstorms of the Huanghe River Valley mainly appear in July-August, the southeast part receives rain earlier than the northwest part. In the Toketo-Sanmenxia reach, rain comes mostly in August. Below the Sanmenxia reach, it occurs mostly between the second ten days of July and the second ten days of August.

References

1. Editorial board of "Physical Geography of China," Chinese Academy of Science: Physical Geography of China (Surface Water), Science Publishing House, 1981.

2. Gao Xiushan, Analysis on Flood of the Middle-Lower Reaches of the Huanghe River in 1761, People's Huanghe River, Vol. 2, 1983.

3. Zhan Daojing, Zuo Jinsang, Probable Maximum Rain Gush and Flood, Publishing House of Water Resources and Electric Power (MWREP), July 1983.

SOME PROBLEMS OF FLOODING AND ITS PREVENTION IN THE LOWER YELLOW RIVER

Liu Changming and Liang Jiyang
Institute of Geography
Academia Sinica
People's Republic of China

ABSTRACT: This paper deals mainly with flood prevention in the lower reaches of the Yellow River. The authors have reviewed the experience in flood control obtained by a number of Chinese engineers and scientists. Finally, an integrated countermeasure is emphasized to solve the problems of flooding in the Yellow River's lower reaches by using systems analysis.

Flooding and Sedimentation

The Yellow River is second among the largest rivers in China, having an area of 752,443 sq.km and a course length of 5,464 km. Located in North China, the river's headwaters are on the Xizang (Tibet)-Qinghai Plateau in the west and the river basin lies to the north of the Yinshan Mt., and to the south of the Qinlin Mt. The Yellow River empties into the Bohai Sea. It flows through a wide area of semi-arid regions with a continental climate. Annual precipitation in the basin averages about 400 mm. The temporal and spatial distribution of precipitation is very uneven. For instance, summer rainfall in most regions accounts for 70% of the annual total. Summer storms occur frequently and rainfall of very high intensity can occur during a short period within a storm. Sometimes, rainfall in one day can equal the average annual precipitation. A heavy storm may in a few hours reach 300-400 mm.

FLOOD GENERATION AND SEDIMENT SOURCES

The floods threatening the downstream areas result mainly from big storms in the middle reaches. According to hydrological records the maximum peak discharge of 22,300 cms was measured in Shaanxian County in 1933 and 22,300 cms was measured at Hauyuangkou, downstream from Zhengzhou, in 1958. The 45-day flood volume of the river at Shaanxian County was about 22

L. M. Brush et al. (eds.), Taming the Yellow River: Silt and Floods, 223–241.

billion cu.

The Yellow River travels through the loess plateau in its middle reaches, where there is a lack of vegetation and the soils are loose. Erosion is very severe when storm rainfall takes place. In some regions the soil erosion modulus can reach 20,000 tons per sq. km per annum. During storms a very high sediment content in the main course of the river occurs due to the confluence of many tributaries with hyperconcentrations of sediment. Sometimes the sediment content in the river and its tributaries exceeds 600 kg per cu.m. and may reach 1000 kg per cu.m. Average annual sediment content of the river in Shaanxian County is 38 kg per cu.m. and the annual sediment runoff amounts to about 1.6 billion tons.

As a result of severe soil erosion, sediment transportation and sedimentation, the river channel in its lower reaches has been silted up over the years and many sections of the riverbed have risen. Consequently, a so-called "suspended river" has been formed. This is the main cause of flooding and overflowing along the downstream reaches. The main dikes for flood prevention have to be heightened and strengthened as the river bed rises.

At present, the riverbed in its lower reaches is generally 3-5 m higher than the adjoining lands behind the levees. In some sections of the Yellow River's lower reaches the height of the bed is higher than both banks by about 10 m.

EMBANKMENT AND PREDICTED FLOODS

Dikes for the prevention of flooding have been built along the lower reaches of the Yellow River from Zhengzhou to the estuary. In sections, the distance between the dikes on both banks in Henan Province is about 10 km., except for a few narrow sections in Shandong Province having a width of 5 km. The narrowest reach is only 0.3 km and the channel capacity for flood water discharge is greatly decreased.

Embankments on the lower reaches of the Yellow River have been renovated several times since 1948 and a great achievement has been attained by avoiding summer flooding in terms of dike breaches for the last 40 years. There are 7 large-scale reservoirs on the main course of the river's upper and middle reaches, including the Sanmenxia reservoir having a dam height of 105 m, with a designed storage capacity of 35.4 billion cu.m at an elevation of storage level at 350 m. Because of siltation, the reservoir's active capacity declined at an annual rate of about a billion cu.m. during the first few years after construction. The role of the Sanmenxia dam had to be reconsidered and reservoir operation changed to discharge sediments and regulate floods. As a result of flood regulation, the Sanmenxia reservoir plays an important role in controlling floods from two of the three major regions of flood sources, i.e., from the areas between Shanxi and Shaanxi Provinces, and the middle reaches' tributaries--Jing He, Luo He and Wei He. Therefore, stormwater from the remaining areas of the Sanmenxia--Huayuankou reaches, including Yi-Luo-Chin Rivers, would be a major threat for the lower reaches of the Yellow River.

For estimating the extreme floods for the reach downstream from Huayuankou there are several approaches. Historically, according to flood investigations, an extreme value of peak discharge of 37,000 cms was observed in 1761. Analytically, the frequency-based peak discharges determined were as follows: 55,600 cms for a recurrence interval of 10,000 years; 42,300 cms for a recurrence interval of 1,000 years and 29,300 cms for a recurrence interval of 100 years. Meteorologically, on the basis of PMP analysis, the PMF would be 43,000-57,000 cms. Finally, an extreme value of flood-peak discharge would be 55,000 cms or so.

THE PROBLEMS OF CHANNEL SILTATION

According to investigation and estimation, about 400 million tons of sediment, which accounts for 1/4 of the annual sediment discharge (1,600 million tons), are deposited in the main course of the river and the rest is transported to the river's estuary and into the Bohai Sea. The channel siltation is formed due to the decline of the channel gradient and the decrease of sediment transporting capacity. The annual rise of the riverbed is about 10 mm resulting in a great decrease in floodwater discharge capacity of the channels. Table 1 shows the rise of water levels in the channels with the same discharge of 5,000 cms for different periods.

Table 1. Rise of Water Levels in Channels

Period	Water level rise (m)			
	Huayuanhou	Gaocun	Aishan	Lijing
1950-1969	1.23	2.48	2.29	2.19
1970-1979	0.51	1.54	1.62	1.53

It is estimated that with the increase of water use in the upper and middle reaches of the river basin, the amount and processes of sediment sources would be changed to increase the sediment content and siltation in the downstream river on

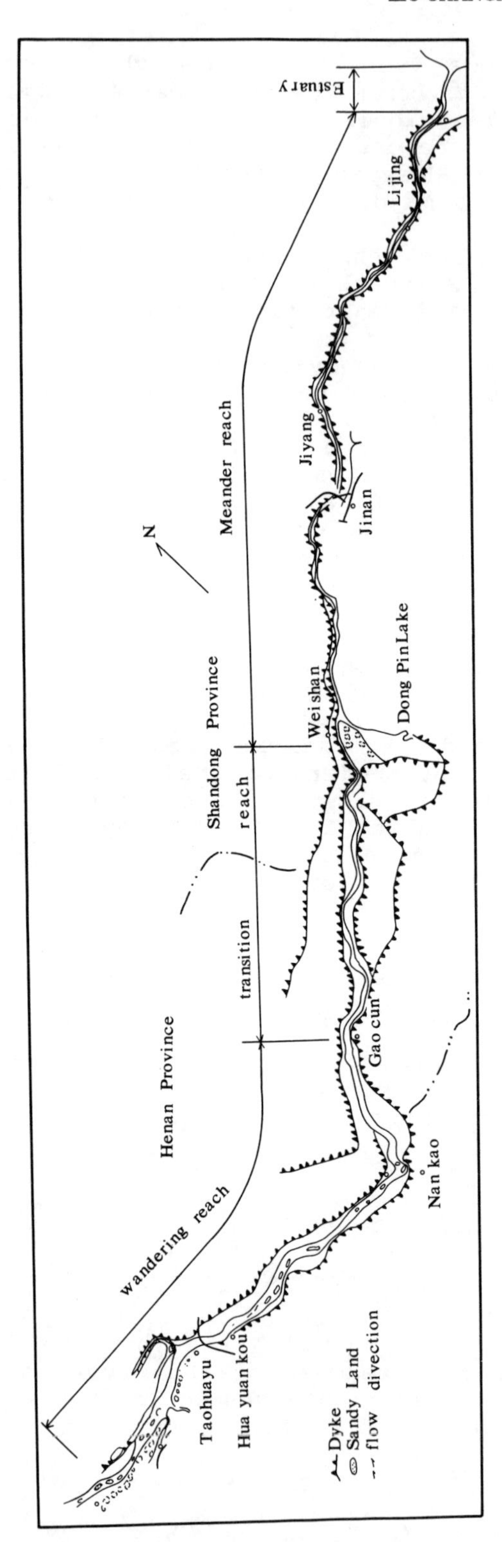

Figure 1. Plan of the Lower Yellow River.

other hand, the siltation would spread upstream owing to both the decline of flood peaks and decrease of sediment-carrying capacity. Consequently, sediment transport into the sea would also be cut down.

CHANNEL DRAINAGE CAPACITY OF FLOOD PEAK DISCHARGES

Recently, the main course of the river's lower reaches can be roughly divided into two major sections: the Henan and Shandong sections. The former is wide and the latter is narrow. So, in general, the channels are characterized by being wide upstream and narrow downstream (Fig. 1). Accordingly, the dikes designed for flood prevention are as follows: 22,000 cms at Huayuankou for the Henan section, and 10,000 cms for the Shandong section. In accordance with these flows the embankments have been heightened and strengthened. At present, the peak discharge with a 100 year reccurrence interval can be reduced to 20,000 cms by conjunctive operations of three existing reservoirs--Shanmenxia on the main course, Luhun on the Yihe River and Guxian on the Luohe River. Moreover, Dongping Lake, if used as a flood retardant area, can retard 2.0-2.58 billion cu.m of flood water with a retention capacity of 10,000 cms of flood-peak discharge. In addition, the Beijindi retardant area has a retention capacity of 2 billion cu. m of floodwater and 10,000 cms of peak discharge which can be used in case of emergency. The floodpeak discharge can be controlled at 10,000 cms downstream of Aishan in Shandong Province.

It must be pointed out that a high flood-peak discharge of 40,000 cms is expected to occur in the lower reaches and flooding will be uncontrolled with the existing engineering works. Many people have worried about an extreme flood that may take place in the near future that would lead to breaching of the Yellow River in its lower reaches. The estimated maximum probable loss caused by such a breach resulting in a big flood would be very high (see Table 2).

Table 2. Maximum Probable Loss due to the Low Yellow River's Breach Resulting from an Extreme Flood

Breach location	Affected area (sq.km)	Affected population (thousand)	Output value in Affected Areas: Agriculture (billion Yuan)	Output value in Affected Areas: Industrial (billion Yuan)
North Bank	75,080	34,879	10.88	26.75
South Bank	100,695	55,187	14.34	12.16

Source: Mr. Cao Quifa, Geo-information Systems Lab, Institute of Geography, Academia Sinica

Historic Review of the General Plan for Harnessing the River

As a natural hazard, flooding and overflowing of the river's lower reaches can be traced back to antiquity. According to Chinese legends, the Great Yu was a representative who conducted work on harnessing the Yellow River by floodwater diversion around the year 2000 B.C. The earliest record of a river course change found in historical documents was in 602 B.C. After the Xihan Dynasty, in about 200 B.C. a dike system along the lower reaches of the Yellow River was built for flood prevention. Since that time a number of changes in the course of the river have been recorded. For the past two thousand years the Yellow River's course in its lower reaches has changed between Tianjiu in the north and the estuary of the Huai He River in the south in the alluvial fan of the lower reaches of the Yellow, Huaihe, and Haihe Rivers. Sometimes the river emptied into the Bohai Sea and sometimes into the Huanghai Sea. Historical statistics show that two dike breaches every three years and one course change in a hundred years on average have resulted from floodings during the past two thousand years. Frequent flooding, dike breaches and course changes have resulted in successive disasters for the people in the river's lower reaches.

There were a number of outstanding persons who did successful work on harnessing the Yellow River. Great Yu (about 2,200 B.C.) paid attention mainly to drainage in terms of floodwater diversion. His tactics were to use the land gradient for water discharge by removing obstructions in the river channels and enlarging the outlet of the river with multi-bifurcations. In addition, he used depressions to store and to retard flood water. These measures were feasible and achieved success, so a story of the Great Yu's contribution, as a legend, was handed down from ancient times. After more than 4,000 years, however, the Great Yu's tactics against the Yellow River's flooding is now restricted by the dense population along both sides of the river course.

Jiarang (about 6 B.C.) developed three countermeasures to harness the river. He thought that the best way would be to discard the old dikes and to direct the river channel northward to empty into the sea. The second best way would be a floodwater diversion to scatter flows into depressions. He thought that the use of the old dikes to defend against floods would be an undesireable measure. Jiarang's idea of harnessing the river has persisted for more than two thousand years. But the best countermeasure that he suggested is also restricted by the large population.

Wangjing's (69 A.D.) proposal for harnessing the river was recognized by the Hanming Emperor in the Han Dynasty and the proposal could have been implemented by 100,000 people.

He dredged the waterways, removed the obstructions and carried out the renovation of about 500 km of dikes. In addition, he reconstructed the Bianqu canal and built up new sluice gates for the canals. After a year of implementation of his project, a fairly good result of controlling the river was obtained owing to the shortening of the waterway towards the sea and by increasing flood water drainage. Obviously, Wangjing's focus on embankment works coincided with the recent measure of strenghtening the dikes, which can be adopted to control the flooding for a certain period.

Panjixun (16th century) carried out work on harnessing the river four times during the period 1565-1592. This was one of the larger scale projects in the history of harnessing the Yellow River.

Panjixun was knowledgeable regarding the sediment content and the sediment-carrying capacity of the flows. He estimated that the Yellow River had a large amount of sediment with a small amount of water and a sediment concentration that would reach a very high value (60%). Therefore he reasoned that the limited water flow could not make all of the sediment move. He suggested a dike system for correcting and concentrating the river flows to carry the sediment. In order to increase the sediment-carrying capacity of the river flows Panjixun attached great importance to embankment works. On the basis of carefully summarizing the experience of embankment engineering he created the dike system for river training (Figure 2): the interior levees for controlling the channels (increasing sediment carrying capacity); the exterior dikes for prevention of big floods; the transverse and arched levees for strengthening the systems. In order to raise the silt carrying capacity of the Yellow River, Panjixum adopted a measure to divert clear water from the Huaihe River to the Yellow River for transporting the sediment. All of these measures enabled the river channels to approach stability.

Obviously, the concept of the inbalance between sediment and water was presented by Panjixum 400 years ago. Even now it still can be of theoretical importance for us to harness the Yellow River.

To summarize the historical experience in harnessing the Yellow River conducted by the ancient Chinese during last four thousand years, the following countermeasures were used:

1. Dredging the river.
2. Floodwater diversion and retention.
3. Establishment of a levee system.
4. Water-sediment regulation including increasing the silt-carrying capacity.

All of these came from practice and were tested by the ancient Chinese people.

In recent times Mr. Liyizhi, a distinguished scientist,

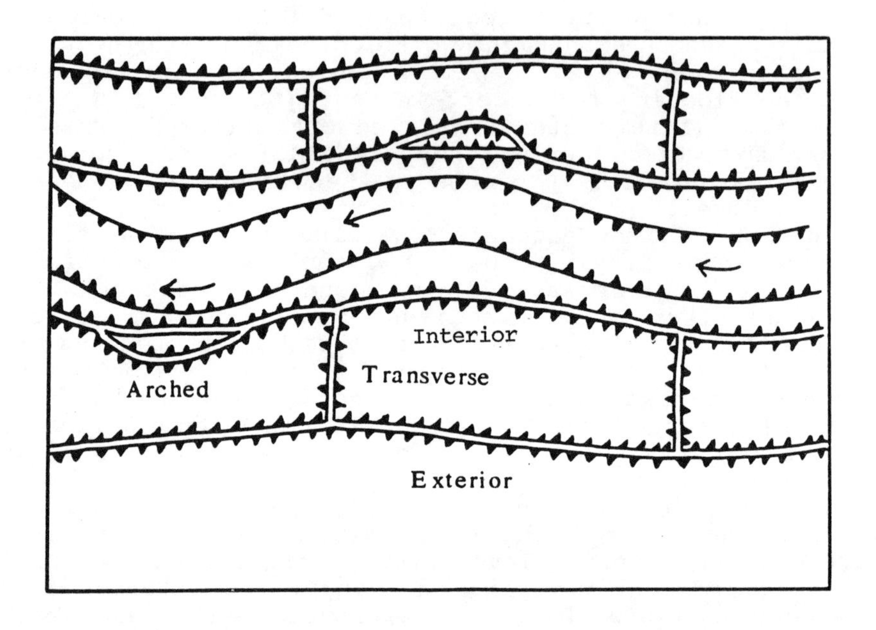

Figure 2. The Dike System for River Training Engineering, after Panjixun

(reprinted with permission from Regulated Rivers: Research and Management, 1989, John Wiley & Sons Ltd.)

did much work on harnessing the Yellow River. To carry forward the ideas suggested by the ancient Chinese people, Mr. Liyizhi made every effort to introduce modern science and technology from the Western countries into China including the measurement of floods and sediment, hydraulic engineering modelling and applied hydraulics. Meanwhile, some foreign experts in hydraulics came to help in the Yellow River's studies and encouraged a measure of afforestation in the upper-middle Yellow River.

From this we learned that the ancient Chinese scientists were concerned mainly with the problems in the lower reaches of the Yellow River. Foreigners suggested that soil erosion be controlled in the upper and middle reaches of the Yellow River. But Mr. Liyizhi stood for an overall plan for the upper, middle and lower reaches of the Yellow River.

For the upper and middle reaches he recommended:

1. Reducing sediment yields by afforestation.
2. Building flood-retention reservoirs to regulate peak flows.
3. Developing canals for irrigation associated with flood diversion.

For the lower reaches:

1. Digging drainage canals for reducing peak flows.
2. Training the river and forming stable channels for normal floodwater levels.

The above-mentioned suggestions have been of significance in planning and harnessing of the river.

An Analysis of the Countermeasures

WATER AND SOIL CONSERVATION

The erosion area is located in the middle reaches of the river. The soil erosion modulus averaged over a large area can reach 37,000 tons/sq. km. a year. A basic reason for flooding occurring in the river's lower reaches is channel siltation resulting from deposition of sediments produced and discharged by the river's middle reaches. Therefore, some people think that the basic way to harness the river should be water and soil conservation in its middle reaches. According to the analysis of sediment grain sizes, the deposited silt in the downstream course mainly consists of the coarse particles, larger than 0.05 mm. This implies that most fine particles smaller than 0.025 mm could be transported to the sea with the flows.

Survey data of sediment sources have shown that about 80%

of the coarse particles come from an area of about 100,000 sq. km in the middle reaches in which about 60% is in an area of 50,000 sq. km. Therefore, sediment control should emphasize countermeasures of water and soil conservation in the areas where the coarse particles of sediments originate.

A number of small watersheds such as Dali R. and Wuding River, etc. were treated by soil conservation works and sediment control was successful. The soil conservation works used on these watersheds comprised mainly terracing, afforestation, planting grasses, building sediment control dams and small-scale reservoirs.

On the basis of our study the following problems in evaluating the effects of the soil conservation works are worthy of examination

(1). The effects of sediment control on the small watersheds are obvious. For instance, a sharp high hydrograph of the uncontrolled small watershed can be changed essentially into a smooth and low one by these measures (Figure 3). But such effects are not identified on the large watersheds due to lack of large-scale implementation of soil conservation. So the problem is how to extrapolate water and soil conservation effects for large watersheds from the small watersheds.

(2). The small-watershed experimental data obtained shows that the soil conservation works can reduce more sediment yield than water yield (Figure 4) (Liu et al., 1985). But such results may differ on large watersheds because large watersheds have much more channel flow than the small watersheds (Figure 5). So it is not likely to be feasible to use small watershed data to predict exactly how sediment control affects the large watersheds.

(3). Our research has defined that the magnitude of soil conservation effects is a function of water input, namely excess rainfall. It varies with changes in direct runoff. According to an analysis of comparable and paired watersheds this function was determined; when water input is very low, the effect of treated watersheds is almost the same as that of untreated catchments. The effective difference of the pair will increase with greater water input. When the water input is very high, the differences will again decrease rapidly (Figure 6). Obviously, we cannot determine the soil conservation effects on an extreme flood by using the data of ordinary rainstorm events.

(4). Among soil conservation measures, afforestation has relatively high effects on floodpeak flows and on sediment yields.

Because of the uncertainty of soil conservation effects

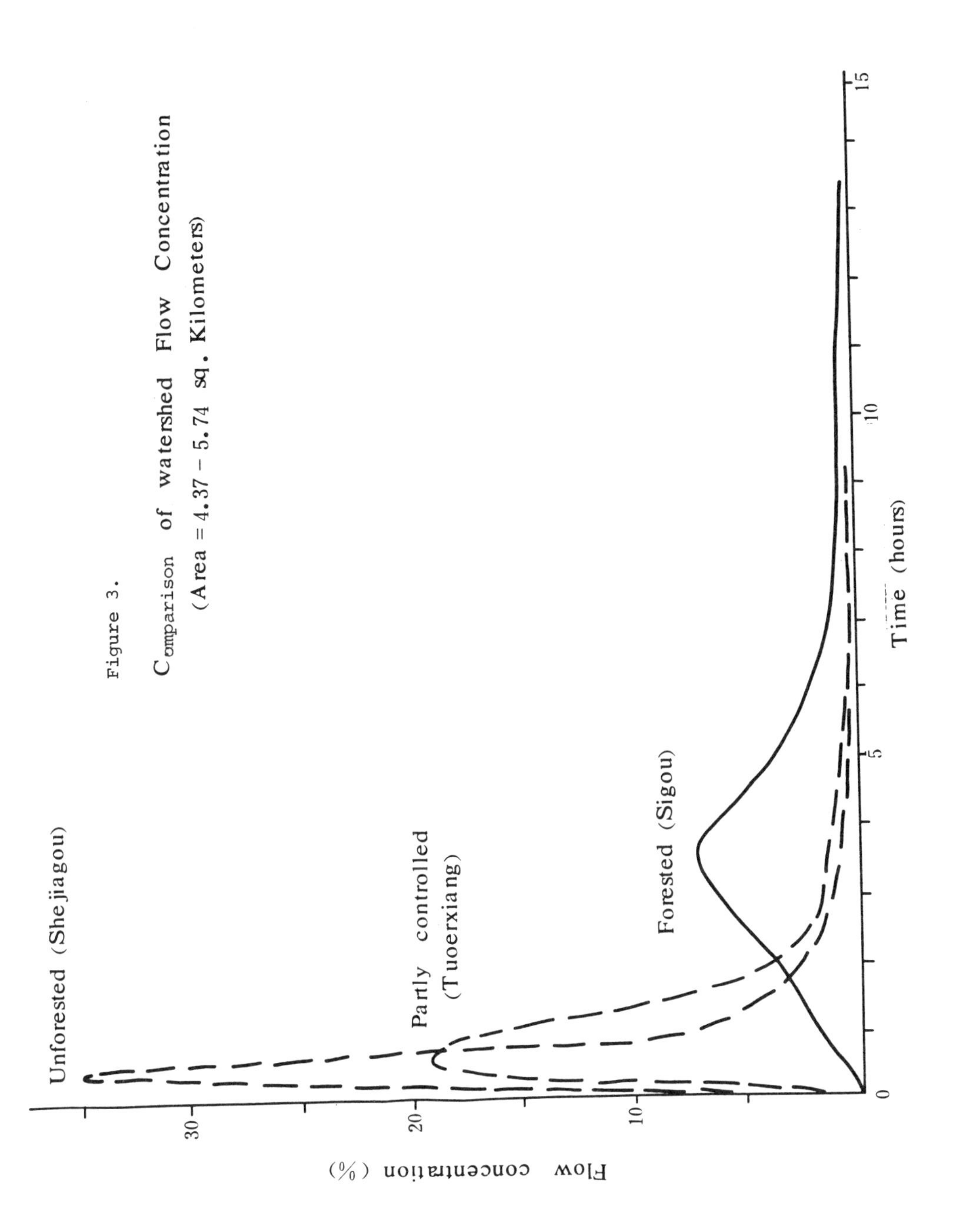

Figure 3.

Comparison of watershed Flow Concentration
(Area = 4.37 - 5.74 sq. Kilometers)

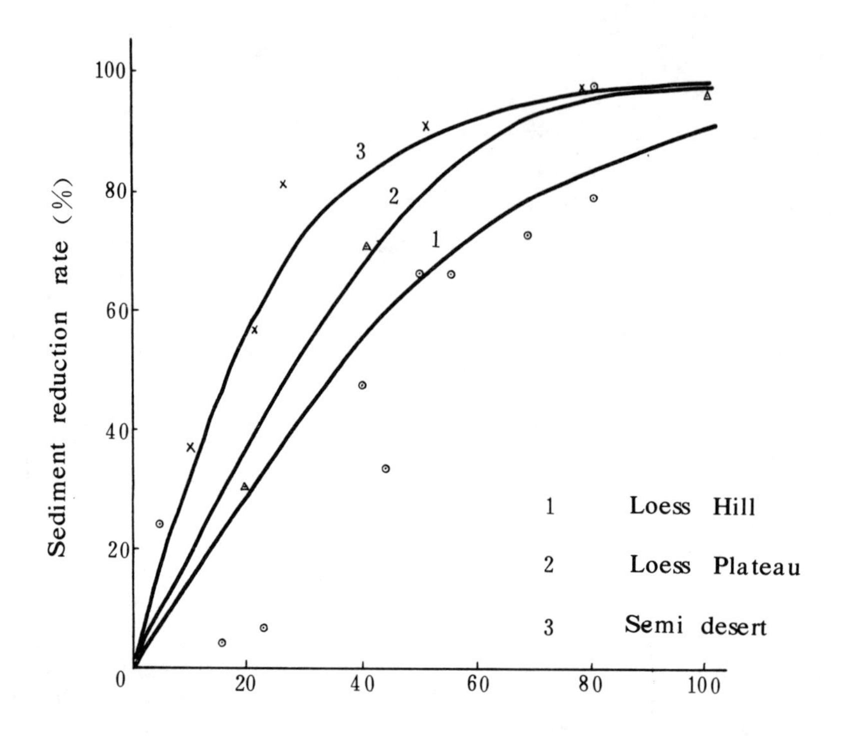

Figure 4. Soil Reduction Rate with Soil Conservation Control

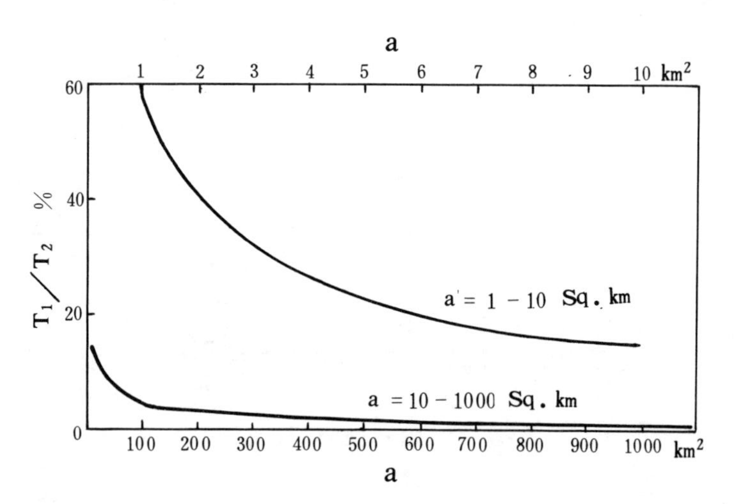

Figure 5. Comparison of Large Watersheds with Small Watersheds

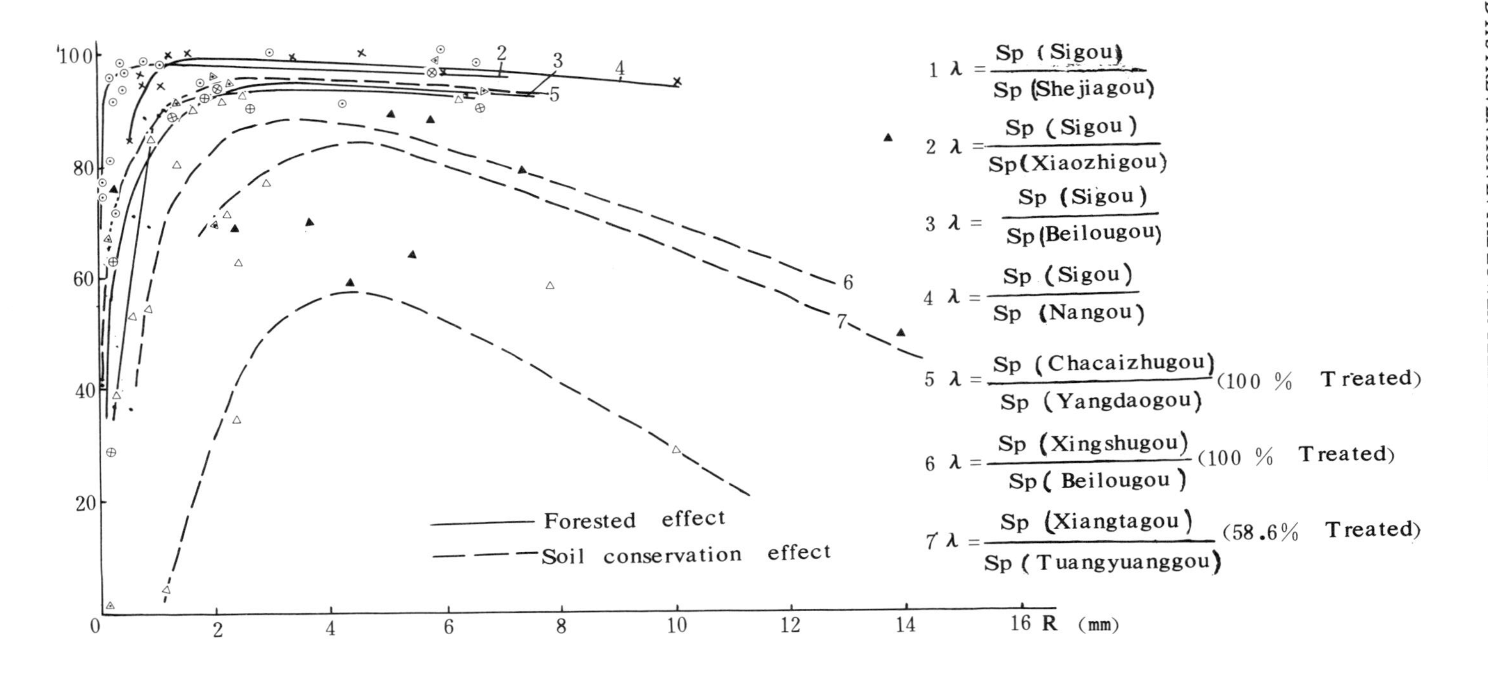

Figure 6. Reduction of Peak Discharge with Runoff

on large watersheds, some people doubt the success of such measures in the short term. Moreover, most areas of gully and valley slope erosion resulting from of gravity are unlikely to be controlled the by above-mentioned measures. Therefore, it would be advisable to consider soil conservation works in long-term planning.

FLOOD PREVENTION IN THE YELLOW RIVER'S LOWER REACHES

Flood prevention in the lower reaches is closely related to the sediment problems. Recently, the river's flooding was mainly controlled by dikes. Because of successive siltation the riverbed rises successively and the dikes have to be raised continuously for flood prevention. Following the founding of the People's Republic the dikes were heightened four times, in 1950, 1955, 1964 and 1976. Although the embankment works have achieved great success in flood prevention, the dikes cannot be heightened indefinitely. In such a case there are two countermeasures to be considered as alternatives:

Building a New Channel Instead of Using the Old Course. This countermeasure has been suggested by some scholars such as Mr. Zhou Dexiu (1979) and Mr. Ye Qingchao (1985). Geomorphologically, Prof. Ye Qingchao predicted that the channels of the Yellow River's lower reaches have evolved to old age and cannot be used much longer. From this point of view a careful study of the possibility of making a new channel was carried out by him. Although this measure is theoretically possible, the construction of a new channel would require resettlement of more than two million people who are living along the area which would be occupied by the new river course. Consequently, some people think that the implementation of this project is not feasible.

Maintaining the Status Quo. This idea assumes that the present river channels will be maintained for about 50-100 years. For this countermeasure, apparently, the sediment control in the channels held by the dikes has to be improved and channel dredging is needed to increase the capacity of sediment and flood discharges. On the basis of such an idea there are a number of measures that can be adopted.

1). Sediment and flood-peak reduction by using the large reservoirs. There are some large reservoirs including the Sanmenxia and the planned Longmen, Xiaolangdi and Tiaohuayu reservoirs. The reservoirs not only function as flood-peak discharge control but also have an effect, on the sediment-water relation by adjusting siltation and washing sediment downstream. The Sanmenxia reservoir's operation has provided the experience for such adjustment (Long Yuqian et al., 1979; He Guozhen, 1981; Qian Yiying, 1984).

2). Reduction of Sedimentation by Warping. There are five depressions suggested for warping, namely, the floodplains between Yumenkou and Tongguan, Wenxian-Mengxian in the middle reaches, QuanYantg-Fengqiu, Dongmingxian and Taiqianxian in the lower reaches. It is estimated that the total capacity for sediment amounts to 25 billion tons. A gross reduction of sediment in the lower reaches could be 10.5 billion tons with an annual sediment reduction rate of 290 million tons (Li Baoru, 1985). In addition, the warping also has other beneficial effects because it helps fertilize the soil in the warping area, elevates depressions in the floodplain and helps to strengthen the dikes.

3). Reduction of sediments by suction dredgers associated with water diversion. It estimated that there are about 200 suction dredgers available for use. Mr. Bao Xicheng (1980) expects that about 150 million tons of sediment per annum can be drawn by 500 suction dredges. This amount makes up 37.5% of average annual sediment deposition of the lower reaches.

4). Increase of silt-carrying capacity by transferring clear water from the Yangtze Valley into the Yellow River's main course. The so-called south-to-north water transfer schemes planned comprise three routes, i.e., the West Route, Middle Route and East Route. The West Route having three lines would transfer water from the Yangtze River's upper reaches. The other two routes would divert water from the Hanjiang River, a tributary of the Yangtze, directly from the main course of the Yangtze River's lower reaches northwards to the North China respectively. All of the schemes were proposed originally to supply water for urban, domestic, industrial and agricultural uses. Because the three routes would enter and cross the Yellow River, each transfer route would have a high probability of providing water for solving the problem of transporting sediment in the Yellow River. With respect to making adjustments for sediment, the Middle Route seems to be the best one (Liu Changming et al., 1983). The water from the West Route can be regulated by the existing reservoir cascades in the upper Yellow River and can also be used for carrying the sediment of the downstream channels. It is estimated that an increase of 20 billion cu.m of water will occur and will help to avoid siltation in the channel.

5). River Training. A series of engineering measures for river training have been studied by the hydraulic engineers of the Yellow River Conservancy Commission. Among the engineering works, the establishment of a stable channel at normal water stages is likely to be the principal measure providing a base on from which the river would be controlled effectively.

From the above-mentioned description we can see that there are many countermeasures for harnessing the Yellow River. Although different scholars have different opinions as to the tactics proposed, all of these countermeasures are likely to be coordinated according to their own effects on flood prevention. On the basis of such coordination an integrated effect will be obtained.

Some Suggestions

On the basis of the review of the issues of flood prevention we would like to propose some suggestions of countermeasures for discussion.

1. Different measures concerned with flood prevention of the Yellow River have different effects and such effects can be divided into two types, namely long term and short term. The former comprises water and soil conservation works and the construction of large reservoirs upstream of the river's middle reaches as well as water transfers. Obviously the implementation of these measures is a long-term measure and the effects on flood prevention will be felt in the distant future. The latter consist of strengthening and heightening of dikes, water and sediment regulation by the existing engineering projects, river training works, flood diversion and retention, dredging and warping. These measures are relatively easy to implement and will come quickly into effect. But the effects of such measures are uncertain in the long-term period and the engineering works have to be carried on constantly. In such cases, the short-term and the long-term countermeasures should be planned in a unified way.

2. The problems of flooding and sediment in the lower reaches of the Yellow River are related to many influencing factors, which mainly comprise the natural conditions and social and economic foundation. Therefore, harnessing the Yellow River is so complex an issue that a wide field of problems must be considered thoroughly and systematically. In view of such complexity any single measure of the those mentioned above seems to be limited in its capability of solving the problems thoroughly, although it may have great effectiveness in harnessing the river. From this point of view, the authors of this paper would like to stress that all the countermeasures available and acceptable for harnessing the river should be combined and coordinated with each other. In general, the ultimate solution of the problems will be reached in a integrated way.

3. Based upon careful and serious identification of the various measures desirable for the Yellow River, an integrated solution should be worked out by the systems approach. From this point of view a descriptive evaluative model can be suggested as follows:

(a). The objective function

$$\text{Min } Z = \sum_t \frac{1}{(1+a)^t} \sum_i C_{i,t} M_{i,t} \qquad \forall\, i,t$$

where Z is objective function, namely, minimizing of costs resulted from the measures; $M_{i,t}$ is reduction of flood loss for the measure i at the time of t. $C_{i,t}$ is the unit cost of the measure i at time of t; a is the interest rate.

(b). The constraints

1) water balance

$$W'_t = W_t - W_t^T \; M_t$$

$$M_t = \sum_{i=1}^{n} M_{i,t} = (M_{1,t} + M_{2,t} + \ldots + M_{n,t})$$

$$W_t^T = (W_{1t}, W_{2t}, \ldots, W_{nt})$$

$$Q'_t = F\,(Q_t, W'_t)$$

where: W'_t is flood reduction volume depended on flood runoff (W_t) and on flood reduction effect resulted from the measure M_t; Q'_t is the decreased flood flow being a function of Q_t and W_t

2) sediment balance

$$S'_t = S_t - S^T_t \; M_t$$

$$S^T_t = (S_{1t}, S_{2t}, \ldots, S_{nt})$$

$$B_t = F(S'_t, Q'_t)$$

where: S'_t and B_t are sediment reductions and index of channel maintaining against siltation.

3) sediment transport capacity

$$\sum_{i,t} S_{i,t} \leq TS$$

$$TS = F(Q'_t, B_t)$$

where: TS is channel silt-carrying capacity, a function of Q'_t and B_t.

4) discharge capacity of the channel

$$\sum_{i,t} q(M_{i,t}) \leq q\ max$$

where: $q(M_{i,t})$ is discharge capacity of the channel; q max is the designed flood peak flow reduced by the measures.

5) values of the variables should be positive

$$M_{i,t} > 0 \qquad \forall_{i,t}$$

The above-mentioned descriptive, evaluative model is formulated on the basis of comprehensive consideration in solving the problems of flooding. The model presented here is simply constructed and some parameters may not be involved in it. Mathematically, the model seems to be solvable. The difficulty, however, will be in the determination of the exact parameters. Therefore, a careful research on the parameters must be carried out.

Conclusion

There are many countermeasures possible for flood prevention in the Yellow River's lower reaches. Most measures seem to be relevant to the problems as identified because of effects of each on flood control as observed in practice. However, the problems of flooding are affected by the influence of many factors. Therefore a single measure may not satisfy the overall solution of such complicated issues. In such a case an integrated approach is likely to be advisable. To meet this challenge, systems analysis appears to be the best way to decision making process to the overall harnessing of the Yellow River.

References

Baoru Li, A comparison of sediment reduction of various measures in the lower reaches of the Yellow River, Study and Practices on the Yellow River, Water Conservancy Press, 1984.

Changming Liu, et al., The effects of forests on water and soil conservation in the Loess Plateau of China, China Geographer, No. 12: Environment, W.Pannell et al. (ed), Westview Press, 1985.

Changming Liu, et al., Interbasin water transfer in China, Geographical Review, Vol. 73, No. 3, 1983, American Geographical Society.

Guozhen He et al., Water and sediment regulation by the Sanmenxia reservoir and the characteristics of sedimentation, The People's Huanghe, No. 6, 1984, The Yellow River Conservancy Commission Press (YRCCP), Zhengzhou.

Qingchao Ye, The characterestics and tendency of sedimentation in the lower reaches of the Huanghe River, Management and Development of the Huang-Huai-Hai Plain, Vol. 1, Science Press, 1985.

Xicheng Bao, Strengthening of dikes by warping, Flood Prevention on the lower reaches of the Yellow River. 1980, YRCC press, Zhengzhou.

Yiying Qian, Channel regulation by the Sanmenxia reservoir's operation in downstream reaches of the Yellow River, The People's Huanghe, No. 6, 1984 Yellow River Conservancy Commission Press, Zhengzhou.

Yuqian Long et al., Reform and operation of the Sanmenxia project, The People's Huanghe, No. 3, 1979, Yellow River Conservancy Press, Zhengzhou.

SIMULATION OF FLOOD PROPAGATION IN CHANNEL OF THE YELLOW RIVER BY CONVECTION-DIFFUSION EQUATION

Liang Jiyang and Liu Changming
Academy Sinica
Institute of Geography
State Planning Commission
People's Republic of China

The lower Yellow River bed is higher than the ground surface on both banks. It is known as the famous suspended river. Because of this, every breach during flooding causes more disasters than any other river in China. Flood threats to the lower Yellow River come from storms in the middle reaches. The main storm regions are located between Shanxi Province and Shaanxi Province in the basins of its major tributaries of Jin, Wei, Lou, and around the main course from Sannmenxia and Huayuankou, including the Yi, Lou, and Qin tributaries.

Because of the deposition of sediment, channel capacity in the lower reach of the Yellow River has been reduced year by year.

The capacity of the channel is lower than the peak discharge of 22,300 cu.m/sec. which occurred in 1958. It is very important to make accurate real-time flood forecasts for dike protection and for decisions regarding the operation of flood diversions and flood detention.

The middle reaches of the Yellow River lie in the Loess Plateau, covered by erosive and loose material. Erosion takes place during storms and sediment is moved to the main channel by surface flows.

The Yellow River flows have a very high silt-carrying capacity during floods; sometimes the concentration can be as high as 600-1,000 kg/cu.m.

The propagation process of the flood wave in the channel can be simulated by the continuity equation and the momentum equation. According to the boundary and initial conditions, the process of flood movement can be obtained using a numerical method. It is necessary to add a sediment continuity equation when flood processes of the Yellow River are simulated. Due to hyperconcentrated, unsteady flows in the channel, the riverbed will be changed greatly and more detailed morphological data of the channel are needed. In addition, the complexity results from deriving the sediment

L. M. Brush et al. (eds.), Taming the Yellow River: Silt and Floods, 243–256.

continuity equation so an attempt was made to use the convection-diffusion equation for simulating flood wave propagation in the channel of the Yellow River. The convection-diffusion equation is regarded as a simplified form of continuity and momentum equations.

The convection equation describes the wave propagated by time but the shape of the wave does not change. The diffusion equation describes how the wave is attenuated with time, but it does not move. A combination of both equations can describe the process of wave advancement and attenuation.

This paper uses the convection-diffusion equation to simulate flow movement in the channel of the Yellow River.

An implicit scheme and the Crank-Nicolson differing scheme are used for flood routing. In general, it is necessary to calibrate by using a physical model analogue. Similarly, the parameters of the mathematical model should be determined when using mathematical models to simulate real processes. In this paper, an optimization technique is used to ensure an optimal solution.

Two reaches were chosen for study. The Huayuankou-Jiahentan reach has a length of 106 km. The flood process is a mono-input and mono-output system. The Xiaolangdi-Huayuankou reach has a length of 128 km. Two large tributaries, the Yilou He and the Qin He, converge into the reach. A schematic map is given in Figure 1. The flood processes are a multi-input and mono-output system.

Numerical Method of Convection-Diffusion Simulation of the Channel Flows

MATHEMATICAL MODEL

The principles which describe energy conservation and conservation of matter of the flow in the channel are based on the flow momentum and continuity equations. To simplify these equations, we write the convection-diffusion equation as follows:

$$\frac{\partial Q}{\partial t} + C \frac{\partial Q}{\partial x} = D \frac{\partial^2 Q}{\partial x^2} + C q \tag{1}$$

in which Q = the flow discharge, t = time, x = the horizontal distance along the channel, and q = lateral inflow.

Equation (1) is a parabolic partial differential equation.

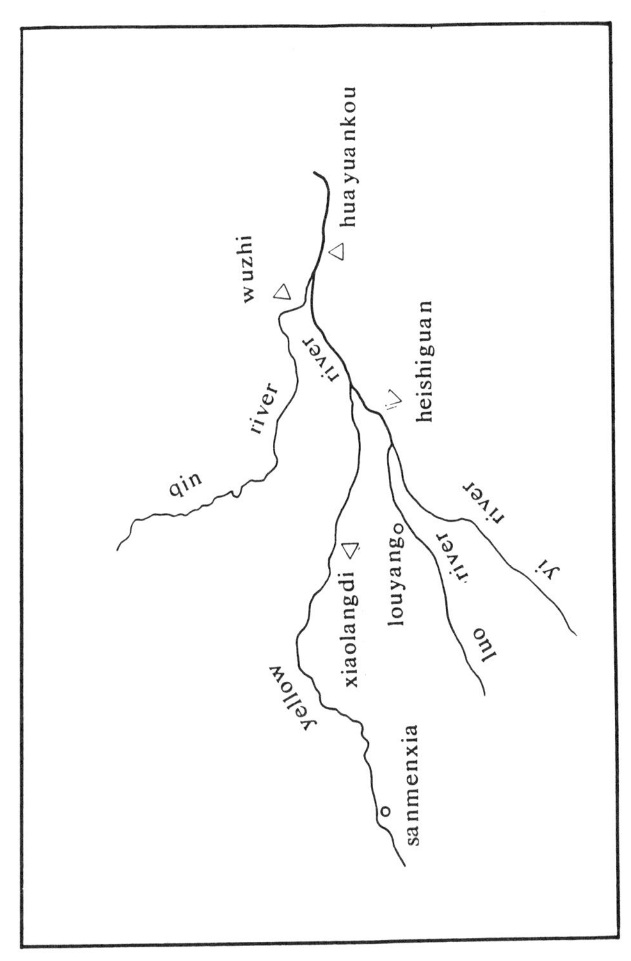

Fig.1 Sketch map of simulation reach

when

$$C\ (x,t) > 0$$

$$D\ (x,t) > 0$$

in the region $R\ (x \geq 0,\ t \geq 0)$

Initial condition:

$$Q\ (x,0) = f\ (x) \tag{2}$$

$$L \geq x \geq 0$$

Boundary condition:

for the mono-input and mono-output systems,

$$Q\ (0\ ,t) = F\ (t) \tag{3}$$

$$t \geq 0$$

for the multi-input and mono-output systems as Xiaolangdi-Huayuankou reach,

$$Q\ (x_x,\ t) = F_x\ (t)$$

$$Q\ (x_h,\ t) = F_h\ (t) \tag{4}$$

$$Q\ (x_w\ ,t) = F_w\ (t)$$

$$t \geq 0$$

in which x_x, x_h, x_w are the distances from the hydrologic station Xiaolangdi (on the mainstream), Heishiguan (on the tributary Yi Lou He) and Wuzhi (on the tributary Qin He) to outlet cross sections, respectively.

Eq. (2), (3) or (2), (4) compose first boundary value problem of the partial differential equation.

DISCRETIZATION OF THE EQUATION

The length of the reach is L. To divide the length into N1 sub-reaches, the length of the sub-reach $\Delta x = L / N1$.

$$X_k = k\,\Delta x, \qquad k = 0,1,\ldots,N1 \tag{5}$$

The duration of flood is T. To divide the time into N2 interval, and the time step $t = T / N2$.

$$T_j = j\,\Delta t \qquad j = 0,1,\ldots,N2 \tag{6}$$

The difference quotient formulas are introduced as follows:

$$Q'_x(k, J+\theta) = \frac{1}{2\Delta x}\left\{Q(K+1, j+\theta) - Q(K-1, j+\theta)\right\} + O(\Delta x^2) \tag{7}$$

$$Q''_{xx}(k, J+\theta) = \frac{1}{\Delta x^2}\left\{Q(K+1, j+\theta) - 2Q(k, j+\theta) + Q(K-1, j+\theta)\right\} + O(\Delta x^2) \tag{8}$$

$$Q'_t(k, j+\theta) = \frac{1}{\Delta t}\left\{Q(k, j+1) - Q(K, j)\right\} - \frac{(1-2\theta)\Delta t}{2!} Q'_{t^2}(k, j+\theta) + O(\Delta t^2) \tag{9}$$

By making some simplifications and ignoring the error of cutoff, the difference equation corresponding to Eq. (1) in point $(k, j+\theta)$ can be written

$$\frac{Q_{k,j+1} - Q_{k,j}}{\Delta t} + \frac{C}{2\Delta x}(Q_{k+1,j+\theta} - Q_{k-1,j+\theta}) - \frac{D}{\Delta x^2}(Q_{k+1, j+\theta} - 2Q_{k, J+\theta} + Q_{k-1, j+\theta}) = C\, q_{k, j+\theta} \tag{10}$$

Eq. (10) is convection-diffusion equation discretized. The error of cutoff is ignored while using the difference equation instead of the differential equation, so the solution of the difference equation must be shown to converge.

DIFFERENCE EQUATION

Implicit Scheme

let $\gamma = \Delta t/\Delta x^2$, $Q_{k,\,j+\theta} = \theta\, Q_{k,\,j+1} + (1+\theta)\, Q_{k,\,j}$

$\theta = 1$, Eq. (10) becomes

$$-\gamma\left(\frac{C\Delta x}{2} + D\right) Q_{k-1,j+1} + (1+2D\gamma)\, Q_{k,j+1} + \left(\frac{C\Delta x}{2} - D\right) Q_{k+1,j+1} = Q_{k,j} + C\Delta t\, q_{k,J+1} \qquad (11)$$

let $j = 0$, then

$$-\gamma\left(\frac{C\Delta x}{2} + D\right) Q_{k-1,1} + (1 + 2D\gamma)\, Q_{k,\,1} + \gamma\left(\frac{C\Delta x}{2} - D\right) Q_{k+1,\,1} = Q_{k,0} + C\Delta t\, q_{k,1} \qquad (12)$$

$$k = 0, 1, \ldots, n-1$$

These are the systems of linear algebraical equations with unknown $Q_{1,1}$ $Q_{2,1}$, $Q_{n-1,1}$

When discretization initial conditions and boundary conditions are given, the solution of Eq.(12) would get the discharge at every point in the reach on the time layer j=1.

Taking the value of time j solved as the initial condition time by time, the discharge of time layer j+1 can be solved by Eq. (12) layer by layer. In order to get discharge process from time 1 to time T, we have to solve number $T/\Delta t$ groups linear algebraical equations.

It is well known that the implicit scheme is stable and convergent, so that it will be a practical calculation scheme.

According to the implicit scheme and direct calculation of parameters, which will be described in the ensuing paragraph, the authors predicated the flood processes of July, 1973, for the Lower Yellow River. The runoff measured at the upstream three stations (Xiaolangdi on the river's main course, Heishiguan on the tributary Yi-Lou He south of the main course, and Wuzhi on the tributary Qin He north of the main course) was employed as the input to the system and the downstream hydrography at Huayuankou was predicted as the output of the system. Fig. 2 shows the predicted hydrograph compared with the observed hydrograph.

Crank-Nicolson Difference Scheme. There is always a problem of approximation when the difference equation is used instead of the differential equation. When explicit schemes and implicit schemes are used, $t \to 0$, $x \to 0$, cutting error for differential equation is replaced differential equation is $O(\Delta t + \Delta x^2)$. However, by using the Crank-Nicolson difference scheme to construct the difference equation, the cutting error will be $O(\Delta t^2 + \Delta x^2)$, and it is steady and convergent without preconditions. So it is a better difference scheme.

let $\theta = 1/2$, Eq. (10) becomes

$$- \gamma \left(C \frac{\Delta x}{2} + D\right) Q_{k-1,\, j+1} + 2\,(1 + \gamma D) Q_{k,\, j+1}$$

$$+ \gamma \left(C \frac{\Delta x}{2} - D\right) Q_{k+1,J+1} =$$

$$\gamma \left(\frac{C \Delta x}{2} + D\right) Q_{k-1,\, j} + 2\,(1 + \gamma D)\, Q_{k,j}$$

$$- \gamma \left(\frac{C \Delta x}{2} - D\right) Q_{k+1,j} + C \Delta t (q_{k,j+1} + q_{k,j}) \quad (13)$$

When Q at the j time layer have been given, for getting Q at j+1 time layer we have to solve the linear algebraical equations system. The difference with the implicit scheme is the right side vector consisting of values of three points in the network. If initial and boundary conditions have been given, we can get full process of the every time layers $Q = f(x,t)$.

Using Crank-Nicolson difference scheme, and the flood process at Huanyuankou station of September, 2-19, 1981 as input of system, the lateral input is zero. The output of the system can be calculated as a hydrograph at Jiahetan downstream. Fig. 3 shows the comparison between observed and simulated hydrographs.

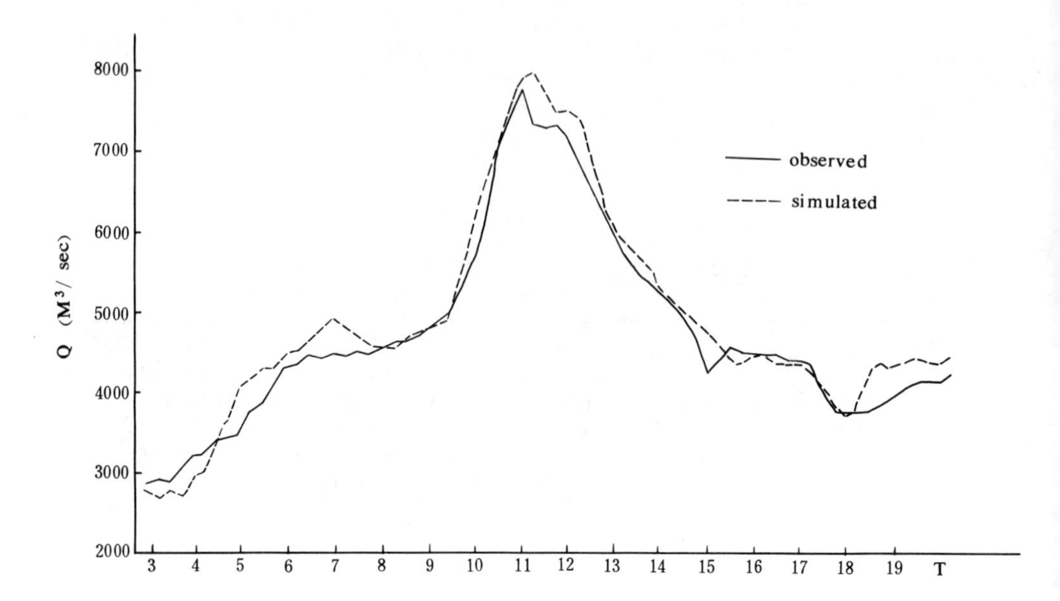

Fig.3 Comparison between observed and simulated hydrograph on Jiahetan station (Sept., 1981)

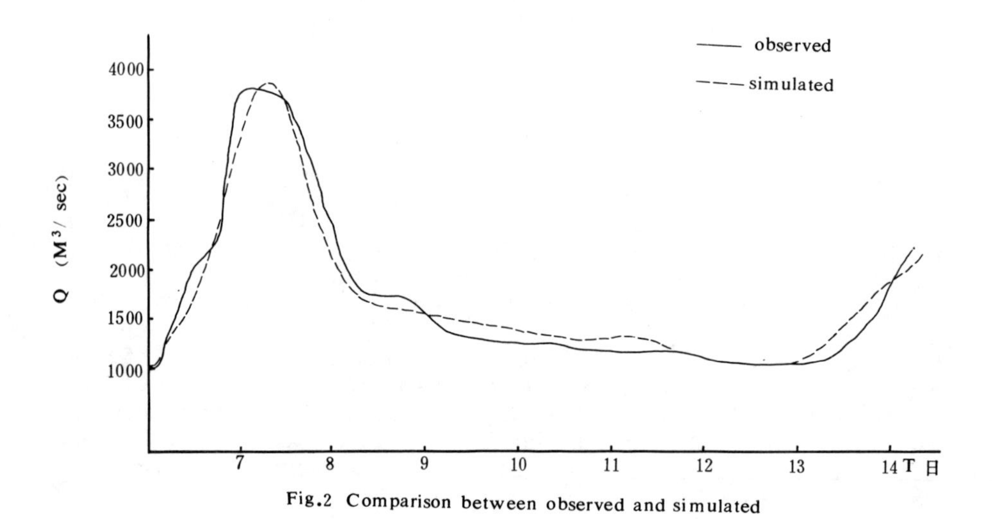

Fig.2 Comparison between observed and simulated hydrograph on Huayuankou station (July, 1973)

SOLUTION OF A SYSTEM OF LINEAR ALGEBRAIC EQUATIONS

Difference equation is obtained after discretizing the differential equation, then by applying specific difference scheme, a system of the linear algebraical equations is obtained. The problem of solving the differential equation becomes the problem of solving the algebraical equations.

As mentioned above, the linear algebraical equations derived from implicit scheme and Crank-Nicolson difference scheme their elements of which aren't zero in coefficient matrix concentrate in three diagonal. The form of matrix and vector are:

$$\begin{vmatrix} b_1 c_1 & \cdots & & \\ a_2 b_2 c_2 & \cdots & & \\ & \vdots & & \\ & & & a_n b_n \end{vmatrix} \begin{vmatrix} Q_{1,j+1} \\ Q_{2,j+1} \\ \vdots \\ Q_{n,j+1} \end{vmatrix} = \begin{vmatrix} Q_{1,j} + c\Delta t\, q_{1,j+1} \\ Q_{2,j} + c\Delta t\, q_{2,j+1} \\ \vdots \\ Q_{n,j} + c\Delta t\, q_{n,j+1} \end{vmatrix}$$

in the above matrix

$$a_i = -\gamma\,(C\,\Delta x/2 + D)$$

$$b_i = (1 + 2D\gamma)$$

$$c_i = \gamma\,(C\,\Delta x/2 - D)$$

$$i = 1, 2, \ldots, n$$

It is known from Eq. (11), (12), (13) that input of the tributary one step earlier than input of upstream main course is required.

The right side vector and b_i from Crank-Nicolson difference scheme have a little difference.

The linear algebraic equation in which the coefficient matrix is three diagonal can be solved by a special form of Gaussian elimination.

OPTIMIZATION OF MODELING PARAMETERS

Calculation of Model Parameters. Convection-diffusion equation simulates the movement process of the natural flood wave in the channel, the parameter C describes the character of flood wave transfer velocity toward the downstream, and parameter D describes the character of flood wave attenuation.

Only if these two parameters respond to the character of the channel, for instance, natural hydraulic conditions such as width, slope, roughness of channel and overbank flow etc., is the simulation result obtained fairly good.

Hayami suggested the diffusion simulation of second order systems. The parameters were determined by formulas as follows:

$$C = \frac{1}{B}\frac{dQ}{dy} \approx \frac{3}{2} V$$

$$D = Q/2BS_o \tag{15}$$

This formula should be used for prismatic channels.

Price suggested half-experimental formulas to determine variable parameters in natural channel as follows:

$$\bar{C} = C + Q^* \frac{d}{dQ}\left(\frac{L}{T_p}\right)$$

$$D = \alpha \frac{Q_p}{L}$$

$$\alpha = \frac{1}{2B_o} \frac{\sum (Li/S_{o\,i}^{2})}{[1/L \ \Sigma (L_i/S_o^{1/3}{}_i)]} \tag{16}$$

The method of determining parameters by Eq. (15) is a little simplified. It is difficult to accord with real hydraulic conditions, and Eq. (16) requires too much knowledge of the morphometry of the channel. So, it is necessary to find a better method to determine the parameters.

Rosenbrock Method to Determine Parameters of Model. In recent years, optimization techniques have been widely used. They are powerful tools in system analysis. In the analysis of hydrological systems, Rosenbrock's method is applied preferably. It actually is a step and direction double acceleration method based on optimization of one dimension, which will find out the optimal point on one straight line for every search, then determine next direction according to special rules. The steps are as follows:

a. Given the initial value point 1 of the parameters, as the straight line which passes this point and parallels the horizontal axis optimizes parameters, get a point 2.

b. Optimizing parameters point 3 at the straight line of which passes point 2 and parallels vertical axis, supposing this point number is N.

c. Optimizing parameters point 2N at the straight line of which passes point number 2N-1 and 2N-3; optimizing parameters point 2N+1 at the straight line of which passes point number 2N and vertical to the straight line which connected point number 2N-1 and 2N-3.

d. Repeat above steps, iterate stop by stop and let the objective function reduce gradually until it reaches prevenient of demand or the standard to stop run. Fig. 4 shows a parameter optimization process which simulates floods for the reach from Xiaolangdi to Huanyuankou on the Yellow River. The optimal point is reached after searching on 9 straight lines.

Selection of Objective Function Types. In this paper, the least-squares principle was used in optimal parameters of the model.

$$F = \sum_{i=1}^{N} (q_i - \hat{q}_i) = \min \tag{17}$$

in which,

F -- objective function

q , $\hat{q}$ -- the values of observed and calculated

N -- the number of data

the objective function is sensitive for parameter variance in diffusion simulation of channel flood. Fig. 5 shows the distribution of the objective function, which simulated flood propagation in the reach from Huayuankou to Jiahetan of the Yellow River in September, 1981.

(figure 5 here)

As a standard in evaluatiing precision, the objective

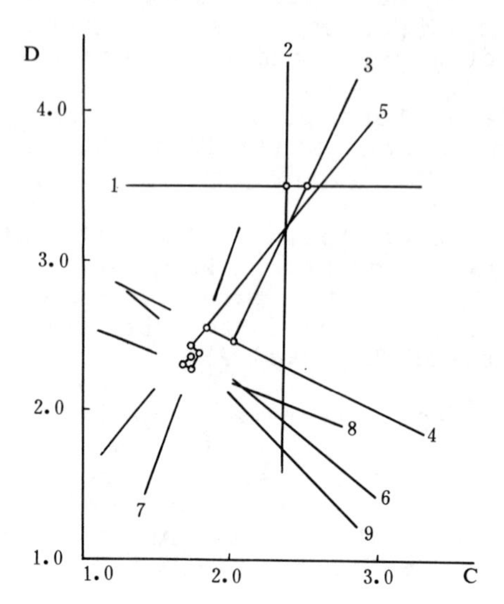

Fig.4 Sketch map of searching process for optimal parameters

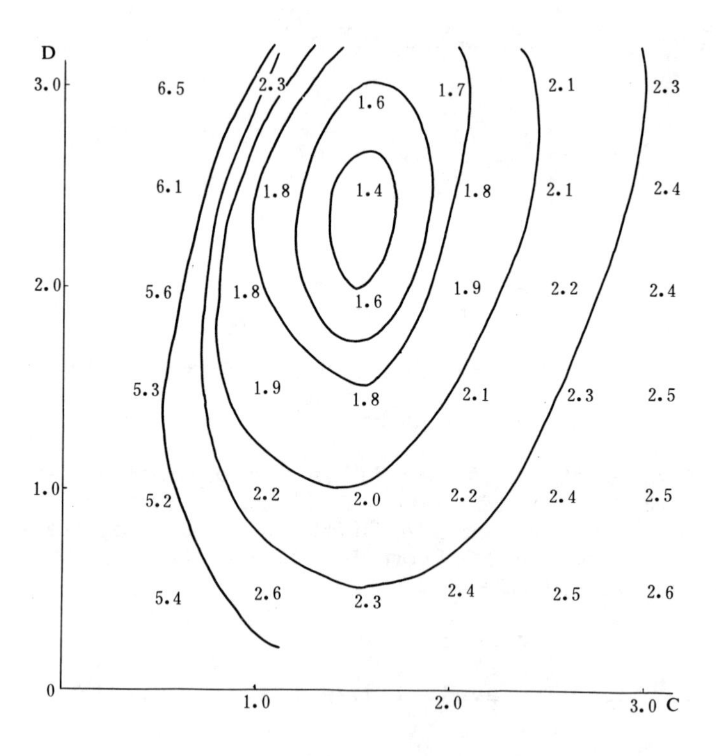

Fig.5 Distribution map of objective function

function has various forms. Kutchment and Koren have suggested the integral form of the absolute value error, Nash and Sutclif have suggested a standard of dimensionless objective function.

The selection of an evaluation standard often depends on the purpose. If the research puts emphasis on water volume, the objective function for a volume form should be adopted. In some cases, a logarithmic form is applied to avoid the inconvenience of large magnitudes in the calculating process. Sometimes, for considering the difference for a series of points, for example, the emphasis is on flood peak, or a few great values of series in frequency analysis, Eq. (17) can be modified by adding weighting factors.

Different forms of objective function may influence the predicted value of the model parameters, and different methods of optimization technique do not have too much influence upon parameter value itself. However, there are considerable differences in the convergent speeds and degree of approximation.

This paper took two reaches of the Yellow River as examples to show the process of unsteady flow in the Yellow River channel by using the convection-diffusion equation.

To a discretized convection equation, we can get the flood routing formula of Muskingum, so the Muskingum method is a special form of solution of the convection-diffusion equation. But the convection-diffusion equation has a flexibility applied to flood routing in the channel.

It will have a great help for flood routing in terms of understanding the influence of sediment on flood wave propagation. Since the channel bed of the Yellow River changes intensely during floods and the sediment content is much higher than other rivers of the world, it will be necessary to add sediment continuity equation in addition to the continuity and momentum equations to achieve a better result.

Because there is so much detailed morphological data of the channel in the lower Yellow River, and there are not any tributaries that come into the course for a long distance, it provides a good condition to analyze unsteady flow in the channel.

References

(1) Feng Kang, Method of Numerical Calculation, National Defense Industrial Press, 1978.

(2) Liang Jiyang, System analysis of flood wave, in Channel Yellow River, 1985, 5.

(3) Ali Osman Akan, Ben Chie Yen, Diffusion - Wave Flood

Routing in Channel Network, J. of Hydraulics Division, ASCE, June, 1981.

(4) Price R. K., Comparison of Four Numerical Methods for Flood Routing, J. of Hydraulics Division, ASCE, July, 1974.

(5) Fleming George, Computer Simulation Techniques in Hydrology, New York, Elsevier, 1975.

A GENERAL SURVEY OF PROBLEMS IN FLOOD STUDY ON THE LOWER REACH OF THE YELLOW RIVER

Hua Shi Qian
Professor and Chairman of the Academic Committee
Nanjing Research Institute
of Hydrology and Water Resources
The People's Republic of China

The Yellow River downstream of Huayuankou hydrometric station is the lower reach which runs across the North China Plain over a total length of 785 km to the sea. Owing to channel accretion over a long period of time, the riverbed is higher than the ground surface on both banks, thus forming a "Suspended River" which depends on dikes to contain the water. The river is actually a watershed of the Haihe River. There are no large tributaries feeding the Yellow River except a small tributary of Wenhe River on the lower reach. Major floods on the lower reach mainly come from three intervening areas. These are:

(1) the catchment for the reach between Tuoketuo and Longmen, with an area of 111,691 sq.km;

(2) the reach between Longmen and Sanmenxia, including those tributaries Jinhe, Luohe, Weihe etc., with a watershed area of 161,671 sq.km;

(3) the drainage basin of main river between Sanmenxia and Huayuankou, including those tributaries of the Yihe, Luohe and Qinhe Rivers, covering an area of 41,687 sq.km.

The storms always occur in the areas (1) and (2) mentioned above. These areas simultaneously gave rise to major floods in 1843 and 1933. They are characterized not only by high peak flows, but also by heavier concentrations of sediment due to severe soil erosion on the Loess Plateau in the middle reaches which brings channel shifts on lower reaches during the flood period by silting or scouring of the riverbed. Area (3) is situated in another storm area which also yields major floods such as those in 1761 and 1958. During flooding little time is available for carrying out flood prevention work after

L. M. Brush et al. (eds.), Taming the Yellow River: Silt and Floods, 257–260.

getting the flood forecasts, and this constitutes a serious threat to the lower reaches. The regions above Tuoketuo contribute a discharge of 2000-3000 cu.m/s during this period and constitute the base flow of the flood downstream. Because floods contain a large amount of sediment, studies of the Yellow River are more complicated than other rivers. Although the Yellow River Conservancy Commission has done a great deal of work on this subject, some problems need to be solved to prevent floods on the lower reaches.

The following problems need further study.

The first problem is to research the compound probability of a flood from Sanmenxia meeting with floods between Sanmenxia and Huayuankou. From the record and the historical literature for the past thousand years, the floods from (1) and (2) and from (3) do not generally meet simultaneously, but the runoff from the area above Sanmenxia comprises over 30-50% of the total volume flowing through the gaging station at Huayuankou.

But, as is known, the floods from the area above Sanmenxia can be retained by a reservoir with a flood capacity to retain about 6 billions cu.m, while the floods from the intervening area can be controlled by the Luhun reservior on the tributary of the Yihe River and Guxian reservoir on the Luohe River. However they can only control 21% of the intervening area, and have a total capacity for flood retention of only about 1.1 billion cu.m. Therefore, different compositions of natural floods will have different flood peaks and volumes in one day, three days, five days and twelve for a certain frequency of design flood at Huayuankou after the regulation by reserviors. In the same manner, computations have been made of hydrographs of different frequencies such as 0.1%, 1% and 2% etc. at Huayuankou. Hydrographs of different frequencies at hydrometric stations downstream such as Gaocun, Sunkou, Luokuo and Lijin can be determined simply by the method of flood routing. These are the data necessary to decide whether to open retarding basins during extraodinary floods. Some say that deposition of silt in the backwater area of Sanmenxia reservoir aggravates the problem, but I think these deposits can be dug out after floods as was done in the lower reaches of the Columbia River where mud and debris flows due to the explosion of St. Helen's volcano in the United States in 1980 were deposited. Of course, this may cost a large amount of money, but it may be worth it to prevent flood damages on the lower reaches. Moreover, the annual cost is not high in consideration of the benefits.

The second problem is to realize the ability of flood prevention works such as the dikes, reservoirs and retarding basins now and in the future. As a result of sediment deposition, the ability of flood prevention works to be effective will be reduced year by year. To maintain the protection capacity in the future, what engineering measures and non-engineering measures should be done in our planning? Can we solve the problem by hydrology, hydraulics and river dynamics?

The third problem is to research sediment accumulation in reservoirs and channels, in particular at the terminal of backwater of Sanmenxia reservoir and the channel at the river mouth. The objective is to reduce the risk of flood damages on the lower reaches. This problem is not only scientific but also deals with the flood damages in the upper or lower provinces as well as the lives of inhabitants and their environment. Therefore, an optimized operational schedule for Sanmenxia should be the focus of further study.

The fourth problem is to study the proposals for optimization of engineering measures. This includes raising the dikes, constructing new reservoirs in tributaries or constructing the Shialongdi reservoir on the main stream. By systems analysis, a mathematical model of a flood prevention system can be designed which will optimize the function of multi-objectives. Although many inferior solutions can be obtained, a compromise solution can also be determined by the decision makers. Of course, the cost must also be considered.

The fifth problem is to measure the benefits yielded by non-engineering measures. Flood information and forecasts must be made so that the retarding basin can be used to allow inhabitants to move to a safety zone. In the Yellow River, because of the rapid change of the cross sections during the flood periods, the water stages are reduced due to the scouring of the riverbed during the peak discharge. Even if forecasts can be made for the peak discharge, it is difficult to actually estimate the crest stage. The width of the river above Gaocun is from five to twenty km, so discharge measurements take four hours or more. Based on these data, it is obvious that it is difficult to predict floods accurately.

The sixth problem is to develop flood insurance by retarding basins by computing the annual average flood damage with return periods of 20, 50 and 100 years. In combination with risk analysis, calculations for the payment of flood insurance should not reduce the burden of the government, but benefit the inhabitants.

The seventh problem is that current changes in the channel affect the discharge passing through the railway bridge near Zhengzhou, the major railway from Beijing to Guangzhou. The variation of discharge is a function of an angle of direction of flow with the bridge opening. If the main current flows perpendicular to the bridge, it can pass Q cms through the bridge, but it can only pass .5 Q cms for an angle of 30°. Therefore, the flow direction under the bridge must be controlled. This problem not only deals with the safety of bridge, but also threatens the dikes above the bridge. The problem can be studied by simulation techniques either by physical or by mathematical models. The object of simulation is to determine the maximum discharge that can be passed through the bridge during peak discharges.

The last problem is to study the effect of sediment accumulation at the mouth of the Yellow River. The channel of the river mouth to the sea shifted on a fan-shaped estuary in the past, but now is limited to a narrow area to protect the Shengli oil field. Therefore, the sediment accumulation in the channel at the river mouth will be faster than before and will reduce the slope of river more quickly in the future. The result is a decrease in the margin of safety for the reach in Shantong province. The inevitable consequence is probably to open breaches on dikes at a certain time in the future. The sedimentation rate at the estuary must be carefully estimated to plan for flood prevention systems on the lower reaches of the Yellow River. Certainly, breaches of dikes can't become a permanent change of waterway on the lower reach since the founding of Sanmenxia reservoir. It is easy to close the sluice gates of the reservoir in a dry season, and the base flow from the intervening areas below Sanmenxia is small in dry seasons. This is a favorable condition under which to stop breaches in dikes.

Reference

(1) "Flood Prevention on the Lower Reaches of the Yellow River," Yellow River Conservancy Commission, Zhengzhou, China, 1980

FLUVIAL PROCESSES OF THE LOWER YELLOW RIVER AND ESTIMATION OF FLOOD CONDITIONS

Ye Qingchao
Institute of Geography
Chinese Academy of Sciences
People's Republic of China

ABSTRACT: The Yellow River is a famous river that carries sediment at hyperconcentrations. On average, the runoff volume at the Huayuankou Hydrographic Station is 47 billion cu.m annually, only one-twentieth of that of the Yangtze River, and the annual mean sediment discharge is 1.6 billion tons, 3 times as much as the Yangtze River. The disequilibrium between runoff and sediment discharge seems to be the crux of the problem which causes sedimentation and a rising bed in the lower Yellow River, gradually making the Yellow River a "Hanging River," and causing frequent abnormal or catastropic floods, avulsions and diversions. In the last 30 years, the annual mean accretion rate in the lower Yellow River has been about 10 cm/y, resulting in a situation where the difference in height between the nearby ground surface inside the banks and outside the banks is 6-8 m in the reach from Huayuankou to Dongbatou, and 3-5 m in the reaches downstream from Dongbatou. It is very important and urgent to study the processes of the evolution of the Yellow River further and to estimate flood conditions in advance.

The History of Avulsions and Diversions of the Lower Yellow River

Historically, for natural and social reasons, the Yellow River was in a very dangerous situation with frequent avulsions occurring about twice every three years. For the 2,500 years when the Yellow River flowed in the region to the north of the modern Yellow River course, about 70% of the period was from 602 BC to 1946. Since 602 BC, when the first great diversion occurred, over 1,500 avulsions and 7 great diversions have taken place (see Fig.1). The extent of diversions, avulsions and inundations cover an area of 250,000 sq. km and reached Tianjin to the north and the drainage basins of the Yangtze River and the Huaihe River to the south. It is obvious that the Yellow River is one of the rivers which possesses the

L. M. Brush et al. (eds.), Taming the Yellow River: Silt and Floods, 261–274.

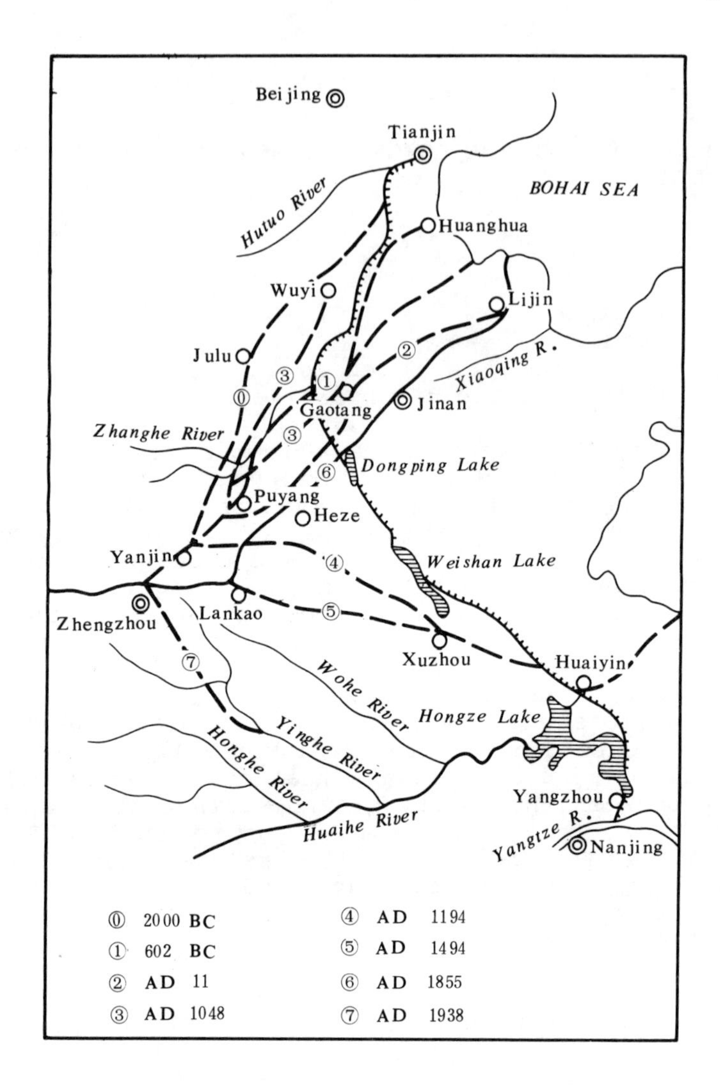

Figure 1. Historical Migration of the Huanghe

strongest sedimentation and the most frequent avulsions and diversions in the world (see Table 1).

The Relationship Between Deposition Rates and Environmental Changes

The lower Yellow River flows through a broad region which is known tectonically as the Huabei Depression. For a long time, this region had been sinking and filling with thick sediments. The Quaternary sediment thicknesses are 150-200 m on the average and 400-600 m at a maximum. From the Tang and Song Dynasties, the sedimentation of the Yellow River has made it a "Hanging River," restrained only by dikes. An example of one of the ancient courses of the Yellow River from 1194 to 1855 is named the Abandoned Yellow River. At present it has the following parameters. Its length is 873.1 km; and the difference between the bed surface and the ground surface outside the dikes is 3.68 m on an average and 6.1 m at a maximum. The difference in elevation between the bar surfaces and the ground surface outside the dikes is 6.8 m on average and 9.8 m at a maximum. According to these data, it can be calculated that the average deposition rate was 0.026 m/y during the period. The channel bed had been aggrading gradually and the incoming sediment could not be carried by the river. Thus, the Yellow River had its sixth great diversion in 1855, and the course of the Yellow River during the Ming and Qing Dynasties was abandoned.

According to field investigations and data analyses, the sedimentation in the modern Yellow River course during the years from 1855-1985 is stronger than that in the Abandoned Yellow River course. The following parameters give a general idea of this sedimentation. The length of the course from Huayuankou to the river mouth is 726.9 km; sediment deposited in the course from 1855 to 1985 was 18 billion tons. The difference in height between the bed surface and the ground surface outside dikes is 3.75 m on average and 10.72 m at a maximum. The difference in height between the bar surfaces and the ground surface outside the dikes is 3-8 m on average and 11.0 m at maximum. The age of the modern Yellow River is only one-sixth the life span of the Abandoned Yellow River, but the sediment deposition rate of the former is larger than that of the latter. Thus the danger that the modern Yellow River will have avulsions, diversions and floods has increased, compared to the Abandoned Yellow River.

The average sediment thickness in the Huayuankou-Lijin reach during the period 1855-1954 was 5.37 m, i.e., an annual mean rate of 0.06 m (see Fig. 2).

From 1954 to 1982, the lower Yellow River passed through different stages in the development process. The changes can be summarized as follows.

Table 1. The Situation of the Great Diversions of the Lower Yellow River

Sequence of diversions	Occurrence time	Avulsion sites	Outlets into the sea
First	602 B.C.	Suxukou northeast of Hua County	Cangzhou & Huanghua Counties, into the Bohai Sea
Second	A.D. 11	Weijun Prefecture northwest of Puyang county	Bin & Lijin Counties, into the Bohai Sea
Third	1048	Shanghusao, Chanzhou (modern Changhuji east of Puyang)	The north one flowed via Tianjin into the Bohai Sea; the south one flowed via Duma River (the modern Maja River) into the Bohai Sea
Fourth	1194	The ancient dike in Guanglu village, Yangwu (modern Zhangdafuzhai northwest of Yuanyang)	Yuntiguan, Qingjiang-Kou, into the Yellow Sea
Fifth	1499	Kaifeng & Jinglongkou	Via Huaihe River course into the Yellow Sea
Sixth	1855	Tongwaxiang	Lijin, into the Bohai Sea
Seventh	1938	Huayuankou	Via Huaihe River into the Yellow Sea

Table 1 --- Continued

Bifurcation situations	Reasons for divertions
Bifurcated in the place near river mouth	Natural
id.	Natural and man-made
Bifurcated into two courses	Natural
Single course	Natural and man-made
Separated into several courses via Guohe River, Yinghe River, and Sihe River into the sea	Natural and man-made
Separated, then single course	Natural and man-made
	Man-made, after 1947, via Daqing River into the Bohai Sea again

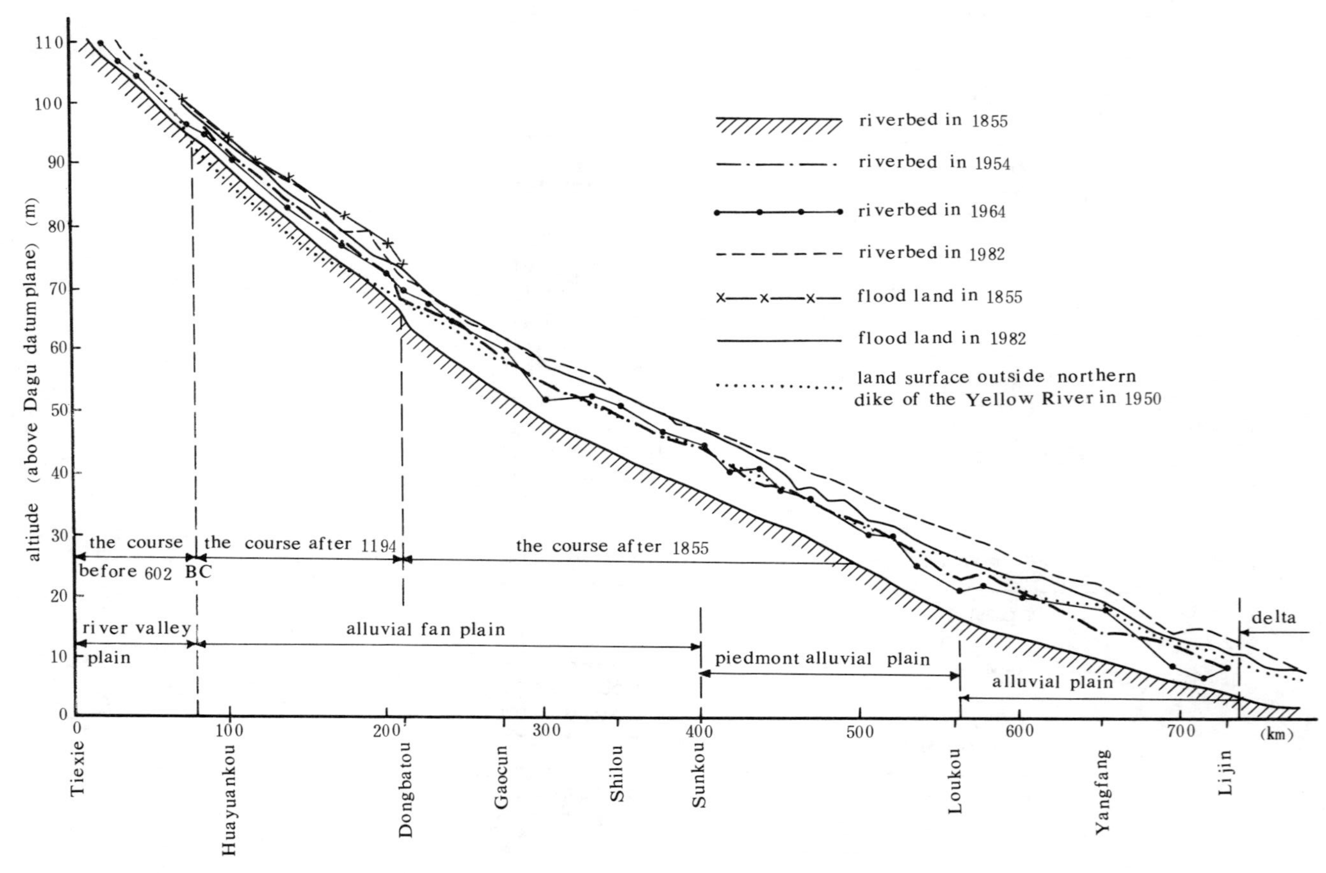

Figure 2. The Longitudinal Profiles of the Lower Reaches of the Huanghe River (from Management and Development of the Huang-Huai-Hai plain, No. 1, with permission)

From 1954 to 1959, the river was in a stage of natural adjustment, the annual mean accretion thickness was 0.14 m in the Huayuankou-Lijin reach. From 1960 to 1964, the Sanmen Gorge Reservoir had been constructed on the Yellow River to store water, retain floods and releasing sediments. In this stage, the reservoir drained clear water from the dam, which caused scouring downstream, and the average erosion rate was 0.26 m/y in the same reach. From 1965 to 1970, the reservoir functioned to store clear water and release high sediment-laden water in different seasons. Thus, the riverbed accumulated again at an average rate of 0.083 m/y. From 1970 to 1974, deposition on the riverbed increased and the sediment deposition rate was 0.11 m/y. From 1975 to 1982, owing to the decrease of rainstorms in the regions of the middle reaches and profits gained from water and soil conservation, the deposition in the channel decreased slightly and the annual mean sediment deposition rate was 0.05 m/y. From 1965-1982, the annual mean sediment deposition rate was 0.10 m/y in the Huayuankou-Lijin reach. Such a strong sedimentation in the channel is a threat to the flood control works in the lower Yellow River (see Table 2).

The reasons for accelerated accumulation in the lower Yellow River follow: (1) The water and soil loss in the Loess Plateau has become more and more serious, especially with regard to accelerated erosion induced by humans. Up to the 1970's, the sediment yield reached 2.2 billion tons per year, (2) Since 1949, the Yellow River has had no avulsions and diversions, hence the accumulated amount of sediment reached 7 billion tons from 1950 to 1982, (3) The larger the incoming sediment coefficient, the stronger the deposition (see Fig. 3), (4) Most of the coarse-grained sands ($D \geq 0.05$ mm) have been deposited in the channel, which increases the accumulation and accelerates bed aggradation, and (5) Rising of the base level of erosion at the river mouth also affected sedimentation in the channel.

At the Sanmen Gorge Station, the coarse sediment discharge was 0.546 billion tons annually during the period 1965-1980, or about 34% of the total sediment discharge of the Yellow River. The coarse sediments come mainly from the tributaries of both sides of the Yellow River gorge in the Longmen-Hekouzhen reach, the upper reaches of the Luohe River and the Jinghe River. The total drainage area is 158,000 sq. km and the sediment yield is 0.456 billion tons annually, in which 0.155 billion tons come from the bedrock areas. These coarse sediments are the main deposits that have accumulated in the lower reaches. Therefore the drainage areas mentioned above are the main regions in which to practice water and soil conservation.

The Possible Avulsion Sites and Related Disaster Conditions in

Table 2. Statistics of the sediment thicknesses of the lower Yellow Rive course during the years from 1954 to 1982

Period / Item / Thickness / Reach	1954 - 1959		1960 - 1964		1965 - 1969		1970 - 1974		1975 - 1982		1965 - 1982		1954 - 1982	
	a*	b**	a	b	a	b	a	b	a	b	a	b	a	b
Tiexie-- Huayuankou			-1.91	-0.37	0.70	0.14	0.23	0.05	-0.10	-0.13	0.96	0.054		
Huayuankou-- Dongbatou	1.03	0.17	-1.08	-0.22	0.49	0.10	0.21	0.04	0.27	0.034	1.67	0.093	2.10	0.074
Dongbatou-- Aishan	1.00	0.07	-1.00	-0.20	0.27	0.05	0.49	0.10	0.40	0.05	1.78	0.099	2.86	0.102
Aishan-- Lijin	0.46	0.077	-1.85	-0.37	-0.52	0.10	0.97	0.19	0.52	0.065	1.45	0.08	1.98	0.071
Tiexie-- Lijin			-1.46	-0.29	0.50	0.098	0.475	0.095	0.27	0.034	1.47	0.082		
Huayuankou-- Lijin	0.83	0.14	-1.31	-0.26	0.42	0.083	0.557	0.11	0.40	0.05	1.63	0.096	2.31	0.082

* a Sediment thickness
** b Net sediment deposition rate (m/y)

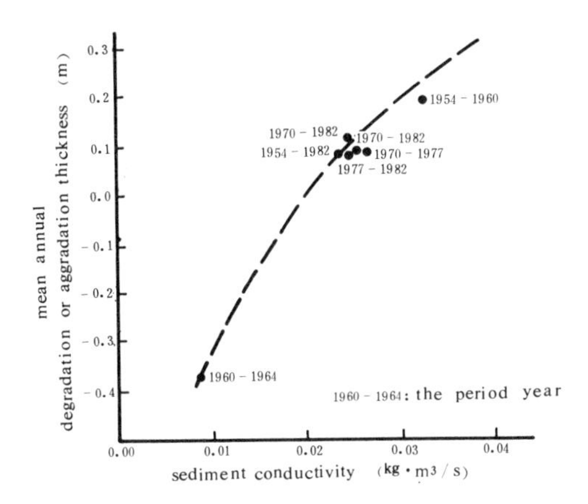

Figure 3. The Relation Between Degradation or Aggradation Thickness and Sediment Conductivity in the Lower Huanghe River.

(from Management and Development of the Huang-Huai-Hai plain, No. 1, with permission)

the Lower Yellow River

According to the evolutionary history of the Yellow River, the geological and geomorphic conditions and the tendency of the possible avulsions along the river, it has been predicted that the Qinhekou-Yuanyang reach and the Yuanyang-Gaocun reach will be possible avulsion sites in the future.

Based on the different discharges of flood peaks, i.e., 55,000 cms, 49,000 cms and 22,300 cms, and on the roughness coefficient, 0.04, at the Huayuankou Station, computer simulations were made to predict floods. The calculated results showed that the northern boundary of the possible flood extent would not go beyond the Majia River, the Weihe River, the Weiyun River and the Zhangweixin River, the southern boundary would reach the Linhuang (which means "face to the Yellow River" in Chinese), the dike west of Jiyang and the zones on the north of the Tuhai River east of Jiyang (see Fig. 4).

Considering the real landforms and the conditions of the surface features, it is possible to estimate the inundated area. The results are as follows. If an avulsion takes place under the discharge of flood peak of 55,000 m3/s, the maximum flood area will be 22,082 sq. km in the Qinhekou-Yuanyang reach, and 21,190 sq. km in the Yuanyang-Gaocun reach (see Table 3). The economic losses, including surficial comprehensive losses would be losses for delays on the railways, lost cultivated lands destroyed by sediments, the lost production in the oil fields and losses suffered in the towns and cities. This would probably cause losses of several ten-billions of yuan.

A Tentative Idea for River Management

At present, because the erosion in the Loess Plateau hasn't been controlled, deposition in the lower Yellow River is continuing and the rising process of the river bed cannot be stopped. Thus, the more the bed rises and the more the height of the banks increases, the critical situations become more and more ominous. This vicious cycle provides the favorable conditions for avulsions and diversions.

Yellow River management is an important problem and has been studied by many specialists at home and abroad. At present, the policy is to hold back sediment in the middle and upper reaches and release sediment in the lower reaches and to conduct flood detention on both sides. In addition, many profitable suggestions to manage the lower Yellow River have been put forth, such as the construction of Xiaolangdi Reservoir, increasing the dike heights, separating the flood waters for sedimentation in certain places, making floods in order to scour the bed, separating the flow to release

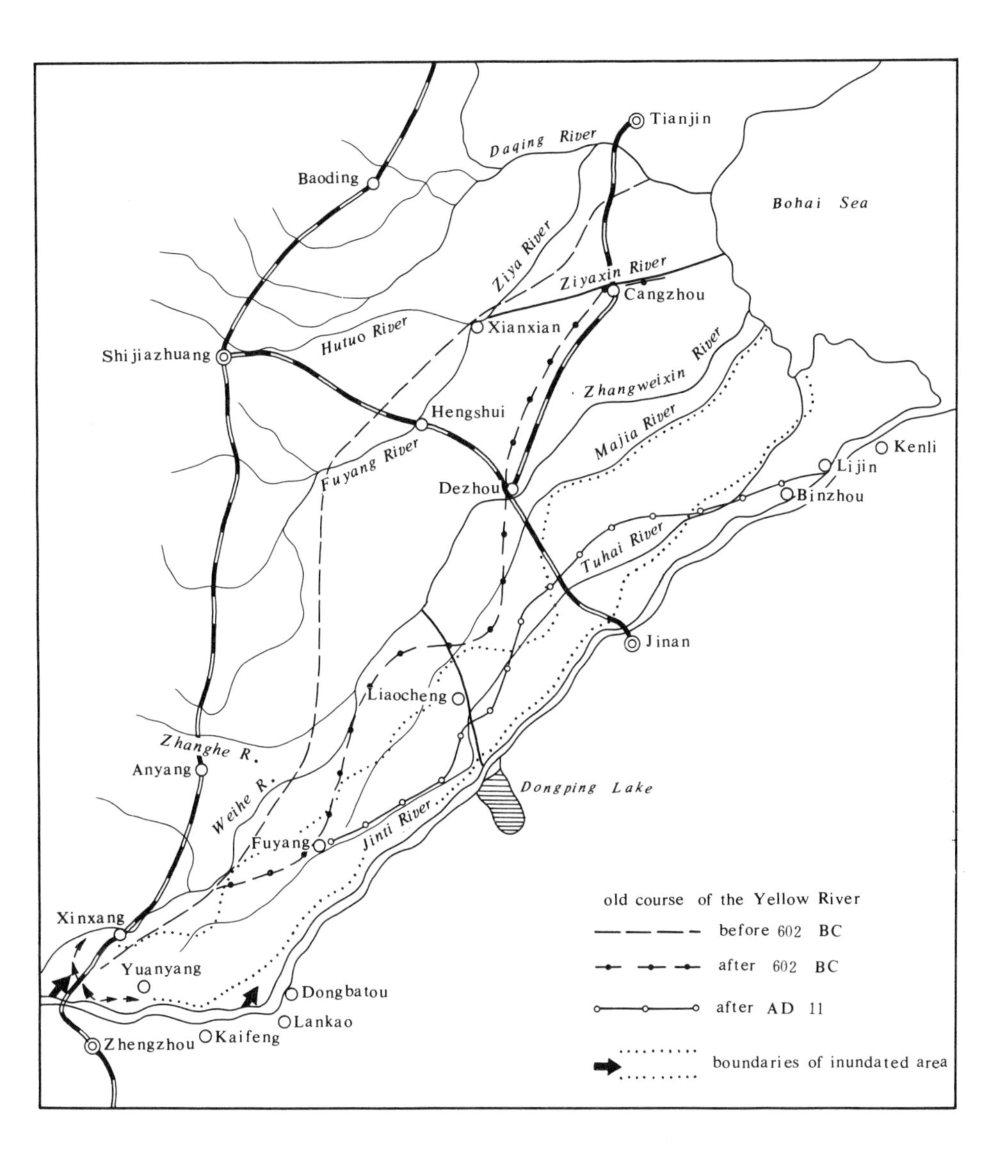

Figure 4. Flooding Caused by A Burst in the Northern Side of the Lower Huanghe River.

Table 3. The areas inundated in different flood discharges

Site of avulsions	Flood peak discharge (cms)	Inundated area (km^2)	The area of the inundated cultivated lands (x1000 mu*)
Qinhekou--	55,000	22,082	2208.2
Yuanyang	49,000	20,928	2092.8
(Yuanyang avulsion)	22,300	15,316	1531.6
Yuanyang--	55,000	21,190	2119.0
Gaocun	49,000	20,172	2017.2
(Caogang avulsion)	22,300	14,916	1491.6

* 15 mu equal to 1 hectare (after Lu Zhongchen)

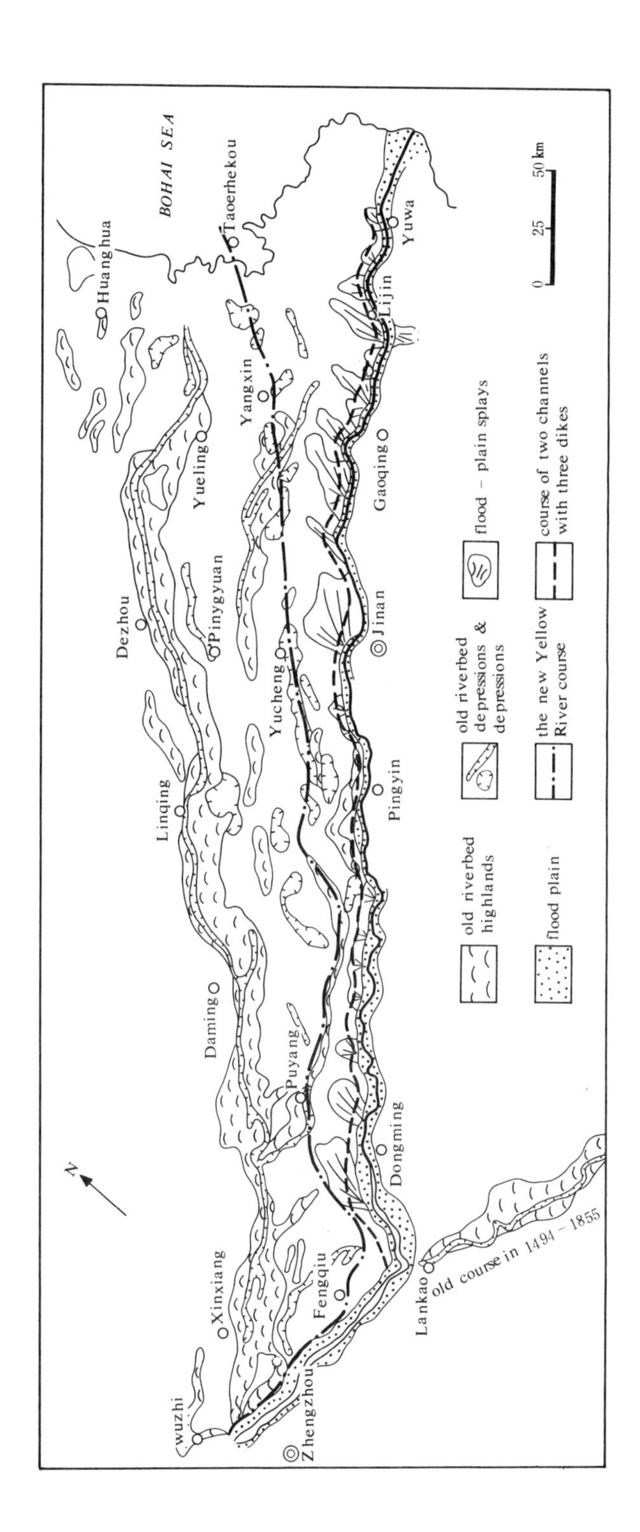

Figure 5. The Fluvial Landforms and General Plan of the River Management of the Lower Yellow River.

sediments, and realigning the channel.

Based on the features of the environmental changes of the Yellow River basin and the evolutional regulations of the course of the lower Yellow River, it is suggested that we should make the best use of the situation and that two kinds of Yellow River management plans may be proposed: (1) In the reaches between Jiahetan and Lijin, a new dike should be constructed at the north side of the Yellow River. Thus, there would be two channels with three dikes. The purpose would be to use the existing channel to drain the flow with bigger discharges. This measure has been shown to be capable of separating flows and releasing sediment in the lower Yellow River making the channel suitable for discharges of 46,000 cms (see Fig.5). (2) In consideration of a long-term strategic policy, the existing lower channel should be abandoned if it cannot meet the demand to pass the highest flood, and a new one should be cleared, as a measure of artificial diversion.

All of the above opinions have their own merits and demerits. They should be studied in depth in the general program of Yellow River management. Both the scientific and technically feasible aspects should be considered and evaluation and comparison of the opinions made in order to provide a much more reasonable scientific foundation for the Yellow River management in the future.

References

1. Lu Zhongchen, 1987, On the possible avulsion sites and the maximum inundated area of the region north of the lower Yellow River: Geographical Research, Vol. 6, No. 4, Science Press (in Chinese)

2. Ye Qingchao, 1985, The sedimentary characteristics and development trend of the lower Yellow River course: The management and development of the Huang-Huai-Hai plain; Vol.1, Science Press (in Chinese)

THE TEMPORAL AND SPATIAL VARIATION OF EROSION AND SEDIMENT YIELD ON THE LOESS PLATEAU

Chen Yongzong
Jing Ke
Lu Jinfa
Zhang Xunchang
Institute of Geography
Chinese Academy of Sciences
Beijing, The People's Republic of China

ABSTRACT: The Loess Plateau in the Middle Yellow basin is located within 34°N--41°N and 102°E--114°E. It covers an area of about 31.9 x 10^4 sq. km, has a precipitation varying from 300 to 700 mm, and an average temperature of 3.6-14°C. This plateau is covered by Quaternary loess with the depth ranging from 100 to 200 m. The relief of the plateau varies from 150 to 300 m. Gully density from 3 to 6 km/sq.km, and the slope is greater than 15 degrees. At present, its population density is about 50 to 200 persons/sq.km in most parts, the vegetation cover is less than 20%, and the cultivation indices 40-60%. Consequently, soil erosion is very severe in this area.

Spatial Variation of Erosion and Sediment Yield

REGIONAL DISTRIBUTION OF SEDIMENT YIELD

Erosion, sediment yield and transport are different but interrelated geomorphic processes. The annual amount of erosion is usually more than the sediment yield which is, in turn, greater than the sediment discharge in the southern and northeastern parts of China. Although there is a time lag, nearly all sediment coming from slope of the Plateau will be carried away by floods in the long run. Because of the greater gradients and higher sediment concentrations during the flood season, the carrying capacities of rivers and gullies are very high, the sediment delivery ratio approximates one and the river channels remain down cutting, except for some parts of the Yellow River within the He Tao plain and from Yumenkou to Tongguan as well as the lower part

L. M. Brush et al. (eds.), *Taming the Yellow River: Silt and Floods*, 275–288.

of Weihe R. where some sedimentation occurs (Gong Shiyang et al., 1979; Mu Jin-ze et al., 1983). Based on the above information, sediment discharge from various rivers can be used to show the spatial variation of erosion and sediment yields of the plateau.

Of the total annual sediment of 1.6 billion tons or so delivered to the lower Yellow River most has been proven to come from the Loess Plateau (Chen Yongzong, 1983). A classification has been made of the average annual sediment discharge of rivers on the Loess Plateau before 1983 and a map of the erosion intensity compiled using this classification. Shown in Fig. 1, the areas with an erosion intensity greater than 5,000 t/sq.km/y are located in the north to the line from Daning, Yanan, to Qing Yang between Liupanshan M. and Luliangshan M., as well as within the upper part of Weihe river basin and Zuleihe R. basin. They cover together an area of 210,000 sq.km which makes up 65.6% of the total area of the Loess Plateau, and the total annual sediment discharge from this region is 1.42 billion tons which accounts for 90.1% of that of the Yellow River. The areas with erosion of more than 10,000 t/sq.km/y lie mainly to the north of Yanan between the Luliangshan and Liupanshan mountains. With an area of 36,000 sq.km, 11.3% of total area of the Loess Plateau, and an annual sediment discharge at 0.373 billion tons this area accounts for 23.8% of the Yellow River sediment discharge. Within this area the Huangpuchuan, Gushanchan, Jialuhe Rivers have annual sediment discharges of more than 15,000 t/sq.km/y. It has been proved that severe erosion occurs mostly in those where active hydraulic, gravitational, and aeolian erosion occur on land covered by sandy loess, loose shale and sandstone. Since population density in this region is small and features of precipitation are similar to those of Yanan, among the factors affecting erosion unfavorable natural ones still play a major role.

Fig.1 The map of erosion intensity on the Loess Plateau in t/sq.km/y.

1, $S < 1,000$
2, $S = 1,000 - 5,000$
3, $S = 5,000 - 10,000$
4, $S = 10,000 - 15,000$
5, $S = 15,000 - 20,000$
6, $S > 20,000$
7, The rocky mountains with good vegetation cover.

SPATIAL VARIATION OF THE COARSER PORTION OF THE SEDIMENT YIELD

The late Prof. Qian Ning (1980) called sediment with a size greater than 0.05 mm coarse sediment and noted that it is these sizes that endanger the channel of the lower Yellow

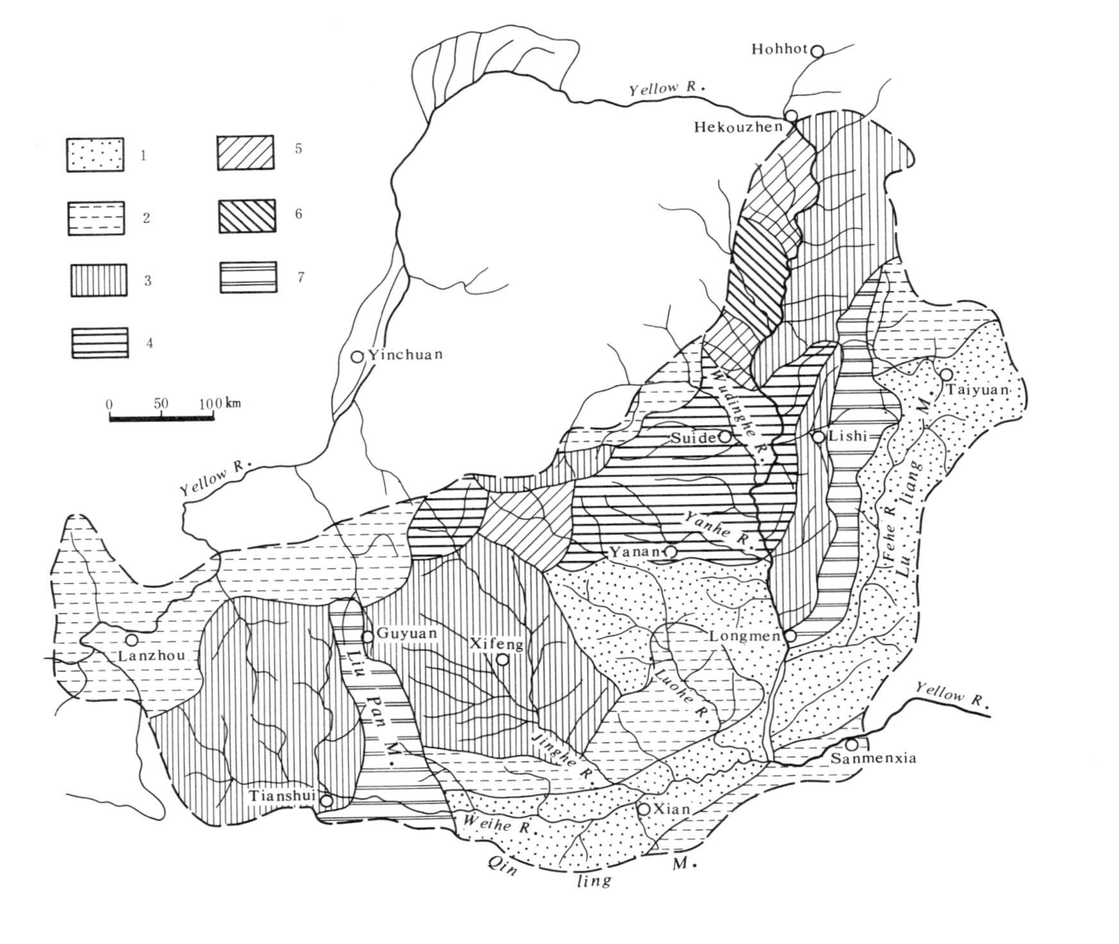

Figure 1. The Map of Erosion Intensity on the Loess Plateau in t/sq.km/y.

River.

Average annual discharge of coarse sediment at Longmen on the Yellow River, Huaxian station on Weihe R., Hejin on Fenhe R., and Zhuangtou on Luohe R., before 1983 were 0.551 billion tons per year with an average coarse sediment load of 824.8 t/sq km/y. Using the discharge of coarse sediment and sediment discharge coefficient as indices, the source areas of coarse sediment can be defined, where the two indices are greater than 825 t/sq.m/y and 1.0 kg/cu.m respectively. Based on this, the map of coarse sediment source was prepared (see Fig. 2).

Fig.2 Division map of sources of coarse sediment in the middle reaches of the Yellow River in t/sq.km/y.

1, S < 850
2, S = 850 - 1,500
3, S = 1,500 - 4,000
4, S = 4,000 - 8,000
5, S > 8,000

As compared with Fig.1, it can be seen that the major source areas of coarse sediment coincide well with those where the sediment yield is greater than 5,000 t/sq.km/y. The amount of coarse sediment from this area accounts for 94.4% of that of the Yellow River. Within the area most coarse sediment comes from the Yellow River basin between Hekouzhen and Longmen, the Luohe R. basin above the Jinfeping, the Malianhe R. basin and the Zhouhe R. These basins with a total area of 130,000 sq.km and have the highest coarse sediment yield amounting to 77% of the total. Among them the river basins of Huangpuchan, Kouyehe, Gushanchan, Tuweihe, Jialuhe, Wudinghe, Qingjianhe, Maliahe, Lower to Jinfeping of Luohe, and Zhouhe make the greatest contribution, providing 0.324 billion tons of coarse sediment per year or 58.8% of the total yield of the Yellow River from 81,000 sq.km. If the amount of coarse sediment from this region can be reduced by a half, the discharge of coarse sediment in the Yellow River would decrease by about 30%. This would greatly reduce the aggradation of the bed of the Yellow River.

SPATIAL VARIATION OF EROSION AND SEDIMENT YIELD IN SMALL GULLY CATCHMENTS

Small gully catchments are the basic units relavent to sediment production and erosion on the Loess Plateau and generally have areas less than 10 sq.km. Sediment production in small catchments shows vertical variation (Chen Yongzong, 1983) (see Fig. 3). Hydraulic erosion dominates and gravitational erosion is seldom seen in intergully areas, but

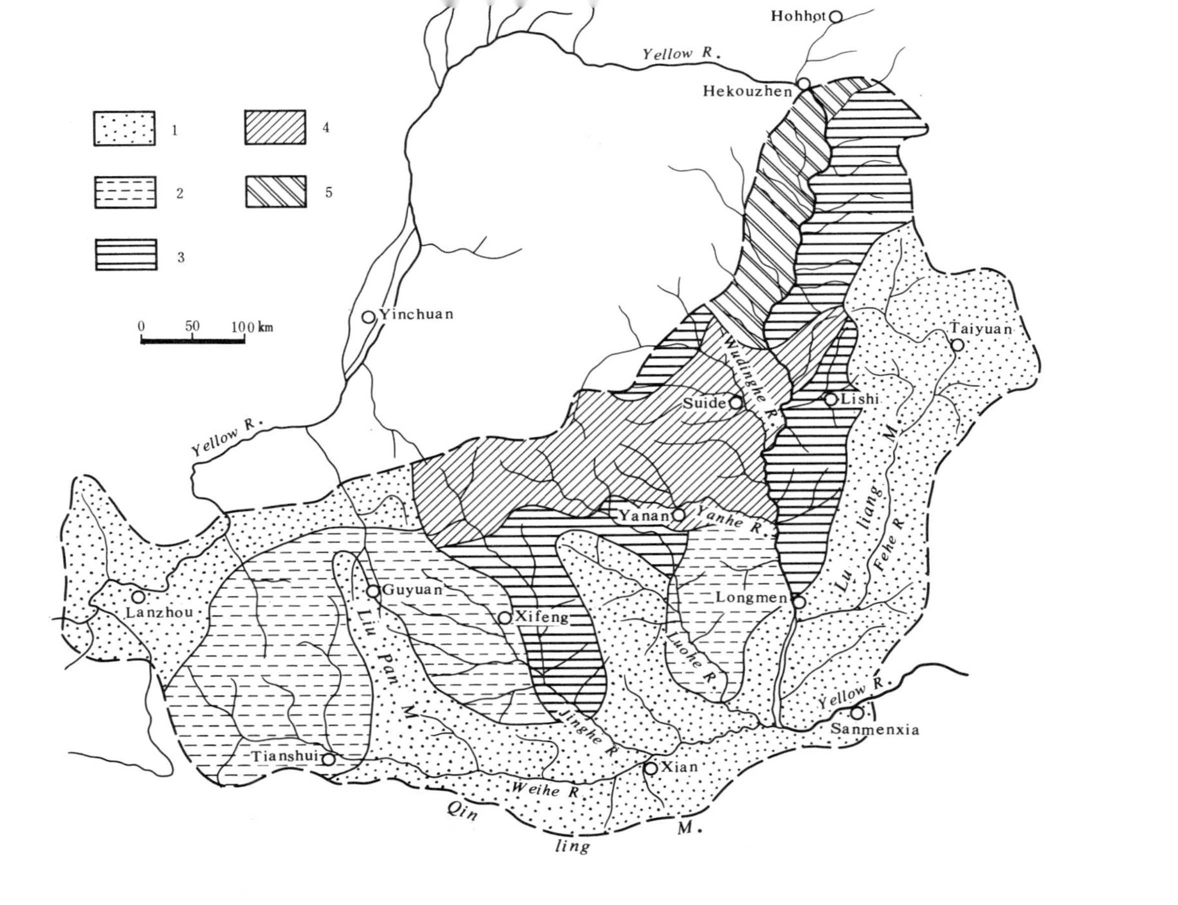

Figure 2. Division Map of Sources of Coarse Sediment in the Middle Reaches of the Yellow River in t/sq.km/y.

both are very active in the gully. Thus a great difference in erosion intensity exists between the gully and intergully areas.

Fig.3 Vertical changes in the amount of sediment yield in a small catchment on the Loess Plateau.

I Area of hydraulic, aeolian, and human erosion on intergully hillslope.
Ia Sub-area of sheet erosion on upper Liang and Mao hillslopes.
Ib Sub-area of sheet and gully erosion gully erosion on middle Liang and Mao hillslope.
Ic Sub-area of gully erosion on lower Liang and Mao hillslopes.
II Area of hydraulic and gravitational erosion in gullies.
IIa Sub-area of hydaulic erosion, landfall, landslips on upper gully slopes.
IIb Sub-area of hydraulic and gravitational erosion on lower-middle gully slopes.
IIc Sub-area of sediment deposit on lower gully slopes.
III Area of erosion on gully-bed.

On the basis of analysis of data obtained at water and soil conservation stations (Table 1), the amount of erosion of sediment per unit area of gully catchment is greater than that of the intergully area by 41.0-76.8% in the hilly region of the Loess Plateau, and by 94.8% in Loess Yuan. If the rainfall-runoff of intergullies is interrupted, the amount of erosion of sediment from gullies will be reduced by 77.7% in the hilly region of Loess Plateau (Zong Beiqing, 1980) and by 85% or so in Loess Yuan (experimental station of Xifen, 1979). Since the area of the intergully region is larger than that of gully in most parts of the Loess Plateau, the sediment yield from intergully regions is greater than that from gullies. For example, the intergully area in the Tanshangou catchment, accounts for 74% of the total catchment, and contributes 15,000 tons of sediment but the gullies only contribute 9,100 tons. The erosion intensity of rainfall-runoff increases from the divide downward on Liang and Mao hillslopes of intergully areas. Sediment yield from erosion on the lower part of Liang and Mao hillslopes can be 1-10 times greater than that on the upper and middle parts. The most active erosion often occurs in the upper part of the gully catchment.

Temporal Variation of Erosion and Sediment Production

VARIATION IN HISTORICAL PERIOD

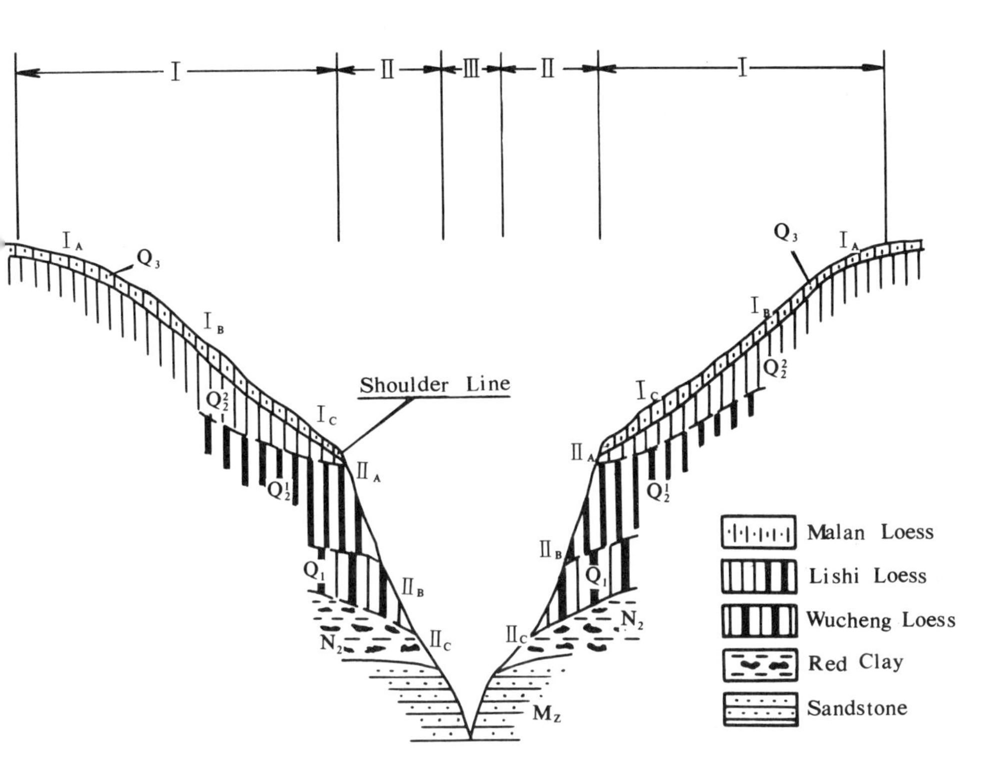

Figure 3. Vertical Changes in the Amount of Sediment Yield in a Small Catchment on the Loess Plateau.

Table 1. The erosion yield of different geomorphic types in catchment

Location		Intergully						Gully				
erosion yield (t/sq.km/y) / Geom.type and place		Upper part of Liang and Mao slope	Middle part of Liang and Mao slope	Lower part of Liang and Mao slope	Coneave slope above gully head	Top and slope of Yuan	Average	Loess gully slope	Red soil gully slope	Average	Average for catch.	Period (yr.)
Loess hilly region	Yong-Daogou catch	509.4	1500.4	7432.8	12403.6		5065.2	26614.8	18748.9	21799.4	20700.0	1963-1968
	Tanshan gou catch	11538.1	1989.0	4.8	21404.3		20284.2	17107.3	35018.5	34375.6	23948.1	1963-1969
Loess Yuan	Nanxiao hegou catch					791.8				15198.4	4946.4	1955 1969-1971 1974

Owing to human activities the sediment yield was as high as 1 billion tons per year in the Loess Plateau three thousand years ago. The excessive increase of population exerted the most important influence on sediment yield. From the Han (206-220 D.C.) to the Ming Dynasty (1368-1944 B.C.), the population was around 60-70 million, and not greater than 100 million until the Qing Dynasty (1661 B.C.). The process of population growth in the Loess Plateau was similar to the country as a whole. For instance, the population in Shaanxi Province remained below 10 million for a very long period and went beyond 100 million just during the Qing Dynasty. As the population increased, pressure on the land and human destruction increased and, as a result, erosion become severe.

Cultivation on steep slopes exerted the most important influence on erosion and sediment production in the Loess Plateau. Three thousand years ago farmland was mostly confined to fertile river valley plains and flat Yuan surfaces. The vast loess hilly area was covered by forest and grass, and erosion was less. Before the Qin Dynasty (200 D.C.) the Yellow River was named Dahe (Great River), its present tributary Yanhe named Qinghe (Clear River), Fenghe named Sufenghe (White River) and Jinhe was also clear. Because of migration to develop border areas, very many people immigrated into the Loess Plateau, during the Qin and Han Dynasties, causing broad destruction of vegetative cover. At the beginning of the West Han Dynasty the name of Dahe was changed to the Yellow River (Shi Nianhei et al., 1985). At the end of the Dynasty the Yellow River became very muddy, but some vegetation still remained. From the East Han Dynasty to the Sui Dynasty the Loess Plateau was mainly utilized by herdsman and vegetation cover was well restored. Soil erosion was reduced.

In the Tang and Song Dynasties, as the woods of the Loess Plateau were greatly destroyed and most grassland was cultivated, erosion increased again. In the Ming and Qing Dynasties, the woods and grass were destroyed thoroughly and sediment concentration in the Yellow River increased dramatically. The Yellow River is well-known for its frequent flooding and changes of channel course throughout history. Although there are many reasons for this, intensified erosion and sediment yield in the Loess Plateau is the major one. Historically the frequency of flooding of the Yellow River reflects the evolution of erosion and sediment yield from the Loess Plateau. Disastrous flooding occurred only once every 200 years before the Qing Dynasty, once every 100 years in the Song and Yuan Dynasties, but twice a year since the Ming and Qing Dynasties.

RECENT VARIATIONS

Analysis of the average annual sediment discharge during the past 60 years, shows that great variations of sediment discharge in the Yellow R. occur that can be divided into three periods.

The first period is from 1919 to 1953, with a sediment discharge of 1.52 billion tons per year. The second one is 1954-1970, with a sediment discharge of 1.86 billion tons per year. The third is 1971-1985, with 1.067 billion tons per year. Because no measures for water and soil conservation were carried out, sediment discharges can be considered roughly equivalent to the erosion sediment yields before 1953.

Although some measures of water and soil conservation were carried out in the second period, due to increased human destruction and more precipitation, the sediment discharge increased by 18.3% compared with the first period. There were leveled terrace fields of 5.1 million mu and dam-land (the land behind the dam) of 1.22 million mu. The construction of dam-land of one mu requires 2,500 tons of sediment and a leveled terrace can reduce sediment yield by 70%. The annual average sediment trapped in the whole region can be calculated to be 0.17 billion tons. The sediment trapped by reservoirs and used by irrigation has been estimated to be 7 million tons annually. Thus the erosion of sediment increased during the second period by 25% compared with the period 1919-1953.

The Loess Plateau has been relatively dry since 1971 with the annual mean precipitation on decreasing by 10-15%. Precipitation during the rain season in the major source area of sediment has been reduced by 5-10%. The frequency of daily rainfall greater than 50 mm and 100 mm has been greatly reduced. Therefore, sediment discharge of the Yellow River has been reduced by 0.79 billion tons per year amounting to 42.6% of that during 1954-1971. In addition, the effect on sediment reduction by newly-built reservoirs is not negligible (Zhang Shen Lie et al., 1986, Xiong Gueshu, 1986). It has turned out that the amount of sediment trapped by the reservoirs, either on the trunk river or tributaries, together with that being used by irrigation, is about 0.112 billion tons per year. The amount of sediment, 0.216 billion tons per year, reduced by level terraced fields and dams amounts to 64.5% (Table 2). This means that the amount of sediment reduced by measures both for water conservancy and soil and water conservation represents 42.2% of the reduced amount of sediment in the Yellow River. Among these measures dams make the biggest contribution to the reduction of sediment.

YEARLY VARIATION

Owing to the influence of the southeastern monsoon, 60-70% of the annual precipitation on the Loess Plateau is concentrated during June to Sept. Therefore, rainfall-runoff erosion mostly takes place in this period. The same thing happens to sediment discharges of the Yellow River and its tributaries.

Table 2 Amount of Sediment Trapped by Level Terraced Fields and Dams on the Loess Plateau

item / number / location	Level terrace (A)				Land behind the Dam (B)			
	Area sq.km	Sediment yield on slope t/sq.km/y	Amount sediment reduced t/sq.km	Total amount of sediment 10^4t	Area (km2)	Amount of Sediment trapped (10^4t/sq.km)	Total amount of sediment (10^4t)	A + B total (10^4t)
above Hekouzhen	3086.7	1115.0	780.5	240.9	134.7	210.0	2828.7	28527.9
between Hekouzhen and Longmen	3340.0	5352.0	3746.4	1251.3	188.0	450.0	84600.0	85851.3
Fenhe R. basin	1163.7	1500.0	1050.0	122.2	488.0	210.0	102480.0	102602.2
Weihe R. and Luohe R. basins	5829.6	5000.0	3500.0	2040.4	185.3	450.0	83385.0	85425.4
Total	13420.0	13420.0		3654.8	996.0		298752.0	302406.8

Based on the analysis of data of the experimental plots at Tianshei, Suide and Lishi soil and water conservation stations, the erosion of sediment during such periods amounts to over 90% of the total for one year. It snows little during winter and snow-melt runoff is seldom seen in the Loess Plateau. Rainfall-runoff erosion mainly occurs in several rainstorms from June to Sept. Fifty-four (54) rainfalls caused runoff at plot No.3 on the Tianshei experimental station from 1945 to 1957. The amount of sediment yielded by 11 rainstorms among them accounted for 87.0% of the total. Fifty rainfalls relating to runoff were recorded on plot No.18 of Suide experimental station. Sediment yielded by eleven rainstorms among the fifty amounted to 83.3% of the total. Erosion from twenty rainstorms on plot No.7 of Tanshanfou from 1961 to 1969 made up 93.3% of the total amount.

The erosion duration is relatively short during every rainfall- runoff process. The sixty one rainfall-runoffs causing erosion were taken place on plot No.7 at Tanshangou from 1961 to 1969 revealed that 80% of the rainfall lasted less than 30 minutes. The amount of runoff (Qi) and eroded sediment (Si) during 5, 10, 15, 20, 30, 60 minutes after the beginning of runoff-production were related to the total volume of runoff (Qm) and total erosion of sediment during the rainfall on plots No.7,9,12 and 3 at Tanshangou. As shown in Table 3, the relationship for 15-20 minutes is best. Erosion intensity at the beginning of runoff is greater than at the end. For example, under the same runoff of 0.02 cms, the sediment concentration at the beginning is over 300 kg/cu.m, but 50 kg/cu.m at the end on plot No.3 of Tanshangou. It has been proved by experiment that this has much to do with splash erosion at the beginning and earth-crusting during the rainfall. If the crust has been broken and rills formed, the amount of erosion sediment will increase sharply.

Conclusion

The area in which sediment yield is more than 5,000 t/sq.km/y covers 210,000 sq. km on the Loess Plateau, within which the area with sediment yield greater than 10,000 t/sq.km/y occupies only 36,000 sq. km. Much attention should be paid to this region for the control of sediment in the Yellow River.

Two major changes of sediment discharge caused by human activities in the Yellow River have taken place in the past sixty years. In any year sediment production is concentrated during a few thunderstorms in summer, especially within a 15-20 minute period. This makes harnessing of erosion more difficult. But if the proper measures and methods can be taken, the sediment discharge of the Yellow River could be reduced greatly.

Table 3 Relationship between runoff amount and erosion yield on plots of Tanshangou, Zizhou county, during different rainfall

No. of plot	Duration (min.)	Qi - Qm			Si - Sm			N
		a	b	r	a	b	r	
7	10	9.5366	1.7481	0.55	11,090.77	1.3613	0.54	32
	15	5.8716	1.4701	0.81	8,191.64	1.2976	0.77	19
	20	4.4442	1.2476	0.92	4,605.57	1.2448	0.92	10
	10	10.9396	1.9237	0.71	29,251.08	1.9188	0.61	32
9	15	16.6600	1.0843	0.76	17,080.29	1.2164	0.84	27
	20	16.6771	1.0445	0.83	17,200.73	1.1073	0.90	18
	10	1.8428	2.6843	0.71	2,816.71	2.4600	0.71	62
12	15	0.5529	1.7910	0.80	2,082.23	1.5854	0.79	53
	20	8.0593	1.2030	0.86	3,445.03	1.1867	0.87	42
	10	1.3457	1.2265	0.74	2,103.06	0.1543	0.43	35
3	15	2.1532	0.9542	0.81	540.28	0.9901	0.98	20
	20	1.9077	0.9926	0.77	442.46	0.9850	0.98	20

References

Chen Yongzong (1983), A preliminary analysis of the processes of sediment production in the small catchment on the Loess Plateau, Geographical Research, No.1, Vol.2.

Gong Shi-yang et al.,(1980), The origin and transport of sediments of the Yellow River, Proc. 1st Int. Symp. River Sedimentation, Beijing.

Liu Dongshong et al.,(1964), The Loess in the middle reaches of the Yellow River. Science Press.

Mo Jinze et al., (1983), Sediment delivery ratio as used in the computation of the watershed sediment yield, Sedimentary Research, No.1.

Qian Ning et al.,(1980), The source of coarse sediment in the middle Yellow River and its effect on the siltation of the Lower Yellow River. Proc. 1st Int. Symp. River Sedimentation, Beijing.

Shi Nianhei et al., (1985), Immigration of woods and grassland on the Loess Plateau, People's Press of Shaaxi province.

Xifen water and soil conservation station (1979), Preliminary study on the ways of harness in the region of loess yuan from the example of Nanxiaohegou, People's Yellow River, No.3.

Xiong Guishu (1986), Reduction of sediment load by water and soil conservation works in the middle and upper reaches of the Yellow River, People's Yellow River, No.4.

Zeng Beiqing (1980), Laws of water and soil loss as well as beneficial result of harness in hilly region, western part of Shaaxi province, People's Yellow River, No. 2.

Zhang Shenli et al., (1986), Preliminary analysis of effect of sediment reduction through improvement of the tributaries and soil and water conservation on the upper and middle reaches of the Yellow River, People's Yellow River, No. 1.

STUDY OF COUNTER MEASURES FOR THE YELLOW RIVER REGULATION AND FLOOD PREVENTION

Wang Wenkai, Wang Tingwu, and Ma Min
Henan Institute of Geography
The People's Republic of China

ABSTRACT: The Yellow River, the cradle of the Chinese nation, is the second largest river of China. It is known throughout the world for its abundant silt, severe sedimentation at the lower reaches, and the high frequency of flooding with tremendous damages. Since liberation, great achievements have been gained in flood prevention and utilization of water and silt of the Yellow River, but flood and silt sedimentation have not been fully controlled. A study of counter-measures for regulation of the Yellow River and flood prevention is one of important probjects to fight against the threat of floods.

The Characteristics of Silting and Flood Disasters at the Lower Yellow River

The lower Yellow River has become a world-famous "hanging river" due to long-term silting. According to the data of the Yellow River Project Committee, during 1952-1979 the height of the river bed increased 1.5 m, with a maximum of 3.0 m. The river floodland is 3-5 m higher than the land back of the levee, and 8-9 m high at Caogang village, Fengqiu county. Flood-carrying capacity of the Yellow River decreases with the increasing height of the river course. The river regime of each stretch is described as follows.

The stretch from mouth of the Qinhe River to Dongbatou: The distance between the two banks is 3-9 km. During flood periods, floods reach the high floodplain due to the increasing height of the river silt, so the high floodplain seems lower.

The stretch from Dongbatou to Taochengfu: The river width is 5-6 km with a maximum width of 20 km. Dikes were built at this stretch to prevent floods from overflowing to the floodplain, so most of the silt deposited in the river channel, where with cross gradients as high as 1/2,000 to 1/3,000, it forms a hanging river over and above the floodplain.

L. M. Brush et al. (eds.), Taming the Yellow River: Silt and Floods, 289–298.

The stretch from Taochengfu to Xihekou: The river width is 1-5 km, and the narrowest place is 400 m wide. There are many curves and dangerous sections in this stretch, so ice floods often occur. The stretch near the river outlet has changed its course many times, swinging left and right of the fan.

Since the beginning of recorded history, breaks at the lower Yellow River have occured 1,593 times with 26 major course changes. The affected area of the flood is 250,000 sq. km, spreading west to Taohuayu, north to Tianjin, south to the Changjiang-Huaihe Rivers and east to the beaches of the Bohai Sea and the Huanghai Sea. Every break brought a severe disaster to industrial and agricultural production, and the life and property of the people, disturbing natural river systems and intensifying the harmful effects of drought, flood, salinity and sand on the Huang-Huai-hai plain, which is an important base of agricultural and industrial production of China.

At present, the flood protection work is not effective. When a severe flood occurs, even with the Sanmenxia reservoir in place, the flood peak at Huayuankou will still be 46,000 cms, which is well above the current capacity of the flood prevention works. Prevention measures should be strengthened to ensure safety along the lower reaches.

Discussion of Projects of Yellow River Regulation and Flood Prevention on the Basis of the Characteristics of the Yellow River

The Yellow River, which passes over the largest Loess Plateau of the world, is known worldwide for its abundant silt. Its silt load is ten to scores higher than that of any other river of the world, so its regulation and flood prevention are most complex. Silt and water of the Yellow River influence and interact with each other.

Various opinions and projects have been proposed for regulation, development and flood prevention of the Yellow River, such as enhancing soil and water conservation work; changing the river course; opening up flood relief channels at the lower reaches; strengthening the height of the dikes; warping on a large scale; building reservoirs on tributaries and the main stream; and transferring the Changjiang River water into the Yellow River. Not one single measure can solve the complex problems of the Yellow River. The most efficient way to solve the problem is to combine the overall, comprehensive and systematic measures together by considering the upper and lower Yellow River. The evaluation of the feasibility and efficiency of each project should be based on the characteristics of the silt and water changing rules of the Yellow River, the national financial condition and basin economic conditions. Every project has its own strong and weak points, limitations and difficulties. Simple descriptions of several projects follow.

Projects of soil and water conservation: This method is accepted as a basic measure for the regulation of the Yellow River and it needs to be strengthened. According to the historical record, once there was a luxuriant forest and clear water which played an important role in the ecological environment. On one hand, the work of soil and water conservation is still efficient in reducing sediment. On the other hand, it will take scores of years to achieve obvious efficiencies in soil and water conservation on a large scale. The reason is that there is a large population living on the Loess Plateau; drought, low precipitation, severe soil erosion, sparse vegetation, extensive cultivation and a weak economic basis are unfavorable factors for the implementation of conservation work. Returning cropland to forests and animal husbandry suggests that engineering measures (such as gully treatment, dam building, water storage and silt retention) should be carried out to improve the standard of living and production levels of the people.

Project of Changing the River Course: It is possible to make the Yellow River flow along a new course within the best topography and shortest distance. Although the economic loss of the planned change of the river course is less than that of river course changes by river breaches, it still requires the abandonment of 534,000 ha of cropfields, the movement of 2.5-3 million people and the reconstruction of irrigation systems, drainage and communication. This will cost scores of billions of Yuan. Unfortunately, this project can't solve the problem of river sediments, and this "new" Yellow River will become a "hanging river" again within 100 years. Also the new dikes will become unstable and will not be able to keep the floods in check.

Project of Opening up Flood Relief Channels at the Lower River Reaches: Opening up flood relief channels will reduce the silt-carrying capacity and the power of scouring the main channel. Even worse, the main river channel will undergo deposition again. "Sedimentation must be the consequence of river diversion" -- which is the experience of the Yellow River regulation gained by successive dynasties. Flood diversion projects will cost much as well.

Project of Dike Heightening and Strenthening: These projects are neccesary to prevent floods, but can't bring the floods of the Yellow River under permanent control. Yearly safety along the river can't be ensured by this project. The work of heightening long, lengths of dikes is difficult, and the cost is also enormous--about 1 billion Yuan. The dangerous situation will become greater with the increasing height of the dikes.

Project of Building Reservoirs on the Main Streams and Tributaries: Building reservoirs on the main streams and tributaries may control the floods, regulate silt and reduce river sediments. The floods of the Yellow River, which are different from those of other rivers, mainly come from the upper and middle reaches. There are no flood flows into the lower reaches of the Yellow River. Therefore floods can be controlled by building reservoirs along the upper reaches in valleys and mountainous areas, and in the main tributaries. Flood peaks of the Yellow River are high but of small volume. The flood duration lasts only 12 days and is mainly concentrated within 5 days, so that flood control reservoirs with small storage can reduce the flood discharge effectively. The Yellow River has great silt-carrying capacity during flood periods and high precipitation years. Silt and water rates can be regulated by reservoirs to carry more silt into the sea during these periods.

To solve the problems of flood prevention, water and soil conservation measures should be taken and reservoirs built. In recent years, the problems of the Yellow River have been mitigated by regulating water and silt. These measures have helped avoid the disadvantages of the river, and have served the purpose of reducing sediment, strengthening the dikes, reclaiming soils, and irrigating farmland. For the long run, water should be transferred from the Changjiang River into the Yellow River.

Solving the Problem of the Yellow River by Using its Silt and Water

1. The lower Yellow River becomes a river above ground with both banks 1,369.5 km long. Each year a great amount of earth is used to widen, heighten and strengthen its two banks. The plain along the lower reaches is part of the Huang-Huai-Hai Plain, which has suffered from floods, droughts, waterlogging, saline and sand for thousands of years. There was once about 670,000 ha of sandy saline land, making up 40% of cropfields on this plain. This plain is severely deficient in water: about several million hectares of farmland need irrigation from the Yellow River, a prerequisite for the rational utilization of silt and water resources of the Yellow River. It has been found that there are 0.8-1.5 kg N, 1.5 kg P, and 20 kg K in one ton of silt. The key to the regulation of the Yellow River is managing the balance of water and silt to reduce the sediments at the lower reaches and increase the flood-carrying capacity.

2. Combination of warping and irrigation to fully play the advantages and avoid the disadvantages of the Yellow River. If the combination of silt and water utilization is proper, the silt and water of the Yellow River can be put into full use. There are two methods to accomplish this. One is by

warping the low land outside the dikes, and the other is by soil reclamation through warping. The main measure adopted for strengthening the dikes is to warp the lowland outside the dikes. This has the advantage of less investment, small labor forces and high efficiency. The first step is to draw out the Yellow River water by gravity flow, and then raise the water by engines or suction dredges. A great quantity of water left by warping is beneficial to the irrigation of the plain. It is known that each suction dredge can suck 200,000-300,000 cu.m of silt and leave 1 million cu.m of water for irrigation of 670,000 ha of paddyfields or 1,340,000 ha of dry fields. The scale of warping by gravity flow or by pumping is larger than that of suction dredges, and more water left by warping can be used for irrigation, and groundwater recharge. Sand sedimentation outside dikes and irrigation of the plains should be integrated, otherwise, a severe waste of water resources cannot be avoided.

Warping lowland at the back of the levee and strengthening dikes using suction dredges has achieved obvious efficiencies in flood prevention and sediment reduction in the lower reaches. For example, when 200 suction dredges were operated at the same time at the lower reaches in Shandong province, their total flow was 70-80 cms and silt volume 200 kg/cu.m, ten times higher than the average silt volume of the Yellow River during that same period. The total silt volume of 200 suction dredgers is approximately equal to 800-900 cms of the silt volume of the natural river. It is estimated that the annual silt volume of 500 suction dredges is up to 120 million cu.m, greatly surpassing the average annual river sediment of 0.1 billion tons.

Land reclamation by silt sedimentation is a main method to deal with saline sandy land in the warping areas. Silt sedimentation, land reclamation and land returned to cultivation should be combined to gain high efficiencies in silt deposition. Slight side seepage outside the dikes because of warping may occur. A large area along the lower reaches should be planned as a strip warping area, but the acreage of each strip should not be too large, so that the efficiency of equal deposition and a smooth warping surface can be achieved. Sand of the Yellow River is much more abundant than mud in non-flood season. Sand can be deposited in the bottom of the warping strip, and in the flood season, mud on the warping strip can be used to help the fields.

Table 1. Grain Diameters of Suspended Load in Sanxia and Chincang (in mm)

Station month	1 - 3	4 - 6	7 - 9	10 - 12
Shanxia	0.0462	0.0362	0.0272	0.0375
Chincang	0.0375	0.0292	0.0197	0.0320

According to the investigations of the Yellow River Project Committee, several areas along the lower reaches are suitable for the establishment of large warping areas. These are Yuanyang to Fengqiu, Lankao county to Dongming county, the Taiqian area, and the Wenmeng area. The warping area from Yuanyang to Fengqiu is an example. This area covers an area of 532 sq. m. The elevation difference inside and outside the dikes is 6-11 m. There are large areas of lowlands and saline lands in this warping area. The task of dike protection is arduous during the flood season. Land reclamation by warping and raising rice with the Yellow River water have been carried out for many years in this area. On the basis of this area's planning made by the Yellow River Project Committee, annual sand quantity by warping is 105 million tons, and the annual river sediment reduction is 40-100 million tons. The depth of warping is 5.2 m high, with a total capacity of 2,780 million cu.m. The warping volume is 3,890 million tons. The utilization period of this warping area is 37 years.

In Zhongmu, Kaifeng, Weishi, the eastern part of Henan, there is about a million ha of undulating sandy and saline land, and hillocks alternating with depressions as the "products" of all previous flooding of the Yellow River. On the southern side of the saline sandy area, there is a mixture of warped soil, sandy soil mixed with warped soil, and other agricultural soils. This area, a part of the Zhoukou warping area, is deficient in irrigation water. The lower groundwater level is due to the utilization of well irrigation. In this area, land reclamation by warping on the upper reaches, and field irrigation with water left by warping on the lower reaches will have double efficiencies. It is planned that in this area, the drawing flow of the Yellow River will be up to 213 cms, reclaiming 58,800 ha and irrigating 670,000 ha of land with the water left over by warping. The utilization of this warping area will bring great benefits, improving water and soil conditions of millions of hectares of cropfields.

Transferring the Changjiang River Water into the Yellow River and Improving the Ecological Environment

In the study of agro-climatic resources for planning in China, the gains and losses in water are the differences between precipitation and evaporation. This is the basis for drawing the line between annual water gains and losses. The balance line begins in Shandong province, via the Huaihe River, the Chinling Mountains, to the lower reaches of the Yalung Zangbo River, Xizang. Water-deficient areas are on the north side of this line, and water abundant areas are on the south side. The whole Yellow River basin is located in a water-deficient area. The annual water deficiency in the northwestern part is 600-1,000 mm, 600-800 mm in Inner Mongolia and Ningxia area, and 200-400 mm in the North China Plain.

The water source of the Yellow River is mainly from precipitation with annual runoff of 58,000 million cu.m.

The annual average runoff of the basin is 77 mm, making up 28% of the average runoff of China, and 14% of that in the Changjiang River basin. The runoff is below 50 mm in some areas along the middle and lower reaches, with 4,500 cu.m/ha accounting for 17% of the average value of water quantity per hectare of China, or below 1/10 of that of the Zhujiang River basin and the Changjiang River basin.

In the irrigation areas of China, waters of the Yellow River are allocated to two types of irrigation areas -- unstable irrigation and perennial irrigation. The average annual precipitation can't satisfy the normal water requirements of various crops. In Ningxia, Shanxi, and Shaanxi, natural precipitation less than 50% of the water requirements of crops. In these areas, the annual precipitation change is great with uneven distribution, 70-80% of precipitation of the whole year concentrates in July-September, so only a small amount can be utilized by crops (see Table 2).

Before 1949, there was about 334,000 ha of irrigated land in the whole Yellow River basin, with water consumption over 4,000 million cu.m. Water irrigation has developed quickly with the development of economic reconstruction. After liberation, by the end of 1980, the irrigated area was up to 5,334,000 ha (including 1,200,000 ha irrigated with groundwater, 1 million ha outside the basin withdrawing water of the Yellow River), a score more than that before liberation. At present, the annual water consumption of industry and agriculture is about 27 billion cu.m, of which, above Huayuankou, the annual water consumption is over 17 billion cu.m, and below Huayuankou, the annual water drawn from the Yellow River is 10 billion cu.m. Water requirements

Table 2. Agricultural Water Quantity Gains and Losses

Region		Northwest Region	Ningxia	North China Region
		Xi'an	Yinchuan	Shijiazhuang
1951-70 annual average precipitation		604.2 mm 6042.0 cu.m/ha	205.4 mm 2053.5 cu.m/ha	558.9 mm 5589.0 cu.m/ha
water need of crops	spring wheat		3000-4500 cu.m/ha	
	winter wheat	4050.0 cu.m/ha		3300.0 cu.m/ha
	maize	3750.0 cu.m/ha		3000.0 cu.m/ha
yearly water quantity of gains and losses		-2658.0 cu.m/ha	-2446.5 cu.m/ha	-722.0 cu.m/ha

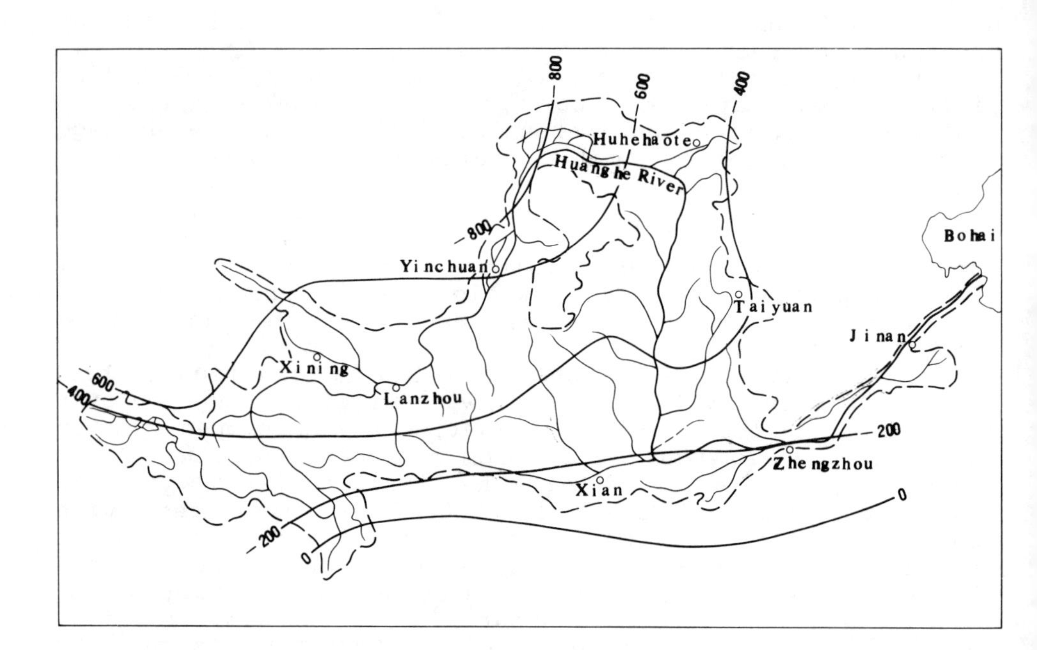

Figure 1. Annual Distribution of Water Quantity of Gains and Losses in the Yellow River Basin.

of industry and agriculture will increase with the development along the middle and lower reaches, which is another factor emphasizing the inadequacy of water flow of the Yellow River.

There are regional differences in runoff distribution, unequal annual allocations and great yearly fluctuations in the Yellow River basin. For example, the controlled basin area above Lanzhou only makes up 26.9% of the whole basin area, perennial runoff is 32.5 billion cu.m accounting for 58% of the whole basin. The basin area from Lanzhou down to Hekouzhen makes up 21.7% of the whole basin with a perennial water yield of 300 million cu.m. In order to rationally use water resources, a series of water projects should be developed in stages. Each stage of development at the middle reaches for irrigation, forestation, or land reclamation will consume a certain volume of water and will reduce the quantity of flow of the lower Yellow River.

According to observational data, from 1969 to 1978 the annual average runoff at Huayuankou was only 37.4 billion cu.m, 10 billion cu.m less than that in the 1950s. The flow into the sea during April-June each year is only 3.5 billion cu.m. The river at the lower reaches during 1972-1981 dried up 8 times for a total of 118 days. The 666 km strech from Gaochun village of Shandong province down to the river outlet dried up for 12-32 days. The flow reduction made the Yellow River water-silt relation unbalanced. It is estimated that drawing 5 billion cu.m of water from the reaches from Lanzhou will increase the silt by 0.1 billion tons in the lower reaches. After Longyangxia reservoir is put into use, the river sediments will increase by 80-90 million tons due to the discharge reduction of the upper reaches. On the basis of the observed data from 1981-1985, yearly duration of medium flow quantity (> 5000 cms) is closely linked with the silt discharge into the sea. In 1981, 3 billion tons of water could transport 0.1 billion tons of silt into the sea. In 1982, 5.5 billion tons of water only transported 0.1 billion tons of silt into the sea (see Table 3).

Table 3 Medium Flow Quantity Duration During 1981-1985

year	>3000 cms duration	>4000 cms duration	>5000 cms duration	flow quantity for 0.1 billion tons of outlet silt
1981	62	43	24	30.1
1982	17	13	7	54.8
1983	86	47	12	48.1
1984	67		11	47.8
1985	38	25	7	51.0
average				45.0

The only way to maintain a scouring force is to increase the flow in the lower reaches.

From the above descriptions, the main problem of the Yellow River is shown to be an imbalance between water and silt. The imbalance trends will be severe with development and the need for increased water supply. Expanding water sources is the key to solving the problems of the Yellow River. Sufficient flows can scour away silt, reduce river sediment, lower the river channel, prevent flood disasters and solve the problem of water needs along two banks of the Yellow River.

Transferring water from south to north is an inevitable way to solve the problem of water deficiency in the north. According to the plan for transferring water from south to north along the middle course, the dam of Danjiangkou reservoir should be 175 m high, drawing flow rate should be up to 1200 cms, and annual outlet water flow 23 billion cms. Thus it is possible to draw the water from the reservoir into the Yellow River to scour away silt during proper periods each year.

If the planning of the Yellow River regulation is combined with transferring water from south to north along the middle course, the project of transferring the Changjiang River water into the Yellow River will be realized. This will basically change the current conditions of the Yellow River, without moving the people along the two banks to other places; occupying crop fields; and affecting the industrial and agricultural production distribution. After the diversion is completed, the proportion of water and silt will be balanced, and the flood prevention and water requirements at the lower Yellow River can be satisfied. Therefore this project is considered to be effective for Yellow River regulation.

A STORY OF SOIL LOSS AND SOIL DEGRADATION ON THE LOESS PLATEAU IN CHINA

Tang Keli, Zhong-zi, and Kong Xiao-ling
Northwestern Institute of Soil and Water Conservation
Academic Sinica
The People's Republic of China

ABSTRACT: The Loess Plateau lies in the middle reaches of the Huanghe or Yellow River. Grievous soil erosion has caused the Huanghe to transport an average annual sediment load of 1.6 billion tons. Of this, about 400 million tons are deposited in the riverbed of the lower reaches, and the rest, or about 1.2 billion tons, empty into the sea, along with the loss of much fertile soil.

Since the sediment problem is the most direct and most outstanding one in controlling the Huanghe River, people have often focused their attention on the amount of sediment as well as the silt component of the sediments on the river bed. Little has been reported on the quality of river sediment and assessments of soil degradation in eroded lands.

Sediment produced at the watershed comes from sheet erosion on the slope surfaces and gully erosion in the watershed. Every year 1.2 billion tons of sediment is carried by the Huanghe to the sea. Although this doesn't constitute a direct threat to the riverbed in the lower reaches, it is evidence of soil degradation in the Loess Plateau. The main content of this paper deals with the quality of river sediment, and assesses the problem of soil degradation caused by soil erosion. Futhermore, this paper probes into the patterns of soil erosion, the location and strata of sediment yield in the sediment-yielding areas.

Introduction

Soil erosion and sediment yield are processes not only of soil dispersion and sediment transportation, but also of adsorption and transportation of certain nutrients and other compounds. The size and characteristics of sediment particles have an obvious effect on the enrichment of the compounds. Therefore, a new approach has been taken in analyzing the sediment

L. M. Brush et al. (eds.), Taming the Yellow River: Silt and Floods, 299–313.

sources. Based on the characteristics of the loess soil and the transportation processes, the available nitrogen presents an unstable index with a large amount lost in the transportation process to the river. The contents of total phosphorus and potassium are not sensitive indices for either mature or non-mature loess soils (Tang et al, 1984). In this paper, the components of particle size, the content of organic matter, total nitrogen, hydrolyzable nitrogen and available phosphate and potassium are used as indices to study the relationship between river sediments and soils from the eroded fields, as well as the problem of soil degradation and sediment sources.

Sediment samples were collected at different discharges with 2-3 typical ones taken during the flood periods every year. The hydrological stations collected the samples at the water depth of 0.5 m (although some were taken at 0.9 m) according to standard specifications. Most soil samples were taken from the surface layers of the sloping farmland in combination with the collection of different components of materials from the eroded ground.

Mechanical composition was determined by the pipette method; the content of organic matter by the Turin method; total nitrogen by seminicro Kjeldahl method; hydrolyzable nitrogen by photoelectric colorimetry with potassium permanganatic ammonia reagent; available phosphorus by the Olsen method and available potassium by flame photometry extracted with ammonium acetate (immersion method).

MECHANICAL COMPOSITION OF THE SEDIMENTS IN THE MAIN TRIBUTARIES OF THE HUANGHE RIVER AND OF COMPONENTS OF THE SOIL FROM THE ERODED FIELDS

The mechanical composition of sediments and soils are shown in Tables 1 and 2, which show that the particle sizes of sediments become gradually finer from the north to the south as arranged in order from top to bottom in Table 2. Apart from the sediments of the Huangpuchuan Stream and the Kuyehe River which contain a considerable amount of coarse sand, the mechanical composition of the sediments in the other tributaries of the Huanghe is similar to that of loess in the eroded areas with coarse silts (0.05-0.01mm) and fine sands (0.25-0.05) predominant. This shows that sediment in most tributaries of the Huanghe originates from silty and fine sandy loess, i.e., the Malan loess (Q3), which covers most of the eroded areas on the Loess Plateau. In general, the content of the coarse silts (>0.05mm) in the river sediment is slightly lower than those of soils in the eroded areas, so that coarse grains will deposit while fine grains will be relatively enriched in the transportation of river sediments. In addition, it is likely that fine grains in sediments come from the loess deposited in the early Pleistocene (Q3) (See

Table 1. Mechanical composition of sediments in major tributaries of the Huanghe River (1983)

Rivers	Hydrological stations	Percentage of various particle sizes (mm)							
		1-0.25	0.25-0.05	0.05-0.01	0.01-0.005	0.005-0.001	< 0.001	< 0.01	>0.05
Huanghe River	Sanmen Gorge	0	20.43	62.09	2.96	2.46	12.06	17.48	20.53
Huanghe River	dam-site silt	0.25	71.61	19.19	1.24	3.19	4.43	8.86	71.86
Huangpuchuan Stream	Huangpuchuan	10.56	31.76	24.64	4.36	9.98	18.72	33.03	42.33
Kuyehe River	Wenjiachuan	34.19	51.70	7.34	0.34	4.05	2.37	6.76	85.89
Jialu River	Shenjiawan	0.21	22.21	54.33	3.93	6.17	13.15	23.25	22.42
Dalihe River	Suide	0.29	12.34	61.78	6.29	4.84	14.47	25.60	12.63
Sanchuan River	Houdacheng	0.87	6.60	63.75	5.15	5.37	17.94	28.46	7.47
Qingjian River	Yanchuan	0	11.79	62.00	6.35	5.11	14.82	26.87	11.79
Yanhe River	Ganguyi	3.36	15.16	56.93	6.19	4.94	13.43	24.54	18.52
Lohe River	Jiankou	0.20	9.24	70.52	5.88	5.09	11.32	22.26	9.44
Fenhe River	Hejin	0.16	7.42	44.21	13.03	13.72	21.46	48.21	7.58
Weihe River	Xianyang	0.36	8.11	44.86	10.98	14.14	21.57	46.69	8.47

Table 2. Mechanical composition of plow layer soils in the major areas on the Loess Plateau

Sampling Location	Catchments	Percentage(%) of various particle sizes(mm) 1-0.25	0.25-0.05	0.05-0.01	0.01-0.005	0.005-0.001	<0.001	<0.01	<0.05
Fugu	Huangpuchuan Stream	1.03	52.04	24.61	2.70	7.84	8.09	18.63	52.07
Shenmu	Kuyehe River	0.38	49.70	33.49	2.70	6.29	7.44	16.43	50.08
Yulin	Wuding River	0.84	42.40	36.50	2.67	4.70	12.90	20.27	43.24
Zizhou	Dali River	0.15	25.63	57.73	2.52	5.04	8.92	16.48	25.78
Lishi	Sanchuan River	0.40	32.75	43.10	6.63	8.23	7.37	22.23	33.15
Zichang	Qingjian River	0.20	25.57	61.98	1.82	2.96	7.48	12.26	25.77
Yanan	Yanhe River	-	20.88	55.67	5.57	8.59	9.90	23.25	20.88
Huangling	Luohe River	1.03	16.04	45.65	9.26	19.01	8.66	36.93	17.07
Changwu	Jinhe River	0.24	4.46	53.14	14.96	21.65	6.55	42.16	4.90
Tianshui	Weihe River	0.25	3.28	48.07	11.78	9.34	27.69	48.81	3.53

Table 3), but this viewpoint needs further study.

By analyzing the components of sediment particles in the Huangpuchuan Stream and the Kuyehe River it is known that the sediment particles larger than 0.25mm cannot come from the loess soils but originate from the local weathering of sandrocks and the moving sand deposits (see Table 3), while the total amount of sediment particles larger than 0.05mm increases correspondingly. In the Huangpuchuan stream watershed, the sandrock outcrop and the schistose sand cover areas together comprise 45% of the total area in the watershed and form an important source of the river sediments.

Table 3. Particle sizes of various ground covering materials on the loess plateau

Sampling Locations	Materials	Percentage of various particle sizes(mm)			Percentage(%) of various particle sizes(mm)				
		1-0.25	0.25-0.05	.05-0.01	0.01-0.005	0.005-0.001	< 0.001	<0.01	>0.05
Zhunger Flag	cultivated weathering sandrock	26.96	58.77	6.20	0	1.65	6.41	8.06	85.73
Fugu	weathered sandrock	47.19	32.54	9.68	0.39	4.14	5.97	10,30	79.73
Fugu	weathered mudstone	1.48	4.45	13.05	20.52	30.03	21.52	81.07	5.93
Zhunger Flag	moving sands	3.52	88.82	1.67	0	0.41	5.57	5.98	92.07
Jingbian	Old loess(Q3)	0	8.43	49.31	7.25	5.22	11.18	23.65	8.43

Coarse particles (> 0.25mm) couldn't be seen in the river sediments of the Sanmen Gorge Section, the deposited sediments of the gorge dam-site, or in those of the lower reaches of the Huanghe. The particles with sizes of 0.25-0.05mm made up 20% of the river sediments, but reached as high as 71% of the deposited sediments on the dam-site of the gorge. It can be seen from this that the sediments with particle sizes of (> 0.25mm) carried by the Huangpuchuan Stream and the Kuyehe River have very strong selective behavior in the process of transportation, so that the coarse particles will deposit in the river course along the way and eventually will disappear in the lower reaches of the Huanghe River. It can be seen from the above-mentioned data that the sediment with particle sizes of 0.25-0.05mm may have the greatest effect upon the deposits of the riverbed in the lower reaches of the Huanghe. They mainly come from the loess deposits of the late Pleistocene of the Quaternary on the Loess Plateau and are usually called the yellow cotton soil and the sandy loess soil. The major region of coarse sediment yield borders Yanan on the north.

AVAILABLE NUTRIENTS OF SEDIMENTS IN MAIN TRIBUTARIES OF THE HUANGHE RIVER

With soils of the plow layer from the eroded farmlands as a check, a comparison of nutrient contents of total N, available P, and available K was made with the results summarized in Figures 1, 2 and 3. The contents of three kinds of essential nutrients appear to have regional characteristics. They tend to become greater and greater from north to south in accordance with the geographical distribution of each tributary, roughly agreeing with the geographical regularity of the mechanical composition of loess. The content of total nitrogen in sediments in most tributaries is close to or higher than the lowest value of those in soils. Some of them are close to the average value, but those in the Weihe River are higher than the value of the upper limit. The content of available phosphorous in sediments is obviously enriched, and most of it is higher than in the soils, even reaching 2-3 times as high. In general, the content of available potassium is close to the average value of soils with some individual samples slightly lower or higher than the soils.

Fig. 1. Comparison of total nitrogen contents in sediments of main tributaries of the Huanghe River and in soils from the eroded regions

1. The Huangpuchuan Stream
2. The Kuyehe River
3. The Jialuhe River
4. The Sanchuanhe River
5. The Dalihe River
6. The Qingjianhe River
7. The Yanhe River
8. The Luohe River
9. The Fenhe River
10. The Weihe River
11. The Huanghe River (The Dragon Gate)
12. The Huanghe River (Tonguan)
13. The Huanghe River (The Sanmen Gorge)
14. The Range of Contents of Total Nitrogen in Soils

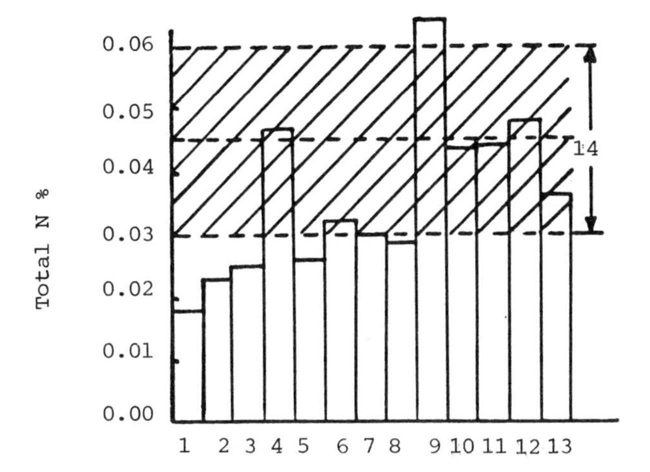

Figure 1. Comparison of Total Nitrogen Contents in Sediments of Main Tributaries of the Huanghe River and in Soils from the Eroded Regions.

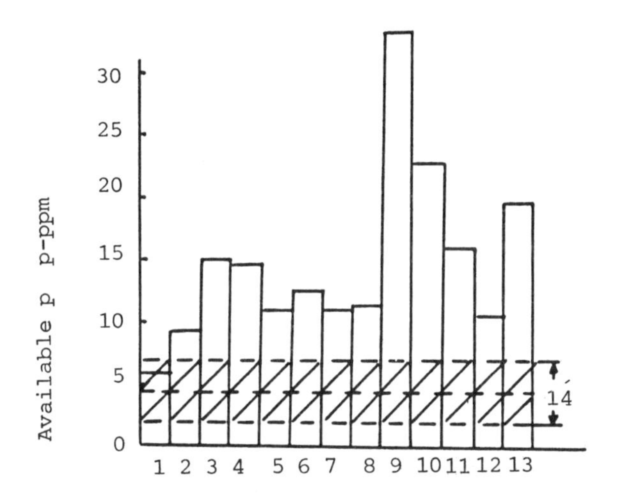

Figure 2. Comparison of Available Phosphorus Content in Sediments of Main Tributaries of the Huanghe River and in Soils from the Eroded Regions

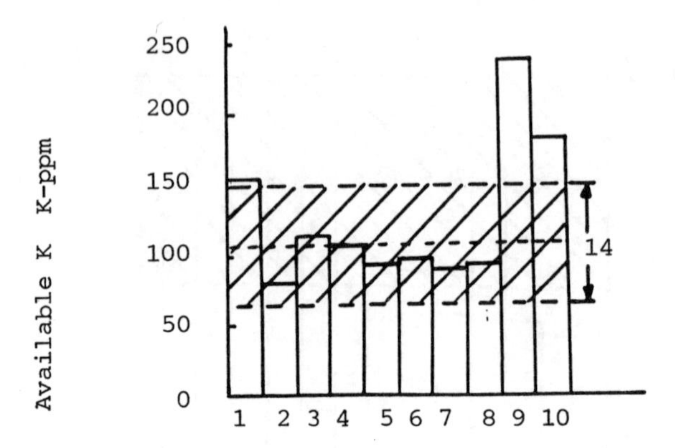

Figure 3. Comparison of Available Potassium Content in Sediments of Main Tributaries of the Huanghe River and in Soils from the Eroded Regions

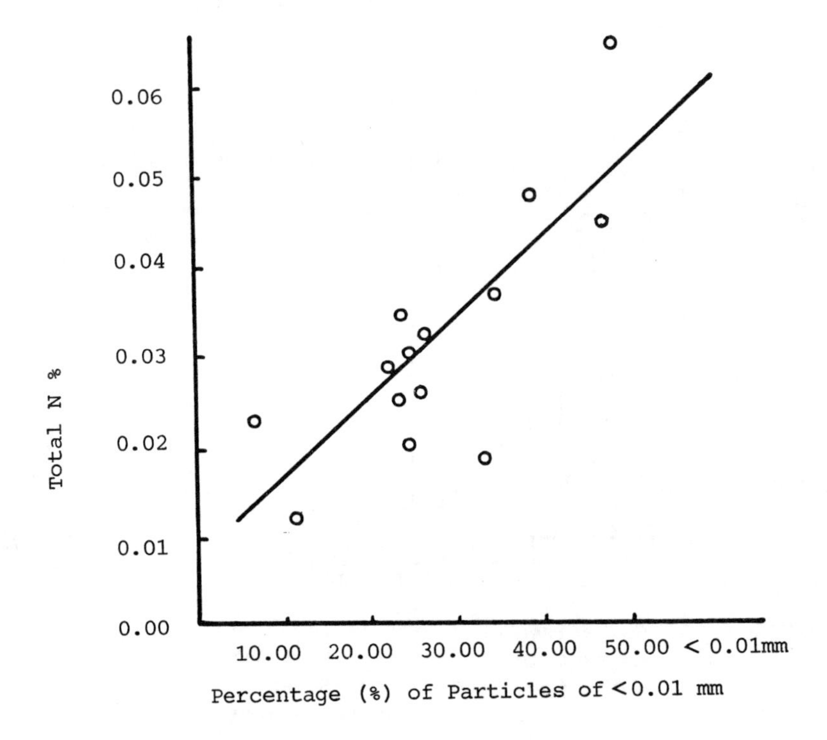

Figure 4. Relationship Between the Content of Total N and the Amounts of Fine Particles <0.001mm

A comparison will now be made between the ratio of the same materials in the river sediment and the soils from the eroded regions, i.e., the enrichment ratio (Table 4). The enrichment ratio of total nitrogen is 0.36-1.24, but most are over 0.5. The enrichment ratio of hydrolyzable nitrogen and available potassium is 1.0 or more. Some of the samples have a hydrolyzable nitrogen content above 2.0. The enrichment ratio of available P is 0.99-4.47--the greatest value among them--but the average value is 2.7. Enrichment of these nutrients appears to be much more obvious in sediments in the lower reaches of the Huanghe at the Sanmenxia Hydrological Station where the enrichment ratio of available phosphorous reaches 3.89. The data show that the closer the river sediments approach to the lower reaches, the more the enrichment of nutrients in the river sediments. It can be estimated that the sediment sources originate mainly from the surface soils of the cultivated farmlands slopes, and the materials are loess soils. Therefore, the cultivated farmland slopes are the priority regions to be controlled.

The relationship between the amount of particles 0.01mm and the content of nutrients in the sediments of the major tributaries of the Huanghe River shows a straight-line relation as illustrated in Figures 4, 5 and 6. As the fine particles increase, the content of total N, P and K appear to have a tendency to increase progressively. By statistical analysis, the following regression equations result: For total N, $y=0.0075+0.0009x$, $r=0.7427$; for available P, $y=0.8784+0.4887x$, $r=0.7724$; for available K, $y=23.40+3.638x$, $r=0.8722$. The correlation coefficients are statistically significant.

It can be seen from the data that although the absolute content of fine particles in sediment is roughly equal to that in the soils from the eroded regions, the quality of the sediment has changed greatly because of the increase in the amounts of various nutrients it contains. For instance, the enrichment ratios of fine particles in the Yanhe River and the Weihe River are close to 1. Nevertheless, the enrichment ratios of available P in the two rivers are 4.47 and 3.40, respectively. Where the content of fine particles is the same, the enrichment of available P in sediment is 3-4 times over that in the soil. If the loss of available P is estimated in terms of sediment, its value will be greater than that measured in soils. The former directly reflects the actual conditions of soil loss and sediment delivery. Based on these researches, the enrichment ratio of available P can be used as an important index to illustrate soil degradation.

Table 4. The enrichment ratio of components of particle sizes and nutrient contents in river sediments and soils from plough layers in eroded regions

Watershed	Particle Size >0.05 (mm)	Particle Size >0.01 (mm)	Total N	Hyd. N	Available P	Available K
Huangpuchuan Stream	0.80	1.77	0.61	1.48	1.30	1.82
Kuyehe River	1.71	0.41	0.58	2.59	3.25	1.24
Dalihe River	0.49	1.55	0.76	1.36	2.17	1.22
Sanchuanhe River	0.23	1.28	1.24	2.46	0.99	0.99
Qingjianhe River	0.46	2.14	1.14	1.11	2.85	1.50
Yanhe River	0.88	1.06	0.75	0.93	4.47	0.96
Luohe River	0.55	0.60	0.36	0.69	3.17	0.69
Weihe River	2.39	0.96	0.75	0.97	3.40	0.83
Huanghe River (Sanmen Gorge)	0.72	0.72	0.82	1.52	3.89	-

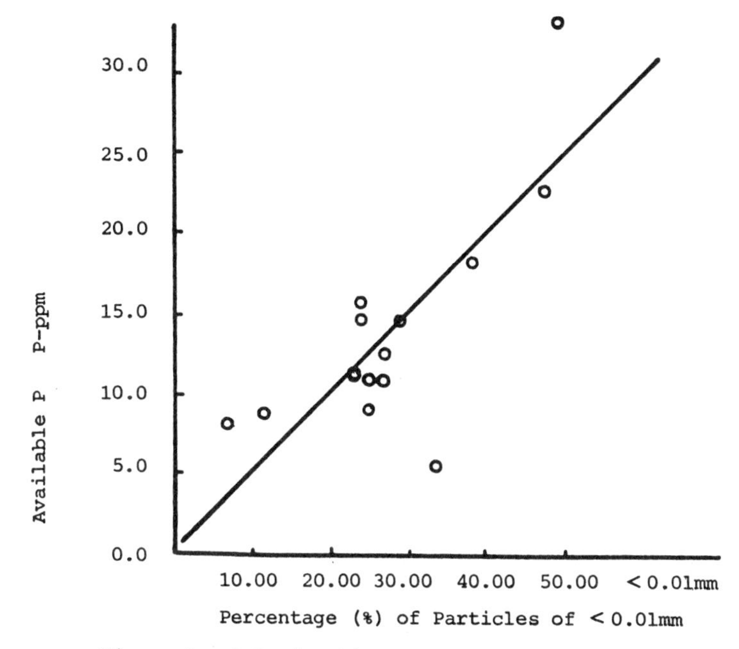

Figure 5. Relationship Between the Contents of Available P and the Amounts of Fine Particles < 0.01mm

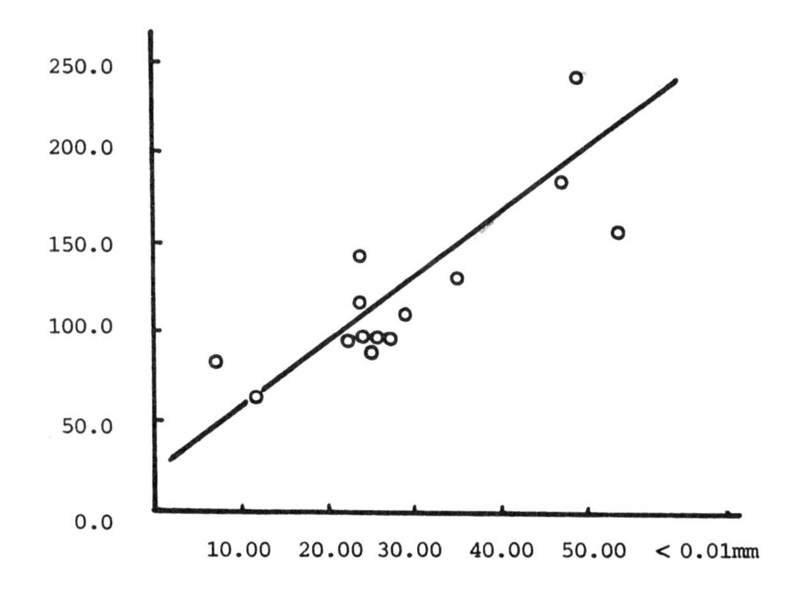

Figure 6. Relationship Between the Contents of Available K and the Amounts of Fine Particles < 0.01mm

Table 5. Variation of nutrient contents at different discharges (1983)

River	Discharge cms	Organic matter (%)	Total N (%)	Hyd. N mg/100g	Avail. P P-ppm	Avail. K K-ppm	Particle size (%) > 0.05	Particle size (%) > 0.01
Huangpuchuan Stream	66	0.30	0.020	4.34	6.2	202.1	24.73	43.27
	306	0.21	0.014	1.04	5.5	110.2	59.93	22.78
Dalihe River	6.5	0.42	0.032	3.16	21.7	132.8	2.07	31.74
	37.5	0.27	0.022	1.30	14.6	91.1	23.18	19.46
Weihe River	951		0.044	2.04	37.1	164.5	6.14	47.29
	2400		0.036	2.90	15.4	168.7	7.91	48.19

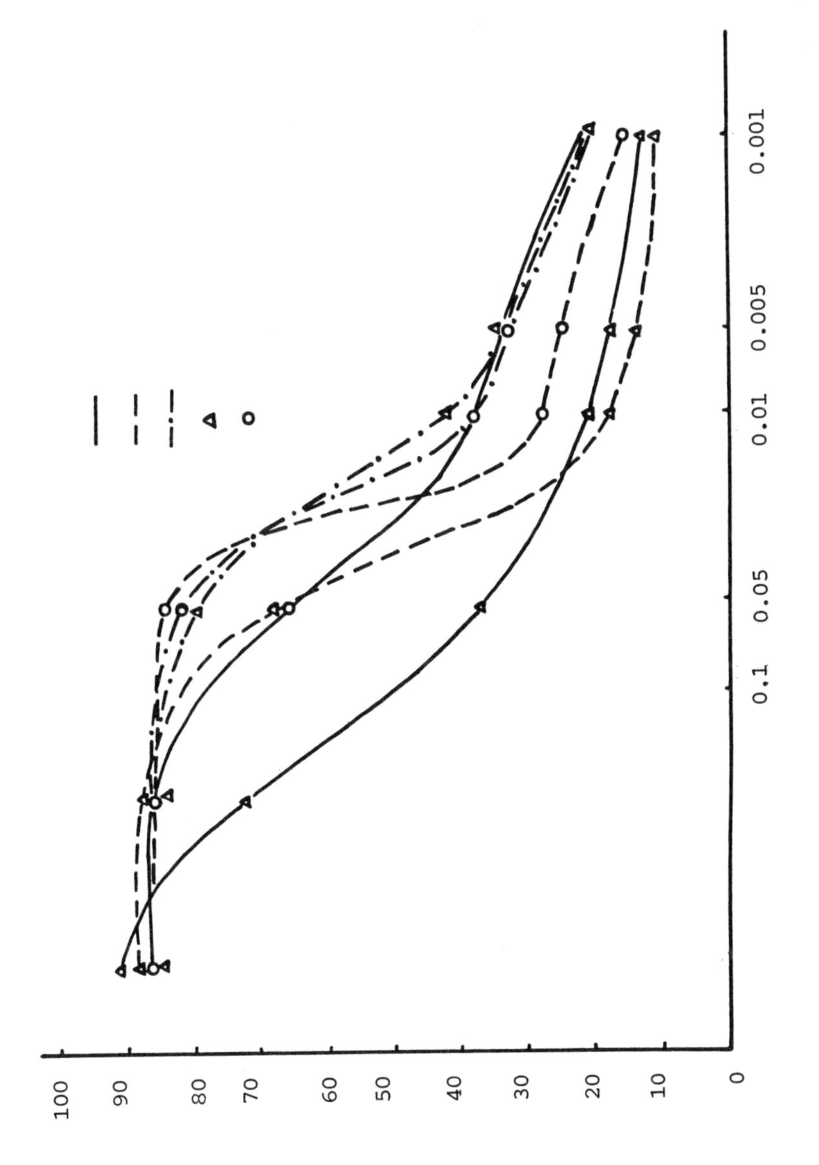

Figure 7. Variation of Particle Sizes with Discharge in Different Streams

THE INFLUENCE OF DIFFERENT DISCHARGES ON THE MECHANICAL COMPOSITION AND NUTRIENT DELIVERY IN RIVER SEDIMENTS

Under the conditions of different discharges in a year, the mechanical composition of the river sediments and the content of nutrients in the same watershed or in different rivers may have great variations as shown in Table 5 and Fig. 7.

In the case of two different discharges, the mechanical composition of sediment in the Weihe River shows basically no change. At the relatively low discharge, available P shows an obvious enrichment. The ratio of the two discharges is 2.41. Total nitrogen has a slight enrichment, and the available potassium shows no change. This illustrates that in spite of different discharges erosion patterns in the eroded regions do not change and the sources of sediments are still predominantly from the loess soils.

In the case of two different discharges in the Huangpuchuan River and the Dalihe River, the mechanical composition the nutrient content show greater changes. When discharge increases, the coarse particles also increase, but the content of available nutrients decrease. In the Huangpuchuan Stream, in particular, sediment particle sizes greater than 0.25mm increase from greater than 1% to above 20%, and the total amount of particles larger than 0.05mm is doubles.

It is obvious that in the case of the higher discharge, erosion occurs mainly in the gullies where weathered material from bedrock crops out. The sediments in the rivers are predominantly the bedload with an indistinct enrichment of available P. Accordingly, at different discharges the quality of the river sediment depends on the erosion patterns of soil loss and composition of the source materials.

In summary, sediments in the Huangpuchuan Stream and the Kuyehe River are predominantly coarse, with particles larger than 0.25mm dominant. They are produced from the weathered materials of bedrock in the eroded regions. When the particles of such a grade are carried to the Huanghe River, distinct deposition appears, but the particles basically disappear in the Sanmen Gorge section of the Huanghe River. Particles from 0.25-0.05mm and 0.05-0.01mm make up most of the sediment in the tributaries and in the lower reaches of the Huanghe. There are very few particles larger than 0.25mm. The mechanical composition of sediments basically agrees with the loess soils in the eroded regions. This clearly indicates that the loess deposits with fine sandy loess soils in particular are the major sources of sediments in the Huanghe River.

The ratio of available nutrient enrichment in the Huanghe and its tributaries is over 0.5 and that of available P is above 2. This illustrates that the farmlands and livestock lands and cultivated farmlands slopes in particular are the major sediment-yielding regions and play an important role in the transportation of sediments as well as large amounts of nutrients to the Huanghe River. So, attention should be paid to erosion control on these kinds of lands.

The amount N in the river sediments is 0.04% on the average and the available P is 10-15ppm. This means that if 1.6 billion tons of sediment are lost every year, 0.6 million tons of nitrogen, roughly equivalent to 1.3 millions tons of urea may also be lost.

References

(1) Miller, M., 1966. Anthropogenic influence of sediment quality at a source. Joint Commission for the Great Lakes, Windsor, Ontario. 61-02.

(2) Tang Ke-li, et al. 1984. A preliminary study on the influence of soil degradation. J.Env.Sci., 6.

(3) Walling, D.E. 1980. Soil erosion and sediment yield.

ON THE STRATEGY OF HARNESSING THE YELLOW RIVER

Feng Yin
Ex-Chief Engineer
Ministry of Water Resources and Electric Power
People's Republic of China

ABSTRACT: The Problem of the Yellow River Control--Proposed Plans to Harness the River.

1. To reinforce the soil conservation works on the tributaries in the middle reaches,
2. To build the Taohuayu Reservoir mainly for flood control,
3. To build three large reservoirs on the main stream of the middle reaches,
4. To provide a floodway in the lower reaches,
5. To provide settling basins along the river for silt deposition,
6. To develop many diversion channels to release the water and silt,
7. To convey 10 billion cu.m of clear water annually for sluicing the river channel, and
8. To improve the river mouth delta.

The Problem of the Yellow River Control

The Yellow River is closely related to the development of the nation and it is well known as the "cradle of Chinese civilization." Whereas the river flood has rampaged for thousands of years over the vast North China Plain, the river has been ill famed as "China's sorrow." Since the founding of the People's Republic of China, under the powerful control of the new government the river has been astonishingly peaceful with no major dike breaches for almost 40 years. But people still worry about harnessing the river in the years to come.

The main problem of the Yellow River lies in the middle and lower reaches. Figuratively speaking, this portion of the river is like a gigantic "sand glass," or hourglass, with the upper bulb collecting the water and silt into the river and with the lower bulb delivering them to the vast plain as far as the sea. Several thousands of years ago, the Yellow River

L. M. Brush et al. (eds.), Taming the Yellow River: Silt and Floods, 315–329.

took a southern route and emptied into the Yellow Sea, to the south of Shangdong Peninsula and deposited a vast delta under the sea, measuring some 50,000 sq. km in area. This gigantic "sand glass" has been working for nearly 200,000 years and will continue to work in the future (see Figures 1 and 2).

The drainage area of the middle reaches amounts to 362,000 sq. km. Within this area, 100,000 sq. km produce three-quarters of the total sediment carried by the Yellow River. The most serious erosion comes from the tributaries in the loess region. The thickness of the loess deposits varies from about 100 m to 300 m. Along the tributaries down to the smallest gullies, water erodes the loess and carries it away as the sediment load of the silt content ranges as high as 60-70%. The loose-packed loess is most susceptible to erosion and the gullies are cut down almost to rock bottom. The length of the gullies measures 3.7 km per sq. km. The modulus of erosion rates are sometimes higher than 25,000 tons/sq.km/y from the tributaries in flood seasons, and the concentrations are sometimes as high as 300-500 kg. per cu.m and more than 1,000 kg. per cu.m has been recorded at the gaging stations.

From the numerous sediment analyses made for the different locations in the drainage basin of the middle reaches, the mean particle size ranges from 0.025 mm to 0.05 mm. It has been well recognized that the coarse particles are the major source of trouble. From the sediment analyses of the runoff in lower reaches, for particles < 0.025 mm, 87.6% of which is discharged to the river downstream of Lijin, while for particles > 0.05 mm, only 44% is discharged downstream. The area of the most serious erosion in the gullied loess regions is about 50,000 sq. km, and the erosion control of these regions should be considered of prime importance.

The building of the North China Plain may have started in the post-Leossic and Aneolithic ages, but the written historical record of the Yellow River only dates back to 602 B.C. For the past 2,500 years, the Yellow River has experienced 1,500 major inundations, 26 significant channel alterations, and 7 course changes which were so great as to move the river mouth from the north of Shandong Peninsula to the south. The vast floodplain extends north to Tianjin and south almost to the Yangtze River. The sediment deposition on land has formed the vast North China Plain. The Chinese people started 2,000 years ago to build levees as a protection against floods. The siltation between the levees tends to raise the bottom of the river and hence the river runs along a ridge which is higher than the land outside the levees. Thus the Yellow River has been called a "suspended river."

The present Yellow River channel was formed in 1855, after a breach of the left dike at Lankao in Henan Province. But

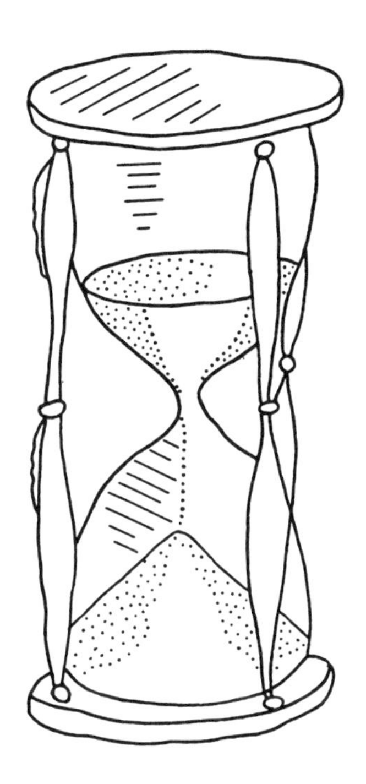

Figure 1. Sand Glass or Hourglass Schematic of the Yellow River

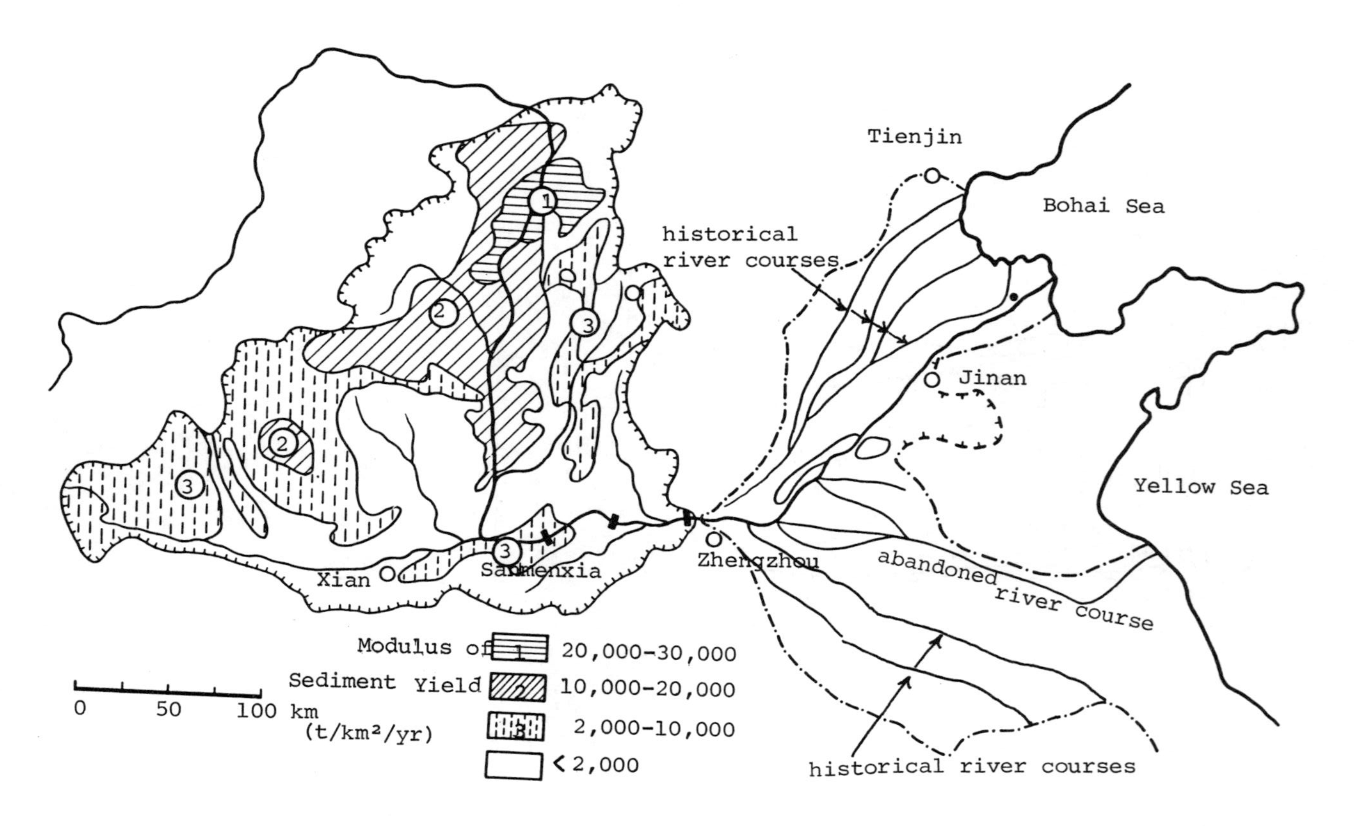

Figure 2. Sediment Source and Sink of the Yellow River Basin

the levees below Lankao were not completed until 1877. During the invasion of the Japanese army the right dike was breached by the Guomindang Army artificially at Huayuankou in Henan, and it was closed in 1946. Therefore the river has flowed in the present channel between the levees only 124 years. In Henan Province, the river dikes are spaced at 5-20 km, averaging 7 km apart, while in Shandong Province the dikes are spaced at 0.4-5 km, averaging 3 km.

There are two detention basins, one on each side of the river channel, and each basin can take 2 billion cu.m of water for storage. But there are 1.5 million people living in the two basins. It would be a serious problem to evacuate the inhabitants if the basins were called upon to store the flood waters.

More than 0.3 billion tons of silt is deposited in the river channel annually. Diversion on both banks along the river takes about 10 billion cu.m of water and more than 0.1 billion tons of silt every year. The rest of the sediment is transported to the river mouth partly to be deposited near the river mouth and partly to be transported to the remote deeper sea.

Comparing the channel profile of 1956 and 1984 with the old channel profile before 1855 (see Figure 3), we may arrive at the conclusion that without a material change of runoff characteristics including the sediment concentration and particle size, the river profile will be approximately of the same gradient. The silt transported to the river mouth is approximately 1.1 billion tons annually. The delta at the river mouth is growing at a rate of 23.5 sq. km annually and the coastline is pushing forward 0.5 km to the sea. With the extension of the river mouth and the siltation of the river channel, the riverbed is rising approximately 6-8 cm every year.

At present, the river bank is 3-5 m higher than the adjoining land behind the levee and at some places the maximum difference in elevation reaches 10 m. In the past 30 years, the levees have been raised and strengthened 3 times and a total volume of 500 million cu.m of earthwork has been accomplished. Now the levees should be rasied a fourth time, and the amounts of earthwork and investment would be still greater. As compared with the old levee system before 1855, it is estimated that the present levees may be raised for the next 60 years. But we are now confronted with the problem of whether there are better ways of controlling the river beside raising the levees, which by that time is too late to work out a better solution.

Fig. 3 Comparison of Floodplain Profiles

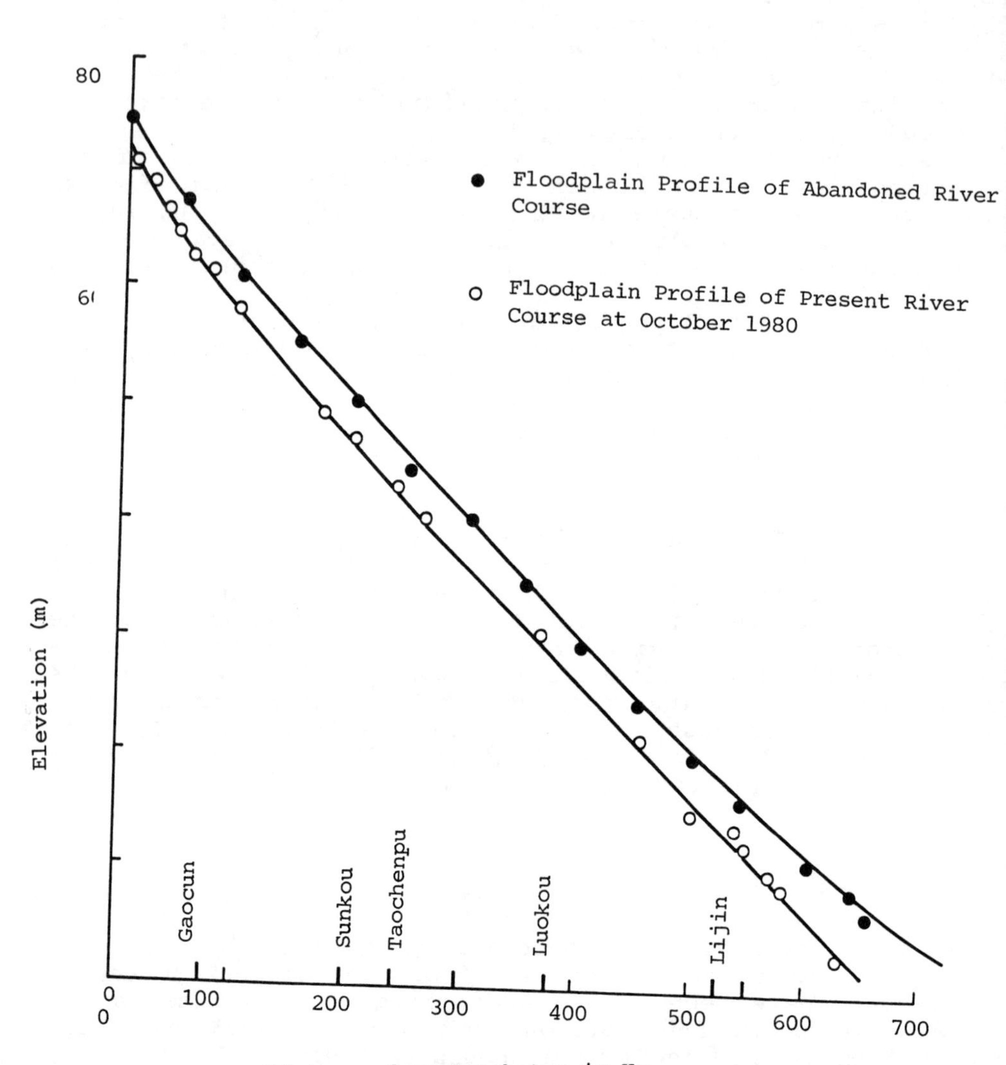

Proposed Plans to Harness the River

The strategy for harnessing the river has been considered the nation's first important problem, hotly debated among the higher government officials and scholars for over two thousand years. The description of the old strategies is beyond the scope of this paper, and only the plans proposed during recent years will be discussed.

TO REINFORCE THE SOIL CONSERVATION WORKS ON TRIBUTARIES IN THE MIDDLE REACHES

The government has pushed forward in many stages soil conservation works such as revegetation, land terracing, gully control works, etc. Besides the controlling of erosion, this helps improve production and living standards of the farmers. A number of experimental stations have been established and the results are promising. A number of large and small-sized reservoirs have been built. According to incomplete statistics, 8,164 "wet filled" dams have been built in Shanxi and Shaanxi provinces. The so-called "wet filled" method is something like the hydraulic fill in western countries. The effective control of erosion in the Fen River and Wuding River basins is almost 50%. Now the benefits from soil conservation are well recognized by the local authorities and the people.

On the other hand, the soil conservation works may be the most difficult task to carry out. There were many ups and downs in the progress of the work. The regions are dry and the precipitation in the rainy season is hardly enough for the growth of vegetation. In some areas, it is sparsely populated, only 20-100 inhabitants per sq. km. The topography is very rugged and communication is difficult.

No soil conservation works can be said to be finished, and even forestation and vegetation require upkeep and maintenance, whereas deforestation, excessive tilling on steeper slopes, road building, and mining have brought about negative results to soil conservation. In the past 30 years, the progress of soil conservation works is not more than 1% annually.

It is estimated that an investment of 10,000 yuan up to 50,000 yuan per sq. km is needed for soil conservation works on lands with various degrees of erosion. Taking these figures for granted, a large sum of government investment is needed. It is obvious that investments, technology and management are necessary and the most important thing is that the soil conservation work must be linked very closely to the promotion of farmers' income in order to bring forth the people's enthusiasm to carry on the gigantic task.

TO BUILD THE TAOHUAYU RESERVOIR MAINLY FOR FLOOD CONTROL

The Taohuayu Reservoir is located at the junction of the middle and lower reaches of the Yellow River (see Fig. 2). It would be most effective for flood control of the lower reaches. The high water level will be raised 15 m above the natural water surface. The total capacity of the reservoir is to be about 5 billion cu.m. The total reservoir area will be 540 sq. km and the total number of inhabitants influenced by the reservoir would be about 100,000. To raise the high water level would cause enormous inundation. The wide river channel will deposit 30-40 milion cu.m of silt annually even if the reservoir is not built. If the reservoir were to be built, the capacity of the reservoir for deposition would be at least 300 million cu.m and a 1933 flood with a recurrence interval of 60-100 years would be controlled by the reservoir. The accummulated deposition would also bring up the silting of Luohe River on upstream and would cause trouble to the city of Louyang.

TO BUILD THREE LARGE RESERVOIRS IN THE MAIN STREAM OF THE MIDDLE REACHES

Sanmenxia is the first large dam to be built in the middle reaches of the Yellow River. Immediately after the completion of the dam, it was discovered that the dam could not work properly with the heavy silt load. Then 2 large outlet tunnels were added on the left abutment and 8 diversion outlets were re-opened to increase the discharging capacity at low reservoir levels. The reservoir is to be drawn down during the flood season to prevent siltation of the reservoir and to store only the clear water during winter to reduce the ice flow and the storage is to be used for irrigation. Only unusual floods from upstream are to be regulated for flood control.

With the experience gained from the operation of Sanmenxia plans have been made to build 3 other reservoirs in succession on the mainstream, Xiaolangdi, Longmen and Qikou (see Fig. 2). The total capacity of the 4 reservoirs would be about 50 billion cu.m. About 30-40% would be available as live storage for flood control, water conservation as well as power generation. During the flood season, the minor floods would be stored in the reservoirs and released as a manmade flood discharging several thousand cms. This discharge is favorable for sluicing the silt in the river channel. Such an operation is considered to be advantageous for the regulation of the runoff together with the silt.

TO PROVIDE A FLOODWAY IN THE LOWER REACHES

At the present stage, the recurrence interval for the projected flood of 22,000 cms ranges from 60-100 years. The provision for a new floodway is aimed at controlling the maximum probable flood (PMF) which amounts to 46,000 cms. The two detention basins can divert a discharge of 20,000 cms with a flood volume of 4.9 billion cu.m. The capacity of the new floodway is planned for the first stage to take a discharge of 5,000 cms and a flood volume of 3.2 billion cu.m. The new dike would serve as a second line of defense. The discharge capacity of the new channel is to be designed to take 15,000 cms. The height of the new dike is to be 15 m. It is estimated that the floodway would be filled up by siltation to the elevation to the present channel bed in almost 30-50 years. At the first stage, the new dike would be 400 km in length. The channel would cover about 100,000 ha of farmland with 700,000-880,000 inhabitants. The construction works would consist of 0.55 billion cu.m of earthwork and 2 million cu.m of masonry work (see Figure 4).

The new channel bed will slope away from the old dike. Heavy training works are required for the protection of the new dike. The construction of the new channel would incude the relocation of 3 cities of one prefecture and 2 districts and the reconstruction of railway, highway, irrigation and drainage systems. Therefore, the alignment and the scheme warrant further study, the results of which can be taken as an alternative plan for comparison.

TO PROVIDE SETTLING BASINS ALONG THE RIVER FOR THE DEPOSITION OF THE SILT

From Longmen down to Taochengpu, 5 locations have been selected for the construction of levees to form settling basins to deposit silt. The thickness of the silting deposition can reach 24 m below Longmen, whereas the thickness of the deposition decreases to only 2 m in the settling basins on the lower reaches and the volume of deposition follows the same order. The total capacity of the 5 settling basins amounts to approximately 30 billion cu.m.

The capacity of the settling basins in the middle reaches is larger than the basins on the lower reaches and the inhabitants per sq. km living in the upper basins are fewer than those in the lower basins. Therefore, the settling basins in the middle reaches would be considered as a first choice.

The construction of settling basins consists of the building of the enclosing dikes which would be relatively high for the upper basins, inlet structures of large capacity, and the outlet works. Also the inhabitants must be relocated, perhaps in stages. The construction of the upper basins should be made after the completion of the reservoirs on the middle reaches.

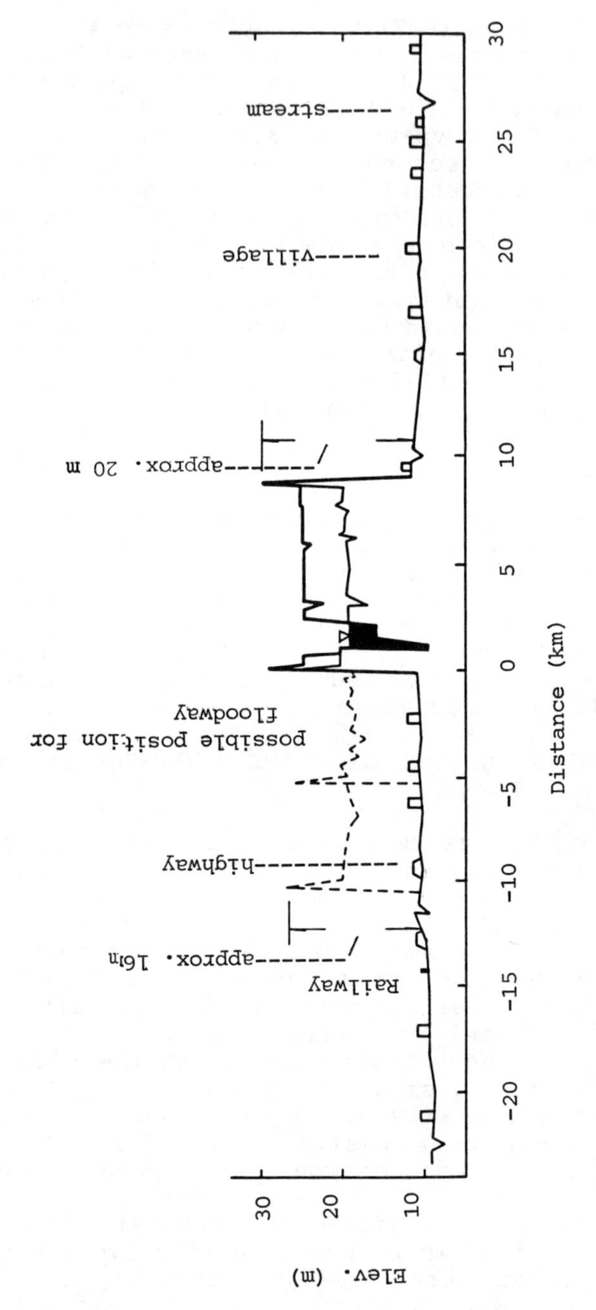

Figure 4. Typical Cross Section of Lower Yellow River

There are 2 ways to release the desilted water: one is to send the water back to the Yellow River, whereas in the lower reaches the desilted water can be either discharged into the river or released to the neighboring streams. In the latter case, it would cause difficulties in flood season.

In my personal opinion, this fifth proposed plan seems to be quite attractive, but I am afraid that a great amount of earth and concrete works would be required and many people would need to be relocated as the location of the settling basin is densely populated. Based on the cost per cu.m of storage, it would be more expensive to build a settling basin than a reservoir.

TO DEVELOP MANY DIVERSION CHANNELS TO RELEASE THE WATER AND SILT

To stop the aggradation of the downstream river channel, a scheme has been suggested to build about 20 diversion channels all the way along the present river channel and to release the water together with the silt through the existing streams. All the relieving channels are to be gated. At the low channel flows, only a few streams would be used for diversion, whereas at high channel flows most channels would be put into use.

The diversion streams are to be made composite in cross sections. The water level in the diversion channel is to be kept flowing full so that the silt may be deposited on the bank, and the main stream channel may be further diverted laterally for irrigation. The suggestion has met with many objections. These diversion streams will be silted up just as the Yellow River, and at last the Yellow River together with the diversion canals would result in a herringbone topography on the North China Plain.

The purpose of this plan is to divert the water and silt evenly all over the vast North China Plain. But the land is not so even, and there are numerous basins and flat ridges. Besides, there are a great number of roads, rivers and streams, irrigation and drainage canal systems and each of the present irrigation canals taking water from the Yellow River is not over 50-100 cms. If about 20 diversions were to be provided, each of the diversion channels should be made with discharge capacity of several hundred cms. It is practically impossible to accomplish this goal.

TO CONVEY 10 BILLION CU.M OF CLEAR WATER ANNUALLY FOR SLUICING THE RIVER CHANNEL

The preliminary studies have been made for the 3 routes--eastern, middle and western--of water transfer from the Yangtze River to the North China Plain. The construction work of the eastern route has already begun, but the water is not to be delivered to the Yellow River. The construction cost of the eastern route is the lowest, the middle route is intermediate, and the western route is the highest. The maintenance and operation cost of the eastern route is estimated to be about 0.12 yuan per cu.m of water. The cost for conveyance of large amounts of water is quite formidable. The water transfer is designed to carry not more than 1,000 cms in the final stage, which is not enough to convey 10 billion cu.m to the Yellow River.

The middle route is to divert water from Han River, a tributary of the Yangtze River. The annual runoff of the Han is 38 billion cu.m. The diversion of 10 billion cu.m from Danjiangkou Reservoir would materially reduce the generation of electric power, whereas deficiency of electricity is a serious problem in central China. The conveyance canal is 482 km in length, and irrigation projects in Hubei and Henan Provinces need a supplementary supply of water. Therefore it is quite uncertain whether this scheme will fulfill the required supply to the Yellow River. Besides, the annual cost of water supply would be very high.

A detailed plan of the western route will be completed in a few years. The transfer project is located at the upstream of the two rivers. The water to be conveyed is expected to be 10 billion cu.m annually. It might be possible but not feasible to carry this out.

TO IMPROVE THE RIVER MOUTH DELTA

Since the last change of the river course in 1855, the coastline has been moved 27 km to the sea. The land formed by the deposition at the delta amounts to approximately 2,200 sq. km. Previously the river was not restricted by the levees at the river mouth and the river channel changed its course once in several or more than 10 years. In recent years the government has been interested in reclaiming the delta. The extension of the river mouth and the siltation of the river channel in the lower reaches causes aggradation in the river channel (see Fig. 5).

There are several diverse opinions regarding this problem. The total amount of silt transported to the river mouth and part of which is deposited at the river mouth is costly to dredge every year. However, a certain amount of the silt deposited at the river mouth may be dredged and distributed over the delta plain as a means of land reclamation. Another suggestion is to employ one main channel for the release of the floods in summer, but to use some smaller channels for release of the low waters. In such a way, the channels should be gated or fuse plugged. But it is doubtful that the channel

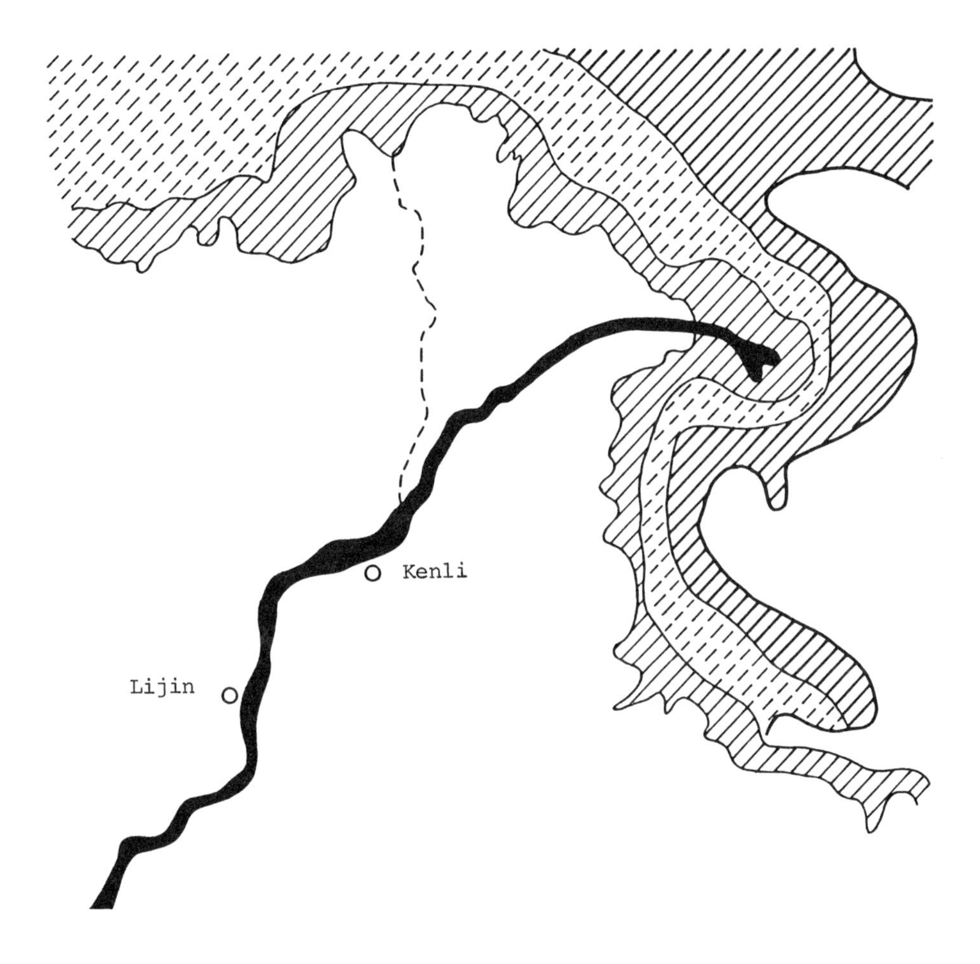

Figure 5. The Extension of the River Mouth and the Siltation Transport into the Sea in the Estuary of the Yellow River.

downstream of the control works might be silted up by tidal action when the channel is blocked. The temporary solution is to provide a new channel to the south of the present channel, and it is expected that the channel may be used for another 10-20 years. Yet another can be provided at the northern part of the delta.

The river mouth problem is bound to be a complicated one. Even if the silt load could be reduced to a certain extent, the degradation of the riverbed would erode a considerable amount of silt from the riverbed which will be deposited at the river mouth. Therefore, the improvement of the river mouth will have to be tackled for many years to come.

Other topics such as the coast protection against the storm tide erosion, fresh water supply, the drainage system of the reclaimed land, etc. are beyond the scope of this discussion.

Concluding Remarks

1. The problem of the Yellow River at the core is the heavy silt load together with relatively small volume of runoff. The erosion of the loess plateau should be controlled, especially the erosion of the coarse material in the most intensive erosion area. The people in tributary areas are very concerned with the silt problem as the development of agriculture and industry depends very much on the soil conservation works.

2. The construction of large reservoirs may require investment, but the reservoirs are multi-purpose projects. New reservoirs of large capacity are near completion on the upper reaches. The runoff characteristics will be changed. The flood peak in the lower reaches will be reduced, whereas the concentration of sediment will be increased. Therefore, regulation of flood and silt released from the reservoirs should be made in the design of the new reservoirs.

3. A possible scheme for the solution of the river mouth problem is to move the vertex of the delta further upstream. In doing so, the river may be shortened in length, steeper in gradient, and more important, a large piece of land may be reclaimed. This scheme should be studied together with that of building a new floodway in the lower reaches of the Yellow River.

It deserves notice that even if the reservoirs release clear water, the river would still carry a certain amount of silt from the erosion of the bed material and therefore the river mouth would continue to extend toward the sea and the improvement of the river delta with land reclamation should be carried on for many, many years to come.

4. A comprehensive study should be worked out for harnessing the Yellow River. Various schemes should be compared on the basis of cost-benefit analyses. The basin planning should be made with long-term arrangement. The planning should be reviewed and revised perhaps every 5 years. The development of planning should be made in stages so we can take steps in working out the general plan.

References

1. Selected papers on researches of the Yellow River and Present Practice, YRCC, 1987

2. A Summary of the History of the Yellow River (in Chinese), YRCC, 1984

3. The Management and Development of the Yellow River (in Chinese).

4. Long Yuqian, Qian Ning (Ning Chien), Erosion Transportation and Deposition of Sediment in the Yellow River Basin, Intern. Jrnl. of Sediment Res., v. 1, no. 1, 1986

5. Zhang Ren, Xie Shunan, Discussion of the Lifespan of the Present Channel of the Lower Yellow River, Jrnl. "People's Yellow River," no. 4, 1987

6. Pang Jiazheng, Yu Limin, Comparison with the Xuhai old course and Assessment of Existing river Course in the Delta of the Yellow River, Jrnl. of People's Yellow River, no. 4, 1987

7. Appendixes I-X of the Report on the Design Assignment of Xiaolangdi Project on the Yellow River, YRCC, 1986

THE PROSPECTS OF FLOOD CONTROL MEASURES ON THE LOWER YELLOW RIVER

Gu Wenshu
Water Resources and Hydropower
Planning Design Institute, MWREP
People's Republic of China

ABSTRACT: Since the beginning of 1950s the flood control work on the Lower Yellow River has undergone several stages. It began with the reinforcement of dikes and construction of Dongping Lake and Shitouzhuang (later reconstructed and renamed Beijindi) flood-detention basins, and successfully handled the large flood of 22,000 cms at Huayuankou in 1958. The Sanmenxia reservoir was built in the late 1950s. It had played an important role in flood protection for the downstream reaches, but excessive sedimentation of the reservoir had caused serious problems upstream, especially silting up of river bed in the lower reach of the Wei River. Reconstruction of the project and a change of reservoir operating rules had to be taken accordingly. In recent years, valuable experience regarding sediment regulation of the Sanmenxia reservoir with the aim of reducing sediment deposition in the downstream channel has occurred. Significant changes in the area of the Yellow River have also taken place due to regulation and control measures occurring in the upper and middle reaches. We must closely follow the new trends and determine the corresponding measures to be taken in the future.

Important Message for the Regulation of the Yellow River--Significant Changes Occured in Streamflow and Sediment Discharge

The sources of water and sediment discharge of the Yellow River are: the water mainly comes from the drainage area in the upper reaches above Lanzhou, while the sediments come mainly from the loess area in the middle reaches. In recent years significant changes have occurred in these areas.

The upper reach of the Yellow River from Longyangxia to Qingtongxia has a length of 918 km and a natural fall of 1324

L. M. Brush et al. (eds.), Taming the Yellow River: Silt and Floods, 331–337.

meters. There are many gorges in this reach. Along this reach, there are many good sites for reservoirs and hydropower projects because of the favorable geology and topography. In addition, there is a relatively small sediment discharge and any inundation which may occur will not significantly affect the sparse population found there. The main purposes of development for this reach would be production of hydropower and water storage for irrigation with incidental flood and ice-jam mitigation benefits.

The proposed development plan would consist of 15 projects with a total planned capacity of 13,700 MW and a total annual energy output of 49,000 GWH. Four of these projects, namely Liujiaxia, Yanguoxia, Bapanxia and Qingtonxia, with a total installed capacity of 1964 MW, have been already completed. The furthest upstream, the Longyangxia project, with a planned capacity of 1280 MW will be put into service before the end of 1988. The Liujiaxia reservoir which has a total capacity of 5.7 billion cu.m. was completed in the late sixties. Since construction a reduction in river discharge during the high-water season has been marked in the Lower Yellow River causing a slight increase of sediment deposition in the downstream channel. The upstream Longyangxia project has a much larger reservoir capacity, totalling 24.7 billion cu.m., and would operate as an overyear storage reservoir. Impounding of this reservior began in October, 1986, and it will take several more years to fill to its normal water level. The regulation effect of this reservoir will rearrange both the yearly and the seasonal distribution of natural flow and would impose important effects on the Lower Yellow River. It will increase the water supply during dry seasons and dry years, but will also create new problems regarding ice jams and sedimentation in the downstream channel.

REDUCTION OF STREAMFLOW AND SEDIMENT DISCHARGE DUE TO WATER AND SOIL CONSERVATION WORK CARRIED OUT IN THE MIDDLE REACHES

Water and soil conservation work in the middle reaches of the Yellow River began in the 1950s, but appreciable reduction of sediment discharge in the river was not marked until the seventies when the technique of constructing dams by hydraulic fill was used, leading to the building of sediment-check dams on an unprecedented scale. There are different opinions about the sediment reduction effect of soil conservation work. Some engineers hold the view that the reduction of sediment discharge in recent years is due to the relatively dry weather with less rainfall and fewer storms. Different kinds of analyses have been made and it has been concluded that besides the factor of lower rainfall, the soil conservation work has also played an important part in reducing sediment. The reduction of sediment discharge mainly occurs in the reach between Hekouzhen and Longmen where most of the coarse sediments are derived, which is very beneficial to the lessening of sediment deposition in the channel downstream.

According to preliminary estimates, the average annual sediment discharge of this reach for the period from 1970 to 1984 was about 200 million tons less than that in the previous twenty years. Comparsion was also made between different years with similar hydrologic conditions. In the past the annual sediment discharge of this reach in slightly dry years had been around 500-600 million tons. In recent years it has dropped to 160-300 million tons under similar hydrologic conditions. Since most of the annual sediment discharge of the Yellow River is concentrated within a few storm floods, analyses were made for several typical storms in the Qingshui River and the Yan River basins. The "1984.7" storm was compared with the "1969.8" storm. The maximum rainfall at the storm center of both storms was around 100 mm. The rainfall intensity and areal precipitation of the "1984.7" storm was slightly larger than that of the "1969.8" storm. The peak flood discharges of the Qingshui River and of the Yan River during the "1969.8" storm were 3520 and 2410 cms, respectively, while the total sediment yield was 54.26 million tons. During the "1984.7" storm peak discharges of these two rivers were 115 and 105 cms respectively, and the total sediment yield was only 0.22 million tons, indicating that a maximum rainfall of 100 mm was practically absorbed. The reason was that soil conservation work has been carried out extensively in this region, therefore the sediment control effect was conspicuous.

Further analysis was made for the extremely large storms in the year 1977. Three very large storms occurred in July and August 1977, each having a maximum daily rainfall exceeding 200mm at the load center. At one particular point the maximum reached 1000mm. During these tremendous storms a number of medium amd small reservoirs and a large number of sediment-check dams were destroyed. An unprecedentedly high silt content was recorded in the Lower Yellow River (the highest concentration of silt at Xiaolangdi station reached 941 Kg/cu.m), and 900 million tons of sediment was deposited in the river channel. This gave the impression that sediment control and soil conservation work seemed to be unstable and unreliable, if the sediment trapped in ordinary years would be entirely washed away during large storms. The actual condition was different from this. The three storms in 1977 were extremely large and the sediment discharge under such circumstances should naturally be abnormal. According to a preliminary survey, the increase of sediment discharge due to collapsing of reservoirs and check dams on the various tributaries amounted to about 100 million tons, while the amount of sediment trapped by the unharmed reservoirs and check dams in the Wuding River basin alone amounted to 200 million tons. The amount of sediment trapped exceeded the amount washed away by floods. The reason was that collapsing of reservoirs and check dams occurred only in a limited area around the storm center, while in the vast region outside the storm center area the numerous reservoirs and check dams could still carry on their sediment retaining functions. Moreover, most of the collapsed check dams were small dams, while the majority of main check dams have remained intact. In the case of collapsed check dams the previously deposited sediment was not completely washed away, but only a channel was scoured having a volume of about 15-25% of the total deposited sediment. These could be repaired easily afterwards. On the whole quite a number of check dams were destroyed in 1977 and 1978, but a majority of them have remained to continue the sediment trapping functions as illustrated in the case of "1984.7" storm which occurred in the Qingshui River and the Yan River basins.

It has been predicted that in the future the reduction of streamflow and sediment discharge will continue during normal and dry years as consumption of water for agricultural and industrial use in the middle reaches of the Yellow River is bound to grow. Decrease of streamflow would inevitably lead to the decrease of sediment discharge. During extremely large floods the reduction would be relatively small and subject to more uncertainties. Therefore, the annual distribution of river flow and sediment would be polarized, i.e. most of the years with quite low river flow and sediment discharges and a few years with relatively high values.

The Prospects of Flood Control Measures on the Lower Yellow River

After efforts over a long time, a flood control engineering system has been established for the Lower Yellow River consisting of soil conservation work in the middle reaches, Sanmenxia reservoir on the the main river, Luhun and Guxian reservoirs on the tributaries below Sanmenxia dam, main dikes along the lower reaches, and Beijindi and Dongping Lake flood detention basins. The Luhun and Guxian reservoirs and Dongping Lake flood detention basin serve only as auxiliary measures. The Beijindi flood detention basin has a population around one million and in recent years the Zhongyuan oilfield has been constructed within this area, therefore the use of the detention basin has become very undesirable. The mainstays of flood control measures now lie on the Sanmenxia reservoir and the main dikes.

The Sanmenxia reservoir has a large capacity and could retain extremely large floods. It was built in the late fifties and some serious problems have been encountered at this project. The eight bottom sand sluices had originally been constructed and used for river diversion during the construction period. They were reconstructed later and are now used for sand sluicing. Used for that purpose for several years, they have been badly scoured and have deteriorated due to erosion and cavitation, and are undergoing elaborate repair and reinforcement. Another sérious problem caused by this project is that if the reservoir backwater level rises above the banks at Tongguan, the upper end of the gorge, excessive siltation would occur at the confluence plain of the Yellow River and the Wei River. The river bed of the lower section of the Wei River was silted up before reconstruction of the Sanmenxia project and dikes have been built to protect the two banks. The drainage paths of the tributaries flowing into the Lower Wei River have thus been hampered, resulting in increased local flooding and waterlogging. Now the reservoir capacity below Tongguan has been reduced to 1.8 billion cu.m. due to siltation, and the reservoir water level would pass over Tongguan during high floods.

In the 41,600 sq.km. drainage area between Sanmenxia dam and Huayuankou, large floods may still occur. The Luhun and Guxian reservoirs can only control a part of this area; therefore there is no control for the area along the main river where severe storms would be very likely to occur.

Since the beginning of 1950s the main dikes along the Lower Yellow River have been reinforced by grouting the embankments with clayey soil, filling up of the depressions behind the dikes with river sediments by hydraulic suction dredges, and other measures. Due to the continuous silting up of the river bed the dikes have been heightened three times, each time with an average rise of one meter. Various problems appeared after the heightening of dikes, including sliding and collapsing of revetments or earth embankments, longitudinal seams on the top of dikes, etc. Since 1980 silting up of the river bed in the Lower Yellow River has been temporarily stopped. Two reasons might account for this advantageous situation. The first reason is the favorable hydrologic conditions in recent years as abundant rainfall occurred in the upper reaches and relatively scarce rainfall in the middle reaches. The second reason is that the sediment control effect of water and soil conservation work carried out in the middle reaches has helped the situation. The silting up of the downstream channel might appear again under unfavorable hydrologic conditions and requires special attention.

Due to the reasons mentioned above, the flood control problem of the Lower Yellow River has not been satisfactorily solved. Should large floods occur, tense situations might appear at both the Sanmenxia Reservoir and the main dikes downstream. The development of Xiaolangdi project has been proposed to ease the situation. The planned Xiaolangdi project is to be located downstream of the Sanmenxia dam and would have an ultimate reservoir capacity of 12.65 billion cu.m. It would offer the following flood control benefits:

(1) It could control a drainage area of about 5000 sq.km. between Sanmenxia and Xiaolangdi where severe floods would be very likely to occur and could diminish chances of using the Beijindi flood-detention basin.
(2) It would somewhat ease the high strain on the Sanmenxia reservoir and the main dikes during large floods. Should a big flood occur above Sanmenxia, it could share the required flood detention capacity with the Sanmenxia reservoir. Should big floods occur below Xiaolangdi or if the main dikes encounter danger, it could reduce its outflow to relieve danger in the downstream. It would also be desirable for the Xiaolangdi to provide the storage required for irrigation and ice-jam mitigation downstream.
(3) It could trap a significant amount of river sediment and also, through rational sediment regulation, would decrease sedimentation in the downstream channel so as to prolong the useful life of the present river course.

How to cope with the continuous silting up of the Lower Yellow River channel has been a problem in dispute. Different estimates of the useful life of the present river course have been made. In regard to the sediment control effects of water and soil conservation work in the middle reaches, the prevailing opinion is that it would last a long time. However, the actual experience in recent years has proved that it would bring successful and promising changes to the loess area plagued by severe erosion. Soil conservation work in the middle reaches consists of three kinds of measures, i.e. biological, structural, and a change in cultivation practice. Planting of grass and trees has been carried out on a large scale only for a few years and so far it is difficult to quantify their sediment reduction effect. The progress of plantation varies with the region. In the relatively flat and broad semi-desert area, it would be faster and bear more fruitful results. But in the hilly and gullied loess area with severe erosion, the land is severed into pieces by numerous gullies and covered with loose soil. The population in this area is relatively dense so that a minimum amount of cultivated land should be maintained to sustain life. Therefore, the progress of planting grass and trees in the gullied-hilly loess area has been subjected to physical and social limitations.

As described before, the significant reduction of sediment discharge since the seventies has been brought about by check dams, reservoirs and terrace fields. These measures increase productivity of the cultivated land and have been supported by farmers. According to a survey, within the period from 1970 to 1984, the average annnual reduction of sediment discharge in the region between Hekouzhen and Longmen was 215 million tons, of which 65% was retained by sediment-check dams, 25% by reservoir and 10% by terrace fields. It is presumed that in the near future the reduction of sediment discharge will still depend on these measures. At the same time, planting of grass and trees should be encouraged, and special emphasis should be placed on the prevention of newly formed soil erosion caused by coal mining, highway construction and other activities destructive to vegetation cover. The soil conservation work in the middle reaches for the period ahead will be very arduous, but it has to be done in order to reduce the sediment loads.

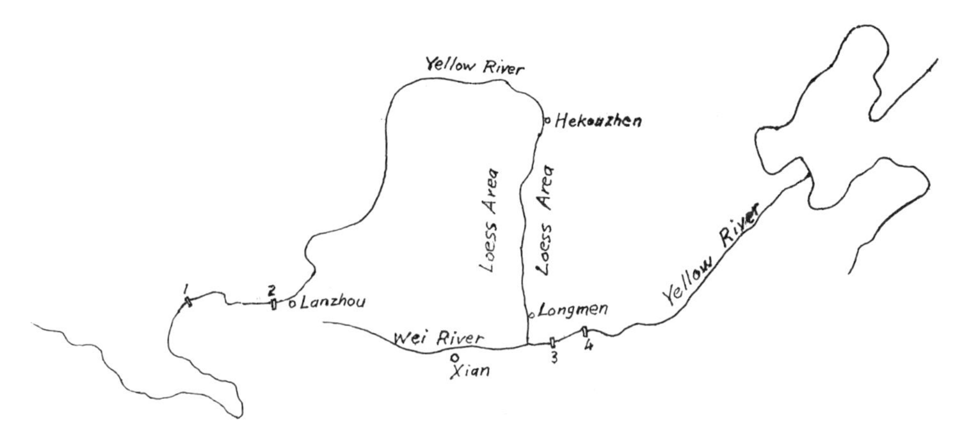

1, Longyangxia Project

2, Liujiaxia Project

3, Sanmenxia Project

4, Xiaolangdi Project

TENTATIVE IDEAS FOR CONTROLLING THE AGGRADATION OF THE LOWER YELLOW RIVER IN THE NEAR AND REMOTE FUTURE

Xie Jianheng
Wuhan University of Hydraulic
and Electrical Engineering
People's Republic of China

ABSTRACT: Changes in the longitudinal profile of the Lower Yellow River and their mechanics have been studied in detail. The suggested regulation measures in the near future are to control the three main factors determining the longitudinal profile, and in the remote future are to divert water and sediments from the river into the low-lying land near the present river course and in the remote regions. Some supplementary measures are also taken into account.

Introduction

The Yellow River is called the second great river in China. In fact, its annual runoff is less than the Pearl River, Songhua River, or Min River, but also less than the larger tributaries of the Yangtze River. The seriousness of the flood control problems in the Lower Yellow River are not due to its large amount of water, but due to its unusually large sediment load. Entering the lower alluvial plain, the Yellow River carries an annual sediment load of nearly 1.6 billion tons with annual runoff less than 50 billion cu.m. A great amount of sediment has been deposited along the river course, forming something like a dividing ridge on which an "elevated river" is flowing high above the great North China Plain, an extraordinary sight among the rivers all over the world (Fig.1). As a result of heavy deposition, the longitudinal profile of the Lower Yellow River has been continuously aggrading and threatens to breach the levees and change the position of the river course. Chinese history records that in the old days there were uncounted flood disasters, and a considerable number of changes in channel position northwards to Tianjin City and southwards to the Yangtze River. Such a phenomenon is nothing but a self-adjustment measure taken by the Yellow River itself in order to distribute the sediments more or less "uniformly"

L. M. Brush et al. (eds.), Taming the Yellow River: Silt and Floods, 339–354.

over the great North China Plain. The present channel of the river was formed after the Tongwaxiang levee breach and has remained there for 133 years.

To make an evaluation of the flood control measures related to river bed aggradation, it is necessary to investigate the present situation of the longitudinal profile of the Lower Yellow River, the laws governing its changes and, based on this knowledge, to seek effective measures that might be used in the near and remote future (Xie 1959,1965,1980).

Present Position

COMMON CHARACTERISTICS OF THE RIVER CHANNEL

The length of the Lower Yellow River from Mengjin, the start of the alluvial plain, to its estuary is approximately 850km. The general channel characteristics and their variation along the main part of the river course are shown in Table 1.

WATER LEVEL CHANGE

The annual changes of the longitudinal profile may be found by plotting the rating curves for different years at the gauging stations and making comparisions among them (Fig.2.) (Xie, 1980, Inst.of Hy.Res.1986). From this figure it can be seen:

(1) The general tendency of the changes in the longitudinal profile is aggradation. The aggrading height for more than 20 years from 1950 to 1974 was 2m or so with a lower value in the uppermost reach. During the next ten years the changes were quite small and even a slight degradation occurred in both the uppermost and lowermost reaches.

(2) Water and sediment input conditions exert a great influence on the silting and scouring processes. During 1960-1964 when Sanmenxia Reservoir, located upstream of the lower alluvial plain, was put into operation with impounding flood water scheme, the sediment input to the Lower Yellow River was reduced radically and the river bed was subjected to a temporary progressive degradation. After that, the Sanmenxia Reservoir was rebuilt for the purpose of enlarging the sluicing capacity and operating with a scheme of impounding clear water and delivering muddy water. Following this period, the river bed turned to aggradation again. Then during 1974-1985, owing to a series of years with small water and sediment loads, no marked changes of the river bed were

Table 1

Station	Huayuankou	Gaocun	Aishan	Lijin Sea
distance (km)	0	189	382	664
slope (0/000)	2.08	1.50	1.16	0.96
bed material d_{50} (mm)	0.092	0.057	0.062	0.062
suspended load d_{50} (mm)	0.0282	0.0196	0.0191	0.0185
river pattern	wandering	transitional		confined meandering
Bd/h	0.19-0.38	0.06-0.09		0.016-0.047
bank full depth (m)	2.0	3.5-4.0		5.0-6.0
coefficient of sinuosity	1.15	1.33		1.20

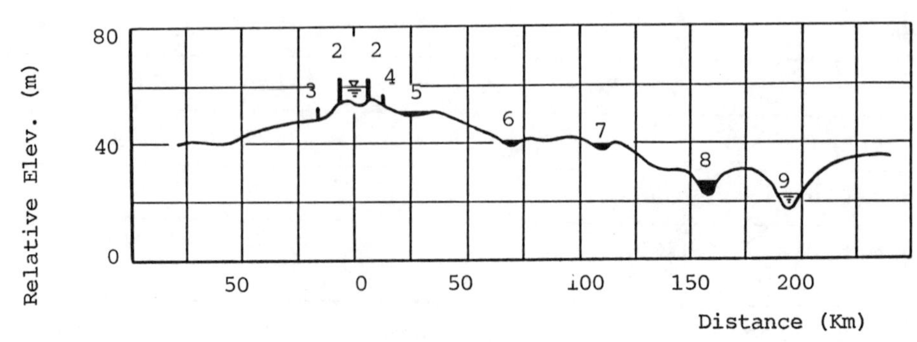

Figure 1. A Great Cross Section of the Lower Yellow River

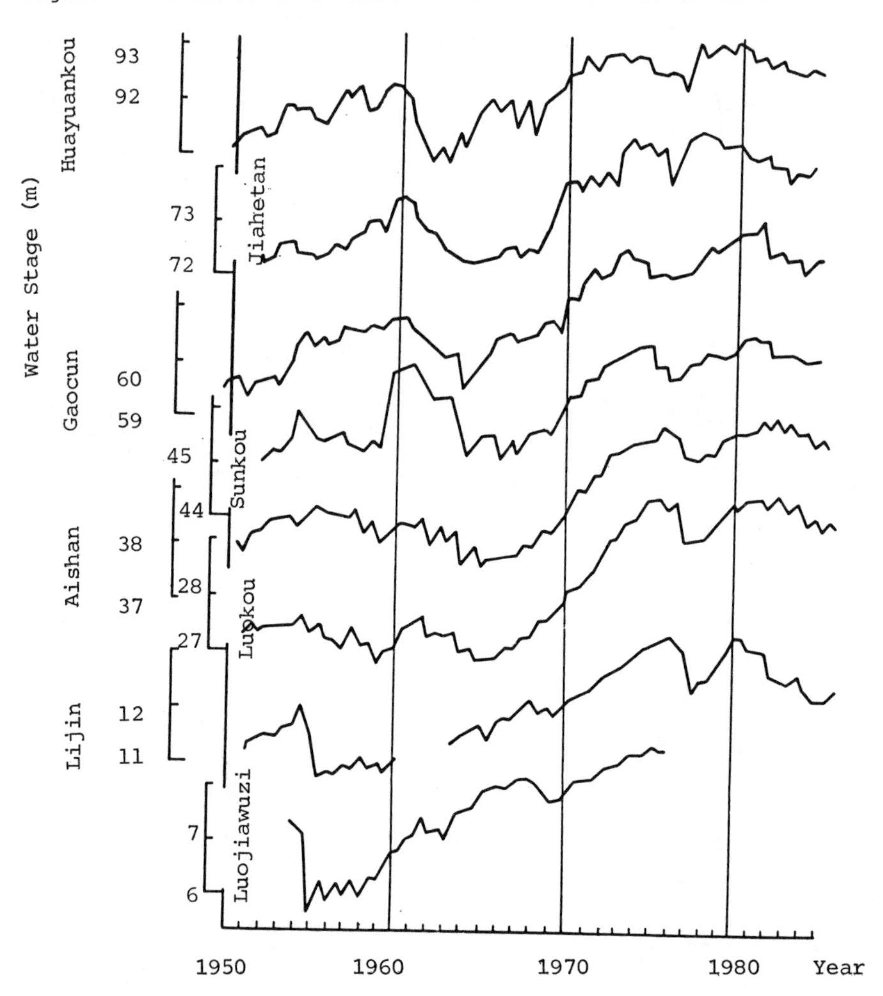

Figure 2. Annual Changes of the Water Levels at Q=3000m/s in Different Stations of the Lower Yellow River

seen.

(3) Changes of the estuary delta also exert an influence to some extent on the river bed deformation. The Xiaokouzi avulsion, a channel position change limited to the estuary region, occurred in 1953 with a river length shortening of 11 km and brought about a remarkable retrogressive scouring of the river bed in the estuary region with its influence reaching up to Luokou, a distance of approximately 280 km from the river mouth. Two avulsions occurring in 1964 and 1976 also influenced the aggradation of the river, though the existence of the local erosion apparently reduced the development of the retrogressive scouring.

DEPOSITIONAL CHANGES

Changes in the longitudinal profile of the Lower Yellow River can also be evaluated by computing the silting and scouring of sediment between different hydrological stations where sediment transport is regularly measured. The results are shown in Table 2.

During 1950-1960 the annual sediment input was 1.795 billion tons and the annual sediment deposition was 0.361 billion tons which reflects the normal situation. During 1964-1973 the annual sediment input was reduced to 1.63 billion tons, but an unusual increase in sediment deposition of 0.439 billion tons occurred. Obviously, this value is a result of extensive bed scouring occurring during 1960-1964. From 1973 to 1984 the annual sediment input was further reduced to 1.115 billion tons. Correspondingly, the annual sediment deposition dropped down to 0.134 billion tons. Being much less than normal, the value reflects the situation of small deposition due to small sediment input. Taking the whole period from 1950 to 1984 into consideration, the annual sediment deposition was only 200 million tons, which is also less than normal because of the influence of the Sanmenxia Reservoir operation. Taking the mean value of annual sediment inputs of 1.56 billion tons from 1919 to 1985 and the situation of sediment deposition from 1950 to 1960 as representive, annual sediment deposition of 314 million tons may be close to the actual mean value for the long run. According to this, of the total sediment input only 1/5 deposits upstream of Lijin. If the sediments diverted for irrigation are added, this increases to 1/4. In addition, approximately 1/2 of the total sediment input deposits on the estuary delta and its near-shore region, and only 1/4 may be carried by tide current and wind flow to the open sea.

DEVELOPMENT OF THE ESTUARY DELTA (XIE, 1965)

The Lower Yellow River discharges to the Gulf of Bohai of which the near-shore region has a shallow bottom (water depth 15m or so) and weak tides (tidal range 1-2m, tidal current

Table 2 Amount of erosion or deposition, 100 million tons

Reach	upstream Huayuankou	Huayuankou Gaocun	Gaocun Aishan	Aishan Lijin	upstream Lijin
1950.7 1960.6	6.20	13.70	11.70	4.50	36.10
1960.7 1964.10	-7.60	-9.24	-5.00	-1.28	-23.12
1964.11 1973.10	8.55	18.17	6.66	6.12	39.50
1973.11 1984.10	-4.58	5.59	10.48	3.25	14.74
1950.7 1984.10	2.57	28.22	23.84	12.59	67.22

velocity 0.5-1.5m/s). The great amount of sediment delivered to the estuary region inevitably causes the estuary to extend and river bed to aggrade. When the river bed rises to some extent, flood waters breach the natural or artificial levees and find a new shorter way to the sea over the low-lying land. After this takes place, the process of the extension of the estuary and aggradation of the river bed will be repeated again. The present estuary delta is currently being formed and developed following this periodic process.

During a small cycle, the extending-avulsion-reextension described above, the water level rise caused by the estuary extending and river bed aggrading after each avulsion will be cancelled by the water level fall caused by the river length shortening and river bed degrading at the next avulsion. On average there are no increases of the river length and, thus, no permanent river bed aggradation and water level rise.

However, after several great cycles of avulsions sweeping over the whole delta has occurred, the river length extension will develop if the shoreline of the whole delta has progressed forward. At that time the next avulsion will be unable to shorten the river length any more. As a consequence, the river bed aggradation and water level rise caused by the estuary extending cannot be fully cancelled and a permanent aggradation of the longitudinal profile will occur. According to analysis of the field measurements, it is believed that great cycles have been completed twice, the first in the middle of the 20s and the second at the end of the 60s (Fig. 3). It is clear that on the delta, the slower the estuary extends the smaller the permanent aggradation of the longitudinal profile will be. The reduction of the estuary delta area from 5400 sq km in the beginning of the 50s to 2200 sq km at present is not helpful.

MODEL OF THE LONGITUDINAL PROFILE CHANGES (XIE, 1980)

Based on the above analysis, the changes of the macroscopic longitudinal profile of the Lower Yellow River may be summarized in brief as follows: The main part of the longitudinal profile has a tendency to aggrade in parallel. For the uppermost reach where the bed material sizes are reduced from gravel to medium and fine sands, the corresponding aggradation height gradually increases downstream. As to the estuary reach, owing to the present delta conditions, the longitudinal slope recently has a tendency to become slightly milder. What is described above only indicates a comprehensive result for a relatively long period. The actual process is quite complicated. The upper reaches are characterized with progressive deposition or scouring alternately, according to the variation of the sediment input conditions, but always with deposition being dominant. The lower reaches are characterized with retrogressive deposition or scouring periodially, according to the variation of the erosion basis conditions, also with deposition dominant, but aggradation follows a special form like undulate steps (Fig.3). Finally aggradation of the middle part of the longitudinal profile is formed by a combination of progressive and retrogressive deposition.

MOVING-EQUILIBRIUM LONGITUDINAL PROFILE (XIE, 1980)

The configuration of the longitudinal profile of the Lower Yellow River has not changed much during the last thirty years. No deformation in the slopes of its main part has been discovered. The reason is that the aggrading height at the inlet of the Lower Yellow River caused by progressive deposition is nearly cancelled by the same height required due to continuous extending of the river length in the estuary and, thus, there is no possibility of building a steep slope with sufficient sediment-carrying capacity to transport all the sediment load. A moving-equilibrium longitudinal profile has been established which is still aggrading, but with the deposition heights more or less uniformly distributed along the course. So the present longitudinal profile of the Lower Yellow River may be seen as a mature one which has passed the stage of rapid deformation. It should be noted that the longitudinal profile of the old channel which existed before the Tongwaxian levee breach in the Ming-Qing Dynasties is approximately parallel to the present one, but with an elevation 3-4m higher. This situation supports the idea to some extent that the present river course is aggrading parallel in its main part. The fact that the lower reaches of the old river course have a milder slope may be explained mainly by the confluence of the Yellow River with the Huaihe River carrying sediments with low concentrations. It should be emphasized that the fact that the Yellow River can approach its mature stage quickly is closely related to its heavy sediment load and strong self-adjustment ability.

Five equations are needed to describe the longitudinal profile of the Lower Yellow River quantitatively:

sediment continuity equation

$$\frac{\partial Q_s}{\partial l} + \gamma' B \frac{\partial Y_o}{\partial t} = 0 \tag{1}$$

sediment-carrying capacity equation

$$S = K_1 \gamma_s \frac{U^3}{a g h w} \tag{2}$$

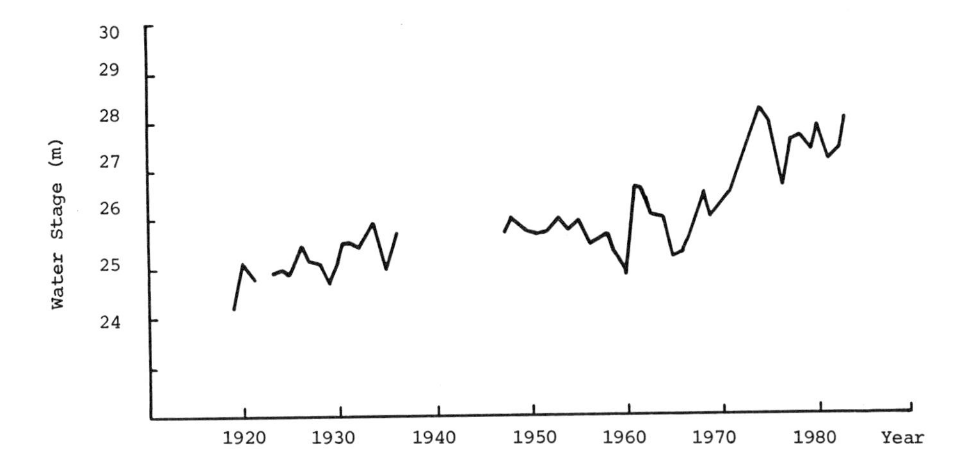

Figure 3. Variation of Annual Mean Water Level in December at Laokou Gauging Station.

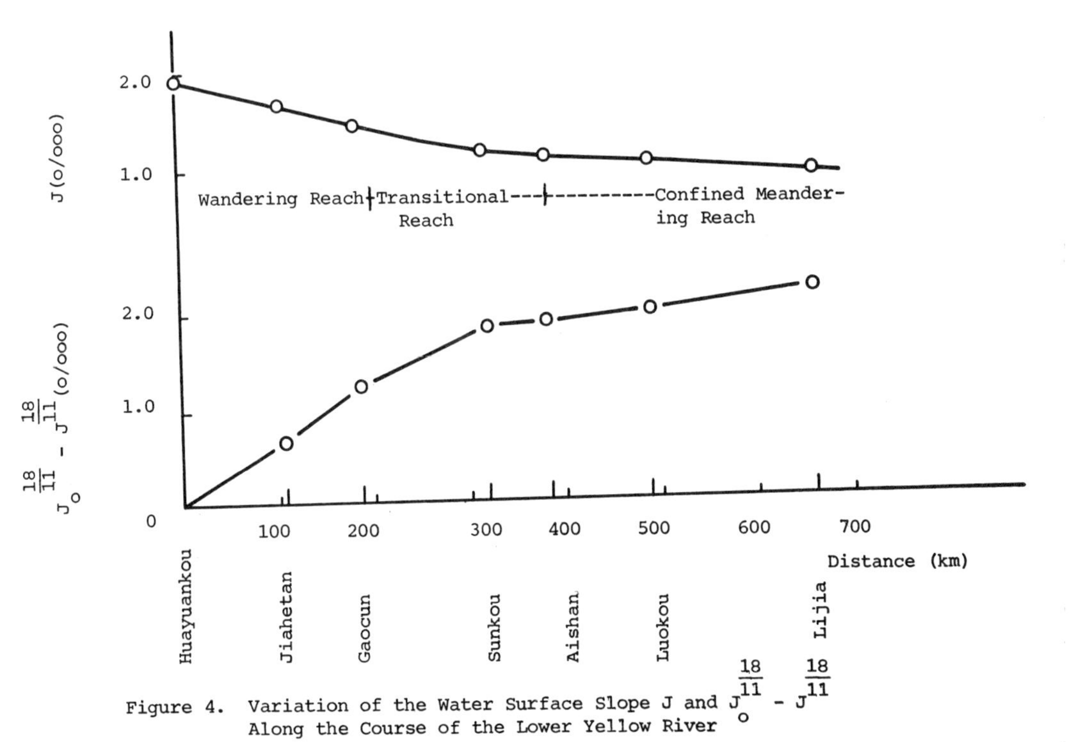

Figure 4. Variation of the Water Surface Slope J and $J_o^{\frac{18}{11}} - J^{\frac{18}{11}}$ Along the Course of the Lower Yellow River

hydraulic geometry equation

$$\frac{\sqrt{Bd}}{h} = K_2 \qquad (3)$$

flow resistance equation

$$U = K_3 \left(\frac{h}{d}\right)^{1/6} \sqrt{g\ h\ J} \qquad (4)$$

flow continuity equation

$$Q = B\ h\ U \qquad (5)$$

in which Q is channel-forming discharge; S is sediment concentration; γ' is bulk weight of deposits; Y_o is river bed elevation; l,t are distance and time respectively; B is river width; h is average water depth, w is average fall velocity of the bed material load; γ_s, γ are specific weights of sediment and water respectively; $a = \frac{\gamma_s - \gamma}{\gamma}$; d is mean diameter of bed material, and K_1, K_2 and K_3 are coefficients.

Combined solution of these equations with the assumption that $\frac{\partial Y_o}{\partial t} = V_y =$ const and given boundary condition at inlet cross section $J = J_o$, at $l = 0$ yields

$$J^{\frac{18}{11}} = J_o^{\frac{18}{11}} - Ml \qquad (6)$$

$$M = \frac{6\ K_2^{\frac{6}{11}}\ \gamma' a\ w\ V_y}{5K_1K_3^{\frac{36}{11}}\ \gamma_s\ g^{\frac{7}{11}}\ d^{\frac{2}{11}}\ Q^{\frac{8}{11}}}$$

The formula shows that for a certain value of J_o the slope decreases gradually along the course following an exponential function. The larger the value M, the more the decrease in slope will be (Fig. 4).

For convenience of discussion, Eq. (6) may be written as

$$J^{\frac{18}{11}} = J_e^{\frac{18}{11}} + M\,(l_e - l) \qquad (7)$$

in which l_e is distance from the inlet to the outlet of the considered reach; J_e is the slope at the outlet which is assumed as a known value.

Tentative Ideas of Regulation

DIFFICULTY OF CHANGING CHANNEL POSITION

In the historical past, the lower Yellow River has undergone a number of channel position changes. Starting from the first one in 602 B.C. there have been 26 such events, on average once per 100 years. The river channel which formed after every event had a lifespan of either several hundred years for the long-lived ones, or several decades for the shorter and even several years for the shortest. Some of the channel position changes in the Lower Yellow River have been natural and some artificial. The natural factors for change have been stated above. The artificial factors deal with the strength of flood control works which may bring about some probability of channel position change and determine the frequency of their appearance. It can be said that the short time changes in river channel position are a result of poor flood control works, and the long time changes in river channel position are a result of the natural development of the river.

The present river channel has existed for more than 100 years which is the average value, but much less than the value of the long-lived ones. At present the water and sediment input conditions during flood season, especially in a year with plentiful water and sediment load, have not changed much. So heavy deposition on the river bed and the natural tendency of channel position change will continue.

However, owing to the rapid development of economic activities, both the natural channel position changes and artificial ones will cause tremendous destruction. To change the channel position artificially would be very difficult at present and much more difficult in the future.

REGULATION MEASURES IN THE NEAR FUTURE (INST. OF HY. RES.,

1986)

Regulation measures suggested for the near future should be chosen from the point of view of not changing the channel position. The basic idea for regulation is to control the river bed aggradation through controlling the three main factors which determine the changes of longitudinal profile and to restrict the aggradation within allowable limits.

(1) The most radical measure would be to emphasize the water and soil conservation in the loess plateaus of the middle Yellow River in order to control the sediment, especially coarse sediment with grain size d $>$ 0.05 mm, delivered to the lower alluvial plain. Analyses based on field measurements reveal that the most harmful sediments to the Lower Yellow River channel are coarse sediments d $>$ 0.05 mm because they represent half of the total deposition (see Tab.3). So, if we can detain all coarse sediments carried by river flow amounting annually to an average of 312 million tons, the deposition process in the Lower Yellow River could be fully stopped. Even if that is impossible, it would help if we could detain half the coarse sediment input, and thereby reduce the river channel deposition by almost a half. For this purpose the regions from which the coarse sediments occur must be regulated by conservation efforts.

(2) The sediment storage capacity of the estuary region must be fully utilized in order to reduce the rate of movement of the shoreline of the delta. Acccording to a simple estimate, aggradation of the original great estuary delta by 1 m may increase the sediment storage capacity by 540 million cu m and, taking the water depth of the near-shore region to be 15m, on average, and the smooth full length of the shoreline 154 km long, the advance of the shoreline of the delta by 10 km may increase the sediment storage capacity by 2310 million cu m. Summing up, we have the total sediment storage capacity in the estuary region 285 x 10^8 cu.m or 370.5 x 10^8 tons. Provided that 3m aggradation of the estuary delta and 30 km advance of the shoreline are allowable, the total sediment storage capacity gained may be used for 140 years, if the annual sediment deposition of 78 million tons is unchanged. In fact, if the estuary avulsions can be controlled to modify the shoreline so that it becomes relatively smooth during a 100-year period there is no necessity to advance the shoreline to 30 km with its full length. To achieve the goal mentioned above, the area of the estuary delta should be enlarged as much as possible and the avulsions should be put in control.

(3) The area of the river channel between the two great levees built on both banks of the Lower Yellow River is 4600 sq km. Every 1m aggradation will provide 46 x 10^8 cu.m or 59.8 x 10^8 tons of sediment storage capacity. In accordance with the allowable 3 m aggradation in the estuary region, the

Table 3

Grain size d (mm)	percentage of total input	percentage of total deposition
$>$ 0.05	20	49
0.025 - 0.05	26	33
$<$ 0.025	54	18

sediment storage capacity of the river channel is only 179.4 x 10^8 tons which can be used only for 57 years if the annual deposition of 3.14 x 10^8 tons is unchanged. So there exists a great gap of sediment storage capacities between the estuary delta and the river channel area upstream of it. This gap may be supplemented, first, by increasing the water and sediment diversion from the Lower Yellow River for irrigation which has been increased to 2 x 10^8 tons per year (1973-1979) and, second, by strengthening the dredging work to enlarge the levee cross section which currently has reached 0.33 x 10^8 tons per year in Shandong province.

For the same purpose another measure may be also used to reduce the aggradation of the river channel. If the sediment input and grain size do not change markedly, the slope J_e in Eq (7) will be unchanged. To reduce J as a measure to reduce the aggradation of the river channel, especially in the upper reaches, the only measure is to reduce the coefficient M. In the functional relation of M values having a great influence and possibly subject to change are K_2 and K_3. The figures given in Tab. 1 show that the value K_2 for the wandering reach is 3-4 times that for transitional reach. It is reasonable to think that a small reduction of the K_2 value for the wandering reach by narrowing the river width in a proper way may result in considerable decrease of the original slope 2.08-1.500/000. The other value, K_3, is also worth examining. Without changing the grain size of the bed material the roughness coefficient n is inversely proportional to K_3. For large discharges the value of roughness of the Lower Yellow River channel is quite small in a range n = 0.007-0.010, corresponding to K_3 = 30-20. But some reaches have a larger value n and smaller value K_3. As an example, for Luokou station n = 0.02 or K_3 = 10 have been observed. Therefore, under specific conditions of the Lower Yellow River, decreasing the roughness for large discharges by increasing the K_3 value to 20-30, where possible, may be effective in reducing the slope. The corresponding measure is to adjust the river training works for bank protection to decrease the roughness as much as possible.

Besides, in order to avoid the unfavorable side effects from controlling the river course, it is necessary to regulate the flood plain by measures such as eliminating the gullies, channels along the levee as well as the large transverse slopes, and utilizing the sediment storage capacity of the flood plain between the two great levees by strengthening the levees to provide much larger cross sections, thus forming a river channel with relatively low elevation which is also preferred. This is especially attractive if done by dredging, where a fair amount of coarse sediment would be removed from the river bed.

Among the three measures stated above, the first, i.e. controlling the coarse sediment input, is of fundamental significance, but in the near future the main effort will be laid on consolidating a foundation for further expansion of its control and the effect is not known. The solution of the problem will mainly rely on the application of the second and third measures. Suppose, after taking such measures, that in 100 years the advance of shoreline of the estuary delta does not exceed 30km, the aggradation of the river bed in its whole length will not exceed 3m. The great levees on both banks can be strengthened to have a top width several 10m with a sufficient free board, and the river bank protection works will form a completed system along the whole river course to make it more stable. In addition to this, there will be put forward a lot of other measures for flood control not discussed in this paper. In this case a better situation without large problems happening may be expected.

REGULATION MEASURES IN THE REMOTE FUTURE

The key problem requiring a solution in the long term is to seek a way to change the historical fluvial processes of the Lower Yellow River which, as noted, distributes a great amount of sediment over the great North China Plain by changing channel positions. At first glance this seems to be impossible. Such misgivings may be justified when the average depth of alluvium over the great North China Plain of as much as 150-200 m is taken into account. Nevertheless, according to the point of view of the author, the possibility of changing such an historic process at least for a long, long period can not be ruled out fully. The following three measures are suggested:

(1) Water and soil conservation works should be further developed in an extensive way. Not only all coarse sediments $d > 0.05$ mm, but also some part of medium sediment $d = 0.025-0.05$ mm should be detained in the watershed of the middle Yellow River. If this can be done, the aggradation of the river bed may be eliminated.

(2) Suppose water and sediment are diverted in great quantities at appropriate points from both sides of the river channel and are transported for a long distance, say several 10km, even several 100km. The fine sediments deposited may be used to raise the low-lying lands and improve them for cultivation. The seriousness of the flood control problems in the Lower River is not due to a large amount of water, but to the unusually large sediment load. As a result of serious deposition, the longitudinal profile of the Lower Yellow River has been continuously aggrading threatening to breach the levees and change the position of the river course. Chinese history records that in the old days there were uncounted flood disasters, even after a considerable increase in capacity as a result of decreasing discharges. But this problem can be solved if all coarse sediments are detained upstream of the lower alluvial plain. In this case the sediment-carrying capacity will be increased considerably in comparison with the present situation and it may be further increased relatively by regulating the water and sediment input through proper operation of large reservoirs in the Middle Yellow River and diversion water intakes in the Lower Yellow River so as to decrease deposition in some cases and increase scouring in others.

(3) Owing to the lack of water resources in North China, looking at it from a long-term point of view, it might be wise to consider diverting water from the South to the North. However, the cost for such a diversion is very high. Furthermore, the water would be more useful in solving personal, industrial and agricultural water supply problems rather than in scouring sediments. So it is difficult to discuss the feasibility of measures at present. However, owing to the fact that big floods in the Yellow and Yangtze Rivers may not happen at the same time during the flood season, taking flood water from the Yangtze River and its large tributary the Hanshui River might not cause harmful effects to those rivers, and might bring a great benefit to the Lower Yellow River. There is no doubt about the possibility of its practical application.

Concluding Remarks

1. The longitudinal profile of the Lower Yellow River has reached a mature stage. Its river bed is still aggrading, but its configuration remains unchanged.

2. The regulation measures in the near future are to control the three main factors determining the deformation of the longitudinal profile. Enlarging the area of the estuary delta and regulating the flood plains between the two levees to increase the sediment storage capacities and using them in full may restrict the whole river channel, especially in

Shandong province, to within allowable limits. Furthermore, narrowing the river width of the wandering reach in a proper way and reducing the roughness, where possible, to adjust the slope may decrease the slope and thus, aggradation of the river bed in Henan province. It is expected that during the next 100 years the progression of the shoreline of the estuary delta will not exceed 30 km, and aggradation of the river bed 3 m. As to the regulation of the coarse sediment source region, the main goal is to consolidate the initial basic work also taking into account further possible control. Similarly, the effects of constructing reservoirs attempting to control sediment should be taken into serious consideration.

3. Regulation measures in the remote future should seek alternatives to distribute the great amount of sediments over the great North China Plain without channel position changes. Under the assumption that coarse sediment has been controlled, such a replacement measure may be formed by diverting water and sediment in great quantities into low-lying lands near the river course and in the remote regions to improve them for cultivation. Construction of Xiaolongdi Reservoir can considerably raise the water level which is valuable for long distance transport of water and sediment. During the flood season utilizing the waste flood water diverted from the Yangtze River and its big tributary the Hanshui River is also an effective measure to reduce the sediment deposition in the Lower Yellow River.

References

1. Xie Jianheng, Longitudinal profile of Alluvial Rivers, Scientific Report of Wuhan University of Hydraulic and Electrical Engineering (WUHEE), 1959

2. Xie Jianheng, Pong Jiazheng, Ding Liuyi, Zhang Guangquan, Zhang Xinhrong Liu Baison, Preliminary Report on the Basical Situation and Basic Laws of The Yellow River Estuary, Scientific Report of WUHEE AND YRCC, 1965.

3. Xie Jianheng, Morphology of Longitudinal Profile of the Lower Yellow River and its Variation, Yellow River, 1980.1.

4. Institute of Hydraulic Research, YRCC, Analysis of the Silting Situation in the Lower Yellow River Channel from 1974 to 1985, Scientific Report of Institute of Hydraulic Research, YRCC, 1986.

THE PROPER MEASURES FOR FLOOD MANAGEMENT AND WATER RESOURCES EXPLOITATION ON THE YELLOW RIVER

Chen Jiaqi
Water Resources Office
Ministry of Water Resources and Electric Power
Beijing
People's Republic of China

ABSTRACT: The outstanding problem in harnessing the Yellow River is flood control. The water resources issues of the Yellow River are also very important. Both these problems are very closely correlated to the sedimentation problems of the Yellow River.

The characteristics of the floods from different parts of the basin of the Yellow River are different, hence different measures for flood control should also be considered. However, we have to recognize that under existing engineering measures, there are still certain risks for ensuring the safety of the lower reaches of the river from flood hazards. Therefore, we should strive for the best to improve the conditions of flood control and to mitigate the threat from floods.

For the water resources issues, we should strongly emphasize the comprehensive utilizaiton of water, and the rational development and scientific allocation of water resources on the Yellow River. For these purposes, rational principles of water resources exploitation on the Yellow River must be studied.

Introduction

The Yellow River is the second largest river in China. Its drainage basin is about 750,000 sq.km, i.e. about 8% of the total area of China, while the long-term mean annual streamflow of the Yellow River is 56.3 cu km, i.e. 2% of the total sum of the annual amount of streamflow of all of the rivers in China. The cultivated land area in the Yellow River basin occupies 13% of the total cultivated land in China, while the population in the basin is 8% of the total. Precipitation, stream runoff, cultivated farmland and population are distributed in the basin quite unevenly. Much of the area where the river flows is arid or semi-arid. Mean

L. M. Brush et al. (eds.), Taming the Yellow River: Silt and Floods, 355–361.

annual precipitation over the whole river basin is approximately 460 mm. The precipitation is concentrated in the flood season, i.e. from June to September, and the variation of precipitation from year to year is also very significant. Therefore these areas are often threatened alternatively by disasters from floods or drought, especially in the middle and lower reaches of the Yellow River. Since the Yellow River flows through the loess plateau, where water loss and soil erosion are very serious, the floods often carry a tremendous amount of sediment and the mean annual sediment runoff amounts to 1.6 billion tons per year at Sanmenxia. This may be the highest value among all large rivers in the world. The riverbed between the two dikes in the lower reaches of the Yellow River has risen gradually due to sedimentation, and the elevation of the river bottom is several meters higher than that of lands beside the river behind the dikes. Therefore the Yellow River is known as the "suspended river." A dike break would create a very difficult problem with regard to water resources in the river basin, the national economy, and the livelihood of the inhabitants along the river. The flood control problems and the exploitation of the water resources in the Yellow River basin are closely correlated with the sediment problem. Due to these features of the Yellow River the counter-measures that should be looked for in attempting to harness the river must take into account the real situation in the river, with some help from the experience gained on other rivers either at home or abroad.

The Flood Control Issues

The biggest threat to the achievement of flood control in the lower reaches of Yellow River is brought about by the large floods occurring in hot summer or autumn. In the past, serious disasters from bursting dikes or changes in the course of the river channel as a result of flooding have brought great suffering to the people. The ice-floods in the thaw period in spring may lead to the rise of river level due to plugs from an ice-dam, and accordingly disasters of inundation in local areas may occur. Although this can bring about some hazard to people, the areas influenced are comparatively small compared to those from the summer-autumn floods. The summer-autumn floods may be divided into upper-reach floods occurring in the area beyond Lanzhou city, and the mid-reach floods occurring in the area between the towns of Hokouzhen and Huayuankou. Floods from these two areas form the principal sources of floods of the Yellow River. As for the area from Lanzhou to Hekouzhen the river flows in an arid zone and floods occurring in this area are usually very small. In most cases the upper-reach floods when routed downwstream have their flood peaks reduced significantly. In the lower reaches of river from Huayuankou to the estuary the Yellow River flows between dikes and even the riverbed is higher than the ground beside the river, no other tributaries except the river Dawenhe join the Yellow River at this section, hence no local floods can be formed there.

The upper-reach floods are mainly formed by successive periods of rainfall of long duration covering a large area. The floods have low peaks and large volumes. The mean duration of one flood may be as long as 40 days. The upper reaches of the Yellow River run through the mountainous areas and the protection objects from flood hazards are mainly the large cities along the river and some points of the main lines of railroad. At present the Longyangxia and Liujiaxia reservoirs are also responsible for flood protection for the city of Lanzhou, and therefore some volume for flood control is being considered in these reservoirs. The upper-reach flood peaks will be reduced after routing from Lanzhou to Hekouzhen and will form the basic flow of the mid-reach floods which also may be divided further into 3 parts. The first are floods occurring from Hekouzhen to Longmen, the second one, those from Longmen to Sanmenxia, and the third one those from Sanmenxia to Huayuankou. These three sections are specified by their storm rainfall centers, and rainfall that may or may not occur simultaneously. Thus the mid-reach floods of the Yellow River may have different patterns. Generally, the floods which form due mainly to the storm rainfall in the mid-reach of Yellow River are more threatening to the lower reaches, such as the floods in 1843 and 1933. After construction of the Sanmenxia Project floods forming in the first and second areas of the mid-reach of the river are controlled partly by the Sanmenxia reservoir which reduces the threat to the lower reaches. But large floods forming in the section of the river from Sanmenxia to Huayuankou may still threaten the lower reaches. For instance, the flood of 1958 was formed in this area and the peak flood value at Huayuankou amounted to 22,300 cms. This is the maximum observed flood since 1919, the flood peak at Sanmenxia in 1958 was only 6400 cms. The Sanmenxia reservoir is unable to control the floods with this rainfall pattern. To solve this problem it has been suggested that a new dam, the Xiaolangdi Project, be constructed to control the floods from the 5,700 sq km of the Yellow River basin between Sanmenxia and Xiaolangdi where the flood of 1954 occurred. The Xiaolangdi Project will not be able to control the floods occurring from the main tributaries in sections of the Yihe, Luohe and Qinghe rivers. At present the Luhun and Guxian reservoirs are set up on the Yihe and Luohe respectively and can, in part, control the floods on these tributaries. Thus, for the floods occurring in the upper and middle reaches, the flood detention reservoirs can be used to decrease the flood peaks and to mitigate the threat to the lower reaches of Yellow River.

Downstream of the Sanmenxia, Xiaolangdi, Luhun and Guxian engineering projects to Huayuankou there are 27,000 sq.km of drainage area where floods still cannot be controlled. After realization of all the reservoirs mentioned, the 100-year flood peak at Huayuankou may be reduced to less than 20,000 cms without using the flood detention area of Beijinti. In addition, the peak at Sunkou may be reduced to less than 13,500 cms, and at Aishan to less than 10,000 cms after flood diversion into Dongping Lake. That means for both sections of Yellow River in Henan and Shandong provinces the 100-year flood can pass through safely. On the other hand, if a 1,000-year flood were to occur in the area between Sanmenxia and Huayuankou with the 1983 pattern, this would be considered as a dangerous pattern for the lower reaches of Yellow River. The peak flood at Huayuankou might be 22,500 cms after flood regulation by the reservoirs, and the flood peak at Sunkou would be 18,050 cms. At Aishan the peak would still be less than 10,000 cms if the Dongping Lake flood detention area is used.

The floods of the Yellow River can be characterized by their abrupt rising and falling hydrographs, but the flood peak is the key for resisting floods in the lower reaches. As the Yellow River is located in the northern part of China, the flood volume in one flood is comparatively less than that of the rivers with the same drainage area in southern China. So it is feasible to bring most floods under control by the compound measures of reservoir detention, establishment of flood diversion-detention areas and the dikes, and in addition, still able to deal with an emergency by the armed forces and the inhabitants. Integrated planning of flood control work is necessary, but so too is organizing and training work related to flood control. Storage of floods by reservoirs often leads to a reduction, or even to the disappearance of the flood-control ability of reservoirs if the floods carry an unusually large amount of sediment. Therefore more attention should be paid to effective measures for regulating sediment in order to maintain the capacity of reservoirs. During recent decades, several sets of measures for regulating runoff and sediment have been worked out, including reservoir water level lowering in flood season. In addition, the mid-reach floods, especially the storm floods from the area between Sanmenxia and Huayuankou, may occur very abruptly and there will not be enough time to adopt effective measures for the safety of inhabitants in the downstream areas. Therefore, automatic real-time flood forecasting and flood warning systems are urgently needed.

In recent times people are very much concerned about the future situation in the lower reaches of Yellow River. Will the rise of the riverbed reach a limiting value? Can the present river course be maintained forever without any catastrophic change of course? Certainly there are different answers from different points of view. It is difficult to give a categorical response to these questions, and there is risk, as well, in drawing either positive or negative conclusions. We cannot treat these problems with an unrealistic optimism, nor unrealistic pessimism. We should prepare for the worst possibilities and do our utmost by taking scientific, realistic and effective measures to assure safety while striving for the possibility of improving the flood control conditions for the sake of mitigation from flood threats. At Huayuankou, Yellow River floods with peak values higher than 30,000 cms may possibly occur. Therefore we are running significant risks if a flood peak of only 22,000 cms cam be safely discharged. A flood detention area as a means of flood control in the lower reaches seems necessary on the Yellow River. To assure effective operation of these flood detention areas, political measures should be studied. We should do the best to maintain the stability of the present river channel course. Meanwhile we should also be prepared to meet the challenge of the probable change of the river course. In brief, we have to do more with great attention and effort at controlling the development of situations in order to mitigate the probable disasters from floods.

The Water Resources Issues

The drainage area above Lanzhou occupies one third of the total basin area of the Yellow River, but the annual streamflow volume at Lanzhou is 56% of the total. The principal agricultural areas are located downstream of Lanzhou. The density of population is also larger downstream. In these areas local water resources are very inadequate and the streamflow of the Yellow River is the only reliable water source. On the upper reaches of the Yellow River, from the gorge at Longyangxia to the gorge at Qintongxia, the fall of the river is about 1500 m, and the exploitable capacity of hydropower nearly 13,000 Mw. Therefore, this reach may be considered for hydropower development.

From Qingtonxia to Hekouzhen the streamflow of the Yellow River declines due to high evaporation and seepage losses. During the irrigation season a large amount of water is delivered to the Ning-Mong irrigation district. Sometimes at the peak irrigation time the Yellow River at Hekouzhen dries up. Downstream from Hekouzhen in provinces Shanxi, Shaanxi, Henan and Shandong provinces, water from Yellow River plays an important role in water supply for agricultural, industrial, environmental and urban uses. Among them, Shanxi, Shaanxi and the northern part of Inner-Mongolia possess many coal mines which are important energy bases for China, but they are also very short of water. Rational utilization of the streamflow of the Yellow River is the only way out for these areas. For the Haihe River plain and the large cities such as Beijing and Tianjin, there is a plan to divert water from the Yellow River to augment water supplies, but the Yellow River possesses only a very limited annual runoff. Hence the potential water supply capacity is not very high. The annual yield of water per sq km in the Yellow River basin amounts to only 77,000 cu.m, which is about 14% of that in the Yangtze River basin. Thus, we should treat this problem properly and with great attention, otherwise conflicts over contending uses of water will occur among these areas and water users. The long-term mean value of annual stream flow of the Yellow River is about 57 cu km, excluding withdrawals. In fact, in recent decades the mean value of the observed annual streamflow at the hydrological station at Huayuankou is only 47 cu km, instead of 57. Therefore, further development of the Yellow River water resources should emphasize comprehensive utilization, unified planning with due consideration for all concerned, proceeding from consideration of the whole, directed toward optimal social gross benefits as a goal. Thus the scientific allocation and rational distribution of water resources can be undertaken.

In order to develop industry there must be corresponding agriculture must also be assured. However in these areas water resources are the restricting factor for economic development. The allocation of the Yellow River water among different districts along the river is such a large problem that it should be studied very carefully. In this case, no matter how strained the demands for water are, a principle of water allocation must assure a certain amount of streamflow draining into the sea for the purpose of maintaining the ecology, environment, and for estuary protection including the transport of soluble and soluble matter to the sea. It is a wrong principle of water allocation "to drink off Yellow River water." We must keep in mind that the amount of water diverted from the Yellow River should be restricted within certain limits and over-drafts of the river flow should be forbidden. At present the annual surface water consumption is nearly 27 cu km, and groundwater consumption in the Yellow River basin is 6.4 cu km. Not only at Hekouzhen in the middle reaches, but also at the estuary the Yellow River has "dried up" several times. So we should put development of water use in the Yellow River basin under rigid control.

As mentioned previously, the upper reaches of the Yellow River have advantageous conditions for hydropower development. However, in the construction of large reservoirs for hydropower development, multipurpose necessities should be taken into account such as flood control, ice-flood control, irrigation, industrial and municipal water uses downstream. In view of the fact that in the upper reaches of Yellow River, above the Heishanxia gorge, agricultural water use is comparatively low and the sedimentation situation is also less severe than that in the middle and lower reaches, there are opportunities to arrange reservoir operation to optimize energy production. But at the Heishanxia section, it will be better to set up a reservoir of reverse regulation in order to meet the needs of flood and ice-flood control, water supply and sediment scour in the middle and lower reaches of the Yellow River.

For the plain areas of the Haihe River in North China where the important metropolises of Beijing and Tianjin exist, the shortage of local water resources has led to a very tense situation. Although in long-term planning, consideration has been given to setting up an inter-basin water transfer engineering project to divert water from the Yangtze River to the north, the transfer can scarcely be realized in the very near future. To solve the problem of tense relations between water supply and demand in Beijing and Tianjin, it should be noted that the Yellow River has some surplus water in winter. It would be feasible to divert water from the Yellow River to Beijing and Tianjin to cope with the difficulties of water supply. Certainly this is only an interim measure. From a long-term point of view the ability of the Yellow River to supply water for Beijing and Tianjin is very restricted since the water resources of Yellow River are needed to meet the water uses within the basin itself. In a few years the Yellow River might have no surplus water to support regions out of its basin. However, the Yellow River at present may play an important role in the transition period mitigating the tense situation of water in North China.

References

Research Group, YRCC, The Harnessing and Development of the Yellow River, Shanghai Education Press, 1984

Editorial Board of the Journal PEOPLES YELLOW RIVER, Research and Practice on the Yellow River, Water Resources and Electric Power Press, Beijing, 1986

Chen Jiaqi, Rational Development of Water Resources for Permanent Utilization, Journal OF NATURAL RESOURCES, NO.2, 1987, Science Press

YELLOW RIVER SEDIMENT PROBLEMS AND SYNOPSIS ON THE STRATEGY OF HARNESSING THE RIVER

Dai Yingshen
Bureau of Geology and Mineral Resources, Henan Province
People's Republic of China

ABSTRACT: The history of harnessing the Yellow River could be traced back to the legend about regulated rivers and watercourses by King Yu. Since then 4000 years have passed, but the problem of flood threats has not been thoroughly solved. What is the reason? Because the Yellow River contains more sediment than most rivers in the world, it presents difficulties in harnessing and developing the river. So, the first step is to control the sediment. If the problem of sediment cannot be thoroughly solved, it will be impossible to bring the river under permanent harnessing. After examining the problem of sediment, this paper will discuss the strategy of harnessing the river and the strategic decision of flood control in the lower reaches of the Yellow River.

The Geological Environment of Producing Sediment by Erosion in the Yellow River Basin

The Yellow River from its source flows toward the east, through the Qinhai plateau, desert and loess plateau, and enters into the North China Plain near Zhengzhou. The relief of the basin from west to east can be roughly divided into three steps. The first step is mainly formed by high mountains which are about 5,000 m above sea level. The second step is composed of loess plateau and desert plateau, 1000-2000 meters above sea level. The third step is a plain with an elevation less than 2,000 m. The changes in topography and relief of the Yellow River basin are generally controlled by the geological conditions. For example, in the upper reaches to the west of Xunhua the earth crust rises extensively, the mountains are high, the valleys are deep and the river currents are rapid. In the upper middle reaches from Xunhau to Zhengzhou the different vertical movement of the earth's crust is obvious, the elevated area is a plateau, and the lower parts are basins. In the Yellow River canyons alternate with

L. M. Brush et al. (eds.), Taming the Yellow River: Silt and Floods, 363–375.

basins. The width of the river courses varies and the flows are different. In the lower reaches to the east of Zhengzhou the geological structure of the floor is complex with different vertical movements. The lower parts have become troughs, and the higher elevations have become uplifted. The river course in the trough is wide with wandering and less winding. The river course in the uplifted parts is narrow with more windings, and belongs to a type of meandering river (see Fig.1).

Historical Facts of Rich Sediment of the Yellow River

It is well known that the Yellow River contains rich sediment. As early as at the formation of the river, about 1.5 Ma ago, sediment produced from erosion of large areas of the basin began to deposit in the basin. Later, with the development of the ancient river system, regional erosion with large sediment production appeared. All of these left obvious traces in the strata of loess in the plateau. The loess in the middle reaches of the Yellow River basin is very thick, and the stratigraphy complete. The loess accumulated from the Pleistocene to the Holocene. According to the stratigraphic sequence, from old to young, it is divided into Fossil loess; Old loess (divided again into upper and lower members); New loess and Latest loess. The geological times of the loesses are, respectively, Early Pleistocene, Middle Pleistocene (the lower member is early middle Pleistocene), late Pleistocene and Holocene. According to the results from paleo-magnetism and other methods, the absolute ages of these loess strata are: Fossil Loess 2.43 Ma, Old loess 1.15 Ma (the per member of Old loess 0.47 Ma), New loess 0.1 Ma and Latest loess about 0.01 Ma. However, these sets of loess are widely spread in the basin. There is evidence of water erosion for a long period on the top of each set of loess. These indicate the erosion characteristics and sediment-producing intensity of each of the sedimentary periods. We know from these that four cycles of sediment erosion have taken place in the course of ancient history. A climax period of sediment production from erosion occurred during in each cycle. The main characteristics of each climax period are as follows:

THE CLIMAX PERIOD OF SEDIMENT PRODUCTION OF THE FIRST EROSION CYCLE

This took place from the end of early Pleistocene to the beginning of middle Pleistocene between 1.15-1.5 Ma B.P. (before present). There are clear signs of fluvial erosion on the top of the Fossil loess in the loess plateau, especially

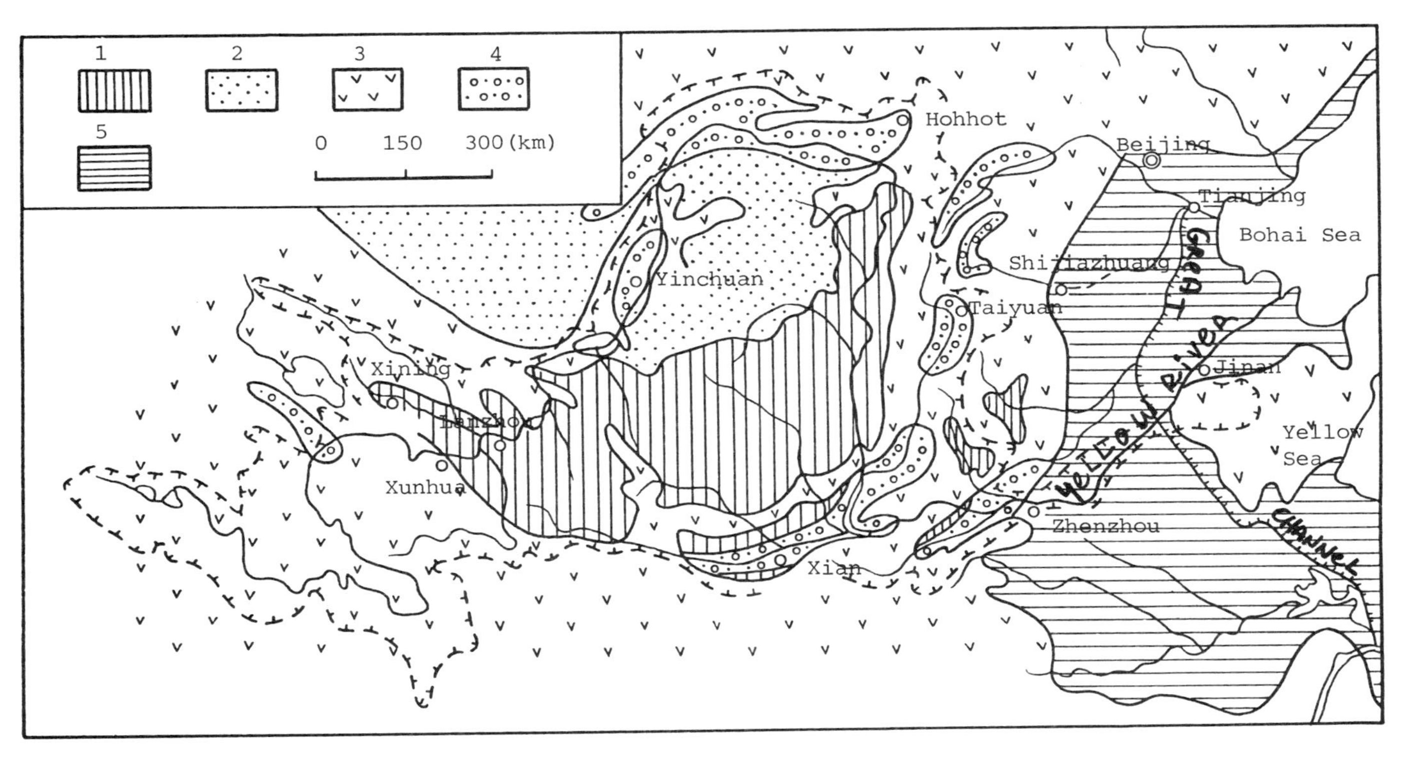

Figure 1. Sketch of Recent Earth Crust Movement and Geomorphic Type in the Yellow River Basin

Key: (1) unevenly upheaved loess plateau; (2) gently upheaved desert plateau
(3) intensively uplifted bedrock mountain area; (4) intensively subsided rift basin
(5) unevenly upheaved and subsided accumulation plain

in zones of upheaval and the uplands where this set of loess has been thoroughly eroded away. Investigations have not been able to locate fossil valleys during this period. It was thus clear that at that time the paleo-water system was limited to the mainstream and main tributaries of the Yellow River. Denudation was the main source of erosion on vast areas and sediment-producing intensity was weak.

THE CLIMAX PERIOD OF SEDIMENT PRODUCTION OF THE SECOND EROSION CYCLE

At the end of early middle Pleistocene to the beginning of late middle Pleistocene, 0.47-0.59 Ma B.P., the fluvial erosion in the paleo-Yellow River basin was very intense. Not only did terraces form in the river and its tributaries, but also paleo-valleys and fossil erosion depressions can be found everywhere on the top of lower member of the Old loess, which were filled by the upper member of Old loess which accumulated in a later period. Meanwhile the paleo-valleys were cut deeper, over ten or several tens of meters in general. From this we know that the sediment-producing intensity of soil erosion within the basin was moderate.

THE CLIMAX PERIOD OF SEDIMENT PRODUCTION OF THE THIRD EROSION CYCLE

At the end of late middle Pleistocene to the beginning of late Pleistocene, 0.09-0.2 Ma B.P., the sediment production from erosion in the paleo-Yellow River basin was more intensive. Not only was a new terrace added in the river, but also a fossil valley system covered by new loess can be found everywhere in the loess plateau. In the deeply cut valleys paleo-landslides of loess formed during this erosion period can be found. This shows that from this time the erosion in the loess plateau entered into a new development stage. Apart from sediment production from fluvial erosion, sediment production by gravity erosion also appeared. Meanwhile, the development of natural sediment-transporting systems gradually became apparent. A great quantity of sediment produced by erosion accumulated in the lake basins and a lot of large, paleo-lakes gradually diminished or disappeared because of deposition.

THE CLIMAX PERIOD OF SEDIMENT PRODUCTION OF THE FOURTH EROSION CYCLE

The end of late Pleistocene to the beginning of the Holocene, 0.009-0.03 Ma B. P., was the great developing stage of the water system of the paleo-Yellow River. Apart from a new terrace that was added, a fossil valley system developed rapidly. Not only did the fossil gully system in loess develop rapidly, but also the development of channels was rapid. Old landslides of loess formed on the edges of valleys in great numbers and were of a large scale. These signs fully show the intensity and the pattern of the sediment production in that period.

The most interesting question in the above-mentioned climaxes of sediment production and intensive erosion in the fourth cycle just discussed was the prosperous period of the ancient peoples in the Yellow River basin. Archaeologists have found fossils and old stone implements and other historical relics of several eras in tributary terraces at different places. The activity of the ancient people was closely bound up with the development of the river. Along with the historical course of development of the Yellow River, the pattern and intensity of sediment production, and soil erosion also increased progressively . The repetition of the appearance of above-mentioned sediment-producing cycles of erosion were not simply repetitious, but developed into higher forms from the primary stages. At present the Yellow River basin is beginning its fifth sediment-producing cycle of erosion.

Sources of Sediment and Characteristics of Sediment Production in the Recent Yellow River

In the Yellow River basin today the precipitation is less and the climate more arid than in the past. From north to south the area can be divided into three climatic zones: arid, semi-arid and semi-humid. The perennial average precipitation in each zone is shown in Fig. 2: the arid zone is less than 400 mm; the semi-arid zone is 400-600 mm and the semi-humid zone is 600-700 mm.

Because the climate of the basin is inclined to the arid side and growing conditions for natural vegetation are poor, the percentage of forest cover in most areas is low and is higher only in the bedrock mountain areas of the southern semi-humid zone and the central semi-arid zone. In the northern part of the loess plateau where the mounds and hills are bare, the area is almost a desert. However, the loess there has more coarse grains, a loose texture and has a lower anti-erosion capacity, and therefore is responsible for a large area of sediment production. In addition, the aeolian sand mainly consisting of silt and fine sand in the desert zones is a drift sand layer without internal cohesion, and is easily transported with strong winds to become the source of the newest loess in the loess plateau. It is the main source of sediment production in the basin.

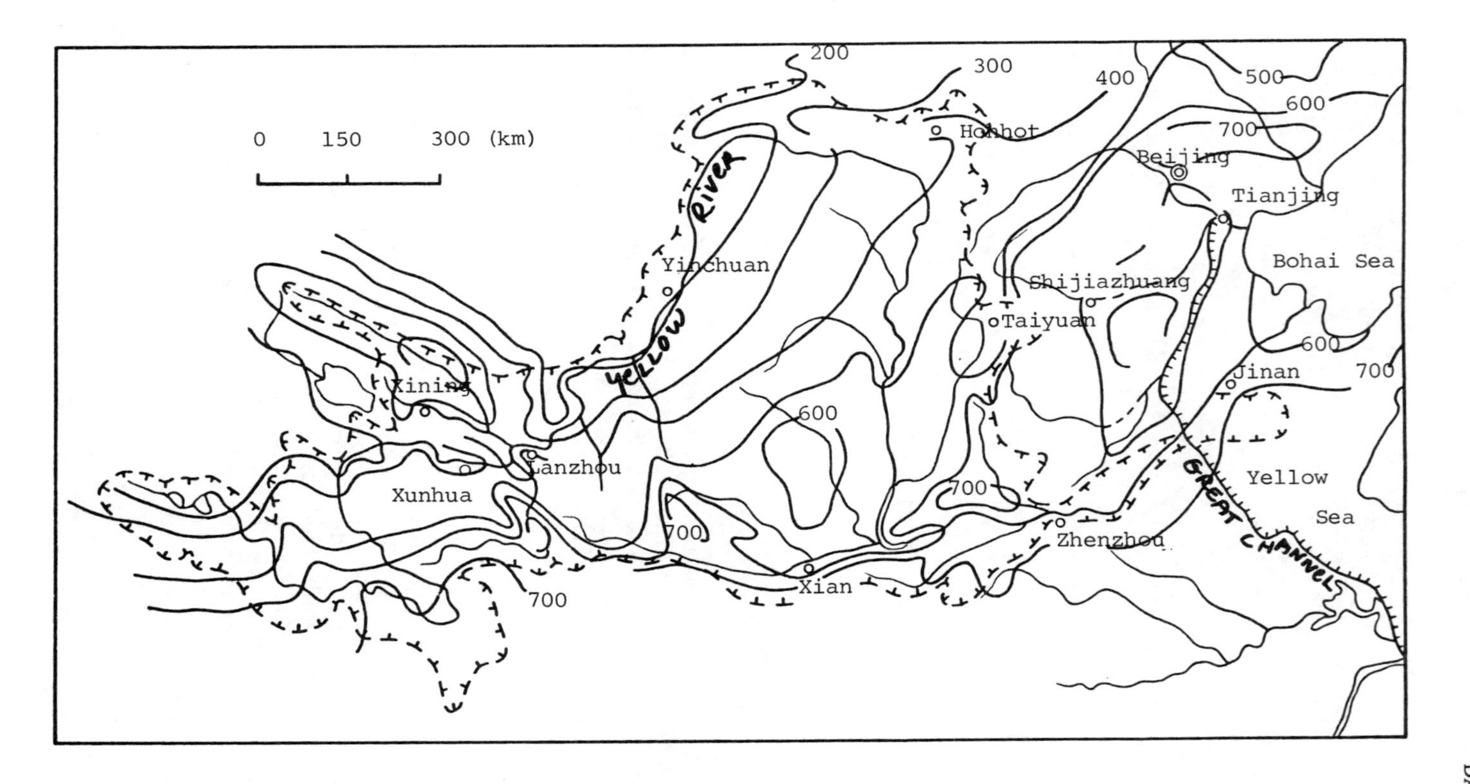

Figure 2. Isogram of Annual Average Precipitation in the Yellow River Basin.

Because of differences in the natural geological, geomorphological and climatic factors in each area of the basin, the pattern of sediment production by erosion also varies. Fluvial erosion is dominant in the bedrock mountain areas, there is fluvial and gravitational erosion in the loess plateau and wind erosion is dominant in the desert plateau. But there are depositional areas which are the fault basins in the upper middle reaches of the Yellow River and in the plains in the Lower reaches (see Fig. 3). However, the difference in intensity of sediment production in each erosion zone is very obvious, as shown in Fig. 4. The annual average sediment discharge (t/sq.km) in the basin is as follows: bedrock mountain areas, 1000; loess plateau area, from 1000 to 10000. But it could be reduced as the annual average erosion of the surface soil by wind erosion is up to 5-7 cm in the desert area. It is clear that sediment production from wind erosion is serious. As for sediment production by erosion in the loess plateau, it varies within the region. In the loess hills of Northern Shaanxi, northwest of Shanxi, south of Longxi and west of Longdong, the annual load discharge rates are more than 10,000 (t/sq. km.). The total area of the above-mentioned sediment-producing region is about over 70,000 sq. km., which accounts for 9% of the total area of the Yellow River basin.

Unity of Sediment Controlling in the Yellow River Basin and Water Regulation in the River Course

The high sediment load of the Yellow River and water regulation in the river course and sediment controls in the region represent two aspects of the strategy for decisions regarding the harnessing of the Yellow River. There are both differences and similarities between them, but intensity is a leading one. Thus, in practice, permanently harnessing the Yellow River sediment controls in the erosive areas and water regulations in the river course should be regarded as a united whole. The Yellow River sediment is predominatly from the loess plateau, but the regions which centralize the sediment production in the loess plateau are not too large, only slightly over 70,000 sq.km. If most or all of sediment production in the centralized sediment-producing regions could be controlled, the sediment that would enter into the Yellow River could be reduced by about half. However, projects for permanent control represent a long-term strategy. It is impossible to achieve significant results in the short-term. Therefore whichever water and soil conservation method is energetically employed in these regions must be strengthened so as to eliminate floods. This is an imminent and major

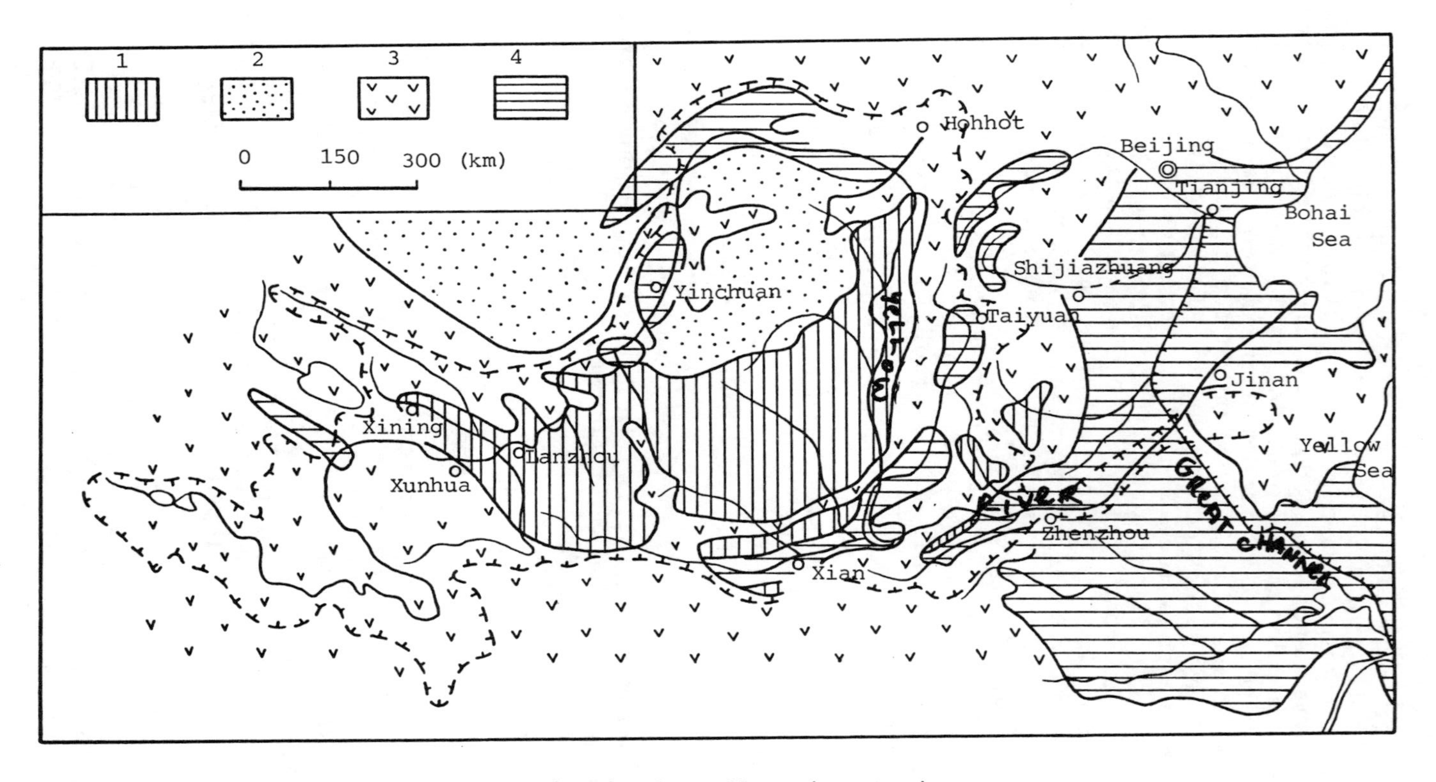

Figure 3. Type of Rock and Soil Erosion in the Yellow River Basin.

Key: Erosion Actions: (1) runoff and gravitation; (2) wind; (3) runoff.
Deposition: (4) fluvial

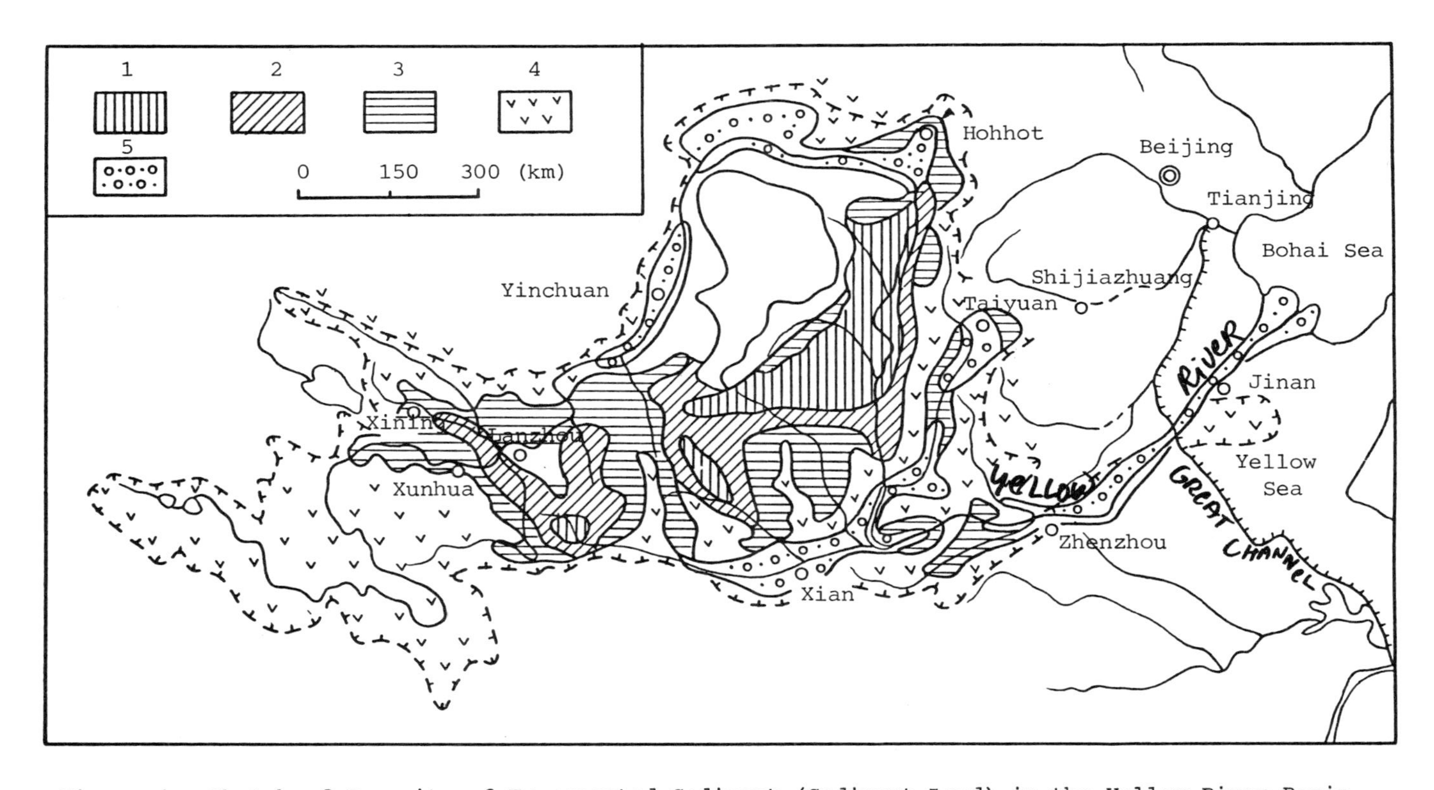

Figure 4. Sketch of Capacity of Transported Sediment (Sediment Load) in the Yellow River Basin.

Key: Load Discharge Moduli (t/km^2a): (1) >10,000; (2) 10,000-5,000; (3) 5,000-1,000; (4) 1,000; (5) deposition.

matter, and the decisions regarding the strategy should be strengthened. In order to retard the speed of aggradation of sediment in the river in the lower reaches and to reduce large pressures due to flooding, reservoirs should be constructed in the mainstream and at tributaries of the middle-upper reaches. These reservoirs would regulate the water and sediment. This would control floods and block sediment, and mitigate flood hazards in the lower reaches of the Yellow River.

Generally speaking, conserving water and soil should supplement one another in order to permanently harness the Yellow River. In detail, this problem is one that divides into two, and the two combine into one. Success or failure would depend on effective operation of systems engineering.

Strategic Decisions for Flood Control in the Lower Reaches of the Yellow River

During the formation of the lower reaches of the Yellow River there have been three stages. The ancient Yellow River before the late Holocene (3000 years ago) flowed northward around the eastern foot of the Taihangshan Mountains and entered the Bohai Sea near Tianjin, and was called the Yuhe River. According to historical accounts, it was mainly a subsurface river (the bed was below ground surface). This was the first stage of development of the Lower reaches of the ancient Yellow River. However, since the Shang and Zhou Dynasties, the withering or disappearamce of many paleo-lakes along the river due to continuous deposition seriously decreased and caused the loss of regulation of water and sediment. Thus, deposition in the river course was intensified, the land on the river was constantly changed, and the undersurface river was gradually turned into a surface river (whose bed was wide, and exposed on the surface). In 602 B.C., the Yellow River broke out of the channel and caused a large migration. This avulsion at Suxukou caused the river to flow northward and enter the Bohai Sea at Huangha County, Hebei Province. This avulsion historically was called the Suxukou migration channel. From then on, the Yellow River has entered into a flooding period, but the surface river still has been reluctantly tied to Zhenggou Period. This is the second stage of development. From the Zhanguo Period (475 B. C.), man began to control the river by dikes, and since the Qin and Han Dynasties, man has been fixing the river course by thousands of dikes. From then on the river has developed into an oversurface suspended river which has turned into an artificial river because of the controlling dikes. This is the third stage of the development of the channel of the lower reaches of the Yellow River. From then on it has been entering a senile stage. Since the Xihan Dynasty, the channel of the lower reaches of Yellow River has been broken through natural causes many times and on large scales. This is the inevitable result of the development of the senile stage of the river.

Because its lower reaches have entered the senile stage, the suspended river is the end result of development, and it cannot be retrieved by manpower. During the drafting of strategic plans to control floods, it is hoped that one would act in accordance with the objective law of its development. Only by having the whole basin in mind is it possible to eliminate the flood threat in the lower reaches of the Yellow River. With this in mind, the following suggestions are put forward:

1. The water and soil conservation in the basin of the upper-middle reaches of the Yellow River should be strengthened. Sediment control in the basin by comprehensive measures must be taken to gradually decrease the sediment entering the river, especially from the central sediment-producing areas.

2. More flood-prevention and sediment-retaining reservoirs in the mainstream and the tributaries of its upper-middle reaches should be built. There are few natural lakes to regulate the water and sediment in the lower reaches. The raging floods cannot be controlled by relying on the dikes as they cannot be used forever. This is the reason for breaches and overflows many times in history. In order to reduce the flood pressure to the dikes of the lower reaches, it would be wise to build more controlling reservoirs upstream. This will increase the safety of the flood-prevention structures of the lower reaches.

3. Flood prevention dikes should be continuously reinforced and extended. Although large-scale floods are protected by the dikes, they are capable of handling more intensive floods. It would be better to continuously reinforce and extend the dikes to ensure their soundness. In addition, geological dangers to the safety of the flood-prevention dikes should be investigated.

4. The waterway should be further dredged to increase the channel capacity. The waterway is seriously silted up in its lower reaches, especially near the seaport where it flows in many directions. In order to avoid that undue deposition, the waterway should be dredged.

5. Vigilant attention should be given to the possible damage to the dikes by earthquakes. The Yellow River basin is a multi-earthquake area. It cuts across eight destructive earthquake zones from its source to mouth. This is an important engineering and geological problem for the Yellow River. Especially from Mengjin to Sunkou, the river cuts through two destructive earthquake zones which pose a clear danger to the dikes. Therefore, anti-earthquake investigations and disaster prevention should be strengthened (see Fig. 5).

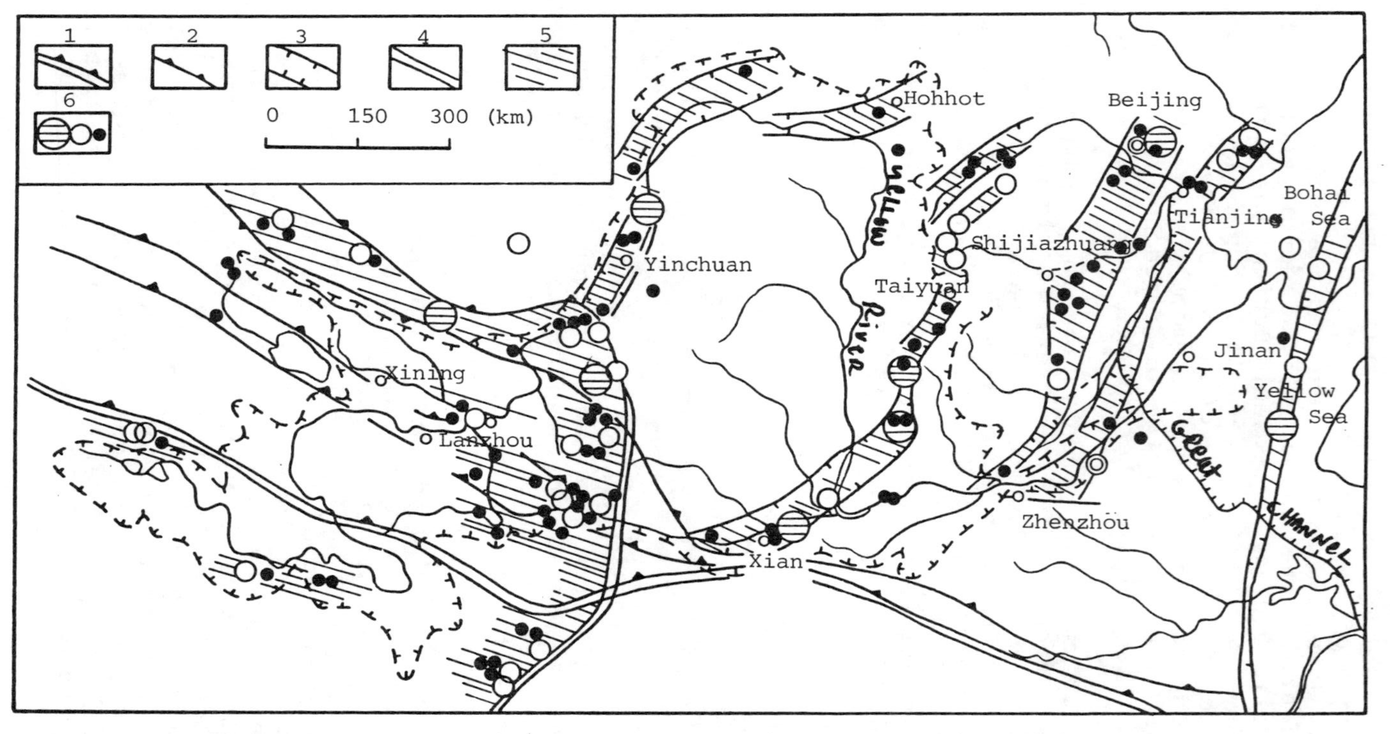

Figure 5. Distribution of the Active Tectonic Zone and the Destructive Earthquake Zone of the Yellow River Basin.

Key: Tectonic Zone: (1) suture lines; (2) subduction zone; (3) Cenozoic rift valleys; (4) Deep fracture.

Earthquake: (5) Destructive earthquake zones; (6) Earthquake epicentra (earthquake magnitude (M): left, M ≥ 8; mid, M=8-7; M=7-6)

6. The Investigation of the Present River Course Development should be strengthened. According to the historic records, the Yellow River has a few times naturally migrated its course in the last 2000 years, which shows that it has developed into the senile stage. The lifespan of the hanging river course is short, even though hundreds of years or a thousand years old. It gets new life by course changes. What is the lifespan of the present hanging river in the lower reaches? This needs investigation and demonstration. It seems that more geological studies are needed to provide scientific basis for the strategic plans ón harnessing the river.

Conclusion

In the upper-middle reaches of the Yellow River basin, loess and aeolian sands are widespread, the climate is arid, the vegetation is sparce and scattered, the precipitation falls unequally during the year, and most rains are torrential. Especially in the flood seasons, the river is swift and violent and carries rich sediment. The water and sediment capacity changes markedly between the wet and arid years. These unfavorable natural factors and the complicated geological environment have undoubtedly created a lot of difficulties for the river harnessing and flood prevention of the lower reaches. For this reason, during studies of the ways of harnessing the Yellow River and probing into the countermeasures, one should never forget that it is a sediment rich river, and possesses a unique natural environment. Therefore, only by taking an overall look at all of the theories and strategic plans of harnessing the Yellow River will it be possible to get good results. Otherwise, efforts will be futile.

DISCUSSION ON THE LIFESPAN OF THE PRESENT CHANNEL OF THE LOWER YELLOW RIVER

Zhang Ren and Xie Shu-nan
Tsing Hua University
People's Republic of China

ABSTRACT: An estimate of the lifespan of the present channel is a strategic problem in harnessing the Lower River. The abandoned channel of the Yellow River in the Ming and Qing dynasties is the best basis for the estimation. The length of the extension of the river course at the estuary and the raising of the river bed elevation are two important parameters determining the possibility of a river avulsion. From the comparison between the conditions of the present and abandoned channels and consideration of the probable hydraulic works built on the Yellow River recently, it can be concluded that the present channel of the Lower Yellow River can be safely maintained for another 100 years. It is not necessary to change the river course artificially in the near future.

Introduction

Since the last change of course of the Lower Yellow River which took place at Tongwaxiang in the year of 1855, the present channel has existed for 133 years. Due to the huge amount of sediment deposited on the bed, it has been aggrading at an alarming rate. Today the floodplain of the lower reaches is 3-5 meters higher than the adjacent lands behind the dikes. As the river channel has become a divide for the Great North China Plain, any breach of the dike will invariably be a great disaster to the country indeed. Thus, how long the present river course will last and when a change of the present course becomes inevitable are the key problems that influence the strategic arrangement of the regulating works on the Yellow River. To examine these points four problems will be discussed.

The Main Cause of the Continuous Aggradation of the Channel Bed of the Lower Yellow River

L. M. Brush et al. (eds.), Taming the Yellow River: Silt and Floods, 377–385.

According to the principles of fluvial processes, the longitudinal profile of an alluvial river depends upon the conditions of the water inflow, the amount of the sediment it carries and the boundary conditions of the river channel. Morphological processes will continue until an equilibrium profile is reached, under which the river is capable of transporting the incoming flow and sediment load. The absolute elevation of the profile depends on the datum of the river estuary. The higher the datum is, the higher the elevation of the channel profile will be.

The Yellow River has brought year by year a large amount of sediment into its estuarine area, only a small part of which can be carried into the deep sea due to the weak tidal conditions of the estuary. As a result, the estuary of the Yellow River has a giant epeirogenetic capacity. Consequently, a high rate of channel extension has resulted and the channel gradient has continuously reduced decreasing the sediment transport capacity of the river. In order to recover the capacity for carrying sediment, aggradation of the river channel must occur. Since the river extension persists continuously, the phenomenon of continuous aggradation on the Lower Yellow River is inevitable.

The Abandoned River Channel Provides a Means of Estimating the Lifespan of the Present Channel

The abandoned course of the lower reaches began at Tongwaxiang in Henan province and ran southeastward into the province of Kiangsu. It served as the main channel of the Lower Yellow River through the dynasties of Yuan, Ming and Qing and lasted for 661 years from 1194 to 1855. During these centuries, the hydraulic geometry of the river was fully adjusted and developed in accordance with the watershed conditions. At present, the abandoned channel still fundamentally maintains the state it was in just before its abandonment in 1855. In addition, there are numerous historical records on its development in ancient books because of its importance on the flood control and navigation of the country at that time. Thus, utilization of these materials will provide a sound basis for estimating the lifespan of the present Lower Yellow River. However, in the historical course of development, past conditions were quite different from the present. For instance, for a long time, no consolidated dike system existed along the Yellow River. As a result, breaching of dikes occurred frequently causing large amounts of sediment to spread over the adjacent lands and resulted in a much lower rate of river extension than occurs at present. In addition, due to the industrial and agricultural developments in the estuarine region, the dike extends toward the sea with the apex of the delta moving continuously downstream. Thus, the freedom for the river channel to shift on the delta is much reduced, bringing about a higher rate of river extension even under the same sediment conditions. If one attempts to estimate the lifespan of the present channel by using the data from the abandoned Yellow River, it is necessary to fully reckon with these changed conditions.

Comparison of the Abandoned and Present Channels

River extension brings about the rise of riverbed elevation with the consequent breach of the dikes when aggradation of the river bed becomes excessive. Difficulties in plugging up openings on the dike will be greatly increased and perhaps become insurmountable. Therefore it may be suggested that the extension of the river course with the accompanying raising of the riverbed are the most important factors determining the behavior of the course of the river. Accordingly, it seems reasonable to assume that the situation of the abandoned channel can be used as a criterion for the potential abandonment of the present river course.

(1) As the ancient Yellow River started to flow southeastward in 1194, its estuary was at Yuntiguan. According to the historical records (Table 1), the estuary moved to a place 155 km downstream from Yuntiguan by 1804. It was roughly equal to the length of river extension in those 600 years. For the present channel, its estuary was located at a place 40 km below Ninghai in 1855 and now it is located at a distance of about 100 km downstream of Ninghai. This means that the length of extension of the present channel is only about 60 km, which is 95 km less than that of the abandoned Yellow River. As a check, we compared the length of these two channels from the same point. Dongbatou is the place near the avulsion in 1855. From Dongbatou to the estuary, the lengths of the old and present river channels are 748 and 649 km, respectively. The difference of channel length is also in the neighorhood of 90 km.

We shall now make an estimate of the river extension rate on the basis of recent data. In the period from 1953 to 1973, the annual ground accretion at the delta was 46 sq. km per year on average. Because of the development of an oil field in the estuarian area the front edge of the delta was contracted to 40-50 km, thus the rate of river extension will be about 1 km each year in the near future. Therefore, it can be concluded that the present river channel will undergo a period of about 90 year to reach the same situation as the old channel on the eve of avulsion.

(2) Comparison of the profiles of the floodplain of the old and present channels is given in Fig.1. For a long period of time, the rise of the riverbed and floodplain were nearly

Table 1 The historical records of the river extension for the abandoned Yellow River reach

Writer	Year	Content of report	Distance from Yuntiguan to estuary	Reference
Jin Fu	1194	"It is not known when the place started to be called with the name of Yuntiguan. Yuntiguan was the location where the Yellow River and Huaihe River pour into the sea. ..."	0 km	The Strategy on Regulation of Rivers, Vol.1
Pan jixun	1587	"In summer, the sixth year of the Wanli reign, Jiang Yilin and I investigated the river regime. It was said that the estuary was located at Sitao(the fourth bend) downstream of Yuntiguan. The river was 7-8, even more than 10(li) wide and 3-4(zhang) deep, ..."	15 km	History of Ming Dynasty, part 7,Vol.84 River and Canal (2)
Pan Jixun	1592	"I made a reconnaissance at the estuary of the river in October of the twentieth year of the Wanli reign. From Jiatao to Siyitao (the eleventh bend) downstream from Yuntiguan. The widths of the river are 5, 7, 8 and 10(li) with the depth varying from 1.5, 2 to 3(zhang). The current run swiftly without any obstruction, ..."	35 km	Xing Shui Jin Jian, Vol.35

Jin Fu	1677	"In the past time, Yuntiguan located at sea side. Now, the delta extends to 120(li) out of Yuntiguan with deposit of about 1(cun) each day. ...."	51.6 km	The Strategy on Regulation of Rivers, V. 6
Dong Anguo	1697	"Yuntiguan was the estuary of the river in the past time. Now the river was silted day by day and Yuntiguan was located 200 (li) from the sea. The discharging of water at downstream reaches become more difficult, so, the deposition in the upper reaches is much increased, ..."	86 km	Xing Shui jin Jian
Chen Shiguan	1756	"Now the distance from Yuntiguan to Er Mu Lou Sea reaches 280 (li), ..."	120.4 km	Xing Shui jin Jian (continued) Vol.13
Sa Zhan	1776	"The distance below Yuntiguan reaches more than 300(li), and the river channel is quite sinuous. ......"	>129 km	
Xu Duan	1804	"From Yuntiguan to the estuary the distance amounts 360-370 (li) along the river channel ......"	154.8 km	Xing Shui Jin Jian (continued) V. 32

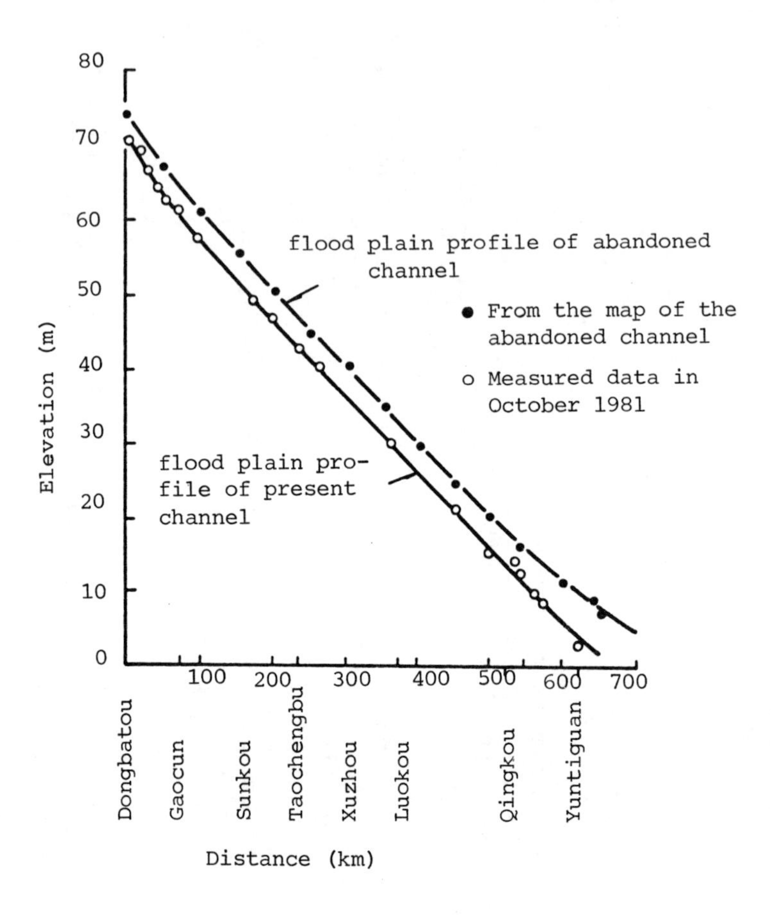

Figure 1. Comparison Between the Floodplain Profiles of the Old and Present Channel

of the same magnitude. Hence, it is possible to use the profile of the floodplain as an indicator for the riverbed. From Fig.1 it can be seen that, in general, the floodplain of the old channel is 4 m higher than that of the present one. During the recent, 1950 to 1983, rate of rise of the riverbed, the water stage for a discharge of 3,000 cu.m/s gives a rough estimate of the rise of the bed level. This was found to increase 4-7 cm annually at six gauging stations along the lower reach. If the effect on sediment detention of the Sanmenxia Reservoir is excluded, the rate of rise in the riverbed will reach 6-10 cm each year. Thus, with respect to the condition of riverbed elevation, it can be said that the present channel will reach the situation of the abandoned channel in 1855 after a period of 50 to 70 years.

(3) In the analysis mentioned previously, the impact of engineering works proposed for the Yellow River in future years has not been included. For instance, after the construction of the Xiaolangdi project, a reservoir with a storage capacity of 12.6 billion cu.m below the Sanmenxia Reservoir, siltation in the Lower Yellow River will be reduced by about 8 to 10 billion tons. This is equal to the benefits derived by maintaining the lower reach in a balanced condition for a period of 20-30 years. If the Wenmeng plain is used to deposit the sediment sluiced from the Xiaolangdi Reservoir and clear water is returned to the river, the balanced period of the lower reach could be extended to about 50 years. In order to meet the requirement for economic development of the country, an additional number of large reservoirs will undoubtedly be built on the Yellow River. Hence, it is reasonable to include these factors in the evaluation of the lifespan of the present channel.

From the facts mentioned so far, it can be seen that the present channel of the Lower Yellow River can be maintained for another 100 years without catastrophic failure of the dike system even if improvements in the dike system and the promotion of flood control techniques are ignored. It may be concluded that to change the present river course artificially to a new one is not deemed necessary at this time.

The Great Consequences of Changing the River Course Artificially

According to the history of the Yellow River, it is well known that breaching the dikes and the switching of the channel to a new course are inevitable due to its heavy sediment load. But under the conditions of a well developed economy and dense population in the Great North China Plain in recent years, extreme care must be taken in considering the results that might follow from a change of river course. Some major problems that may occur are discussed briefly in the following:

(1) The Lower Yellow River is an aggrading and wandering stream. The thalweg of the channel shifts very often from bank to bank in a period of only few days and thus the points of attack by the main current on the dikes change frequently. The well-protected dike system of the present channel was achieved only after a long-term struggle against floods over the years. Although an artificial change in the river course can be carried out in a well-planned way, the new dike system will be difficult to maintain due to the wandering character of the Lower Yellow River. It may be difficult to maintain the relatively nice current situation wherein no breaches have occurred during flood seasons for the 38 years since the founding of new China.

(2) The population in the area along the lower reach is relatively dense with 300-400 people in each sq. km. Due to the large area of the river course covering about 4,000 sq. km, a change of river course will cause a great problem requiring relocation of a population of more than one million people.

(3) The lower course is now, in effect, a main irrigation canal for the Great North China Plain. More than 1.3 million hectares of cultivated land rely on the water supplied by the Lower Yellow River. Because of the dry weather in this area, the water supply is crucial to agricultural production. If the Yellow River changes to a new course the water supply system will be entirely suspended and all of the present irrigation system will be seriously upset. Meanwhile, the water supply from the Yellow River to the Shenli Oil field in the estuarine region and to Qingdao city will also be affected.

(4) In past years, the drainage basins of the Haihe River and the Huaihe River itself were often severely harmed due to the breaches in the dikes of the Yellow River. Once the river course of the Haihe or the Huaihe River was taken over by the Yellow River, siltation in the river channel caused great disasters of flooding and waterlogging of these basins. After liberation, the government has made a great effort to promote high standards for flood control and drainage systems. If the Yellow River changes its course, then the drainage system into which the Yellow River extends will be destroyed.

(5) Many railway and highway bridges and other crossing works have been constructed on the Yellow River. The change of the river course will require the reconstruction of these works.

Consequently, the change in the course for the Lower Yellow River is a very expensive project with many technical difficulties. Therefore, besides safeguarding the existing dike system and exerting every effort to avoid the breaches of the dikes, all measures for prolonging the lifespan of the present channel should be carefully studied and carried out. This would include constructing large reservoirs on the main stem of the Yellow River; depositing the sediment on the floodplains or the lands behind the dike; carrying out the soil conservation program in the coarse sediment source area and the like. Only by reducing the incoming amount of sediment and the rate of aggradation of the river channel of the Yellow River can a stable and secure environment for the development of the Great North China Plain be achieved.

References

1. Zhang Ren, Xie Shunan, "The depositional profile of the abandoned channel and the main cause of continuous aggradation of the Lower Yellow River", Journal of Sediment Research, vol.3, Sept. 1985

2. Xu Fulin, "Study on the changes of the Ming-Qing channel and the present channel of the Lower Yellow River", Yellow River, vol.1 1979

3. Yellow River Conservancy Commission, "Basic condition and the problems on regulation of the estuary of the Yellow River", Selected works of the sediment research of the Yellow River, vol.1, part 2, 1978, Zhenzhou

4. Yellow River Conservancy Commission, "The depositional condition of the Lower Yellow River and the prediction on the developing trend from 1984 to 1995", Yellow River, vol.4, 1985

5. Qian Ning, Zhang Ren, Zhou Zhide, "Fluvial processes", Science Publishing House, 1987

STUDY ON AND ASSESSMENT OF FLOOD PREVENTION MEASURES ON THE LOWER YELLOW RIVER

Wu Zhi Yao
Reconnaissance, Planning and Designing Institute
Yellow River Conservancy Commission
People's Republic of China

Sediment of the River and Fluvial Processes

The Yellow River is known all over the world for its high sediment concentration. The high sediment concentration complicates the problem of achieving flood protection on the lower reaches of the Yellow River. Damage by the Yellow River, formerly called the Sorrow of China, was mainly in the form of breaching and the overflowing of dikes along the lower reaches, essentially due to too much sediment. Sediment carried by the river should be handled satisfactorily through an overall evaluation of the factors involved and various branches of national economy concerned, while dealing with the problem of flood protection.

Unfavorable effects of sediment on flood prevention are mainly as follows:

(1) While large quantities of sediment are being brought down the river, its lower course manifests accretion and the riverbed is steadily rising, having already been "suspended" above the adjacent land surface and becoming a watershed divide for the drainage basins of rivers to both sides. The longitudinal slope of the river and velocity of flow exceed by far hose of rivers on flat-land in general, the maximum velocity being 5-6 m/s at a discharge of 2,000 cms. The same holds true for all old or abandoned channels of the Yellow River now in existence. This is the basic cause of the serious menace of floods on the lower Yellow River. The mean bed elevation is at present 3-5 m above that of land surface on both banks. Flood stages may rise 6-8 m above the land surface, an unfavorable situation which can hardly be obviated within a short period of time. To cope with the floods in a satisfactory way and to prevent breaching and overflowing of dikes is extremely important as the consequences of failing to do so are very serious.

(2) Severe changes in scouring and deposition during floods augment the difficulties of flood protection, and are the most important factor affecting the security of river works and the success of flood prevention. The unfavorable situation mainly takes place in the upper sections of the lower reaches, namely

L. M. Brush et al. (eds.), Taming the Yellow River: Silt and Floods, 387–406.

the reach above Gaocun. As the runoff and silt charges of the Yellow River vary widely both within a year and from year to year, the aforesaid reach is constantly adjusting itself to severe scour and deposition. The result is a broad and shallow channel which is typically wandering in nature. The width of channel in the reach above Gaocun is at present 3-5 km, and deposition amounts to 700 million tons during a flood with hyperconcentrated flows. Channel aggradation within a single year may cause the flood stage to rise 0.6-1.0 m for the same discharge. Extraordinarily high stages may also occur during medium discharge, adding to the burden of flood protection. For instance, the flood stage at Huayuankou Station in 1973, when the discharge was 5,000 cms, much exceeded that in 1958, when flow amounted to 22,300 cms there. All along the lower reach, stages in 1973 were 0.2-0.4 m higher. During floods with hyperconcentrated flows, intensive readjustments of river morphology lead to concentrated flows in the mainstream, the discharge per unit width sometimes reaching 40-50 cms/cm. Shifting of the main current may sometimes bring about abrupt changes in the flow pattern, possibly forming cross currents heading directly towards the main dike and seriously menacing the safety of the dike. Many breaches in the past were a consequence of these cross currents.

It is worth mentioning that the damage due to sediment during flood flows mostly takes place during medium floods, of more or less frequent occurrence. Breaching by cross currents often happens soon after the flood peak has abated. The effect of changes in the sediment regime is weakened during major floods. If a dike breaches upstream of Gaocun, both the scope of inundation and the losses due to flooding are immense. To ensure safe flood protection, consideration of the effects of both major and medium floods should be heeded. Severe aggradation during flood should also be controlled.

(3) The lower Yellow River is steadily subject to channel accretion, so that the dikes have to be heightened step by step in order to maintain the flood-discharging capacity of the channel. The pressure of flood prevention is actually becoming greater and greater. This is a major problem in the long run. Proper evaluation of the prospects of development of channel aggradation bears significantly on long-term policy-making in providing flood protection.

In the past, the lower reaches of the Yellow River witnessed frequent breaches and overflowing, so that the fluvial processes have been very complicated (see Figure 1). Structural improvements and development of the Yellow River in recent years affect the conditions of incoming streamflow and sediment load to a considerable extent. Prediction of channel accretion on the lower Yellow River in the future will be difficult, and opinions by experts differ.

There has been no breaching of dikes on the Yellow River during the past 40 years. The riverbed has been rising at the rate of 3-5 cm/year in the reach above Gaocun, and 5-7 cm/year below Gaocun. Annual sediment delivery to the sea averages 1

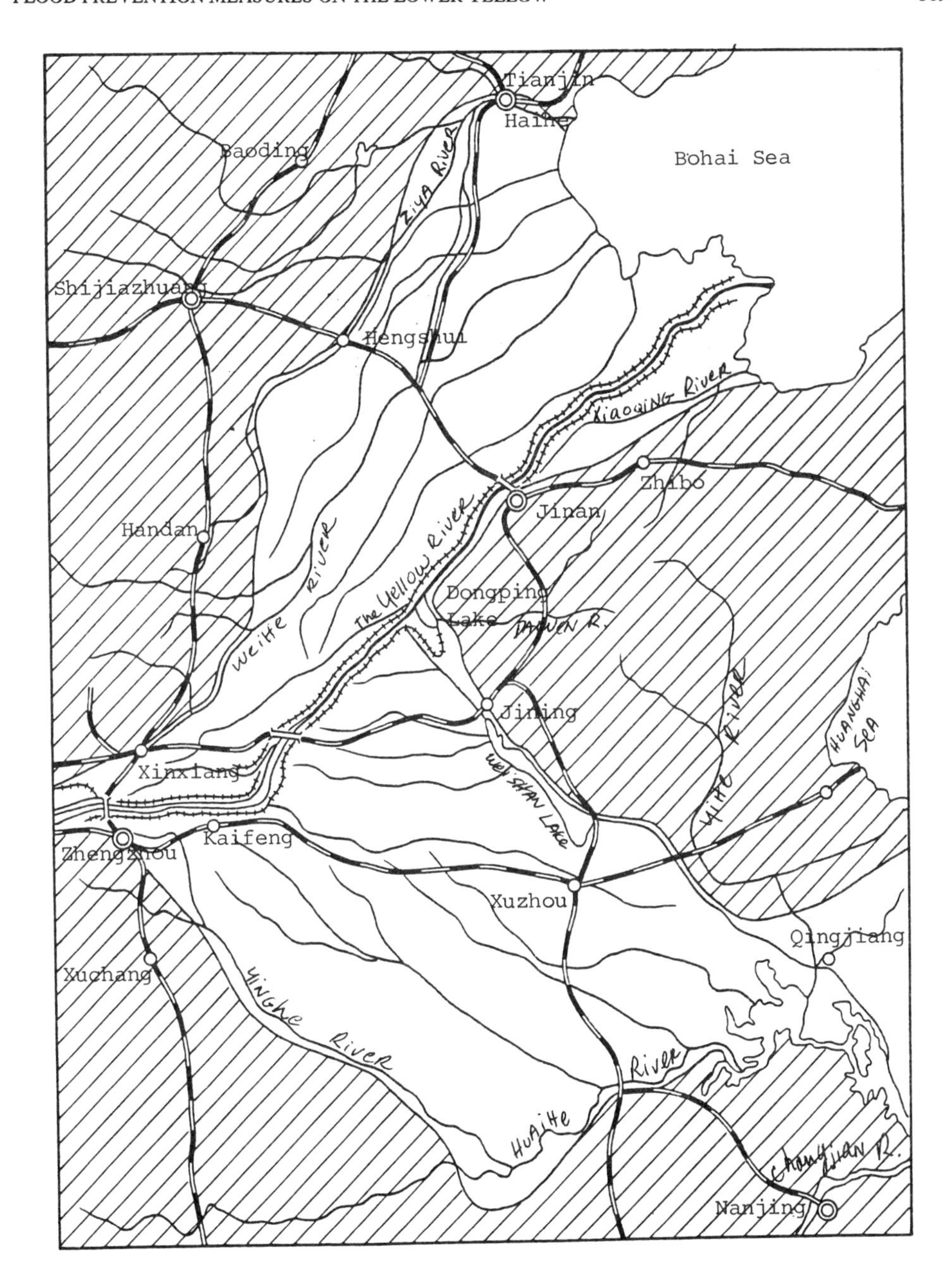

Figure 1. Map Showing Areas Subject ot Inundation During Floods on the Lower Reaches of the Yellow River

billion tons, creating an average of 25 sq.km of newly-formed land. The flow path towards the sea has already changed three times, and the ever-extending channel has not been stabilized. Channel aggradation and readjustment during the past 4 decades have been subjected to effects of previous breaching and overflowing of the dikes, such as in 1933 and 1935. A manmade change of course toward the south was made in 1938, inundating the affected area for almost a decade. Subsequent to 1960, the Sanmenxia reservoir has helped reduce sediment delivery to the lower reaches.

Knowledge of Fluvial Processes: Past and Present

The following ideas have been acquired on the basis of comprehensive analyses of the fluvial processes, past and present:

(1) Fluvial processes in an alluvial river are characterized by self-adjustment toward a state of relative equilibrium of scouring and deposition. The force of adjustment is particularly strong on the lower reaches of hyperconcentrated Yellow River. A new channel essentially adapting to the needs of sediment delivery may be shaped in a relatively short period of time. Capture of Daqing River (below Dongping Lake) over a stretch of 400 km by the Yellow River subsequent to the change of course in 1885 rendered the former to be suspended above the land surface in a little more than 10 years, owing to the large quantities of sediment delivered. Changes of flow path at the river mouth and development of reservoir silting at Sanmenxia have occurred in recent years. The process of channel accretion and rise of riverbed are apparently stepwise. The stage of intensive deposition and readjustment is followed by a stage of slow readjustment resulting from a period of relative stability. When a new channel is being formed, huge deposits effect prompt adjustment of the longitudinal slope, while a channel and adjacent floodplains are formed, the river is soon suspended above the land surface. Channel aggradation along the new course takes place mostly during this period. After a state of relative equilibrium is reached through self-adjustment of the river, the longitudinal slope becomes more or less constant, and adjustments of sediment-carrying capacity are mainly exhibited in changes in river morphology and composition of bed material, with the rate of aggradation markedly diminishing. Evaluation of the average rate of deposition in general terms by the elevation of the deposits and the time interval during which the river is flowing along the path is not sufficient to show the actual conditions during the two distinct stages.

(2) The channel along the lower reach has an adequate sediment-transporting capacity for flood flows. During the past few years, there has been scouring and deposition and the longitudinal slope has tended to stabilize. The general tendency is still continual accretion, though the bed is rising at a lower rate. In the long run, the basic factors

governing channel accretion are the extension of river at its mouth and the deposition of coarse sediment. Based on observed data, each year witnesses the transport of an average of 90 million tons of coarse sediment (grain size exceeding 0.1 mm). As much as 80 million tons, nearly all the incoming coarse sediment, is deposited in the channel, primarily in the reach above Gaocun. It is considered feasible to take measures to detain the coarse sediment and control channel extension at the mouth, to further retard channel aggradation.

Problem of Flood Control

Long-term data are available with respect to floods of the Yellow River. More or less complete hydrometric observations have been carried out at Sanmenxia (Shanxian) since 1919. As early as 1765, gauges for measuring flood stages were established at Shanxian and other places, to effect continuous observation of the rise and fall of river stage. For the most part, calamities of past floods were also recorded in the literature. Through extensive in-depth studies it has been determined that the major floods of the past two or three hundred years are known and represent the significant floods of the Yellow River. In accordance with our present study, it has been found to be more favorable to solve the problem of floods of the Yellow River by controlling the concentration of flood flows.

(1) Major floods originate and are concentrated in the mountainous and rolling areas on the middle reaches. The lower Yellow River, though measuring 800 km in length, being suspended has very few tributaries, and adds only 3% to the total catchment. Besides, the floods on the lower reaches are small and do not occur at the same time as major floods from the middle reaches. Construction of reservoirs on the middle reaches to control the floods will effectively alleviate the threats of floodwater along the entire lower reach.

(2) Floods of the middle reaches result from rainstorms of short duration and high intensity derived from two areas: an area tributary to the Tuoketuo-Sanmenxia reach, and that tributary to the Sanmenxia-Huayuankou reach. The area subject to storms is generally 20,000-30,000 sq.km, and floods of the two zones do not merge. Major floods of the Yellow River are therefore characterized by high peaks and small volumes of runoff. One flood may last 10 to 12 days, but be primarily concentrated in 5 days. It has been estimated that peak flow of 1,000-year recurrence at Huayuankou is 42,000 cms but the total runoff of floodwater exceeding 10,000 cms is only something like 6,200 million cu.m (the designed flood-discharging capacity of channel below Dongping Lake being 10,000 cms). Concentrated control of floodwater through construction of reservoirs will be instrumental in pronouncedly reducing peak discharges without the need of enormous flood-detention capacity (see Table 1).

Table 1 Characteristic values of floods on the lower Yellow River, at Huayuankou station

Flood	Peak flow (cms)	Runoff in 12 days (1 mil.cu.m)	Runoff in 5 days (1 mil.cu.m)	Runoff of flow exceeding 10,000 cms (1 mil.cu.m)
1933, observed	20,400	10,000	5,960	1,160
1958, observed	22,300	8,700	5,190	1,350
P = 1%	29,300	12,500	7,130	3,250
P = 0.1%	42,300	16,400	9,840	6,170

3) Floods from the middle reaches, the Tuoketuo-Sanmenxia reach in particular, have high sediment concentrations. Flood-detention reservoirs will also have to retain sediment. Hence reservoir silting will take place, and large capacities will be needed to cope with the sediments to avoid increased channel accretion downstream.

The Present Status of River Improvement

BASIN MANAGEMENT AND CONSTRUCTION WORK FOR FLOOD PROTECTION

Since the founding of the People's Republic of China, importance has been attached to studies on improving the Yellow River in a comprehensive way. In 1955, a decision was made at the National People's Congress for the purpose of "Eliminating the damages of the Yellow River once and for all, and developing the resources of the Yellow River, through a comprehensive program." Different kinds of studies to design methods for harnessing the Yellow River were carried out through an all-out effort. Through unremitting efforts in the past 30 years or more, marked results in obviating damages and gaining benefits have been achieved.

Seven major projects and hydropower stations have already been built on the main Yellow River, such as those at Liujiaxia and Sanmenxia. The Longyangxia Reservoir on the upper reaches is nearing completion. It will be used for multi-anuual regulation of streamflow. The total capacity of the aforesaid amounts to 40 billion cu.m, and the total installed capacity for generating power, 4 Mkw. On the tributaries, more than 170 reservoirs with a capacity exceeding 10 million cu.m have already been built. Land used for irrigation in the valley tapping water from the river has risen from 12 million mu, before the founding of the People's

Republic of China, to over 70 million mu, almost a five-fold increase. Cropland in basins and valleys on the upper and middle reaches is provided with irrigation water, and a number of high lift pumping stations have been benefitting the arid plateau formerly deficient in water. Beginning in the early fifties, water has been diverted from the lower reaches of the Yellow River for irrigation over an area of 20 millon mu. About 48% of the annual average volume of runoff is now being used to serve industries and agriculture in the entire basin, signifying a rather high rate of water utilization.

Work of soil and water conservation has been developed over a wide area on the Loess Plateau in the middle and upper reaches of the Yellow River. A number of bases for scientific research and experimentation have been established. Different measures of water and soil conservation, each playing its role in an integrated system of technological improvement, have been developed and found to be effective at many places where they have been tried. In recent years, management of small watersheds through contracts with households that assume responsibility for performing the task appropriately has been popularized, and concentrated management of key areas has been stressed, giving a fresh impetus to expediting conservation practice. To date, 100,000 sq.km of land have been managed to some extent improving livelihood of people and conditions of local production to a certain degree. Building reservoirs on the main river and its tributaries and carrying out conservation practices have shown potential for reducing sediment delivery into the Yellow River. In accordance with our analysis, a diminution of 200 million tons of sediment delivered into the Yellow River each year has occurred, taking into account the effect of smaller precipitation during these years.

Flood protection on the lower reaches of the Yellow River has always been the first and foremost task of harnessing the river. During the past 30 years or more, the main dikes have been heightened and strengthened, and additional river training has been carried out. Flood diversion and detention projects such, as at Dongping Lake and Sanmenxia have been completed. It may be said that a complete system of flood protection works has been set up, though preliminarily. In the meanwhile, emphasis has been laid on hydrometric service and flood forecasting, and large masses of people have been organized into a formidable contingent for flood prevention. Based on the functioning of the aforesaid projects and rigorous measures in defending against the floodwater, dike breaches have not occurred during the past 39 years, preserving the safety of life and property in the vast Huanghe-Huaihe-Haihe plains. Altogether 934 million cu. meters of rock and earthwork have been used for which the state invested about 5 billion yuan, including funds for construction, maintenance and repairs. Compared with frequent breaching in the past, the 39 years witnessed enormous economic benefits, obviating direct losses due to possible inundation of 40 billion yuan.

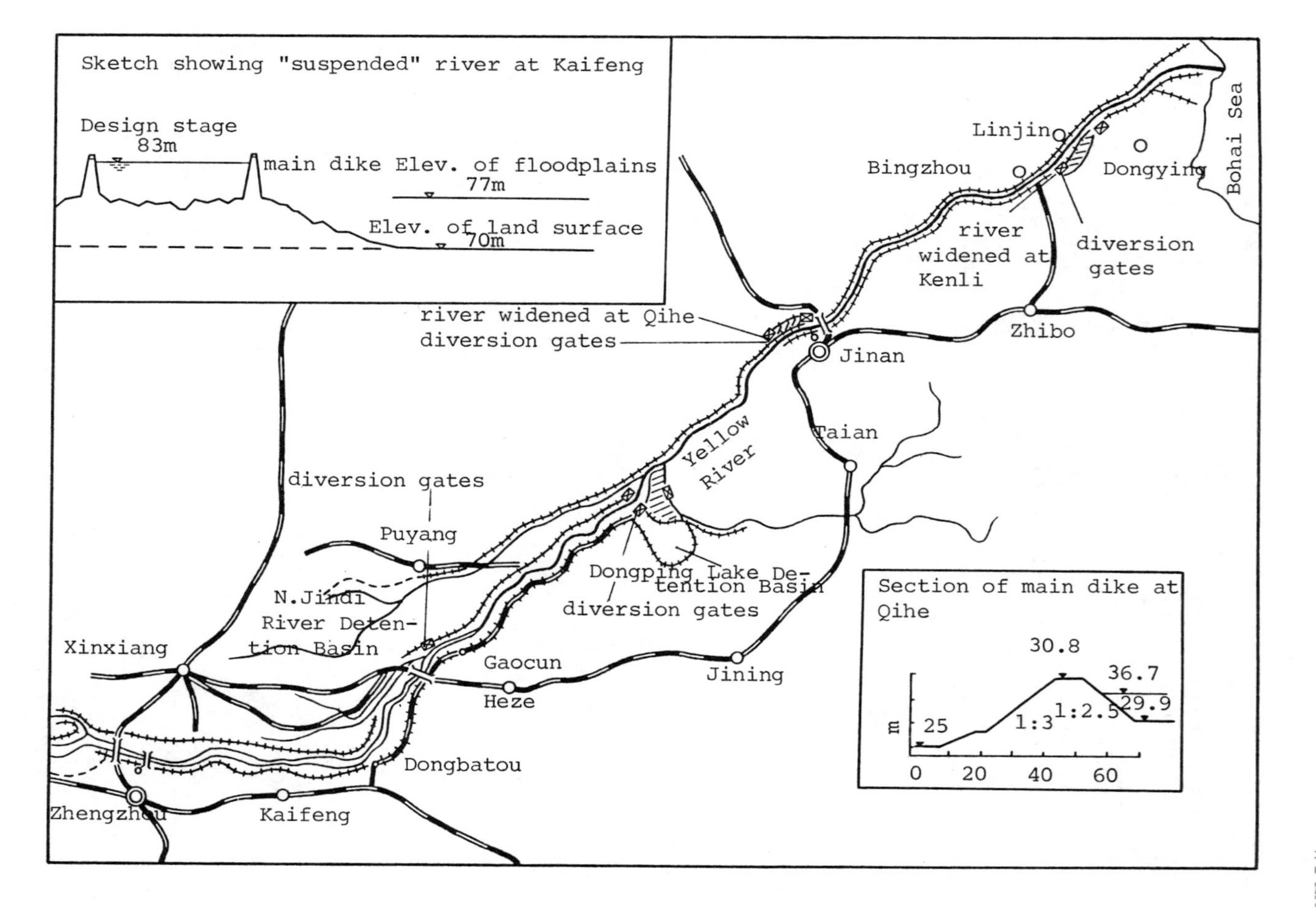

Figure 2. Map Showing Engineering Works for Flood Protection on the Lower Reaches of the Yellow River.

ROLE OF MAJOR FLOOD PROTECTION WORKS AND PROBLEMS THEREOF

The Main Dikes and River Training. Dikes on both banks of the Yellow River measure almost 1,400 km in length. They have been raised three times since 1950, those above Dongbatou gaining 2-4 m and those below, 4-6 m in crest elevation. The dikes are at present approximately 9-10 m in height, with the maximum height being 14 m. The rise of the crest of the dikes exceeds the rise of riverbed due to channel aggradation, so that the flood discharging capacity has been raised, allowing an additional 3000-4000 cms to be discharged into the sea.

As the riverbed is steadily on the rise and the flood stage to be coped with is governed by the discharge at which safety is to be ensured, it has been necessary to raise the dikes accordingly. At present, the dikes are designed to defend against floodwater corresponding to a discharge of 22,000 cms at Huayuankou, as shown in Table 2.

Table 2 Discharges at different stations on the lower reaches against which the dikes are to be defended

Section	Huayuankou	Dongbatou	Gaocun	Sunkou	Aishan and below
Discharge to cope with cms	22000	21500	20000	17500	10000

A number of measures have also been taken to strengthen the main dikes. Probing with pointed rods by mechanical means to detect defects in the dikes and subsequently grouting under pressure has been used to strengthen the dikes. The work has been carried on persistently over the years, and up to the end of 1980 altogether 78 million probing holes had been drilled and the dikes have been strengthened over a total length of 6449 km by grouting with 2 million cu.m of earth.

Making use of the sediment of the Yellow River to form deposits behind the dikes to raise the land surface has been proved effective in checking piping and deformation of foundations of the dike proper due to seepage, thus preventing breaches. Prior to the early 1960s, depressions behind the dikes were silted up as a result of desilting of water tapped from the Yellow River, either by gravity flow or through pumping of the water diverted. Beginning with the 1970s, dredges have been used to move bed material to be laid over a width of 50-100 m behind the dikes, to widen and strengthen them. Up to the end of 1985, a total of 249×10^6 cu.m of earth had been laid over stretches measuring 562 km in all. Practice has proven that reaches with 3-m thick deposits behind the dike have shown no signs of leakage or piping.

Furthermore, strengthening of dikes by construction of sloping downstream clay cores, sand and stone drains, inverted filters, and by berms on either side of the dikes has occurred at some sections.

To prevent direct impingement of the main current on the dikes and consequent breaching, river training has been done together with strengthening of vulnerable spots. Constraints for floodplain protection and defense of the dikes have been carried out in large numbers. At 318 places, vulnerable spots have been reinforced, rebuilt or newly protected and controls and constraints made, effecting protection of the floodplains. Stone revetments measure 474 km in total, comprising 65% of the total length of the river, for which 16×10^6 cu.m of stone has been placed. The flow pattern in a reach of 500 km below Gaocun has been essentially under control and the flow path toward the sea stabilized. All the aforesaid play a significant role in ensuring flood protection.

Through strenuous efforts during the past 30 years or more, the capability of the dikes to prevent floods have been markedly increased. In the past, discharges exceeding 10,000 cms meant breaches or overflowing; one-half of the years with flood flow of 6,000-10,000 cms witnessed dike breaches. During the past 39 years, there were 10 flood discharges exceeding 10,000 cms, and 18 with flow of 6,000-10,000 cms, but no breaches occurred. These suffice to prove the effectiveness of the construction work. It should be noted, however, that the main efforts have been concentrated on heightening the dikes, whereas strengthening of dikes has not been fully accomplished. Dikes of the Yellow River, of considerable length and height as they are, have been rebuilt from the existing embankments handed down by former generations. These were vulnerable and far from free of defects and weak links. Besides, the conditions of the dike foundation are rather complicated and the quality of work of building and strengthening the dikes proper is variable and weak links and vulnerable spots still remain, notwithstanding the improvements already made. The foundations of revetments are comparatively weak, so that a lot must be done in maintenance and rush repairs. Wandering of the riverbed and main current in the reach above Gaocun, and frequent changes in flow pattern will not be remedied within a short period of time. The danger of breaching by ordinary floods still remains. It is therefore important to carry on the work of strengthening the dikes and to lay stress on river training so as to ensure security of flood protection on the lower Yellow River.

PROJECTS FOR FLOOD DETENTION AND STORAGE

Dongpoing Lake Detention Basin. Dongping Lake is situated at the confluence of the Yellow River with its largest tributary on the lower reaches, the Wenhe River, formerly an area of natural detention of floodwater. The lake was rebuilt to store water in 1958, and the reservoir was later stipulated to serve as a detention basin subsequent to 1962.

The Dongping Lake Detention Basin is a most important means of ensuring flood protection in the reach below Aishan. It is intended to deal with floods of both the Yellow River and Wenhe River. As the only major tributary of the lower Yellow River, Wenhe River has a drainage basin of 8,630 sq. km, the maximum discharge ever recorded in history being 8,000 cms. The bed material is mainly coarse sand. As the bed of the Yellow River rises steadily, it becomes more and more difficult for the floodwaters of Wenhe River to be discharged into the Yellow River, thus necessitating detention at Dongping Lake which also serves to cut down the peak flows of Wenhe River and retain the coarse sand to prevent delivery into the Yellow River. In case of a major flood on the Yellow River, the excessive floodwater will be diverted into Dongping Lake at an opportune moment, so as to keep the discharge below Aishan below 10,000 cms. Thus, the amplitude of variation of discharge downstream will be diminished and stability of the channel and safety in flood prevention will be attained.

The detention basin measures 632 sq. km in area, bounded only on the northeast by mountains and rolling terrain, while in all other directions levees are required. The original design stage was El.46 m, but later it was stipulated that maximum storage be controlled at El.44.5 m in the near future, in view of imperfections in dike construction and effects of leakage through the foundation. The corresponding storage will be 3,050 x 10^6 cu.m, of which 400 x 10^6 cu.m will be available for holding floodwater diverted from the Yellow River. There are at present 5 sluices for diversion to pass 7,500 cms (Table 3).

Table 3. Main parameters of detention basins on the lower reaches of the Yellow River

Item	Dongping Lake	N.Jindi River
Area of detention basin, sq. km	632	2,316
Total storage, 10^6 cu.m	4,000	2,700
Effective detention, 10^6 cu.m	2,650	2,000
Diversion, cms	7,500	10,000
Length of enclosing dikes, km	97.4	123
Max. Ht. of dike, m	8.5	13
Cropland, 10^3 mu	457	2,420
Population, 10^3 persons	249	1,255

Taking into account the role of Sanmenxia, Luhun and Guxian Reservoirs in flood prevention, a discharge of 10,000 cms at Aishan corresponds to a recurrence interval of 6-10 years. When floods exceed this value, water is diverted into Dongping Lake and the level of flood protection will be raised to a 60-70 year recurrence interval.

All in all, Dongping Lake plays a significant role in flood protection on the lower Yellow River, its effect being obvious and layout rational. The problems still existing are imperfections of engineering works (the enclosure needs to be further strengthened) and the need of providing a tailwater channel. Measures for satisfactory settlement of immigrants from the flooded area, to improve their livelihood and conditions of production, are to be taken, and safety in operation will be ensured through increasing the capacity of diversion and detention of floods.

Sanmenxia Reservoir. Sanmenxia Reservoir is a key project for harnessing the Yellow River as stipulated in the general plan for improvement of the river. Construction work began in 1957 and the reservoir was first completed in 1960 for impoundment and operation. In the process of construction and operation, however, problems arose so that the size of the project and mode of operation were changed, outlets for flood releases and sediment discharge having been added. The role of the project in obviating damages and gaining benefits is being developed step by step.

Sanmenxia Reservoir commands 92% of the drainage area of the Yellow River, at a key position for regulation of streamflow and sediment regime. Owing to rapid silting and exceedingly large losses due to submersion, the reservoir is used mainly to deal with major floods, while ordinary floods are not detained, thus imposing limitations on its function in flood protection. In accordance with the mode of operation now stipulated, after detention of a major flood from above Sanmenxia, the maximum outflow will not exceed 15,000 cms, thus markedly alleviating menaces of floods in the lower reaches. In case of major floods in the area tributary to the Sanmenxia-Huayuankou reach, regulation of the reservoir approximately will result in detention of 3,000-4,000 x 10^6 cu.m of floodwater, which is of much significance in alleviating the burden of flood protection on the lower reaches (see Table 4).

Table 4. Role of Sanmenxia Reservoir in reducing flood peaks at Huayuankou downstream

Occurrence of flood, %	Peak discharge, cms without reservoir	Peak discharge, cms with reservoir	Runoff, 10^6 cu.m* without reservoir	Runoff, 10^6 cu.m* with reservoir
Frequency of Floods from above Sanmenxia				
1	29,200	17,100	3,690	1,810
0.1	42,100	19,600	7,260	3,580
0.01	55,000	22,000	10,850	5,260
From Sanmenxia-Huayuankou reach				
1	29,200	27,800	3,180	2,100
0.1	42,100	37,800	6,160	3,760
0.01	55,000	46,400	9,680	6,700

*runoff of flood flow exceeding 10,000 cms is denoted

At the initial stage, Sanmenxia Reservoir detained 4,500 x 10^6 tons of sediment and there was no channel accretion downstream over a period of 10 years. Subsequent to 1974, the operation mode was changed to impoundment of water and detention of sediment in non-flood seasons, and discharge of sediment at lowered levels in the flood season, July through October, so that in non-flood seasons, scouring is manifest in the channel downstream instead of deposition. Sediment is concentratedly released in the flood season, so that the ratio of quantity of sediment brought into the sea and incoming sediment load for the whole year actually increases, and a diminution of 50-100 x 10^6 tons of sediment deposited in the channel downstream is witnessed each year, thus effecting a retardation of channel aggradation. Besides, benefits have been gained from multi-purpose development in ice protection, irrigation and power generation among others.

During the process of studying the project in the planning stage and building of the reservoir, mistakes in our work did take place from which lessons have been learned. Success was won through readjustment and rebuilding the reservoir. Since 1974, the canyon reach below Tongguan measuring 120 km in length has been kept in a state of equilibrium with respect to scouring and deposition, so that the effective storage of the reservoir has no longer been encroached upon. Such experience is of value in building of reservoirs in canyon sections of a sediment-laden river.

The existing problems are as follows: It is necessary to study and further clarify the functions and princples of operation of Sanmenxia Reservoir, to carry on reconstruction and maintenance and repairs of the project and further to implement management and utilization of the reservoir area, so

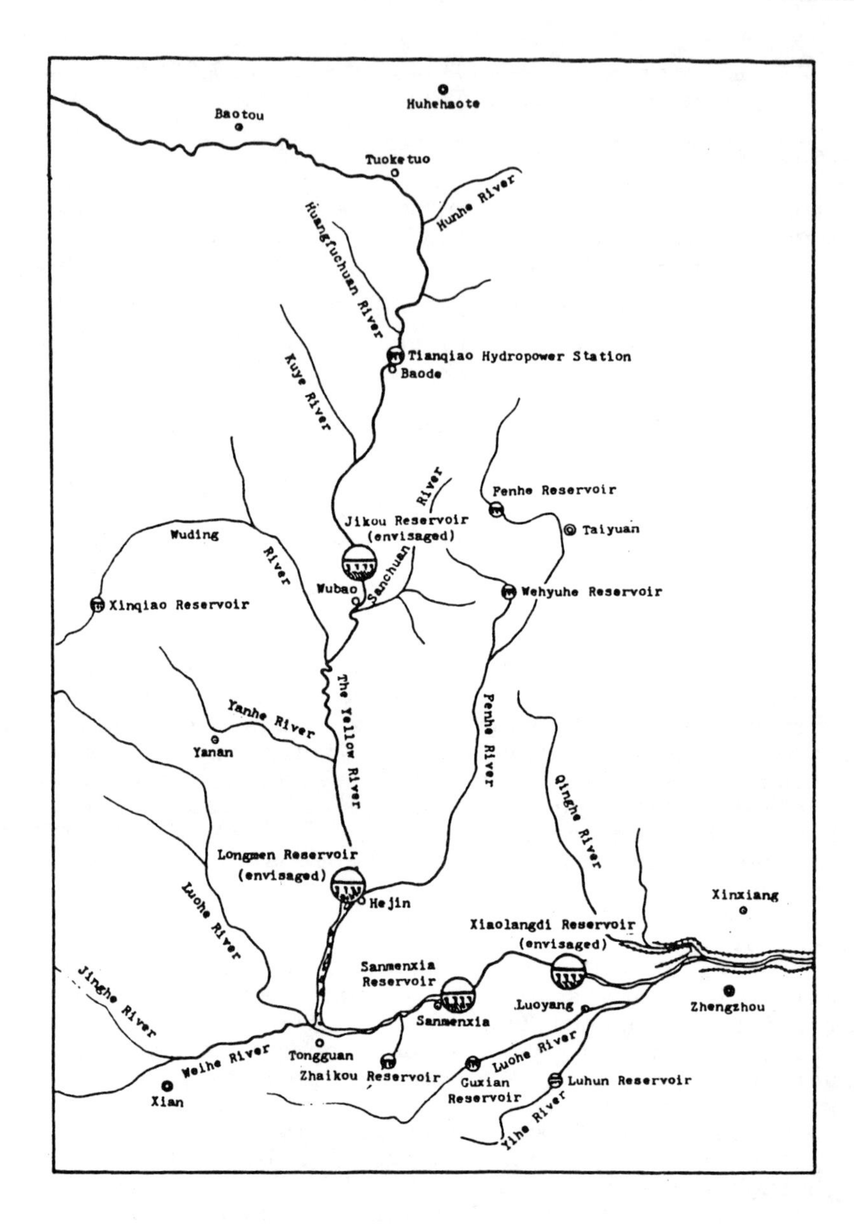

Figure 3. Map showing Location of Reservoirs on the Main River and its Tributaries on the Middle Reaches.

that the project will increase benefits and obviate damage.

Analysis of the Capability of Present Works in Flood Protection. Joint operation of Sanmenxia, Luhun (on the Yihe River) and Guxian Reservoir (now under construction on the Luohe River) to regulate floodwater will provide storage capacity of 4,500-5,500 10^6 cu.m for flood protection. Major floods above Sanmenxia are already subject to control to a remarkable extent, but reservoirs on the tributaries have catchments which are rather too small to effect substantial control, particularly because the capacities are not large enough. Floodwater from the tributaries are therefore not yet subject to adequate regulation. In other words, major floods in the Sanmenxia-Huayuankou reach are not yet effectively controlled. Detention of floods in reservoirs on the main river and the tributaries will lower the frequency of occurrence of a discharge of 22,000 cms from 3.6% to 1.7%, and the dikes downstream will be able to cope with floods of 60-year recurrence interval instead of 30-year recurrence as was formerly the case. The peak flow and total runoff of a flood of rare occurrence will also be reduced to some degree.

In the case of floods exceeding the capacity of the present system of defense, the role of Sanmenxia Reservoir should be given full play, and detention in N. Jindi River basin or at Dagong downstream will be effected at an opportune moment to alleviate damages. Such is our present policy.

In short, the capability of the lower Yellow River system to defend against floodwaters has been pronouncedly raised, thanks to improvements carried out during the past 30 years or more. The danger of breaching or overflowing has been reduced substantially. The work already done serves as an adequate foundation for further reduction of damages. At the same time, valuable experience has been gained. The main problems remaining to be treated are as follows.

The burden of flood protection on the Yellow River is still very heavy. During the period 1950 through 1985, a rise in the riverbed on the lower reaches was as much as 1.2-2.1 m. The unfavorable situation of severe scour and deposition in the reach above Gaocun and shifting of the main current has not been greatly improved. There are still weak links in the long 1,400 km chain of dikes, so that dike incidents during ordinary floods have not been completely ruled out. It is therefore an important task to further consolidate the dikes. Together with rapid development of the economy in the vast plains on both banks downstream and further exploitation of Zhongyuan and Shengli oilfields, still higher demands are being imposed on safety with respect to flood protection and improvement of the estuary section and river mouth. The use of the N. Jindi River Detention Basin, if ever, is being restricted by more and more factors. It is therefore mandatory to work out a policy of raising the capability to protect against floods on the lower reaches and coping with extraordinary large floods.

The flood protection works as a system should be further

perfected. Floods of frequent occurrence are not yet under effective control by existing reservoirs on the main river and its tributaries. The reach above Dongbatou frequented by disasters in the past will be exposed to direct attacks of extraordinarily large floods. With the rise of the riverbed, the probability of old floodplains passing floodwaters is being increased, and the burden of flood protection will be heavier and heavier. Therefore a more reliable guarantee of safety with respect to flood protection is desired. Not much has been done in reducing channel accretion. In the very near future when comparatively small events are to carry larger sediment loads than in previous years, the contradictions will be even more accute as channel accretion may increase. Steps should therefore be taken to cope seriously with this problem.

Table 5. Reducing Flood Peaks at Huayuankou through joint operation of Reservoirs on the Main River and its Tributaries (major flood in the area tributary to Sanmenxia-Huayuankou reach)

Frequency of occurrence of flood, %	Flood peak (cms)		Runoff (Q > 10,000 cms), 10^6 cu.m	
	Without Reservoirs	With Reservoirs	Without Reservoirs	With Reservoirs
1	29,200	25,800	3,180	1,810
0.1	42,100	34,400	6,160	3,200
0.01	55,000	41,700	9,680	5,800

Future Orientation and Measures of River Improvement

ORIENTATION OF RIVER IMPROVEMENT

Many persons show deep concern over policies to reduce damage from the Yellow River and numerous studies have been made. Opinions differ. In the long run, the crux of the problem lies in checking the continual rise of the riverbed. Suggestions and prospects for river improvement may generally be classified into three categories.

First, detention of sediment is supposed to be functional in turning the Yellow River flow into clear water by curbing soil erosion on the Loess Plateau through soil and water conservation. Before the practice of conservation is completely effective, measures are to be taken to detain sediment in reservoirs on the main river and tributaries, by silt-arresters, etc., and through warping as a kind of transition. The aforesaid served as the basis of the planning for harnessing the Yellow River worked out in the 1950s.

Second, warping may be effected on the vast expanse of flat country in the Huanghe-Huaihe-Haihe River basins, to elevate the land surface. Suggestions as such are based on the opinion that all sediment cannot be detained, and the vast plains of North China are capable of accommodating storage of any amount of silt over the long term. Different methods are suggested by different researchers: (1) Widespread development of irrigation with highly sediment-laden water from the Yellow River to effect silting in the irrigated areas, for which large-scale water and silt conveying systems are to be established. (2) Warping through diversion of water from the Yellow River at multiple locations to divert the hyperconcentrated flow separately and to deposit silt all along the paths of flow. (3) Effecting changes of course according to plan, which may be taken as a kind of raising ground elevation through silting in the channels alternately over the long term.

Third, discharging sediment into the sea can prevent the rise of the riverbed. Opinions also differ with respect to this measure. Some hold to the idea of adjusting the streamflow and sediment regimen, modifying the shape of channel sections and raising the transporting capacity of the channel. Others have studied the feasibility of discharging sediment by means of hyperconcentrated flow, to spare the use of water in removing sediment. Still others suggest using water diverted from the southern regions as a means of dislodging and discharging sediment deposited in the channel. Water from the Changjiang River is available and the clear water would be functional in scouring the channel on the lower reaches of the Yellow River.

The following thoughts derive from comprehensive study of the above suggestions, taking into account socio-economic conditions and practical experience in harnessing the Yellow River.

(1) It is possible to reduce channel accretion on the lower reaches and check the pronounced rise of the riverbed.

(2) The problems of both flood and sediment should be taken into account in managing the improvement of the Yellow River. Work in the short run should be coordinated with that in the distant future. Obviating damages should go hand in hand with gaining benefits. The work should be carried out in a comprehensive way, as no single measure, whatever it may be, will satisfactorily solve the problem at present, not to mention harnessing the river once and forever.

(3) Owing to limitations imposed by climatic, geologic, and topographic conditions of the Loess Plateau, soil erosion there can hardly be completely curbed. Loss of soil and water may be alleviated through strenuous efforts over the long term, but it is not realistic to expect marked diminution of sediment delivery into the Yellow River through conservation alone. The Yellow River will remain a sediment-laden river over a long period of time. Improvement of the lower reaches of the river must be based on such a conception in order to gain initiative and avoid serious mistakes in future work.

(4) In order to ensure carrying out socialist construction without a hitch, stability of the lower Yellow River and security of flood protection should be guaranteed in the next few decades. To cope with such an imminent problem, measures

must be taken to exercise better control of floodwater, and to raise the flood-resisting capability of the dikes while checking the continual rise of the riverbed on the lower reaches. Construction of more reservoirs on the main river to exert appropriate control over floodwater and sediment and to detain coarse sediment in particular, are key measures which are practicable and feasible which can change aspects of flood protection on the lower reaches to improve the situation.

(5) The strategic deployment of efforts to harness the Yellow River should be based on a longer time interval, while taking into account limitations of present ideas. It is not practical to think of accomplishing the planning and implementation of construction work to harness the river in one stroke.

Based on studies made by different units of different topics, adoption of measures that are technically reliable and economically rational and feasible can be instrumental in ensuring a stable lower course of the Yellow River in the next 100 years or so. Sound judgments on how things will be in the very remote future cannot be made yet.

(6) From the viewpoint of the situation of flood protection and the trend in its development, it seems unnecessary as well as irrational, to consider manmade changes of the course of the river. Such a measure will not be effective in reducing flood hazards and the burden of flood protection in the near future. There will be serious threats of inundation from the new channel within several decades, and flood prevention work is very difficult to accomplish. Within a few decades, or even in 10 odd years, the new channel will again be suspended. Besides, in doing so all existing drainage networks, communication lines, irrigation systems, water supply and sewerage will be disturbed completely, and several million mu of cropland will have to be acquired and a population of over 1,000,000 relocated. The cost would be tremendous and such attempts would certainly interfere with the social economy in a big way. Envisioning a change of course of the river is therefore impractical.

SUGGESTIONS ON MEASURES OF IMPROVEMENT

Based on consideration of solutions to the problem of harnessing the river during the next 50 or 100 years, the measures to be stressed, whether practiced in the past or not, should be as follows:

(1) The flood protection works on the lower reaches should be further strengthened and adequately rebuilt (including river training, stabilization of flow pattern and appropriate design and direction of flow paths at the estuary section).

(2) Emphasis should still be laid on further conservation and improvement of the middle reaches. The focal point being the source area of coarse sediment.

(3) More high dams should be built in canyon sections of the main river to form large reservoirs, the dead storage of which can be used to retain sediment and provide effective

storage to regulate streamflow and the sediment regime.

(4) Diversion of water from southern regions should be realized step by step, so that an abundant supply of additional water will be available to the lower reaches and made use of at opportune moments to effect scouring of the riverbed.

It is worth stressing the fact that construction of reservoirs on the main Yellow River can be instrumental in solving both the problems of flood protection and sediment diminution on the lower reaches and can be of strategic importance in harnessing the river. It is a key link in quickly changing aspects of flood protection on the lower Yellow River.

Firstly, construction of reservoirs on the main river is functional in quickly and effectively checking further accretion in the channel downstream and in substantially cutting down flood peaks.

Secondly, the reservoirs may be used over the long term for well-coordinated reduction of damage and increase in benefits to attain substantial comprehensive and economic returns.

Thirdly, regulation of streamflow and sediment regime may be effected over the long term. At present the Xiao-langdi, Jikou and Longmen Reservoirs are under study. The dead storage in these will be able to hold about 35×10^9 tons of sediment. Even if we consider reduction in streamflow without corresponding reduction of sediment in the near future, the aforesaid will be functional in keeping the channel downstream from further accretion during the coming 50 years or so. Subsequent to completion and effective operation of the reservoirs, joint operation for regulation of streamflow and sediment regime will be realized through completion of a system of engineering works. These will alter the present situation of inconsistency of water and sediment flow to a marked degree reducing the amplitude of variation of runoff and sediment load. This will favor putting an end to the wandering and ever-changing bed in the broad channel downstream and will increase the sediment-transporting capacity of the channel.

In the long run, sediment of the Yellow River should be further utilized to raise the land elevation behind the dikes, so that the lower course will not be "suspended" anymore, relatively speaking. This is to be coordinated with diversion of water from the Yellow River on both of its banks for irrigation purposes and desilting of water to supply municipalities and industry. Tapping water from the Yellow River which affects silt deposition behind the dikes should be given subsidies by the state. The difficulties of coping with silt in water tapped from the Yellow River for irrigation will soon be overcome, and water courses in the flatland will not be readily silted up which, in turn, favors flood protection in the related area. In so doing, reduction of damage by the Yellow River will be realized once and for all. At present, something like 10×10^9 cu.m of water are being tapped from the Yellow River each year for use by industries and agriculture. Along with the water, some 200 x 10^6 tons of sediment are being carried. If 1/2 of the sediment is laid behind the dikes, radical changes will certainly be effected in several decades.

FLOOD CHARACTERISTICS AND DESIGN FLOOD FOR FLOOD PROTECTION IN LOWER YELLOW RIVER

Shi Fucheng and Wang Guoan
Reconnaissance, Planning & Designing Institute
Yellow River Conservancy Commission
People's Republic of China

Introduction

The Yellow River is the second largest river in China, flowing through Qinghai, Sichuan, Gansu, Ningxia, Inner Mongolia, Shanxi, Shaanxi, Henan, and Shandong provinces as well as autonomous regions. The total length of the Yellow River is 5,464 km, with a drainage area of 752,400 sq.km. The upper reach of the Yellow River is from the river source to Hekouzhen in the autonomous region of Inner Mongolia, the middle reach from Hekouzhen to Taohuayu near the city of Zhengzhou, and the lower reach from Taohuayu to the estuary into the Gulf of the Baohai Sea (see Fig. 1).

There are two main types of floods in the basin of the Yellow River. One is the so-called storm flood, the other, the ice flood. This paper focuses on the former, the storm flood.

The Storm Characteristics

The Yellow River basin is situated in a typical monsoon climate zone. With the arrival of summer, the circulation of the westerly belt pulls back to north, while the subtropical high pressure of the west Pacific moves westward and to the north. At the same time, a large amount of warm-and-wet air, together with the southwest and southeast monsoon, transfers to north and is unceasingly criss-crossed by the cold-and-dry air from north to south. Therefore, a variety of rainfall climatic processes are formed.

CAUSE OF RAINSTORM FORMATION

L. M. Brush et al. (eds.), Taming the Yellow River: Silt and Floods, 407–423.

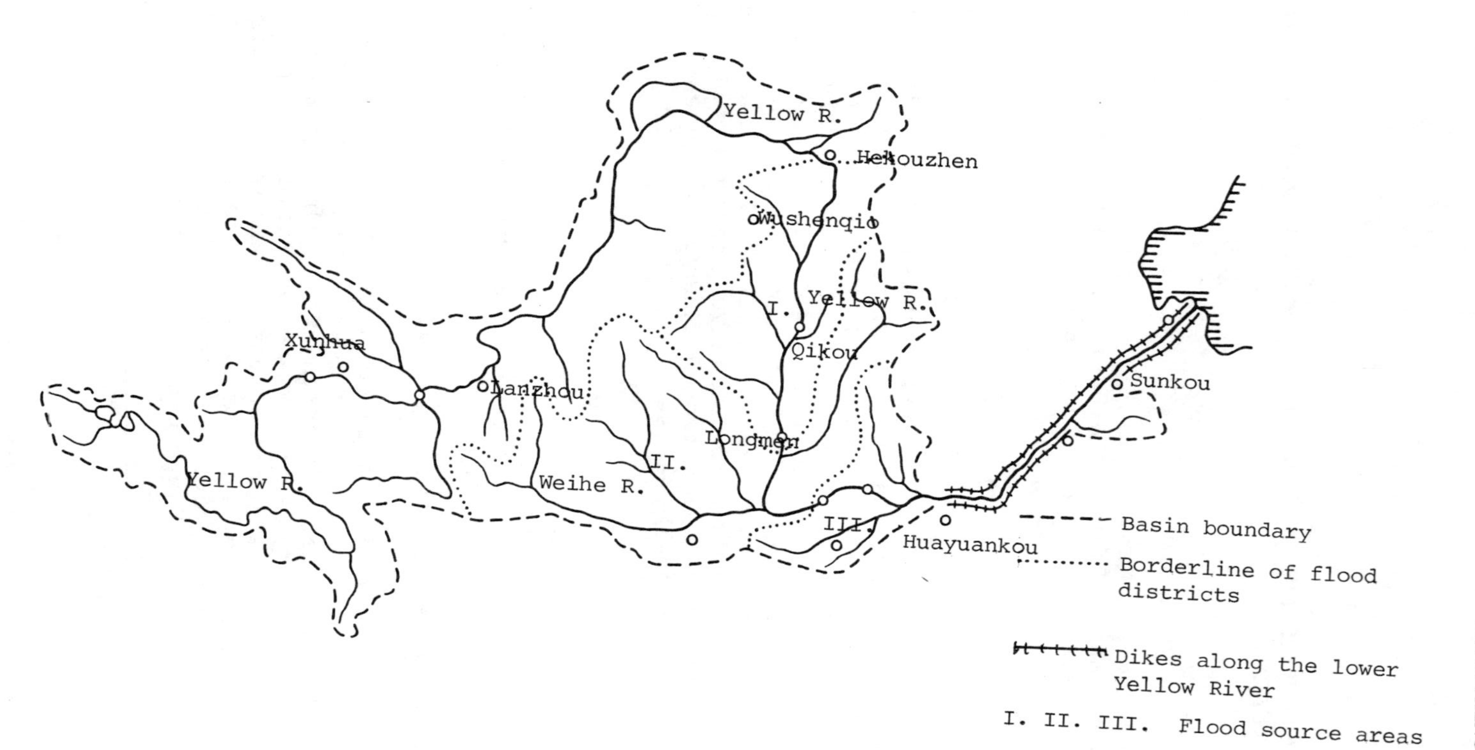

Figure 1. Map of the Yellow River Basin.

The area above Xunhua in the upper reach of the Yellow River is located in the northeast of the Qinhai-Tibet plateau of China. The weather chart of 500 hpa shows that in the summer there is often a shear line having a direction of northeast to southwest between Lanzhou and Changdu. The formation and maintenance of the shear line results from the confluence of the rather cold strong air from northern Xinjiang moving to the east with the strong southwest air-flow moving from east of Qinhai-Tiber plateau to the north. As a consequence, the weather consists of, long-lasting rains or overcast skies.

The causes of heavy rainstorms in the middle reach of the Yellow River can be divided into two types according to the patterns of circulation. One is the longitudinal, the other the zonal type.

The zonal circulation in mid-summer is characterised by the following weather phenomena: the zonal circulation is frequently in the westerly belt; the short wave trough is more active; and the subtropical high pressure is located rather to the north and represents the stable distribution of zonality, east to west. On the weather chart of 700 hpa, there are cold shear, warm shear, westerly trough, and triple point etc. This kind of weather system often forms the heavy rainstorms between Hekouzhen and Sanmenxia in the middle reach of the Yellow River.

Longitudinal circulation, when it is dominant in the westerlies, the subtropical high pressure in the west Pacific is stable and the long wave system moves slowly or sometimes does not move at all. The rainstorm weather system with a shear line from south to north, typhoon and side trough etc., forms the heavy rainstorms which occur between Sanmenxia and Huayuankou on the middle reach of the Yellow River.

RAINSTORM CHARACTERISTICS

Rainstorms of the Yellow River basin occur mainly between July and September. Rainfall above Xunhua on the upper reach of the Yellow River is characterized by large area of rain, long duration, about 20 to 30 days, but with less intensity, seldom over 50 mm, while the characteristics of rainstorms between Hekouzhen and Sanmenxia on the middle reach of the Yellow River may be summerized as heavy intensity and short duration. Strong rainstorms often occur from the middle of July to the middle of August. Using the Yangjiapin hydrologic station, in the Kuyehe river, as an example, it can be seen that the amount of rainfall reached 408.7 mm in 12 hours covering an area of some 22,000 sq.km. The most striking event happened on August 1, 1977, when a heavy rainstorm occured in Wushenqi in the border region between Shaanxi and Inner Mongolia. In the center of rainfall area the amount of rainfall was up to 1400 mm within 9 hours. The rain area within which the amount of rainfall was over 50 mm accounted for 24,000 sq.km. An isohyetal map of the rain is shown in Fig.2.

Large scale rainstorms can be produced within this area under special conditions. In the middle of August in 1933, for example, a heavy rainstorm poured down on the watersheds of all the large tributaries of the Yellow River, from Hekouzhen to Longmen, as well as the Jinghe River, Luohe River, and Weihe River at the same time. The area of rainfall covered over 100,000 sq.km with the rain belt direction being from southwest to northeast. This kind of rainstorm is the typical rainfall pattern causing heavy floods at Sanmenxia.

The rainstorm characteristics in the reach from Sanmenxia to Huayuankou are rather similiar to that from Hekouzhen to Sanmenxia in its rainfall intensity. Usually one rainfall can last two or three days and cover an area of 20,000 or 30,000 sq.km. In the middle of July in 1958, the maximum daily rainfall measured at the Yanqu station was 366 mm and the observed daily rainfall at Rencun reached 650 mm. At the end of July in 1982, the maximum daily rainfall measured at Shiyu station on the middle reach of Yiluohe River was 734 mm with its rain area extending from south to north, as shown in Fig.3.

According to the historical records, there was a large rainstorm in 1761 which covered the whole area between Sanmenxia and Huayuankou and lasted some 5 days.

Because of the vast area of the Yellow River drainage basin, and different storm-forming weather conditions prevailing over different localities, heavy rainstorms do not usually occur in upper and middle reaches at same time. Furthermore, by analyzing the historical literature and recent records, heavy rainstorms from Hekouzhen to Sanmenxia did not occur at the same time as from Sanmenxia to Huayuankou, though both areas are in the middle reach of the Yellow River. These characteristics are very important in dealing with problems of flood protection on the lower reach of the Yellow River.

Flood Characteristics

Floods of the Yellow River in summer and autumn are usually caused by rainstorms that occur during July through September.

FLOOD SOURCE AREAS

There are many lakes and swamps in the source area of the Yellow River. Down the Maduo to Longyangxia there are plains

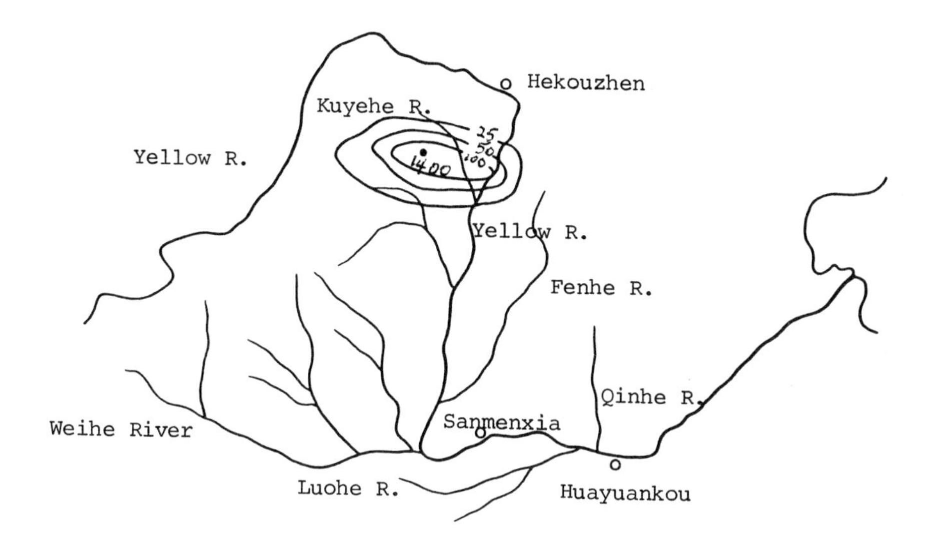

Figure 2. The Isohyetal Line of Rainstorm Occurring in Wushenqi on August 1, 1977.

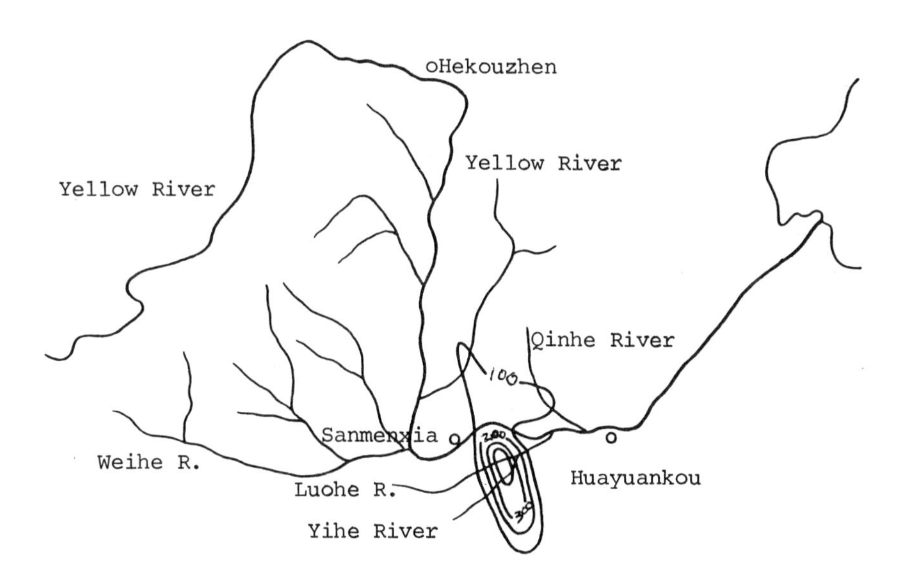

Figure 3. The Isohyetal Line of Rainstorm in the Reach from Sanmenxia to Huayuankou on July 31, 1982.

alternating with canyons. After flowing out of Qintongxia gorge, the river runs through two large plain areas, but at other places, mostly desert or grasslands, the river becomes broader and the slope becomes flatter. Owing to these types of runoff conditions as well as the previously mentioned characteristics of rainfall on the upper reaches, the flood process is characterized by gentle ascending and descending flood waves of long duration and a relatively low peak flow. A flood in Lanzhou usually can last 30 to 40 days and its measured maximum peak discharge has been very seldom over 6000 cms since the Lanzhou hydrologic station was established. The flood wave tends to be lower and flatter as it propagates to Hekouzhen and becomes the base flow of floods from the middle and lower reaches of the Yellow River.

From Hekouzhen to Longmen, the river threads its way between the banks of cliff mountains and drains an area of some 110,000 sq.km. For the most part, except wind-drift sandy lands in the upper part of some tributaries on the right bank and rocky mountains in part of area on the left bank, the river flows through a gullied and hilly loess area which is overlain by loose erodible soil with poor vegetation and heavy loss of soil and water. Those characteristics, together with the heavy rainfall in this area often result in floods with hyperconcentrations of sediment that rapidly rise and fall, yielding high peaks of short duration. A single flood usually lasts one or two days, a series of continuous floods four or five days. The measured maximum peak discharge at Longmen station was 21,000 cms on August 11, 1967, and the measured maximum concentration of sediment was 933 kg per cu.m on July 18, 1966. The maximum peak discharge obtained through investigation of the historical floods of 1842 or 1843 was 31,000 cms.

The reach of the Yellow River from Longmen to Sanmenxia, joined by tributaries such as the Jinghe River, Luohe River, Weihe River and Fenhe River, possesses a drainge area of 188,000 sq.km, most of which is located in the gullied hilly loess area or loess plateau. Other parts are either rocky mountains with forests or alluvial plains. In these reaches sediment concentrations during floods are also very large.

When floods that originate in the area between Hekouzhen and Longmen meet with floods from the area between Longmen and Sanmenxia, large floods with high peaks and a large volume may be formed at Sanmenxia. The flood in 1933 was the highest flood ever recorded since 1919, the begining of the measurement and collection of field data at Sanmenxia gaging station. The peak discharge at that time was 22,000 cms, and the accumulated flood volume during 12 days was 9.07 billion cu.m. The sediment load carried during the flood reached 2.1 billion tons. This number is much higher than the average annual sediment load at Sanmenxia. According to investigation, there was a flood with maximum peak discharge of 36,000 cms in 1843, resulting also from the confluence of peak flows from the two areas mentioned above.

In the catchment from Sanmenxia to Huayuankou, though only 41,615 sq.km, owing to the tight spatial and temporal distribution, can cause large floods. If floods rising from the Yiluohe River, Qinhe River, and from the main stem between Sanmenxia and Huayuankou meet at the same time, a large peak discharge may occur at Huayuankou. In July 17, 1958, for example, a major flood took place in the Huayuankou with its peak discharge of 22,300 cms, which was the largest value ever measured there. Floods which mostly originate from this area, i.e. the area between Sanmenxia and Huayuankou, last about 10 days. Another flood event occuring in 1761 also resulted from the encounter of floods in the main stem with the floods in the tributaries. At that time the peak discharge in Huayuankou was as high as 32,000 cms. Since the founding of the People's Republic of China, floods have occurred there three times with discharges over 15,000 cms. Besides the flood in 1958, there was a flood discharge of 15,000 cms on August 4, 1954, and one of 15,300 cms on July 31, 1982. Flood flows originating from this area constitute the major portion of floods at Huayuankou. Floods in this area often rise and change rapidly. It usually takes only 12 to 14 hours for the flood to flow from major controlling stations on the main stem and the tributaries to the cross-section at Huayuankou, therefore, the forecasting time ahead of an event is rather short.

It has been shown that floods of the Yellow River are derived mainly from the following three areas: between Hekouzhen and Longmen, between Longmen and Sanmenxia, and between Sanmenxia and Huayuankou (as shown in Fig.1. the areas marked I, II, III). Floods originating from the area above Lanzhou form the base flow for large floods of the middle and lower reaches.

COMPOSITION OF FLOODS AT HUAYUANKOU

The cross-section at Huayuankou serves as the general controlling station for floods from different source areas of the upper and middle reaches and is also the entrance to the lower reach. Taking into account that the Sanmenxia reservoir has already been built, it seems very important to flood protection on the lower reach to study the various combination of floods from the area above Sanmenxia and floods from the intermediate area between Sanmenxia and Huayuankou.

It is known that, according to the measured data and historical information, major floods from above Sanmenxia and from Sanmenxia to Huayuankou do not take place in the same year. For the last five hundred years, the years when the floods occured above Sanmenxia were 1632, 1662, 1785, 1841, 1843, 1933, 1942, while the floods from Sanmenxia to

Huayuankou were in 1482, 1533, 1761, 1954, 1958, and 1982.

Taking the floods of 1761 and 1958 as typical examples of floods from the area between Sanmenxia and Huayuankou, and the floods in 1843 and 1933 as typical examples of floods originating from the area above Sanmenxia, the composition of these floods at Huayuankou is shown in Table 1.

It can be seen from the Table that when floods come mainly from above Sanmenxia, both the peak discharge and the flood volume account for more than 90% of those in Huayuankou, while the area between Sanmenxia and Huayuankou contributes a very limited amount of water. It can also be seen that when the floods came mostly from the area between Sanmenxia and Huayuankou, the peak discharge amounted to 70-80% of that at the cross-section of Huayuankou and the volume of the flood in 12 days accounted for only 40-60%. There was still a part of water coming from the area above Sanmenxia. The intermediate area from Sanmenxia to Huayuankou is only about 41,615 sq.km, however, the flood runoff per unit drainage area from this area ranks the highest among all the source areas.

CHARACTERISTICS OF FLOOD ROUTING IN THE LOWER REACH

The river channel in the lower reach is wide and the slope is mild. Here floods tend to have more storage, reducing the peaks. On the basis of statistics of peak flows at Huayuankou for the past four decades, the peak flow in the whole lower reach can be attenuated by 30-60%. Among these, 20-50% in reaches from Huayuankou to Sunkou (the wandering reaches in Henan Province), and 10-20% in reaches from Sunkou to Lijin (reaches in Shandong Province) as shown in Table 2.

It can be seen that the effects are different year by year. The difference has much to do with the following factors: the size of the peak discharge; the shape of the flood hydrograph; the state of river aggradation; and the diversion of floods at the Dongpinhu Lake. Figure 4 shows the flood hydrographs at several stations in the lower reaches in 1958.

Analyzing data obtained since the early fifties on major floods from the Jinghe as well as the Wenhe River, tributaries on both sides of the lower reaches, indicates that they have not met simultaneously with major floods in the Yellow River. Although medium size floods may still occur they will not endanger flood prevention structures.

Design Flood for Flood Protection in Lower Reaches

The design flood for the control section of the lower Yellow River is based upon the cross-section at Huayuankou. However, because of the characteristics mentioned above and the fact that Sanmenxia reservoir has been built, it is necessary to calculate the design flood at Sanmenxia and that in the

Table 1. Flood Composition at Huayuankou Station

flood type	year	peak discharge (cms)			flood volume of 12 days billion cu.m.			% of Qmax at Huayuankou %		% of flood volume at Huayuankou	
		S	S-H	H	S	S-H	H	S	S-H	S	S-H
mainly from	1761	6000	26000	32000	5	7	12	18.8	81.2	41.6	58.4
area between									investigated values		
Sanmenxia &	1958	6400	15900	22300	4.95	3.2	8.18	28.8	71.2	60.5	39.5
Huayuankou									measured values		
mainly from	1843	36000	1000	33000	11.9	1.7	13.6	97	3	86.0	14.0
area above									investigated values		
Sanmenxia	1933	22000	1900	20400	9.19	0.86	10.05	90.7	9.3	91.3	8.7
									measured values		

* H. stands for Huayuankou, S. for Sanmenxia, S-H for intermediate area

Table 2. Effects of attenuation of peakflow in reaches from Huayuankou to Lijin, in percent (%)

year	peak discharge (cms)			percentage of attenuation of peak discharge (%)		
	H.	S.	L.	H.----S.	S.----L.	H.----L.
1953	11200	8120	6860	27.5	11.3	39
1954	15000	8600	7220	42.8	9.3	52
1957	13000	11600	8500	10.8	24.1	35
1958	22300	15900	10400	28.5	24.5	53
1977	10800	4770	4130	56.6	5.2	62
1982	15300	10100	5810	34.0	26.7	62

* H. stands for Huayuankou, S. for Sunkou,and L. for Lijin.

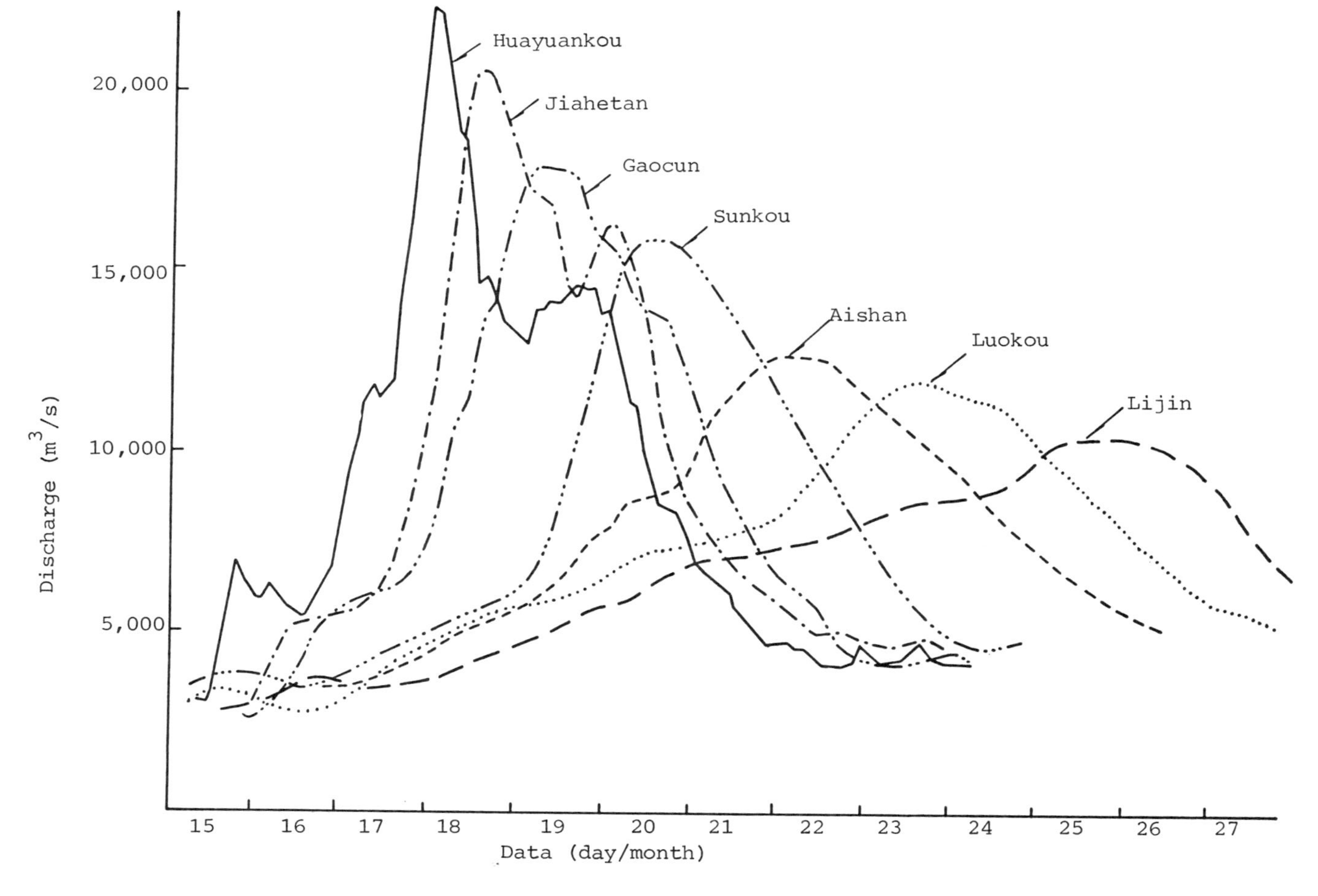

Figure 4. Flood Hydrographs of Stations on the Lower Reaches, 1958.

catchment from Sanmenxia to Huayuankou as well, besides that at Huayuankou itself.

The design flood is determined jointly by statistical analysis, historical floods and an estimate of the maximum probable flood.

HISTORICAL FLOODS

The floods of 1843 and 1761, which influenced the lower reach of the Yellow River very much, are briefly described.

The flood in 1843 was one of the most extreme floods that has taken place. It came mainly from the area above Sanmenxia. The rain belt was distributed from the southwest to the northeast covering the Jinghe River basin, the upper and middle reaches of the northern Luohe River and the large tributaries on the right bank from Hekouzhen to Longmen. The peak discharge at Sanmenxia was estimated to be 36,000 cu.m per s. Because the flood came mainly from the gullied-hilly loess area with severe loss of water and soil, it carried a very large amount of sediment with a maximum of 518 kg per cu.m in concentration. The deposits resulting from this flood are still evident in some reaches from Sanmenxia to Xiaolangdi, 20 meters above the low water river bed, with a thickness of 3 meters.

By analyzing the historical data, investigating the old deposits along the river, and confirming the dates of relics under the silt layer with modern techniques, it can be shown that the flood in 1843 was at least the largest in the last 1,000 years. The flood caused dikes to breach at Zhongmu in the lower reach and inundated over twenty counties.

The flood of 1761 was an extremely large flood and one of the largest in the last several hundred years. It originated from the area between Sanmenxia and Huayuankou. The rain covered a vast area from Huaihe River basin in the south to Fenhe River basin and the Haihe River basin in the north and from Guanzhong area in Shaanxi province in the west to Huayuankou in the east. The rain belt was distributed from the south to the north, with the storm center located in the lower and middle reaches of Yiluo River and Qinhe River as well as the intermediate area along the main stem of the Yellow River. The rainfall lasted about 10 days with high intensity rainstorm lasting only 4-5 days. Based on historical data, it is estimated that the flood of that year had a peak flow of 32,000 cms and a small amount of sediment load. It was the largest flood in the area from Sanmenxia to Huayuankou since 1553, with an estimated return period of 430 years. Breaching of dikes took place at twenty-six sites in the downstream part of the river. The disaster covered about 30 counties in Henan, Shandong, and Anhui provinces.

CALCULATION OF DESIGN FLOOD BY STATISTICAL METHOD

Among stations concerned with the flood protection on the lower reaches of the Yellow River, the Sanmenxia Station (called Shaanxian station before) has the longest series of data with which to calculate flood frequency. Field data have been collected at this station since 1919. The data series at Huayuankou station may be extended to a period comparable to that at Sanmenxia if the field measurements are supplemented by interpolation for the years with no available data. Data for more than 30 years are also available for the area from Sanmenxia to Huayuankou. The data points influenced by operation of Sanmenxia have been restored to the situation without a reservoir. As the historical highest floods have been taken into account in the frequency calculation and the data of various stations have been analysed and tested for its reasonableness in a large scale, the calculated data are stable.

In calculation, the mean value ($\overline{X}$) and the coefficient of variation (C_v) were determined by:

$$\overline{X} = \frac{1}{n} \sum_{i=1}^{n} X_i$$

$$C_v = \frac{1}{\overline{X}} \sqrt{\frac{\sum_{i=1}^{n} (X_i - \overline{X})^2}{n - 1}}$$

The coefficient of skewness Cs was determined by the frequency curve fitting method (C_v was adjusted a little then).

For continuous series:

$$\overline{X} = \frac{1}{N} \left[\sum_{j=i}^{a} X_j + \frac{N-a}{h-1} \sum_{i = L+1}^{n} X_i \right]$$

$$C_v = \frac{1}{\overline{X}} \sqrt{\frac{1}{N-1} \left[\sum_{j=1}^{a} (X_j - \overline{X})^2 + \frac{N-a}{n-1} \sum_{i=L+1}^{n} (X_i - \overline{X})^2 \right]}$$

Where, X_i, X_j stand for continuous series and extreme

variation value respectively. a, L for numbers of historical extreme values and numbers of extreme values in the observed series.

The experienced frequency in the continuous series is calculated by mathematical expectation equation:

$$P_m = \frac{m}{n+1}$$

The experienced frequency of historical extreme value by:

$$P_M = \frac{M}{N+1}$$

Where, P_m, P_M for the experienced frequency in the continuous series and the historical extreme value respectively, m, M for the series numbers in the order from large to small value of continuous series and historical extreme value,and n, N for the numbers of years of continuous series and recurrence period of first term in the historical extreme value.

The results used for each station in planning and design are shown in Table 3.

PROBABLE MAXIMUM FLOOD

After Sanmenxia reservoir was built, the extreme flood which would threaten the lower reaches of the Yellow River is one originating from the area between Sanmenxia and Huayuankou. Therefore, the probable maximum flood from this area was analyzed.

Three methods, i.e., a meteohydrological method, a statistical method and an enlargement of historic flood method, were employed in the analysis. In the meteohydrological method three storm models were used: a local model, a combination model and a model translating storms from a neighboring area.

Water vapor and efficiency were amplified in the first two models and only water vapor was amplified in the last model in maximizing the value for computation. Parameters used in maximizing, such as design dew point and design efficiency, were selected comprehensively by analysis of many methods.

Runoff was evaluated by a storm-runoff model which took into account storage effects and by use of a unit hydrograph.

The historic flood of 1761 was properly enlarged in evaluating the design flood. The multiplier to be adopted was estimated based on the idea used in the meteohydrological method, i.e. from storm --- amplifying water vapor --- probable maximum precipitation. Methods stipulated by national standards were followed in the statistical approach. Results from three methods were combined and a synthetic analysis made. A value of 45,000 cms was finally selected as the

Table 3. Results of flood frequency analysis, cubic meters per second for peak discharge, 100 million cu.m for flood volume

station	basin area (sq.km)	content	mean value	C_v	C_s/C_v	value of frequence as P % 0.1	1.0
Sanmen-xia	688421	peak discharge	8880	0.56	4	40000	27500
		flood vol. 5 days	21.6	0.50	3.5	81.8	59.1
		12 days	43.5	0.43	3.0	136	104
		45 days	126	0.35	2.0	308	251
Sanmenxia to Huayuankou	41615	peak discharge	5100	0.92	2.5	34600	22700
		flood vol. 5 days	9.8	0.90	2.5	64.7	42.8
		12 days	15.0	0.84	2.5	91.1	61.0
		45 days	31.6	0.56	2.5	132	96.5
Huayuan-kou	730036	peak discharge	9780	0.54	4.0	42300	29200
		flood vol. 5 days	26.5	0.49	3.5	98.4	71.3
		12 days	53.5	0.42	3.0	164	125
		45 days	153	0.33	2.0	358	294

Table 4. The needed flood-control storage in lower reaches of Yellow River (based on various standards)

flood standard	typical years	peak discharge (cms)		storage capacity (100 million cu.m) when $Q > 10,000$ cms	
		Huayuankou	Sunkou	Huayuankou	Sunkou
field measured data	1933	20400	16400	16.5	
	1958	22300	15900	17.1	
once in hundred years	1933	29200	21900	36.9	31.5
	1958	29200	21000	31.5	29.4
once in thousand years	1933	42300	31600	72.6	66.0
	1958	42300	29600	62.2	60.4
P.M.F flood	1933	55000	40700	108.5	103.2
	1958	55000	48400	97.9	96.9

maximum probable flood in the intermediate basin between Sanmenxia and Huayuankou. An addition of 1,000 cms released from the hydropower station at Sanmenxia may be added making a total of 46,000 cms for the PMF at Huayuankou.

DESIGNING THE FLOOD HYDROGRAPH AND FLOOD-CONTROL STORAGE

According to the flood characteristics mentioned above, the floods at Huayuankou are mainly due to water-rising either above Sanmenxia, like the flood in 1933, or between Sanmenxia and Huayuankou, like the floods in 1958 and 1982. The flood hydrographs in those years are taken as the typical hydrographs and are then amplified by controlling for the same frequency of the peak and amount into the design flood hydrographs of different frequencies.

The maximum discharge for flood protection downstream on the Yellow River is governed by the reaches in Shandong province. Now the safe discharge in these reaches is 10,000 cms. In order to maintain less than this amount, the storage capacity needed in the reaches above Aishan in Shandong province, based on the data for the typical years, should be as follows: 3 billion cu.m for the floods appearing once in a hundred years, 6 billion cu.m for floods appearing once in a thousand years, and 10 billion cu.m for the probable maximum flood, as shown in Table 4.

In order to regulate water discharge, besides utilization of the existing reservoir at Sanmenxia and Dongpinhu Lake to block parts of the floods, additional measures should be considered, such as building the Xiaolangdi reservoir on the main river or expanding more detention areas, so that the safety of the lower reaches of the Yellow River can be ensured.

References

1. Investigation and analysis of extreme rainstorm in August 1977 in Wushenqi in middle reach of Yellow River, "Yellow River," vol. 2, 1979

2. Flood in August of 1933 on middle reach of Yellow River, "Hydrology," vol. 6, 1984

3. Confirmation of reappearance time of flood in 1843 in Yellow River. "Yellow River," vol 4, 1982

4. Analysis of flood in 1761 on middle and lower reaches of Yellow River. "Yellow River," vol 2, 1983

FLOOD FORECASTING AND FLOOD WARNING SYSTEM IN THE LOWER YELLOW RIVER

Chen Zanting
Hydrology Bureau
Yellow River Conservancy Commission
People's Republic of China

ABSTRACT: This paper deals with the development and problems of flood forecasting and flood warning on the lower reaches of the Yellow River, and some suggestions for improving the work are also made according to experiences obtained in the past 30 odd years and current development and techniques of flood forecasting and flood warning at home and abroad. Influenced by sediment, unsteady river bed and human activities, floods on the lower Yellow River are very complicated in nature bringing a lot of difficulties to flood forecasting. Since the founding of the People's Republic of China, great achievements and many experiences have been obtained in the work of flood forecasting and flood warning on the lower reaches of the Yellow River. Flood forecasting has played an important role in flood protection. However, many problems remain to be solved. In order to improve flood forecasting and flood warning on the lower Yellow River in steps, a system of telemetering precipitation and stage and on-line real-time flood forecasting has begun to be set up in the sub-watershed between Sanmenxia to Huayuankou, at some hydrological stations in the lower reaches and in the floodplain on both banks of the lower Yellow River.

General Introduction to the Development of Flood Forecasting and Flood Warning

A system of flood forecasting and flood warning is an important non-structural measure associated with the success of flood protection. Flood warning information in the lower reaches of the Yellow River began in 1574 (Ming Dynasty). At that time, the hydro-information was sent by riders on horseback at the speed of 250 km a day. In 1765 (Qing Dynasty), water record stakes were set up at Wanjintan of Shenzhou in the main stream of the Yellow River, Gongxian in the Yiluo River and Wuzhi in the Qing River to measure the

L. M. Brush et al. (eds.), Taming the Yellow River: Silt and Floods, 425–449.

rising water levels. Regular hydrological stations were set up at Shenxian and Luokou in 1919, and some hydrological stations, gaging stations and rain stations were established on the main stream and its major tributaries. From these stations messages were sent by means of radio or by cables through the telecommunication department. But the problem was that the stations were sparsely distributed and could not quickly dispatch hydrological forecasts.

The hydro-information work has developed rapidly since the founding of the People's Republic of China. The system of flood forecasting and flood warning established step by step consists of hydro-information stations, a forecasting center, a headquarters for flood protection, administration of reservior releases and water conservancy projects and communication stations (see Figure 1). The Flood Forecasting Center is established in the Headquarters Office of Flood Protection on the Yellow River. Up to 1986, 493 hydrological stations have been set up, including 220 hydrometric stations, 25 stage stations and 248 rain gauge stations. Some hydro-information services were also set up in the Provincial Flood Protection Commands in Henan and Shandong as well as a Reservoir Administration which is in charge of hydrological information and flood forecasts.

The observed hydrological information is sent by special radio sets or telephones through telecommunication bureaus. When it is forecast that the peak flow may exceed the pre-assigned warning limit at Huayuankou station, the forecasting center will notify the appropriate people. The Yellow River Anti-flood Headquarters Offices of Henan and Shandong make complementary forecasts after they receive the information from the forecasting center, and the complementary forecasts are sent immediately to each service and sub-service of flood protection, so that they may be prepared to fight against the flood.

When extreme floods occur, the detention basins must be used. The General Headquarters Office of the Yellow River Flood Protection and the Provincial Command of the Yellow River Flood Protection of Henan and Shandong must send the flood warning to relevant units, such as to people living on the floodplain, along banks and in detention basins. The warnings are sent by telegrams, telephones and radio indicating the amount of rain, stage, discharge and the area which might be inundated.

There are three steps in forcasting: (1) Estimation of possible floods on the basis of predicted rainfall, (2) Forecasting of floods on the basis of actual rainfall, and (3) Forecasting of floods in lower reaches according to floods which have already occurred in upper reaches. The last two steps must be taken in the forecasting of ordinary floods. When a more serious flood occurs, the first step must be made as background only for decisionmakers. Each hydrological

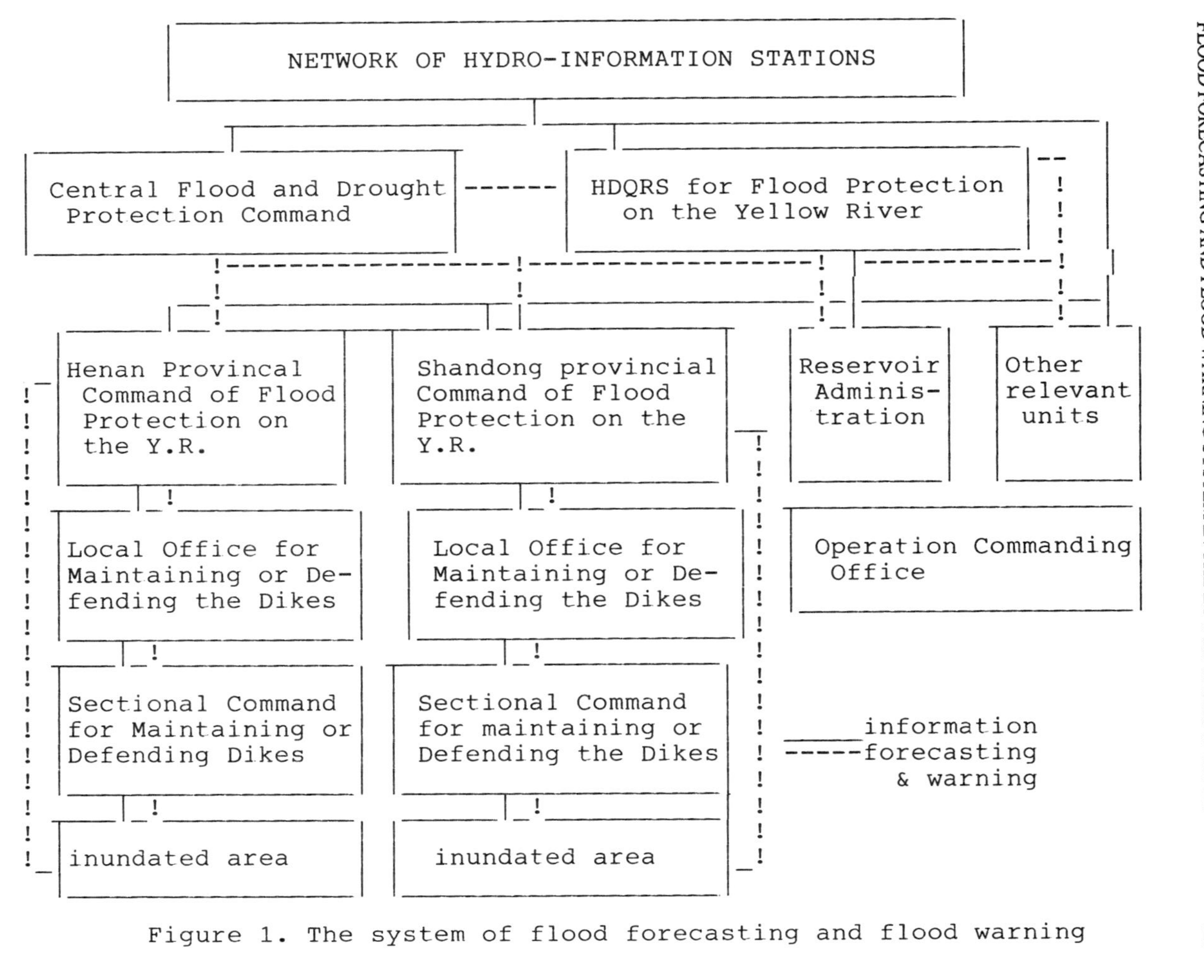

Figure 1. The system of flood forecasting and flood warning

station feeds back hydro-information to the forecasting center while the flood advances. The main reports resulting from forecasts are peakflow, stage, the flood progress and the time of occurrence of peakflow.

This system has made a great contribution to flood protection in the lower Yellow River for more then 30 years. For example, a maximum flood of 22,300 cms at Huayuankou occurred in the lower Yellow River in 1958. During that flood, hydrological information was sent in time, flood forecasts and flood warnings were dispatched without delay. The forecasting was rather accurate throughout the lower reaches. A flood peak of 22,000 cms and water level of 94.40 m was forecast and a flood peak of 22,300 cms and water level of 94.42 m was observed. Because of the forecast, a decision was made to safeguard the dikes and not to use the detention area. Two million people fought their best against the flood and won the victory of flood protection and the inundation of a population of some 1.4 million and 3 million mu of land in the detention basin was avoided.

As a further development of this system, in 1981 a subsystem was set up for real-time telemetering and forecasting from Sanmenxia to Huayuankou. This area is divided into 7 small regions with 233 telemetering stations and one collection station in each small region. A microwave communication system has already been established between the Zhengzhou computer center and Luhun, as well as a pilot region in the upper reaches of the Yihe River between Sanmenxia and Zhengzhou. Telemetering units will be established in a subarea from Sanmenxia to Xiaolangdi and 5 other small regions in near future. Microwave communications and telegrams remain the major tools for sending hydro-information to the forecasting center. The hydrological information received is input to the computer for forecasts. In 1986, several stage telemetering stations were set up on the main stream and the floodplain below Huayuankou, and it is planned to continuously improve the work of real-time monitoring for a prompt grasp of the flood propagation and overbank flow along the river. The flood forecasting and flood warning system will be very much improved after the implementation of these new projects.

Flood Characteristics and Forecasting Methods

FLOOD CHARACTERISTICS

The method of flood forecasting must be adapted to the flood characteristics. Flood characteristics are briefly discussed as follows.

The characteristics of the physiography and floods in the source area of floods are shown in Table 1. In this Table the shape coefficient denotes the ratio of average discharge in a

Tbl 1.Physiographical conditions & features of flood in flood-source regions

Characteristics	flood-source area	upstream from Lanzhou	Hekouzhen -Longmen	Longmen-Sanmenxia	Sanmenxia-Huayuankou	Wen River basin
Geomorphologic (% of total area)	rock mountain area	48.5	11.3	22.8	45.8	66.2
	loess area	9.9	58.8	65.7	44.6	0
	sand drifted area	8.0	29.9	0	0	0
	grassland & plain	36.6	0	11.5	9.6	33.8
rainfall	raining condition	long duration, low intensity	short duration, high intensity	s.d.&h.i. in summer l.d.&l.i. in autumn	short duration, high intensity	short duration, high intensity
	max. daily rainfall (mm)	142.4	1400	358.5	366.5	287.8
runoff production	runoff producing patten	runoff after storage full	runoff in excess of infiltration	runoff in excess of infiltration	runoff after storage full	runoff after storage full
	runoff coefficient	0.03-0.76	0.03-0.40	0.08-0.45	0.05-0.90	0.10-0.79
runoff collection	flood shape	low-flat in shape lasting 22-66hr.	sharp-thin in shape	relatively flat in shape	sharp-thin in shape	sharp-thin in shape
	shape coefficient	0.70-0.84	0.20-0.45	0.34-0.68	0.21-0.48	0.16-0.36
	time	13 days	2 days	1 day	10 hours	-
sedimentation	sediment yield condition	low sediment content	much higher s.c. coarse s.	high s.c. more find sediment	low s.c.	low s.c.
	max. s.c. (kg/mcube)	329(Lanzhou)	933(Longmen)	911(Sanmenxia	103(Heishegou	12.1

flood to peak discharge. Time denotes time required for flood propagation to Huayuankou. It can be seen from the Table that the characteristics of rainfall-runoff and propagation of floods are quite different for floods originating in different areas, and may have different influences on the lower Yellow River. Floods from the upper basin above Lanzhou would have to travel more then 2,500 km lasting nearly half a month to reach the downstream areas. There the flood peaks nearly disappear and the flood merely contributes to raising the base flow. Floods in the reaches from Hekuozheng to Longmen and from Longmen to Sanmenxia (especially in the Jinhe, Weihe and Louhe Rivers) contain a much higher concentration of sediment, and make the riverbed change violently due to aggradation and degradation along the channel. The aggradation value for a flood from Longmen to Tongguan may be 30-50% of the volume of sediment in the flood. Floods with hyperconcentrations of sediment usually result in extremely strong erosion wherein the river bottom may be turned over and extensively lowered. In a flood event in 1970, an average of 9 m in depth of the river bottom was scoured away at the Longmen hydrological station. A long distance below Longmen, the river channel was also scoured greatly and channel storage obviously was changed. Before construction of Sanmenxia reservoir, owing to the relatively deep channel from Longmen to Tongguan, the bankfull discharge was about 12,000 cms, but a flood peak of 10,000-20,000 cms might be reduced by 10-20% through channel storage. In the initial stages of operation of the Sanmenxia reservoir, large volumes of sediment were deposited in the channel, a flood peak of over 5,000 cms would overflow the floodplain, but a 10,000-20,000 cms flood peak might be reduced by 40-60%. After the reconstruction of Sanmenxia reservoir, its capacity for sediment transport has been increased. Discharge over the floodplain and the percentage of reduction of the flood peak have recovered to the original state.

Three other features of floods worth mentioning for the reach from Sanmenxia to Huayuankou are as follows: The first is that when a flood with a hyperconcentration of sediment takes place, the water level may not respond normally. On 7-13th of August, 1977, the flood peak at Xiaolangdi section was 10,100 cms, the maximum sediment concentration reached 941 kg/cu.m. In the process of flood propagation towards Huaguankou no water was added from tributaries. During the rise, the water level suddenly dropped and then rose abruptly after 1.5 hours. The peakflow at Huayuankou didn't decrease but rather increased to 10,800 cms. A second feature is that there are 576 medium or small-size reservoirs on tributaries in this district with a total storage capacity of some 700 million cubic meters. Because of the lower design standard of these reservoirs, they may retain surface runoff for a rainfall with normal intensity, but as heavy rainstorms occur, they might easily collapse and raise the flood peak. Most of these reservoirs are not monitored, and provide no information during rainy days. The third point is that there is a river-cross floodplain at the confluence of two large tributaries, the Yihe River and Louhe River, in this district. When the flood exceeds the limit of flood protection, the dikes may collapse somewhere and part of the flood may flow into the river-cross floodplain resulting in a further decrease of the flood peak. However it is very difficult to accurately estimate the volume of detention, the location, the time and the size of the dike collapses which are all directly related to the detention of floods. After the occurrence of a flood, a great part of the water would return to the channel, which will change the common law of flood movement.

After a flood passes Huayuankou, it flows in the channel between two dikes. The Wenhe River, the largest tributary of the Yellow River in the lower reaches, enters the main channel after passing Dongping Lake. The flow is controlled by the outlet gates of the lake, so the Wenhe River has little influence on the flood of the Yellow River. A tremendous influence may be produced by erosion and deposition as well as by the farmers' production dikes on the floodplain. Along the lower Yellow River, the river channel is wide upstream and narrow downstream. The reach above Gaocun is wandering with a width of 5-20 km between the two dikes. The reach from Gaocun to Taochengpu is a transitional reach, with a width of 1.5-8.5 km between the two dikes. Downstream, the watercourse meanders from Taochenfu to the estuary with a width of 0.4-5 km between two dikes. The capability of flood conveyance in the narrow reaches is less than 50% of that in the wide reaches. The channel shifts continuously especially in reaches above Gaocun. Great changes in aggradation and degradation may take place in an extreme flood. For example, the riverbed was scoured 0.5-1.0 m at Huayuankou and Gaocun resulting in a differece of about 50-100% of the bankfull discharge. This caused a very unsteady relationship between stage and discharge (see Figures 2 and 3), and also changed the storage capacity and channel storage curve drastically.

The floodplain in the lower Yellow River covers an area of 3,070 sq.km, 72% of the total area of the rivercourse, with two-thirds in cultivated land and inhabited by 127 million people. For crop production on the floodplain farmers have constructed many small dikes which are called "production dikes." These production dikes can prevent ordinary floods from entering the floodplain, however, they may reduce the attenuation of the peakflow and hinder the conveyance of major floods. Once production dikes collapse during floods, some of the flood water may also be detained. This changes the laws of flood movement in natural channels and causes a lot of difficulties for accurate flood forecasting.

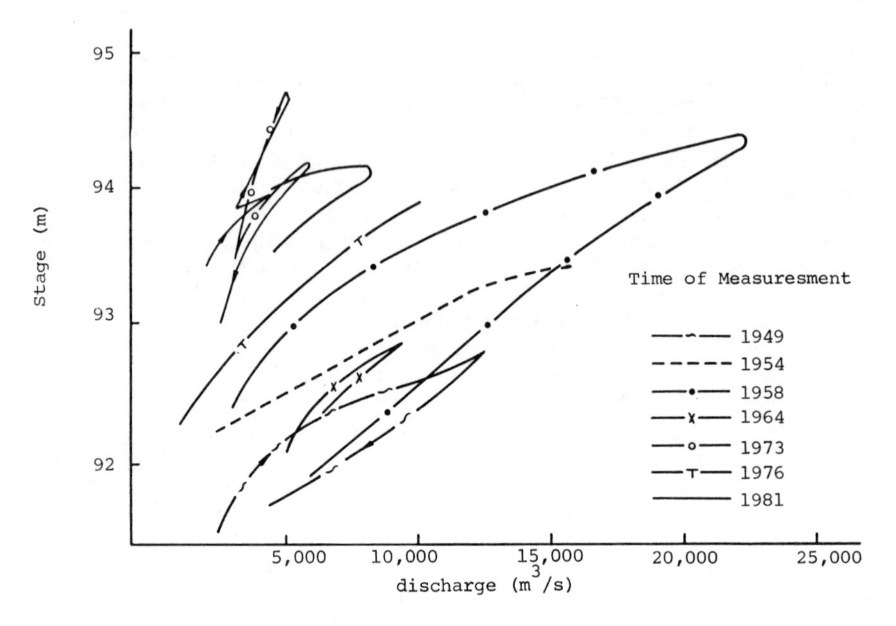

Figure 2. The Relationship Between Water Level and Discharge at Huayuankou Hydrological Station in 1958

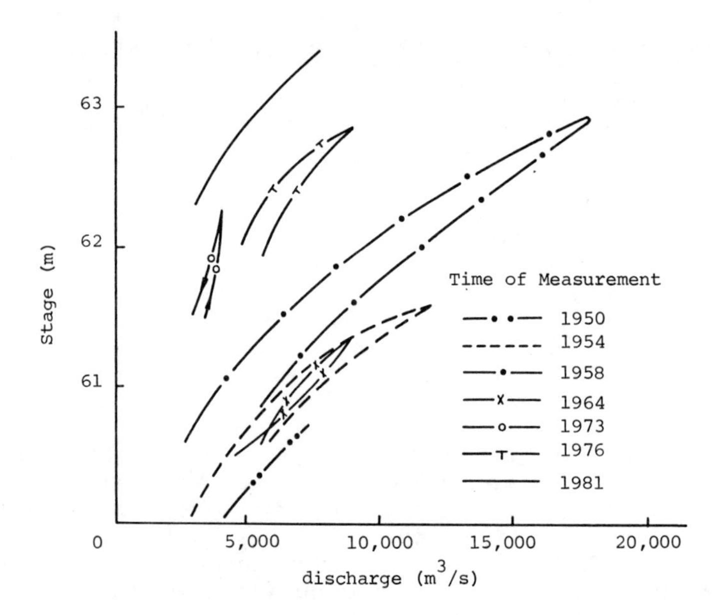

Figure 3. The Relationship Between Water Level and Discharge at Gaocun Hydrological Station in 1958.

FORECASTING METHOD

According to the characteristics of floods mentioned above, the following forecasting methods are used.

Rainfall-runoff Forecasting. For each flood-source area a forecasting scheme has been drawn up, and each scheme may be divided into two parts--runoff yield and flow concentration.

i. Two methods for runoff forecasting have been used depending on the local conditions in each area.

Rainfall-runoff relationship is based on the model of runoff formation with the natural storage.

$$R = f(P, Pa) \qquad (1)$$

in which, R is mean depth of runoff in an area, Pa is antecedent soil moisture, their corresponding relationship is shown in Figure 4.

Two methods are used for an area where runoff will be produced if rainfall exceeds infiltration. One is the relationship between rainfall and runoff with multi-variables (Fig.5).

$$R = f(P, Pa, T) \qquad (2)$$

where T is duration of rain.

The other is the infiltration curve method.

$$f_t = f_c + (f_o - f_c)e^{-\beta t} \qquad (3)$$

or

$$f(t) = f_c + (f_o - f_c)e^{-KP_a t} \qquad (4)$$

in which, f_t is infiltration rate, f_c is steady infiltration rate, f_o is the maxmium infiltration rate under condition of dry soil, P_a is antecedent soil moisture, β , K are coefficients determined by field data.

ii. Calculation of concentration for flow forecasting of outlet flood processes. There are three methods to be applied:

a. Peakflow and flood volume relationships and a generalized hydrograph method.

These methods were used in the 1950s, and are still useful. The peak-volume relationship, or the relationship between total runoff of volume (or mean runoff depth) and the net discharge of a flood peak with parameters of rain duration and

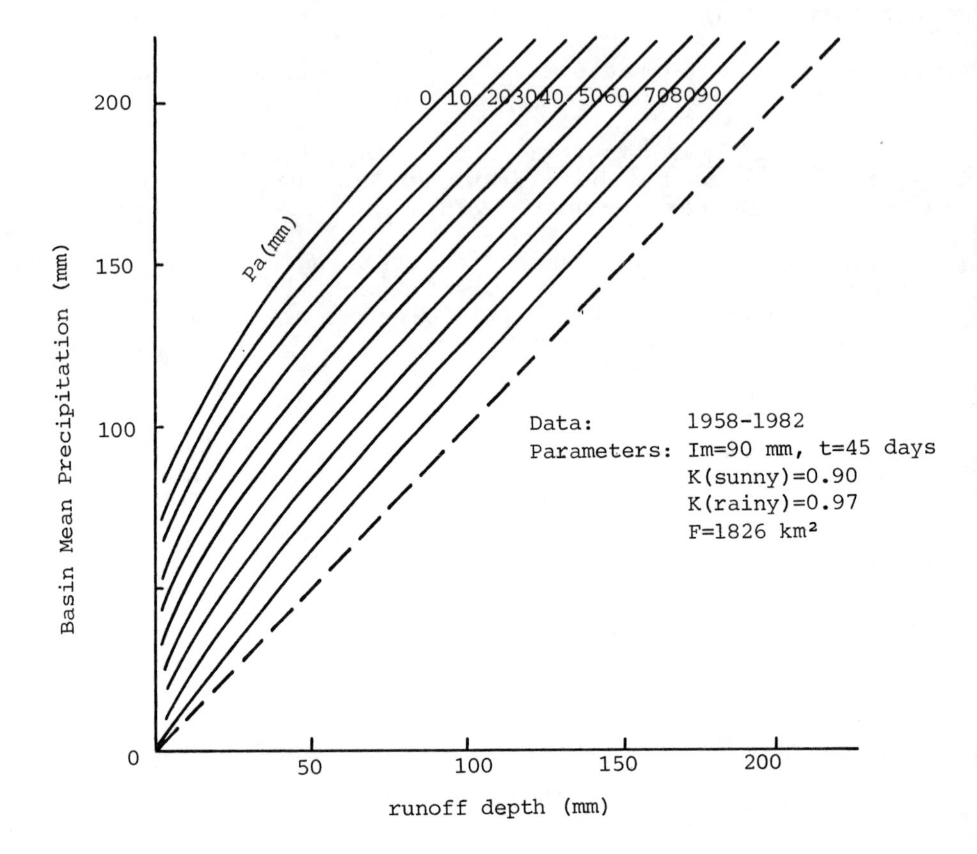

Figure 4. The Relationship Between Rainfall and Runoff in the District from Luhun to Longmenzheng in Yihe River.

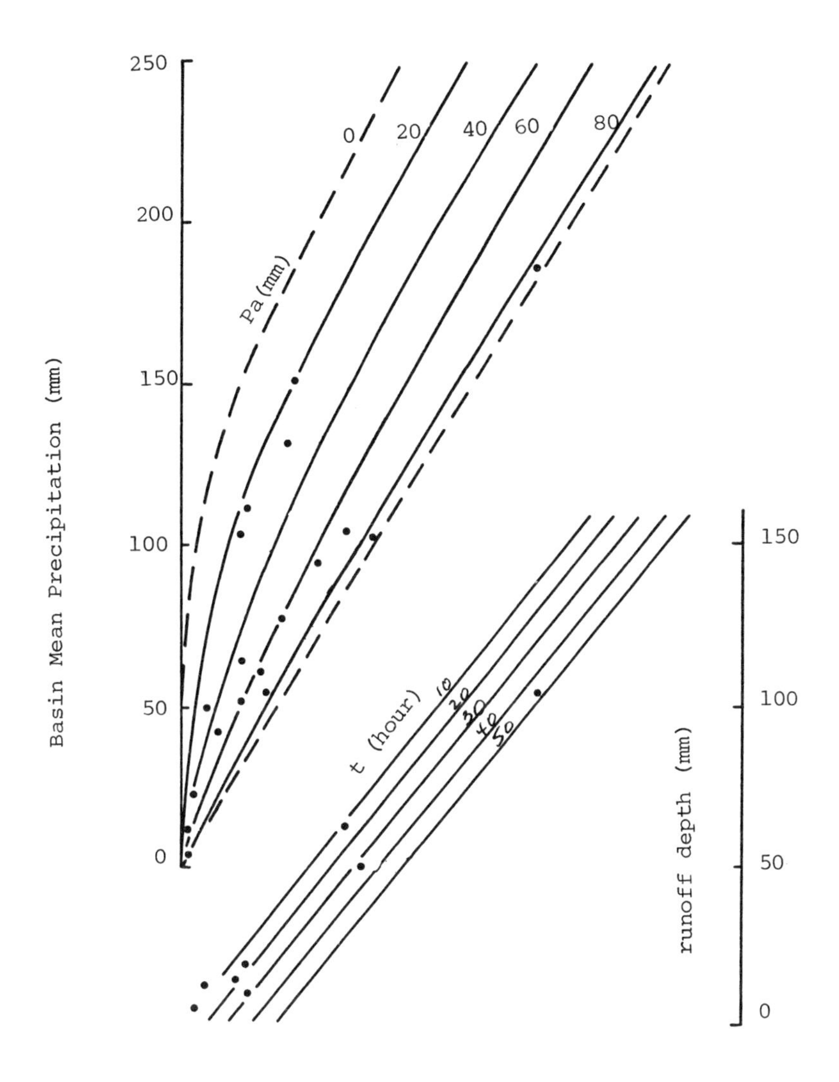

Figure 5. The Relationship Between Rainfall and Runoff in the District from Sanmenxia to Xiaolangdi

coefficient of rainfall distribution (sometimes a factor expressing the weighted area of runoff yield may be used) is shown in Figure 6.

The generalized hydrograph is a curve in which net peakflow and duration of flood are expressed in relative percentages (see Figure 7).

b. Unit hydyograph method.

The unit hydrograph method was often used in the 60s and 70s. Currently the methods of the Sherman unit hydrograph and the instantaneous unit hydrograph are also used. Several hydrographs are prepared according to different patterns of rainfall and one is chosen depending upon actual rainfall distribution.

c. Unit area flow concentration method.

This method has been used from the late 70s to the early 80s. The river basin is divided into many unit area elements for which the instantaneous unit hydrograph is calculated, and its parameters n and k are related to rainfall intensity. The larger the rainfall intensity, the sharper the instantaneous unit hydrograph will be. The outlet flow of each unit area can be calculated based on the rainfall and then routed to the outlet section using the cofficient of channel flow concentration. Lastly, the total runoff at the outlet can be superimposed.

The third method is the best for application, but it is more complex computationally. Chart and computer programs have been prepared for use in routine works.

Flood-peak Forecasting. Because of the violent aggradation and degradation on the riverbed of the Yellow River relating flood water levels at upper and downstream stations cannot be done directly. Considering the influence of the flood shape and the bankfull discharge (reflecting the channel changes due to sedimentation) as major paremeters, a relationship between the relevant flood peaks was established. Denoting the flood shape coefficient by η and bankfull discharge by Q_a, a chart can be drawn for the following types:

i. reaches with no tributaries:

$$Q_{md} = f(Q_{mu}) \tag{5}$$

$$Q_{md} = f(Q_{mu}, \eta) \tag{6}$$

$$Q_{md} = f(Q_{mu}, Q_a) \tag{7}$$

in which, Q_{mu}, Q_{md} are the relevant flood peak discharges at the upstream and downstream stations respectively, η is the flood peak coefficient, Q_a is determined by last flood or calculated by the relationship between stage and discharge at

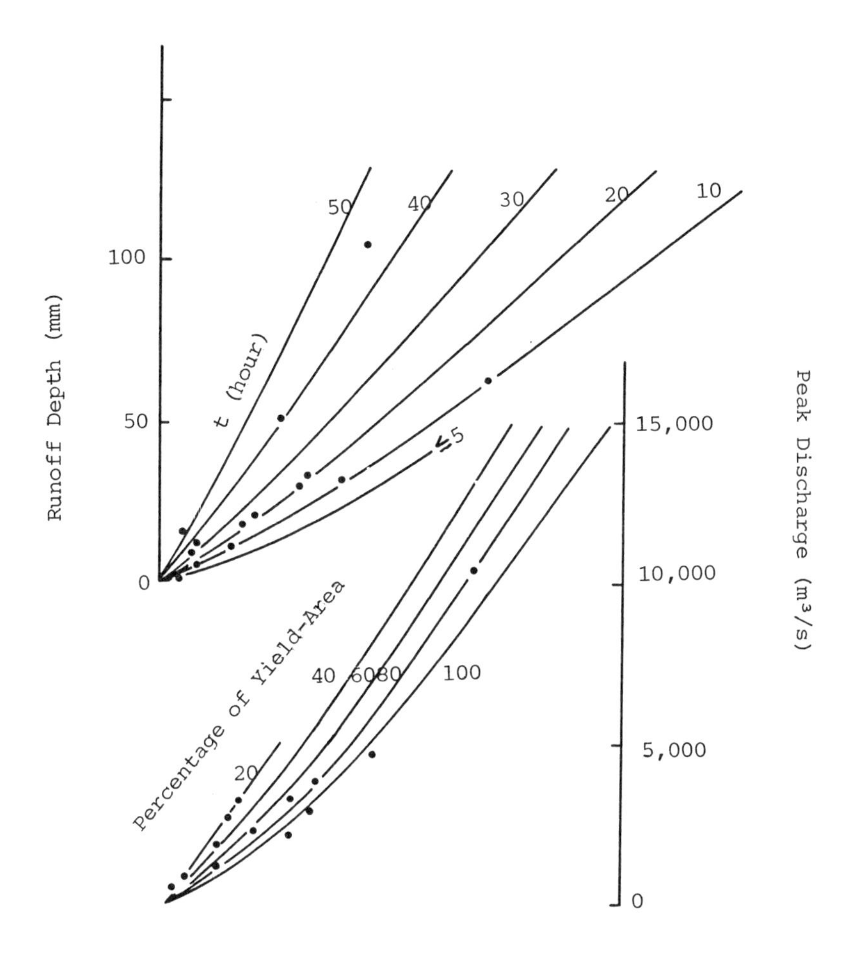

Figure 6. The Relationship Between Runoff Depth and Net Peak in the Region from Sanmenxia to Huayuankou

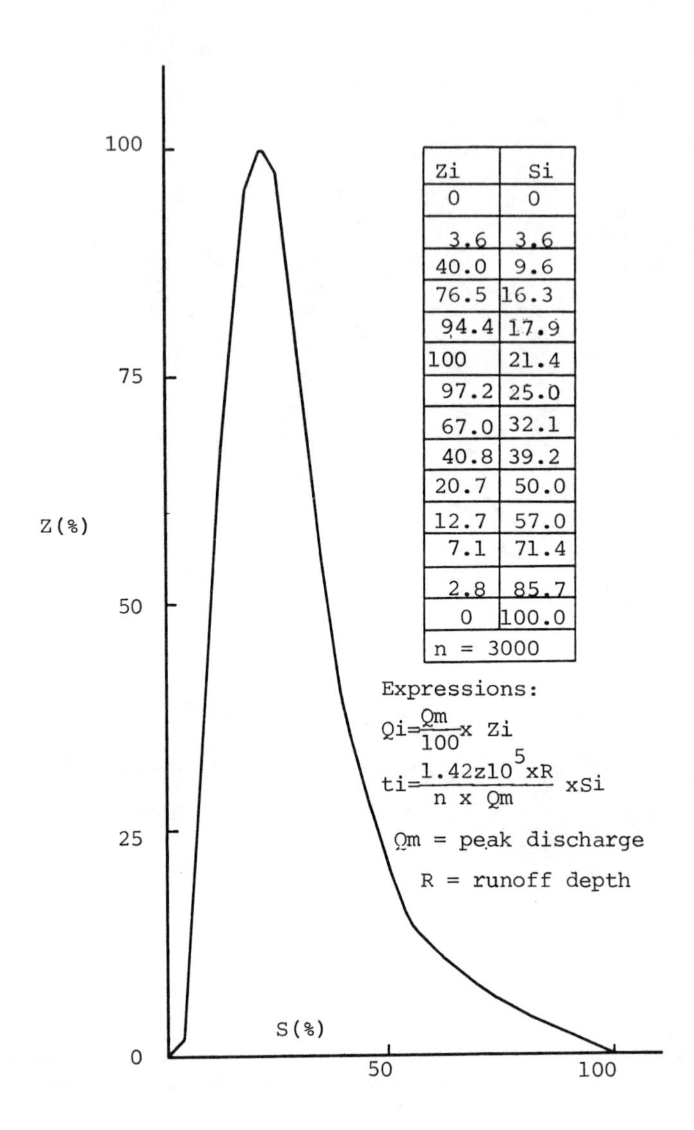

Figure 7. The Generalized Hydrograph of the Flood in the District from Sanmenxia to Xiaolangdi

any time (see Figures 8, 9, 10).

ii. reaches with tributaries:

From Xiaolangdi to Huayuankou the following relationship has been applied:

$$Q_{md,t} = f\left(\sum_{1}^{n} Q_{t-\tau_i}, \eta\right) \tag{8}$$

where $\sum Q_{t-\tau_i}$ is the summation of discharge of stations on the main stream and tributaries from Xiaolangdi to Huayuankou, shown on Figure 11.

Forecasting for Time of Propagation. There are three methods for calculating the propagation time (τ). One is the method of experience. The propagation time can be obtained by the difference in time of the appearance of relevant flood peak at upstream and downstream stations. The second method is to calculate the time of propagation by the wave velocity.

$$\tau = L/\lambda V \tag{9}$$

in which, L is a length of reach, V is an average velocity of flow, λ is a shape coefficient for the cross-sections determined by field data.

The third method of calculating propagation time is channel storage curve method:

$$\tau = \int_{0}^{L} \frac{\partial A}{\partial Q}\, dL \tag{10}$$

Here, A is area of the section, Q is discharge, L the length of the reach. Rewriting (10) as a finite difference equation for n reaches:

$$\tau = \sum_{1}^{n} \frac{\Delta A}{\Delta Q} \Delta L \tag{11}$$

If mean section area $\Delta \bar{A}$ and mean discharge $\Delta \bar{Q}$ of each reach are substituted for Δ A and ΔQ, then

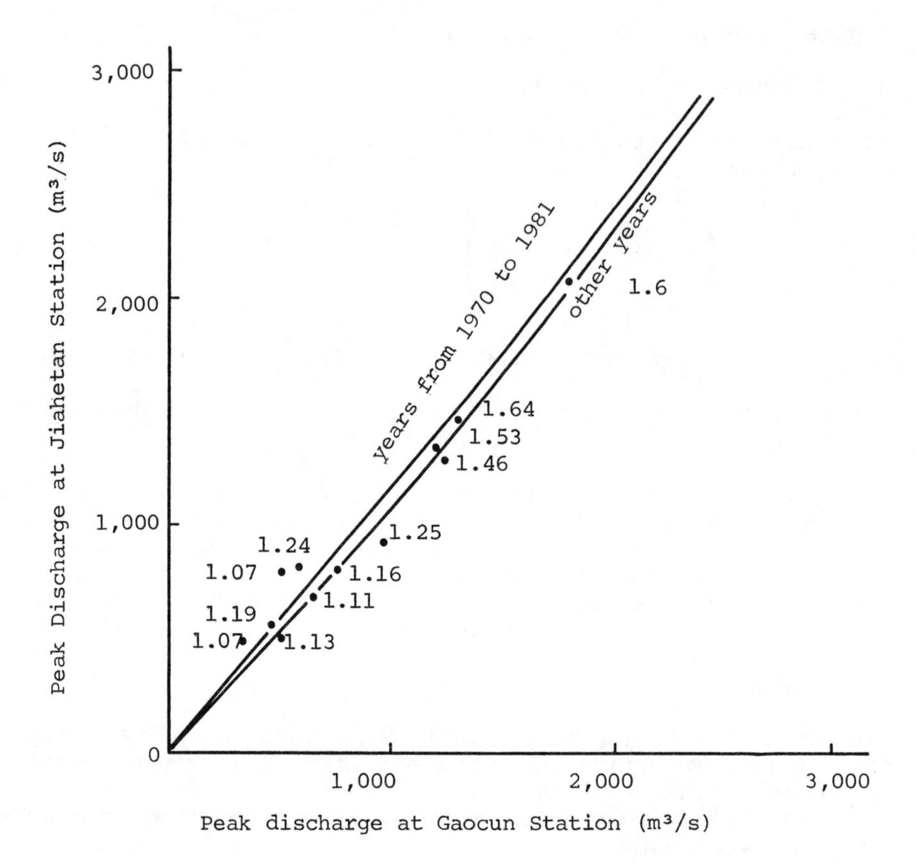

Figure 8. The Relationship of Discharges Between Jiahetan and Gaocun.

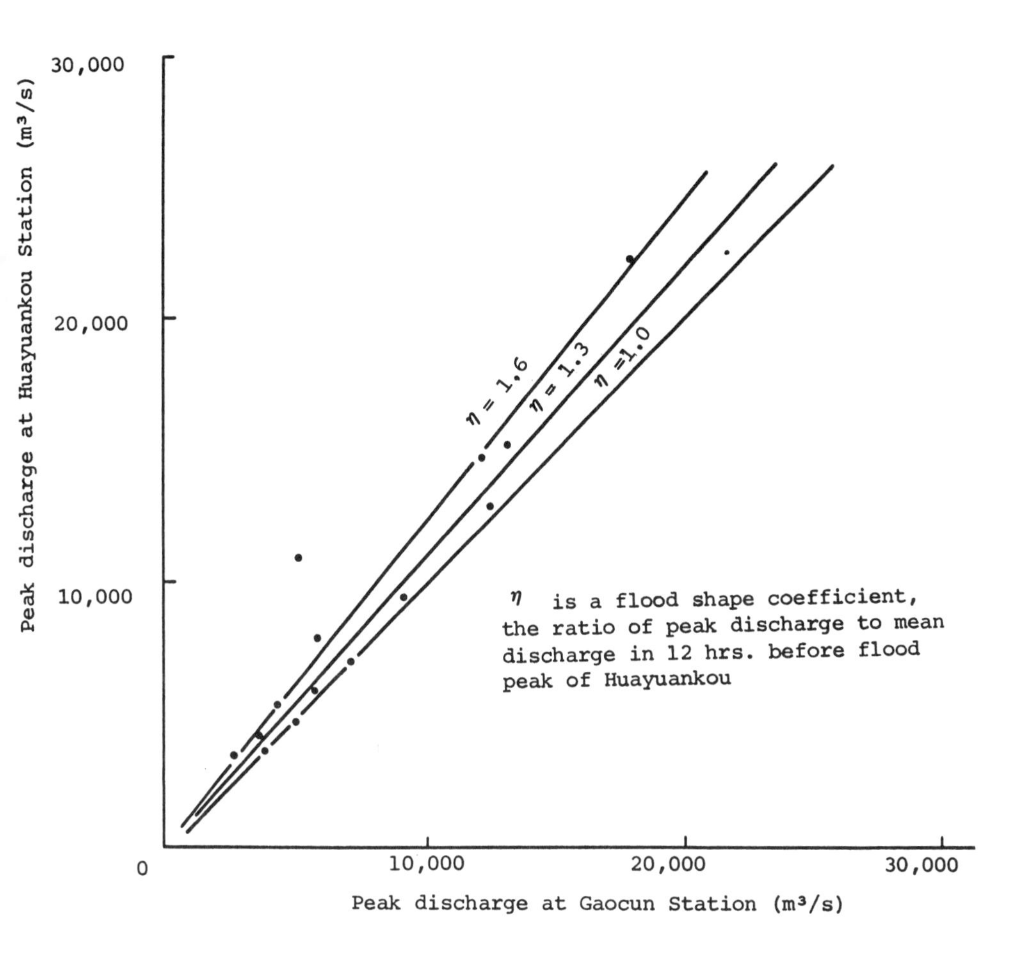

Figure 9. The Relationship of Discharge Between Huayuankou and Gaocun.

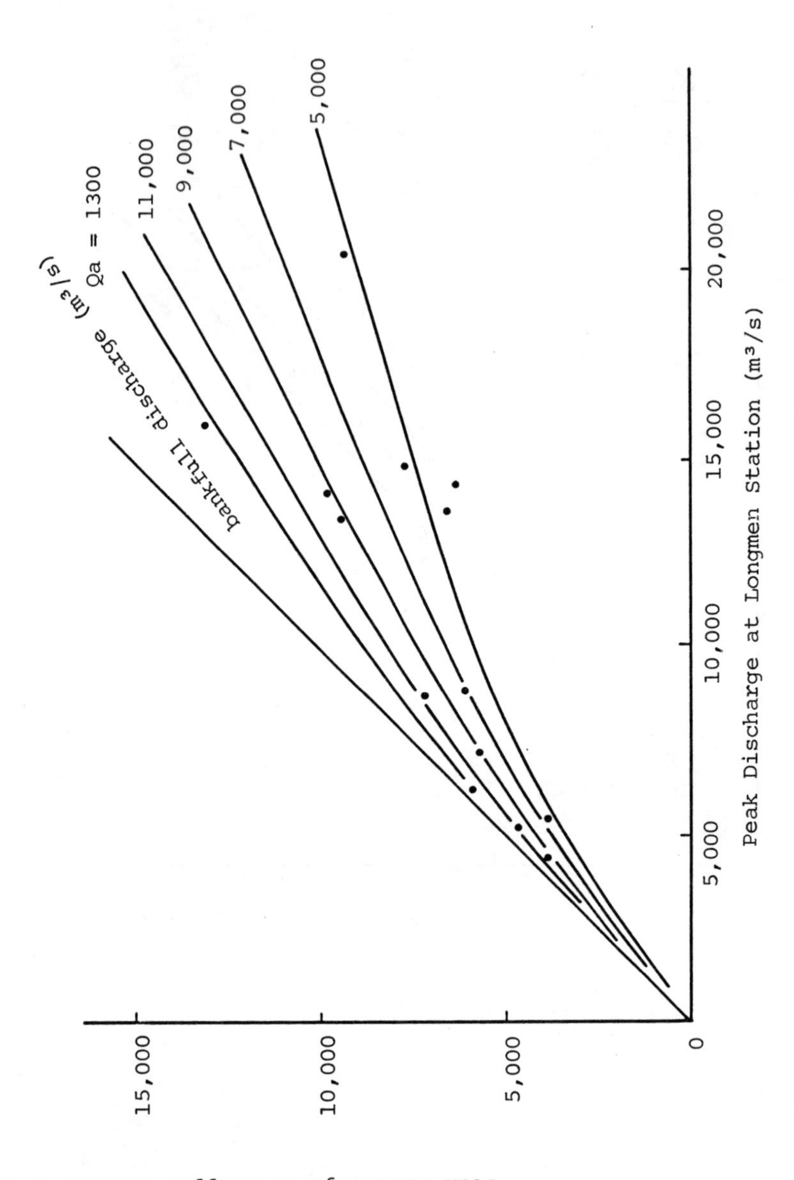

Figure 10. The Relationship of Discharges Between Longmen and Tongguan

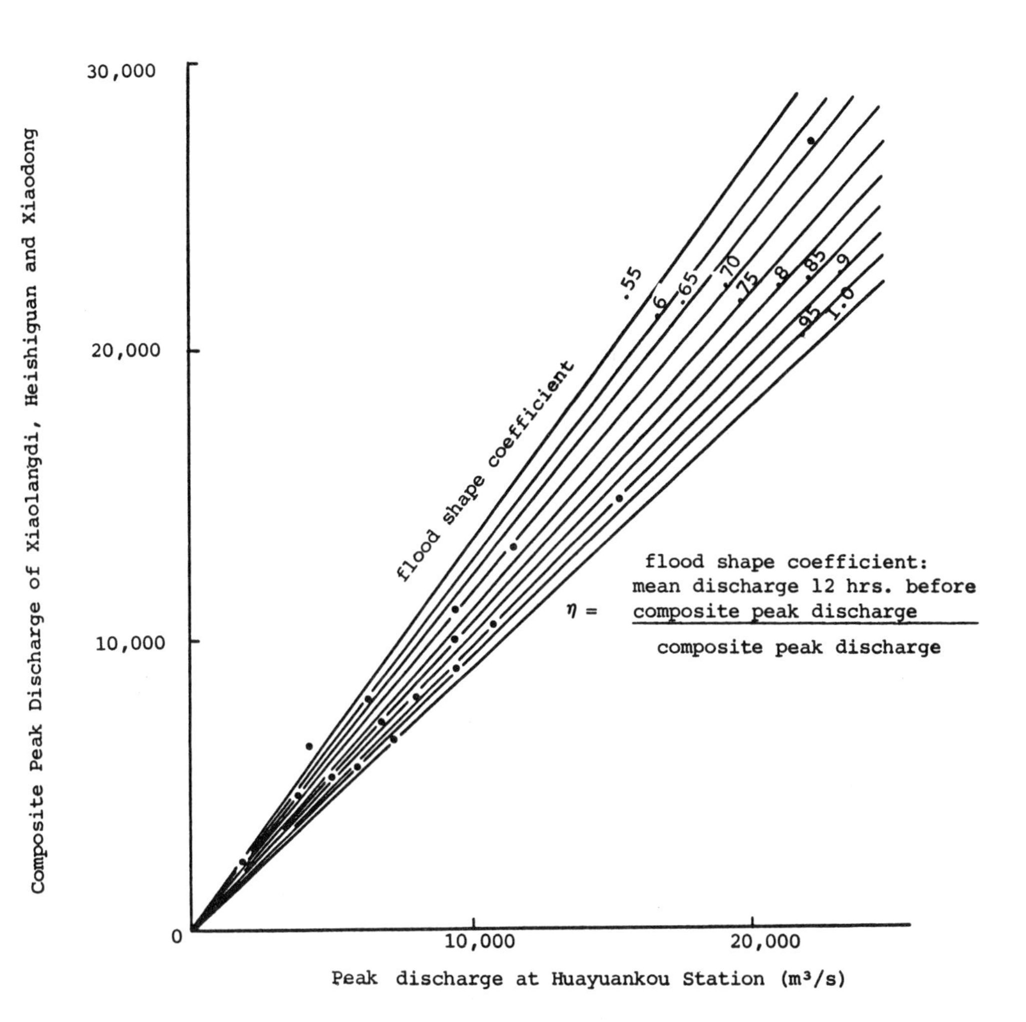

Figure 11. The Relationship Between Flood Peak at Huayuankou and Composite Discharge of Xiaolangdi, Heishigun and Xiaodong.

$$\tau = \frac{\Delta \bar{A} L}{\Delta \bar{Q}} = \frac{\Delta W}{\Delta Q} \quad (12)$$

Here ΔW is an increment of channel storage in successive reaches. Thus τ can be obtained from the channel storage curve $W = f(Q)$.

There are two forms expressing time of propagation, $\tau = f(Q)$ and $\tau = f(Q, \eta)$.

Because the propagation speed is greatly influenced by channel morphology and the flow pattern of the river, it needs to be revised according to the changes that occur.

Stage Forecasting at Flood Peak. There are two methods for determining the stage:

i. Modify the relationship between stage and discharge using field data of stage and discharge to obtain the relationship at any time. Then the stage corresponding to a flood peak can be obtained from the forecasted peak discharge.

ii. Compute erosion or deposition at a section and modify the relationship of stage and discharge. The cross-sectional area is determined after computation of erosion or deposition at the section. Velocity is determined from the relationship of Q and V. The relationship of stage and discharge is then modified so that the relevant water level can then be determined for the discharge forecasted.

Erosion or deposition at a section in the main channel may be computed as follows:

For the rising part of the flood hydrograph:

$$\delta_{Ae} = 0.14\ \delta Q_r \quad (13)$$

$$\delta_{Ad} = 0.00036\ \bar{\rho}^{2} \quad (14)$$

$$\bar{\rho}_d = K \bar{Q}_u^{\alpha - 1}\ \bar{\rho}_u^{\beta} \quad (15)$$

in which δ_{Ae} and δ_{Ad} are rates of section A for erosion and deposition respectively, δQ_r is a varying rate of discharge, $\bar{\rho}_u$ and $\bar{\rho}_d$ are mean sediment concentration at upstream and downstream stations respectively, $\bar{Q}_u$ is a main discharge at an upstream station, α, K and β are parameters determined from field data.

For the falling part of the flood

$$\delta A_f = K^{0.4} \delta A_r \tag{16}$$

Here δA_f and δA_r are rates of change of cross sectional area A respectively for falling and rising flood periods, K is a coefficient. The mean volocity is determined by a relationship between mean current velocity and discharge. Each formula above gives a result for modifying the relationship of stage and discharge on the main stream.

Deposition not scour would take place on the floodplain; the amount can be evaluated by the following formula:

$$W_f = f\ (W_r,\ T) \tag{17}$$

in which W_f is the amount of deposition on the floodplain, W_r is the amount of sediment entering the floodplain, T the time duration of flood flow onto the floodplain. Depth of deposition on the floodplain may then be estimated and used to modify the cross-section over the floodplain. The velocity on the floodplain is then calculated by the corrected section of floodplain, and the relationship of stage and discharge on the floodplain is modified and superimposed on the stage-discharge relationship of the main channel to obtain a stage-discharge relationship for the whole section. Then the relevant stage can be obtained using the forecast discharge.

Flood Process Forecasting. Flood routing is chiefly used:

i. Forecasting without the influence of production dikes. Using the Muskingum method, groups of parameters k, x are evaluated for floods which overflow the floodplain or floods contained in the main channel, and modified by the latest field data at any time in accordance with changes of the riverbed. When forecasting is required, the first step is to determine the bankfull discharge, and then to determine the parameters suitable for flood routing.

For the sharp flood or when reaches are very long, the method of continuous flood routing for unit-reaches should be used. Previous work must be done to divide the reaches into unit-reaches.

The coefficient of flow concentration of continuous flood routing can be calculated by the Muskingum method.

$$P_{on} = C_o{}^{n} \qquad (m = 0) \tag{18}$$

$$P_{mn} = \sum_{i=1}^{n} B_i C_o{}^{n-i} A^{i} C_2{}^{n-i} \qquad (m > 0,\ m-1/2 >= 0) \tag{19}$$

$$A = C_1 + C_o C_2$$

$$B_i = \frac{n!\ (m-1)!}{i!\ (i-1)!(n-i)!(m-1)!}$$

in which, P is a coefficient of flow concentration, m the number of unit-reaches, n is the number of periods, C_o, C_1, and C_2 are three parameters of the Muskingum method.

The other way to evaluate the coefficient of flow concentration of continous flood routing is based on the method of Characteristic Length. Flow concentration curves are given as follows:

$$P_{mn} = \frac{\Delta t}{K_L \ \Gamma(n)} \left(\frac{t}{K_L}\right)^{n-1} e^{-\frac{t}{K_L}} \tag{20}$$

in which, P is a coefficient of flow concentration, Δt a unit period of computation, K_L a propagation period for the characteristic length, Γ is the Gamma Function, n is the number of characteristic reaches, t the period evaluated from the middle of t, and e the base of natural logarithms.

ii. Forecasting with the influence of production dikes. The Muskingum method is also used with parameters of k, x, when the production dikes are sound and no overflow into the floodplain takes place.

There are two conditions for breaching of production dikes. Breaching points on production dikes may already be in existence before the flood. When floodwater begins to overflow onto the floodplain, it will automatically flow into the production dikes. Two methods can be applied for this condition: (a) the sub-process of the flood overflowing into the floodplain at the breached point is determined by using empirical relationship of floods entering the floodplain, routing this sub-process and the flood process of the main stream to the downstream section respectively and superimposing them, (b) routing the whole flood process on the main stream first, and then the results of routing are used in a computation similar to routing through reservoir regulation.

Another condition is that the production dikes breach suddenly in the process of a flood. For this condition, the part of the discharge that exceeds the bankfull discharge will be deducted from the process at the beginning of breaching until no more water can be held over the floodplain. The remaining flood is routed to the next station using parameters of flood routing in the main channel. To consider the other part of flood process on the floodplain, if there is a breach point at the lower part of a production dike, the flow may be routed to the exit section and superimposed to the flood conditions in the main stream. Otherwise the flood water retained on the floodplain will return to the main channel only after a flood is over.

For each method mentioned above, the first important thing is to estimate whether the production dikes will collapse. If no definite answer can be made, two results should be calculated; with and without breaching of the dike as upper and lower limits of the forecasts. For floods exceeding the protective ability of the production dikes, all production dikes may collapse, and forecasting should be made under the condition the the flood overflows on the floodplain.

The Application of Hydrological Models. Since the late 70s, hydrological models have been widely used at home and abroad. These are the Luhun, Xianjiang, Sacramento, CLS, Tank and Linear Perturbation models. These models have been tested in some regions, and parameters of the models are being calibrated in the hope of finding suitable models to be applied in various regions.

Problems and Suggestions for Improvement

PROBLEMS

There are two problems at present. One is that the lead time is too short, so that measures of flood protection cannot be brought into full play on time. The other is that forecasting methods are imperfect and accuracy of the forecasts is unstable. Occasionaly large errors may occur which defeat the demands of flood protection.

(1) Retaining floods in Sanmenxia reservoir is an important flood protection measire for the lower Yellow River. At present it takes 15 hours to close the gates. It takes 24 hours for floodwaters to flow from Sanmenxia to Huayuankou. An earlier forecast of discharge at Huayuankou is needed if a major flood occurs in the intermediate area below Sanmenxia and all gates have to be closed on time. The lead time for forecasting is only about 10 hours from Sanmenxia to Xiaolangdi, if more water is added in the district from Xiaolangdi to Huayuankou, the lead time in the forecast is still less then 10 hours. After real-time telemetering system of flood forecasting from Sanmenxia to Huayuankou is established, data collection and transmission will be improved and the time required for forecasting will be shortened. In other words, the lead time will be increased. There are many people living on the floodplain of the lower Yellow River and in pre-assigned detention basins. It is necessary to send flood forecasts and flood warnings accurately as early as possible to facilitate the evacuation work and to minimize possible damages due to floods, and to allow decisionmakers

time to determine whether the detention basin will be used or not.

(2) The accuracy of flood forecasting usually meets the needs of flood protection. But owing to the influence of groups of medium and small sized reservoirs, production dikes and an unsteady riverbed, accuracy of flood forecasting for each flood event cannot be assured, especially in forecasting large floods. In 1982, for example, floodwater was detained on river-cross floodplain at Yi-Luo River confluence leading to a relative forecasting error of over 20%.

SUGGESTIONS

(1) Besides establishment of an automatic telemetering system for the on-line real-time flood forecasting, meteorological weather forecasts should be improved. Quantitative heavy rain forecasting should be developed, and more radar should be set up to monitor the distribution of rainfall. Equipment for cloud atlas receiving and digital processing should be installed in order to improve the accuracy of rain forecasting and to coordinate closely meteorological forecasts with flood forecasts.

(2) Further studies should be made to improve the methods of flood forecasting. There are two ways to solve the problems due to the influence of production dikes. One is to establish an auto-regression equation for real-time correction or the Karman filtering method. But this method can be applied only when the actual hydrological information is at hand, so the lead time is much shorter. The other is to find a way to predict the possible collapse of production dikes. But at present not enough data are available for such a study. Automatic telemetering systems for determining the water level on the floodplain should be established. Aerial remote sensing should be used to obtain information on the collapse of production dikes and the diversion of floods onto the floodplain.

The methods below can be used to solve problems due to the influence of scour and deposition:

(i) By modifying the parameters of flood routing using corrections on a real-time basis, the accuracy of forecasts should be increased, but the lead time for forecasts cannot be increased.

(ii) Establishing a mathematical model of unsteady flow under the conditions of a movable riverbed (actual boundary conditions needed in the computation) is very difficult to do on a real-time basis.

(iii) A more feasible and practical approach is to combine the existing methods of flood routing with the estimation of scour and deposition of the river channel.

Based on the practice of flood forecasting on the lower Yellow River, it is known that the influence of production dikes is more important than that of movable beds, and this characteristic should be considered in the improvement of forecasting methods.

As for the influence of groups of medium and small reservoirs in the tributaries, investigations and further studies should be made, and influence from reservoirs should be considered in each sub-region respectively in the establishment of a hydrological model. The telemetering stage gaging stations should be set up on reservoirs to obtain hydrological information on reservoirs.

References

1. Qian Yiying, "Basic Characteristics of Flow with Hyperconcentration of Sediment," Proceedings of the International Symposium on River Sedimentation, 1980.

2. Chen Zanting, Lou Qinjun, "Flood Characteristics and Flood Control in the Lower Yellow River," Proceedings of PRC-US-Japan Trilateral Symposium/Workshop on Engineering for Multiple Natural Hazard Mitigation, 1985.

ICE FLOODS ON THE LOWER REACH OF THE YELLOW RIVER AND MEASURES FOR ICE FLOOD PREVENTION

Cai Lin
River Engineering Division
Yellow River Conservancy Commission
People's Republic of China

ABSTRACT: Every spring and winter ice floods often occur on the lower reach of the Yellow River, jamming up the river channel and raising the water level with the consequent overflowing of ice floods onto the floodplains. If the ice were to cause breaching of the dikes, a great disaster could occur. The Yellow River changed its course after the dike-breach at Tongwaxiang in 1855. During the following 54 years from 1883 to 1936, there were 21 years with dike-breaches occurring at 40 places along the lower reaches of the Yellow River during the ice-flood period. On average, dike-breaches took place once every other year. The menace of ice flooding has been somewhat reduced since the operation of Sanmenxia Reservoir in 1960. This paper deals with measures for ice flood prevention in the lower reaches of the Yellow River. It also gives a brief introduction to the characteristics of ice floods in the lower reaches and measures for ice-flood prevention currently practiced.

General Introduction to the Ice Floods

The mainstem of the Yellow River in the lower reaches takes a turn to the northeast after passing Dongbatou in Lankao (Henan Province). There is a gain of 3° in latitude when the river reaches Kenli (Shandong province), where it empties into the Bohai Sea. Water in the river channel always freezes in winter. According to statistics of the past 35 years, the probability of freeze-up in the lower reach was 86%. The morphology of the river is such that the channel is wide and shallow upstream and narrow and meandering downstream. The differences in latitude and width, as well as the variation of discharge contribute to the complexity of ice regimes and flooding on the lower reaches of the Yellow River.

In winter there are, on average, 30 days with mean daily air temperature below zero for the reach above Lankao and 48 days

L. M. Brush et al. (eds.), Taming the Yellow River: Silt and Floods, 451–467.

or so for Jinan reach. In the estuary zone, subzero air temperature exists for over 74 days. The long-term mean daily air temperature at Beizheng is 3.3^{o} C lower than that in Zhengzhou. The difference may be still larger in some individual years. For instance, in mid-January 1957, the 10-day mean air temperature at Beizhen was 7.4^{o} C lower than that in Zhengzhou, and on Jan. 18th, it was 15.2^{o} C lower. After mid-December a strong cold wave often prevails which causes freeze-up of the river channel. The freeze-up date in the Henan reach is generally in early January, whereas freeze-up in the estuary zone (in the reach below Luokou in Shandong province) is in mid- or late-December. Because of the difference in intensity of cold air currents, the freeze-up date, the length of the river channel under ice cover, and the ice volume in the river channel differ greatly. The earliest freeze-up date ever recorded was December 12th, the latest February 17th. The shortest section that may freeze is only 40 km, but the longest may be more than 700 km. The minimum volume of ice in the river channel is only several hundred thousand cu.m, but the maximum volume may reach 140 million cu.m. The break-up date in the Henan reach is generally in late January, and that in the Shandong reach may be in late February or early March. Thickness of the ice cover also varies from 0.1 m in the Henan reach to 0.3-0.5 m in the estuary.

In winter the incoming streamflow has a great influence on ice floods. Before the operation of Sanmenxia Reservoir most of the incoming streamflow was from the reach above Sanmenxia during freeze. Because of the intermittent freezing and thawing of ice in the middle reaches of the Yellow River, the incoming streamflow is irregular during the winter season, which has an unfavorable influence on ice floods. Variations of the upstream and downstream sections in width, temperature and latitude as well as great changes of discharge contribute to the complexity of ice regimes on the lower reach of the Yellow River and the severity of ice floods. According to historical records between 1883 to 1936, there were 21 years with dike-breaches during the ice-flood period. In 1951 and 1955, because of severe cold and complex ice regimes, the ice piled up forming ice dams with consequent dike breaches at Lijin county (Shandong Province). Much land and many villages at Zhanhua, Lijin and Kenli were subject to inundation during this severe disaster.

Characteristics of Ice Floods in the Lower Yellow River

Three conditions associated with ice floods are:

1. Small discharges and high water levels: Water levels along the lower reaches of the Yellow River rise substantially during two periods, the freeze-up period and the thawing

period. In the freeze-up period, because of both the ice resistance and the large amount of ice slush that accumulates beneath the ice cover after freeze-up, roughness of the bottom of the ice cover is increased causing the water level to rise. The higher water level is comparable to that when the discharge is 5-6 times larger during non-frozen periods. In the thawing period the melted ice from the upstream reach jams the channel cross-section in the downstream reach and brings about a substantial rise in water level.

Variation in the range of water stage in the ice-flood period depends on the influence of the ice on the flow. Generally, the water stage gradually rises higher due to ice resistance. When ice jams or ice dams occur in the river channel, the backwater effect is seen immediately, and the water stage rises rapidly. Figure 1 shows the stage-rising process when an ice dam was formed at Jinan reach (Shandong Province) on Jan. 27, 1970.

During ice-flood periods, ice jams often lead to a rise in water stage. Maximum values of water stage are near or above those in summer floods. Comparisons of maximum stage during the summer floods and ice floods at Lijin station in individual years are shown in Table 1. It can be seen that the maximum stage of the ice flood in 1951 was the same as the maximum stage of the summer flood in 1958, although the peak discharge of the ice flood in 1951 was 9240 cms smaller than that of the summer flood in 1958.

2. Discharge of ice flood increases gradually as the ice flood propagates along the way. There are no large tributaries entering the main stream below Huayuankou. When a summer flood propagates downstream, the discharge gradually decreases due to the influence of channel storage, whereas in the winter the discharge of ice floods tends to increase as the flood wave propagates downstream. During the freeze-up period the amount of oncoming streamflow is blocked, bringing about a continuous increase of channel storage and a rise in water stage. During break-up, this amount of water is suddenly released, resulting in a flood peak which tends to increase as it propagates downstream. Taking the ice flood of 1957 as an example, the flood peak of 920 cms at Gaocun increased to 1,010 cms when the flood reached Sunkou, and became 1,260 cms at Luokou. When the flood reached Lijin, the flood peak amounted to 3430 cms. Ice jams and the consequent rise in stage are more likely to take place than an increase of peak flow as the flood propagates along the reach.

3. Narrow sections are susceptible to ice jamming. Most serious historical ice-floods in the Lower Yellow River were due to jamming in the narrow and meandering sections. The

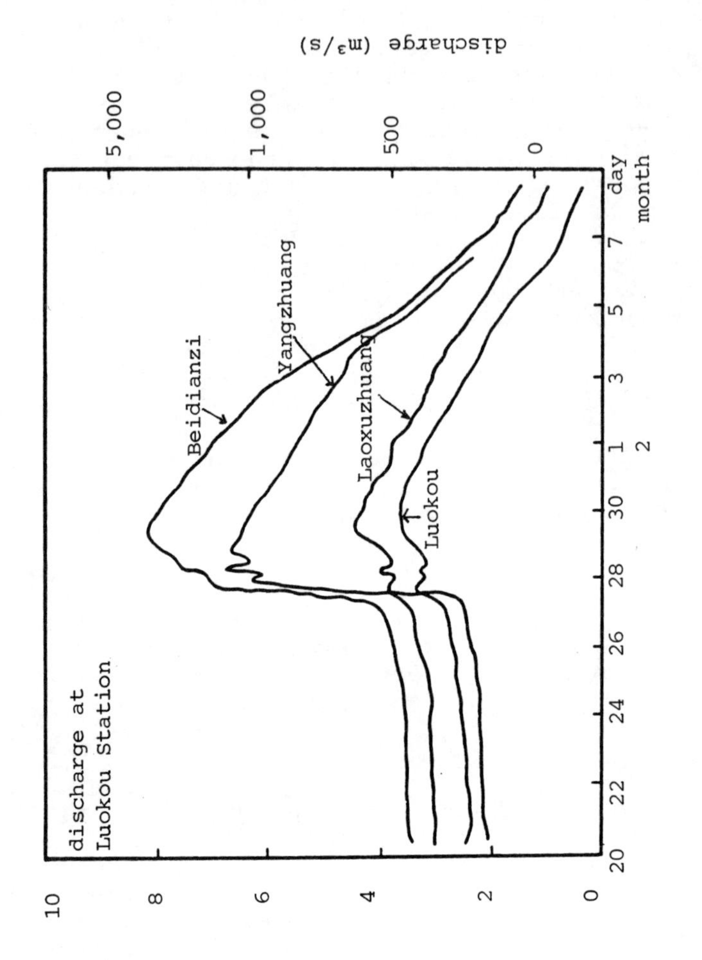

Figure 1. Hydrograph after Formation of Ice Dam at Laoxuzhuang in 1970. (Ice jam began to occur at Laoxuzhuang at 11:30, January 27th.)

Table 1. Comparisons of maximum stage values during summer floods and ice floods at Lijin station in individual years

Years	Summer floods Qm(cms)	H(m)	Ice floods Q(cms)	Hm(m)
1950	6210	12.57	1760	11.08
1951	5780	12.53	1160	13.76
1953	6860	13.30	1060	12.59
1955	5760	12.00	1960	15.31
1956	6050	12.34	2810	12.24
1957	8480	13.25	3430	12.88
1958	10400	13.76	1490	11.34
1959	7180	12.67	595	11.37
1965	5250	12.72	1350	11.14
1968	6800	13.51	1050	13.15
1970	5720	13.47	1130	13.27
1973	3640	13.31	1010	14.35
1979	4090	13.85	906	14.67

Qm - maximum discharge Q - relevant discharge

Hm - maximum stage H - relevant stage

river channel is wide upstream and narrow downstream with more channel storage in the upstream sections. During freeze-up the channel storage in wide shallow sections above Aishan amounts to 60% of the total. During break-up periods blocked water is released suddenly, resulting in a flood peak. At this time the ice in the reach below Aishan is still very hard and ice cover is rather thick and does not readily break. In addition, the capability of conveying the ice flood (the ice-flood discharge capacity) in the narrow sections is very low, and ice dams are likely to form. Among the ten largest events, ice dams occurred nine times on the reach below Aishan. Serious ice jams and ice dams occurred in 1951, 1955, 1969 and 1970, and brought about great disasters before Sanmenxia Reservoir was put into operation. Once when the river froze up entirely in the Ningxia-Nei Mongol reach, it was earlier by 20 days than in lower reaches. Moreover the discharge was relatively large prior to freeze-up and decreased drastically after the freeze-up. It became, of course, gradually larger accompanying the reduction of flow beneath the ice cover. It took nearly 20 days for the low discharge in Ningxia-Nei Mongol reach, in the initial freeze-up period, to propagate down the river. By the time it reaches the lower river course, the lower river would readily be frozen under such a low flow condition. Freeze-up of the river in the lower Yellow River under low flow conditions features a low ice cover, an early freeze-up date and small cross-sectional area under the ice cover. In later stages when the river is still frozen, a larger flow coming down from Ningshia-Nei Mongol reach may force the ice cover to lift, resulting in the sudden or violent breaking of the jammed ice which may cause great damage.

Measures for Ice-Flood Prevention in Current Practice

The most serious problem of ice floods in lower reaches of the Yellow River is the occurrence of ice jams or ice dams on the river channel.

Experience obtained in ice-flood prevention in the past can be classified as ice flood prevention by reservoirs, diversion of water and ice, and increase of discharge capacity of ice floods.

The reservoir can be operated in two ways for ice-flood prevention: one is to store part of the ice flood in the reservoir and control the outflow so as to reduce the threat of an ice flood in the lower reach. The other is to release comparatively high-temperature water from the lower layers of the reservoir or from the turbines to change the ice regimen

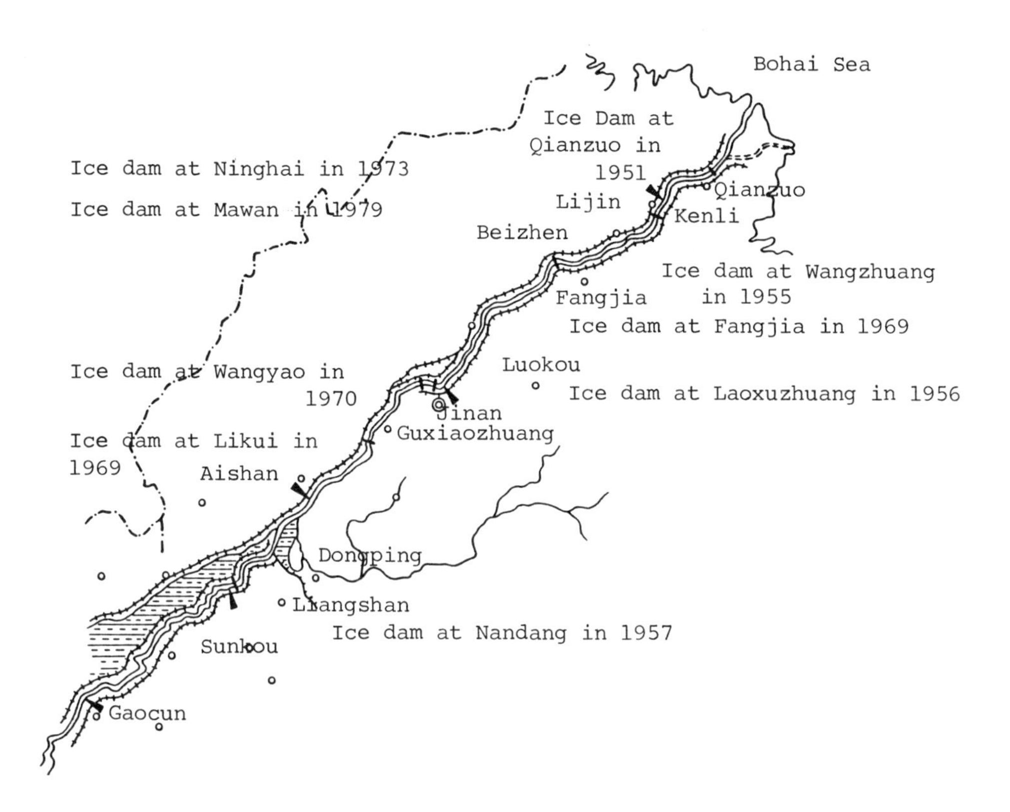

Figure 2. Map of Places where Ice Dams Occurred on the Lower Reach of the Yellow River.

downstream of the reservoir.

CONTROLLING THE OUTFLOW FROM THE RESERVOIR

Twenty years of experience in the operation of Sanmenxia Reservoir for ice-flood prevention shows that the threat of ice floods in an ordinary year can be reduced by storing some of the runoff in the reservoir and releasing flows at the most opportune time. At present the capacity available for ice flood storage in Sanmenxia Reservoir is about 1.8 billion cu.m. The operational modes adopted at Sanmenxia Reservoir are discussed as follows:

Regulation During Break-up Period: The purpose is to control the outflow prior to break-up and reduce the channel storage increment to avoid occurrence of a "violent break-up." This method was used very often before 1973. Prior to the date when break-up on the lower reach is forecast, the sluice gates of Sanmenxia Reservoir are closed so that limited outflow or even no outflow from the reservoir is allowed. The menace of ice floods can be reduced if the outflow from the reservoir is executed at the right moment, which is determined on the basis of accurate forecasts of the break-up date. During an ice flood period in 1967, for instance, a length of 616 km of river channel was subject to freeze-up, involving 140 million cu.m of ice and a channel storage increment of 1.12 billion cu.m. The ice flood was very serious. The channel storage and ice flood peak were greatly reduced after the sluice gates were closed on January 20th. Twenty-five days after closure of the sluice gates, 1.14 billion cu.m of water was detained in the reservoir.

This mode of operation is exercised only during the break-up period. During other periods of ice flood oncoming streamflow is not regulated. This may cause unfavorable freeze-up situations due to lower temperatures in the lower reaches. In addition, after the Sanmenxia Hydropower Station was put into operation, there is a conflict between discharge required for power-generation and the cut-off of discharge by closing of the sluice gates.

All-time Regulation During Ice-flood Period: This operational mode is to impound a certain amount of water in advance of the freeze-up in the lower reach so as to increase the volume of water released during the initial stages of freeze-up and increase the flow conveyance below the ice cover on the premise that no ice jam will occur. In accordance with the magnitude of channel storage capacity, the volume of water released from large to small discharges should be controlled in steps. By the time of break-up, the volume of water released should be further reduced so that channel storage can be kept to the minimum needed for safe conveyance. A typical practice is shown in Fig.3.

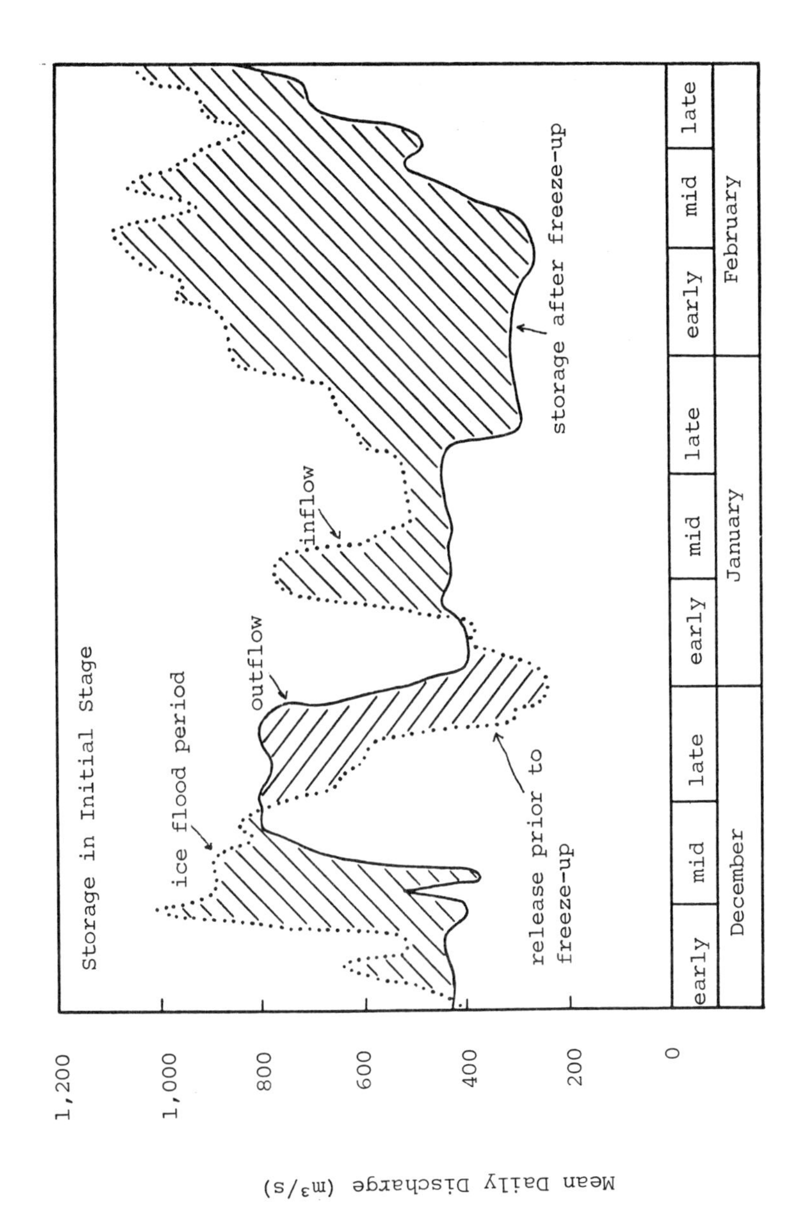

Figure 3. Hydrograph of Inflow and Outflow of Sanmenxia Reservoir During Ice Flood period (1976-1977).

This operational mode has the advantage of the aforesaid modes and avoids the disadvantages of the freeze-up of the river channel due to small discharge. In addition, it can also make it easy to determine the opportune moment for releasing water from the reservoir even if the forecast on break-up is not accurate. In an ordinary year there is no need to use all of the water from the Sanmenxia Reservoir thus normal operation of some of the turbine units in the hydropower station can be maintained. During years with abundant flows and a severe ice regimen, the reservoir capacity available for ice flood control is rather limited. The scope of regulation and the right time for exercising the regulation are not easy to determine.

THE INFLUENCE OF OUTFLOW FROM THE RESERVOIR ON SHORTENING THE TIME OF ICE-COVER AND THE DURATION OF FREEZE-UP

In large reservoirs, the water temperature under the ice cover is still above 0° C. The temperature of water released through turbines or from other hydraulic structures such as bottom sluices usually remains above 0° C for some distance of travel before it gradually decreases to 0° C. Before the construction of Liujiaxia Reservoir on the upper reach of the Yellow River, the river channel near Lanzhou froze up every year and ice jams often occurred. After construction of Liujiaxia Reservoir there have been no occurrences of such phenomena, even the appearance of border ice or ice slushes within 100 km downstream have been reduced. Before the river freezes up, water released from Sanmenxia Hydropower Station is generally at a discharge of 500 cms. From the comparative analysis made in 1974, the water temperature in the Huayuankou reach was increased substantially. The distance influenced was more than 270 km. because the Sanmenxia Reservoir is far from the reach where the initial freeze-up occurs each year, and the river channel in-between is wide and shallow. Therefore, the amount of comparatively high-temperature water released from Sanmenxia Reservoir in winter has little influence on reaches initially frozen up. Nevertheless, some influence on the nearby Henan reach may be observed.

Diverting Water and Ice

a) Using diversion gates or sluices along the Yellow River to divert water. Nn effective way to prevent ice jams or decrease the ice flood peak is to use diversion gates on both banks to dispose of the water and ice detained in the channel, or to establish ice flood detention areas along the reaches with a high occurrence of ice jams to dispose of the water and ice and to reduce channel storage. The existing capacity of diversion gates or sluices on both banks of the lower Yellow

River can amount to 2,000 cms. Around 1960, diversion gates or sluices on both banks diverted large amounts of water during the ice-flood period and the channel discharge was reduced to less than 150 cms, greatly alleviating the menace of an ice flood.

During an ice-flood period in 1977, more than 900 cms of water was diverted along the reaches above Aishan for ice-flood prevention as well as for irrigation during break-up period. Along with the operation of Sanmenxia Reservoir both channel storage and peak flow were reduced during break-up period. As a consquence, the frozen ice broke up tranquilly, section by section.

In practice, those diversion gates or sluices chosen should be capable of diverting water on a relatively large scale into an adequate drainage system. Preparation for diversion should be done in advance, including maintenance of diversion gates or sluices and construction of the necessary structures in the canals. Canals should be emptied prior to freeze-up so that the water stored in canals does not freeze and form ice jams.

b) Widening the river channels for ice floods. Ice jams occurring in the narrow section of the lower reach endanger the dikes. Thus two projects for widening the river channel in the narrow reaches below Aishan were introduced. One is a widening project from Doufuwo to Balizhuang near Qihe County in Shandong Province, 39.2 km in length. The effective flood storage capacity here amounts to 390 million cu.m, and the design inflow discharge is 2800 cms. It is called "Northern Qihe Widening Project." The other project is between Boxing and Kenli in Shandong, 38.6 km in length. Its effective flood storage is 330 million cu.m with a design inflow discharge of 3,440 cms. It is called "Sorthern Kenli Widening Project." These two projects can relieve the menace of ice floods which once occurred in these narrow sections.

The general idea is to construct a new dike outside of the original dike in the narrow reach, and then to connect it with the original dike to form a widened area to store ice floods in case of necessity. Inflow and outflow are controlled by gates. According to the location of the ice jams, the original dike can also be removed by explosion at the right moment to let the water and ice enter the widened area so that menace of the ice flood to the dike system can be relieved.

Increasing the Capability of Discharging an Ice Flood to Prevent Ice Jans in the River Channel. The main measure adopted is to break the ice. Various methods for ice breaking include such options as trenching the ice cover manually, promoting thawing by speading earth or ash on the surface, using an icebreaker to open up the ice cover, bombarding by

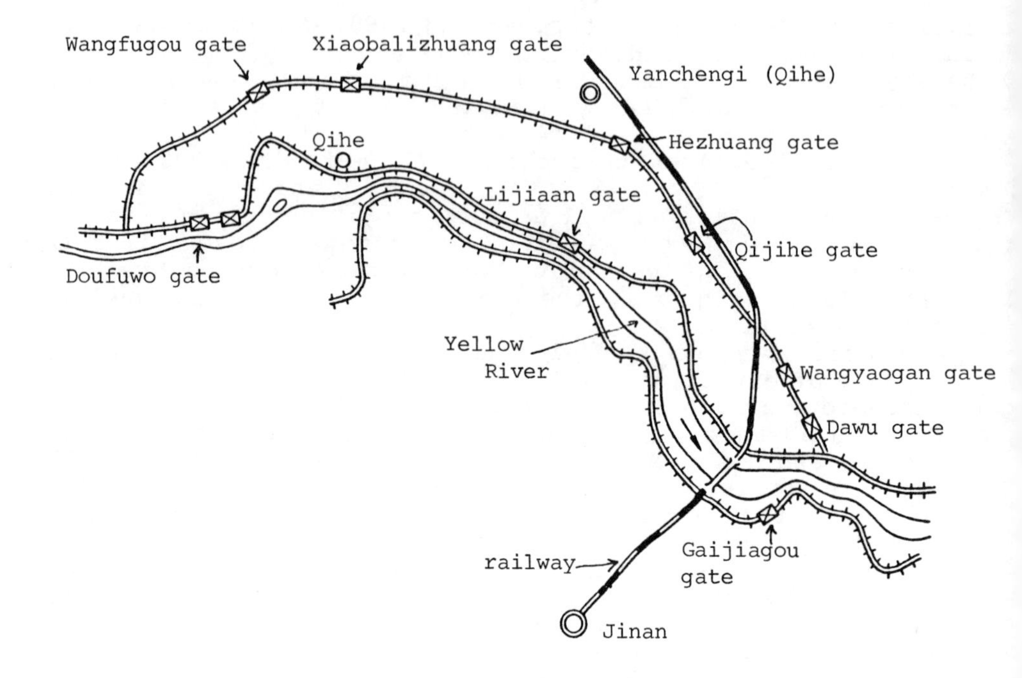

Figure 4. Map of Northern Qihe Widening Project

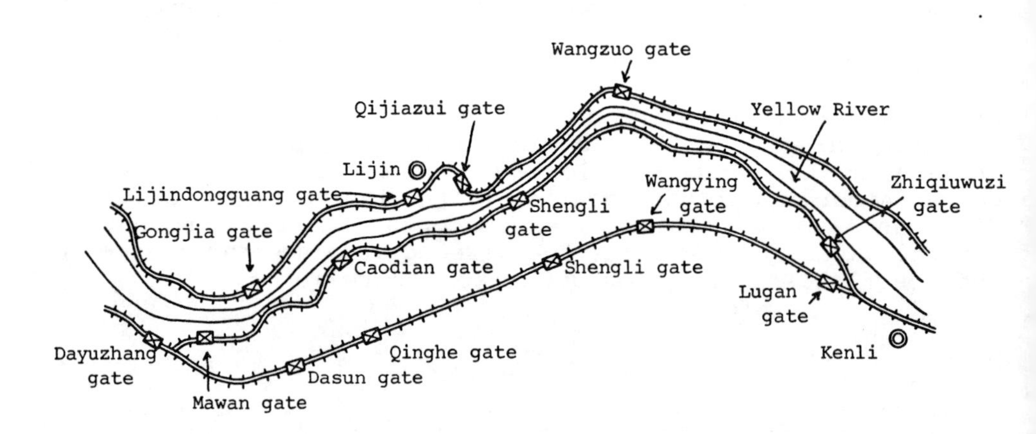

Figure 5. Map of Southern Kenli Widening Project.

means of airplanes, shelling with guns, blasting etc. In recent years breaking the ice by manual efforts has been gradually improved and has become an effective method for ice breaking.

According to the forms of ice to be broken, ice breaking can be classified as ice-cover breaking, ice-jam removing and breaking of large pieces of ice floe.

(1) Ice-cover breaking. If the ice cover is broken on certain reaches prior to break-up, it is possible to form a free flowing channel to let the oncoming streamflow pass smoothly. This is the basic method used for ice-flood prevention in lower reaches of the Yellow River. In order to prevent the occurrence of ice jams caused by the large pieces of ice flowing downstream after ice-cover breaking, the ice cover is usually broken into small pieces which remain floating in original sections or even melt into the water, which favors tranquil break-up once the large discharge comes. In accordance with experience obtained in ice breaking, three guidelines are followed: (a) The ice cover should be broken in narrow instead of in wide sections to avoid the possible occurrence of ice jams in the narrow sections due to the flowing of the ice blocks from the wide sections upstream. Even if ice jams occur in the wide section, ice and floodwater can still flow downstream via the floodplain, and the threat to the dikes is comparatively small. (b) The ice cover should be broken close to the time of break-up. This is a key step for effective ice breaking. If ice breaking is done too early, the river channel might freeze up again and all the efforts would be in vain. If ice breaking were to begin too late, there would not be enough time to properly carry out the work. (c) The length of the ice-covered narrow section to be broken should be determined according to the ice volume upstream. Ice breaking should extend a certain distance downstream of the narrow section in case an ice jam might occur near the exit and extend backwards to the narrow section.

(2) Breaking large pieces of ice floe. Large pieces of the ice floe may get jammed upstream of hydraulic structures across the river or in narrow sections. Thus shelling or blasting may be resorted to to break up the huge pieces of ice in narrow and meandering sections or at upstream hydraulic structures to reduce jamming to assure a favorable break-up.

(3) Removing ice-dams. The bombardment the ice dams by means of airplanes, shelling with guns, or blasting may also be used. Aimed at critical places in an ice dam, consecutive blasting from downstream to upstream of ice dams can be used. If this is supplemented by water diversion through trenches previously enlarged, the efficiency of removal is increased. Large packs of explosives set off in series give very satisfactory results.

On Measures for Flood Prevention in the Near Future

FURTHER STUDIES ON THE OPERATIONAL MODE OF SANMENXIA RESERVOIR FOR ICE-FLOOD PREVENTION

During the ice-flood period, the stage in front of the dam for the operation of Sanmenxia Reservoir is expected to be kept below 326 m to minimize backwater effects. The relevant storage capacity is 1.8 billion cu.m, which is too small to meet the requirements of storage capacity for ice-flood prevention. After the operation of Longyangxia Hydropower Station in the upper reach of the Yellow River, the volume of the water released will be increased during non-flood periods, which will lead to an increase of the oncoming streamflow for the lower reach of the Yellow River in an ice-flood period. In the years with severe ice regimes, the storage capacity of Sanmenxia Reservoir for ice-flood prevention will be far from sufficient. Experience gained in the operation of Sanmenxia Reservoir in ice-flood periods should be studied and reviewed so that the limited storage capacity can be used more effectively for ice-flood prevention. The regulation adopted at present includes the impoundment of water in the reservoir in advance so as to release more water just prior to freeze-up in the lower reach, to delay the freeze-up date and increase the capability of underflow conveyance in ice-covered sections. This regulating mode has been practiced for nearly ten years and it has been found that the effect is limited by the allowable water stage in the reservoir and the available usable storage capacity. The reason is that if the operating water stage is too high, sediment will deposit at the head of reservoir and be sluiced off while the water level is lowered prior to flood season. As the storage capacity is comparatively small and the ability of regulation rather limited, it is necessary and worthwhile to develop an optimization scheme for effective regulation of the reservoir.

In addition, studies should be also made on the principles of flow conveyance of the ice-covered river channel in the lower reaches.

After freeze-up ice slush accumulates beneath the ice cover and takes up part of the cross-section. The flow velocity decreases due to the increased roughness of the bottom of ice cover caused by the ice slush beneath the ice cover. Because of this, the flow of the river channel under ice (underflow of the river channel) will decrease after freeze-up. According to the observed data in the Yellow River and rivers in the Soviet Union cited in the literature, the roughness beneath the ice cover is largest at the initial stages of the freeze-up period. When the freeze-up of the river is steady, the bottom of ice cover becomes smooth due to erosion by the stream current and the roughness decreases accordingly. In Ningxia and Nei Mongol reach, for example, on the fifth day after break-up, the roughness of the bottom of the ice cover was 0.06 and decreased to 0.045 in the following 10 days. After 15 or 25 days the roughness became 0.035. In accordance with this pattern, Sanmenxia reservoir should release less water in the inital stages of freeze-up to reduce the channel storage increment. When the flow conveyance capacity under ice cover (underflow conveyance capacity) becomes normal, more water can be released from Sanmenxia Reservoir. According to conditions in the lower reaches, the outflow discharge from Sanmenxia Reservoir can be controlled at about 300 cms during the initial stage of freeze-up and continued for a duration of 10 or 15 days. Then the outflow discharge can be increased to 500-600 cms. This operational mode of Sanmenxia Reservoir would overcome the difficulties involved in decisionmaking for grasping the appropriate moment for regulation of the reservoir.

This operational mode and the aforesaid mode (impounding the water in the reservoir in advance and then releasing more water in the initial stage of freeze-up period) require different outflows from the reservoir. Further studies are needed in order to choose the suitable operational mode for these releases.

DIVERTING THE WATER AND ICE

Using the present water consumption and that projected in the year 2,000 as basic conditions for computing the volume of inflow to the Sanmenxia Reservoir now and in the future, and taking Longyangxia Reservoir (which will soon be finished and put into operation) into consideration, a prediction of required storage capacity for ice-flood control was made. Hydrologic data from 1950 to 1975 for the prediction of the actual freeze-up date near Aishan to the full break-up date in Shandong reach give the duration for which the controlled release at Sanmenxia Reservoir is required, and these were also used in the computations. If existing oncoming streamflow is used in the prediction, the maximum storage capacity needed is about 4.4 billion cu.m. If the value of oncoming streamflow in the future is used, and the increase of water demand in agriculture and industries along the upper and lower reaches is taken into account, the maximum storage needed will be 3.5 billion cu.m. If the water detained in Sanmenxia Reservoir for ice-flood prevention is still limited to 1.8 billion cu.m, the rest, or 2.6-2.7 billion cu.m of floodwater, must be stored somewhere else in order to assure the safety of the lower reaches. Before the construction of new reservoirs on the main stream, this amount of floodwater can be diverted from the wide river reaches as follows:

Diversion in Combination with Water Transfer Projects for Water Supply to the Northern Part of China. In 1981-1982, temporary measures were implemented to divert water from the Yellow River northward to solve the severe water shortage in the city of Tianjin. There were three routes for diverting the water from the Yellow River to Tianjin. In the winters of 1982 and 1983, 511 million cu.m of water were diverted through two of the previous routes. The diversions played an important role in reducing channel storage increment and alleviating the menace of ice floods.

In order to solve the problem of water shortages in Northern China, studies are being made of the feasibility of diverting water from the Yellow River. According to the preliminary plan, 1.7-1.8 billion cu.m of water can be diverted during the ice-flood period (December to February) each year with a discharge of 200-300 cms. This diversion could reduce channel storage increments and cut off peak discharge, greatly alleviating the menace of ice floods.

Ice Flood Prevention by Means of Dongpinghu Reservoir. In order to assure the safety of the narrow sections below Aishan in Shandong Province and overcome difficulties involved in controlling the release from Sanmenxia Reservoir due to the short lead time and long distance of flow conveyance, Dongpinghu Detention Reservoir can be used to divert the floodwater at an appropriate time to reduce the channel storage increment.

Dongpinghu reservoir is situated at the upstream end of the narrow section in Dongping, Liangshan and Wenshang counties. The total area of Dongpinghu Reservoir is 632 sq.km. The reservoir can be operated in two stages: first, in the former lake area, and afterwards in the newly-added lake area. The former covers an area of 209 sq.km, and the newly-added, 423 sq.km. When the water level reaches 46 m, the storage capacity of the former lake is 1.2 billion cu.m, and that of the newly-added one is 2.8 billion cu.m. The two water transfer projects can divert 1.7-1.8 billion cu.m. There is still about 800 million cu.m of water to be coped with by Dongpinghu Reservoir.

Dongpinghu Reservoir can serve as a supplementary water supply to the reach below Weishan for irrigation in the dry season (in May and June) and for industries in Qingdao. Meanwhile the storage operation of Dongpinghu Reservoir can play a role in reducing the channel aggradation in the non-flood season.

COORDINATION OF OPERATIONS OF XIAOLANGDI AND SANMENXIA RESERVOIRS FOR ICE-FLOOD PREVENTION

Xiaolangdi Reservoir is a proposed large multi-purpose project on the main stem of the Yellow River below Sanmenxia. The design height for the dam is 152 m, forming a total storage capacity of 12.7 billion cu.m.

Coordinated operation of Xiaolangdi and Sanmenxia reservoirs, supplemented by the operation of Dongpinghu Reservoir could eventually control the ice floods. In order to minimize both deposition in Sanmenxia Reservoir and an increase of elevation of the riverbed at Tongguan, and to assure the output of hydro-power during the years with severe ice regimes, Sanmenxia Reservoir is required to store one billion cu.m of ice flood, Xiaolangdi 2.5 billion cu.m, and the remaining 900 million cu.m of ice flood could be diverted by Dongpinghu Reservoir and other diversion projects. If the two water transfer projects for diverting the water from the Yellow River to Hebei and Tianjin were taken into consideration, Xiaolangdi Reservoir could be operated to divert 1.5-2 billion cu.m of floodwater and the flow discharge might be increased by 150-200 cms for power generation.

In the general plan, floods will be controlled first by the operation of Xiaolangdi Reservoir, coordinated with diversion operations in the downstream reaches, and afterwards by diversion of Dongpinghu Reservoir. Safety of the lower reach in the ice-flood period could be assured.

As Xiaolangdi Reservoir is large and deep, the temperature of the water released from the reservoir will be higher than that of the water in the channel before the construction of the reservoir. The potential influence on ice regimes downstream of this temperature difference should be further studied.

References

1. "Ice Flood on the Lower Reach of the Yellow River," River Engineering Division, Yellow River Conservancy Commission, Department of Hydraulic Engineering, Qinghua University Science Press, 1979.

2. "Regulation Effect of Sanmenxia Reservoir on the Control of Ice Run on the Lower Yellow River," Chen Zanting, Sun Zhaochu, Cai Lin, Wang Wencai, "Yellow River," 1980, fifth edition.

CURRENT UTILIZATION AND FUTURE PROSPECTS OF WATER RESOURCES OF THE YELLOW RIVER

Bai Zhaoxi
Reconnaissance Planning and Design Insititute
Yellow River Conservancy Commission
People's Republic of China

ABSTRACT: The Yellow River is the main water resource in North and Northwest China. At present it supplies an area of 850,000 sq.km and would be larger if Tianjian is included. The shortage of water resources has become one of the vital factors affecting the economic growth and people's daily lives of the related area.

Characteristics of Water Resources in the Yellow River

The water resources of the Yellow River basin originate from precipitation. According to the statistics from 1919 to 1980, the average annual runoff reached 58 billon cu.m in the drainage area, equivalent to 77 mm of runoff depth, and was only 27% of the average for the entire country. The volume of water per capita and per unit land area (mu) in the whole basin are 720 and 300 cu.m per annum, respectively, corresponding to only 27 and 17% of the nation's average. Therefore, in comparision with other large rivers in China, it is obvious that the Yellow River is deficient in water resources. However, with an annual discharge of 1.6 billion tons of sediment, the Yellow River ranks first in sediment load over other rivers in China.

The Yellow River is also marked by its uneven distribution of water and sediment. In the area above Lanzhou, 30% of the total catchment yields 56% of the entire water. Covered mainly by high plateaus, grassland, lakes, and swamps with cold weather, the region has a large water-retaining capacity and abundant and stable runoff. With an annual average sediment content of about 3 kg/cu.m, the upper basin yields 6% of the total sediment load and is regarded as the main source area of clear water in the Yellow River basin. Compared with the upper region, the annual runoff at Hekouzhen, outlet of the next reach, is less than that at Lanzhou due to losses in the intermediate arid area and the combination of less

L. M. Brush et al. (eds.), Taming the Yellow River: Silt and Floods, 469–476.

precipitation and high evapotranspiration. The middle reach from Hekouzhen to Longmen, a region covered predominantly by hilly and gullied loess, contributes 12% of the runoff and 58% of the total sediment from an area constituting only 15% of the overall drainage basin. The next region from Longmen to Sanmenxia comprises 25% of the entire river catchment and has 19% of the water runoff and 33% of the sediment. The above two regions are characterized by small runoff and high sediment concentration. The following reach with an area of 6% the total catchment from Sanmenxia to Huayuankou yields 10% of the total runoff and with a small amount of sediment. The Yiluo River with only 30 million tons of sediment each year becomes the second clear water source area of the Yellow River. Regional variation of water resources is shown in Table 1.

The spatial and temporal distribution of runoff are not uniform. In the main river, 60% of the annual water runoff and 85% of the sediment discharge takes place in the flood season from July to October, however the relevant figures are 80-90% and almost 100% respectively in some of the tributaries on the Loess Plateau. From March to June, only 20% of the annual runoff together with a very small amount of sediment occurs in the main river, while flows are entirely absent in some minor or medium-sized tributaries. According to the field records, the ratio of total runoff in wet years to that in dry years is 3-4 to 1 in general for stations on the main stem of the river, while the figure may be as high as 5-12 to 1 for some tributaries, and still greater differences have been recorded in the semi-arid rolling loess regions. In comparision with water runoff, even greater annual variations occur in the sediment load.

Another aspect of the variation in annual runoff of the Yellow River is the sustained periods of succesive dry years. In the 60 years of record, from 1922 through 1932 successive droughts occurred as all-year-long dry periods. Average values of water runoff in these 11 years were 70-75% of the perennial average, while the average sediment load was approximately 20% of the annual average as recorded at Shanxian station.

With a length of 5,464 km and a fall of 4,660 m, the Yellow River has a potential hydroelectric power of 40,550 megawatts. The exploitable annual generating power is about 117 billon kwhrs, with an installed capacity of 28,000 megawatts. To utilize the head of 1,317 m avaiable from Longyanxia to Qingtongxia in the upper reach, a plan for installing 12,460 megawatts in this region has been put forward. Power stations at Liujiaxia, Bapanxia, Yanguoxia and Qingtonxia have already been built and put into operation while the power station at Longyangxia is now under construction. Obviously, a large hydroelectric power base will soon appear in this region.

Table 1. Distribution of water resources along the Yellow River

Reach	Catchment	Area 1000 sq.km	%	Runoff 10^8 cu.m	%	W (cu.m) /mu	W (cu.m) /capita
upper	above Lanzhou	223	30	326	56	2250	4730
	Lanzhou-Hekouzhen	162	21	-8			
middle	Hekouzhen-Longmen	112	15	72	12	280	1190
	Longmen-Sanmenxia	191	25	114	19	130	310
	Sanmenxia-Huayuankou	42	6	60	10	400	570
lower	Huayuankou-Bohai sea	22	3	21	4	130	200
	total	752	100	585	100	300	720

W is volume of water per mu of cropland or per capita

Present Utilization of Water Resources

Although the Yellow River enjoys a long history of water resources utilization in its basin, only a very small amount of water was used before 1950. Rapid development has occurred since then. By 1980 the total water supply soared to 35 billion cu.m, of which 27 billion cu.m was used by industry and agriculture, including 7.6 billion cu.m used in the Huaihai and Haihe basins along the lower reach (see Table 2). Ninety-five percent of the billion cu.m water is allocated to agriculture because industrial and domestic uses are relatively limited.

The utilization factor of water resources for consumptive use in the entire river basin approaches 47%, which is high compared to other rivers of China. In the past three decades, the exploitation and utilization of water resources prompted changes in the river basin and brought about an historic alteration of the Yellow River from one of disaster to a profit river, showing some remarkable benefits.

THE DEVELOPMENT OF IRRIGATION

To increase crop yields, nearly 80 million mu of land have been irrigated since 1950, including land irrigated by groundwater and tapped along the lower reaches. This doubled the average irrigated land per capita and is 5 times more than the irrigated land of 1949. Notable benefits have been shown by a 2-4 fold increase of crop yield over the original. Though only one-third of the cropland in the Yellow River basin is irrigated, the land yields half of the total crop output of the river valley. Numerous examples may be cited to illustrate the remarkable effects of irrigation on the semi-arid land in the basin. For instance, the crop output of 1982 for an irrigation district in Ningxia was 1.05 million tons, about 6.5 times than that in 1950; the crop yield per unit area of the arid and desert plateau in middle Gansu reached 250 kg per mu after being irrigated by water pumped to high land in stages. In the Jingtai irrigation district of Gansu province, cereal production had increased by 9.2 times compared with that in 1971 prior to irrigation, and the economic output from agriculture increased by 12.5 times. In Shaanxi province, cereal output in Sinhe and Weihe irrigation districts amounted to 500 kg/mu. Along the lower Yellow River, according to the statistics in 1981 from 34 counties, the average cereal production per unit of land had increased by 2 times over the past three decades. In the People's Victory Canal irrigation system, production per mu exceeded 500 kg for cereal and 50 kg for cotton, an increase of 4 and 2 times respectively. As a consequence, the conclusion can be drawn that the irrigation in the Yellow River basin promises a bright future for the valley's agriculture.

Table 2. Water consumed by industries and agriculture in the Yellow River basin in 1980 (in billion cu.m.)

Reach		Water consumed by industries & agriculture: Agriculture	Industry	domestic uses of cities & rural country	Total	total water withdrawn from the river
upper	above Hekouzhen	11.8	0.2	0.1	12.1	16.3
middle	Hekouzhen-Huayuankou	5.9	0.1	0.1	6.1	8.6
lower	in basin	1.6	0.05	-	1.65	2.4
	out basin	7.0	0.5	0.1	7.6	7.6
	total	26.3	0.85	0.3	27.45	34.9

WATER SUPPLY FOR INDUSTRIAL DEVELOPMENT AND DOMESTIC USE

Industries in the Yellow River basin have been developing at a rapid rate. Before the founding of the People's Republic of China, there were a few coal, textile, match-producing and flour mills and very few domestic industries in the cities of Lanzhou, Xian, Taiyuan. Water consumption was restricted and could basically be satisfied by the supply from groundwater. Now in the Yellow River basin, various sorts of industries such as coal, material, metallurgical, machine-building, power, chemical, pulp and paper, textile, food, medical instruments and medicine production as well as some well-developed domestic manufacturing plants have been established. Total industrial water supply of 1980 was 2.8 billion cu.m (including groundwater) and the total output value from industry in the basin amounted to 25 times that of 1949. At the same time, water supply for domestic use in cities and towns has been provided, and water shortage problems facing 64% of the population in mountain areas and 56% of the stock have been solved. Since 1980, water has been intermittently transported to Tianjin a long distance to supplement the serious shortage of its industrial and domestic water supply.

POWER DEVELOPMENT FOR INDUSTRIAL AND AGRICULTURAL DEMANDS

Great achievements have been made since 1949 in hydropower development in the Yellow River valley. At present, taking into account the power stations with 500 kw or more in the river valley, a total capacity of 2,540 megawatts of hydropower has been installed, generating 12 billion kwhr each year. The six hydropower stations on the main river have together supplied 82 billion kwhr of electricity. All these stations and plants not only provide power to the local areas, but satisfy the demands outside of the river basin. With the development of hydropower on the upper and middle reaches, a number of irrigation projects with high pumping-heads have been set up due to the available power for the development of irrigation on the high plateau.

Many problems remain to be solved in the development and utilization of water resources of the Yellow River. Runoff regulation in the Yellow River basin is inadequate. The total capacity in the existing reservoirs available for regulation of surface runoff constitutes only 18% of the annual runoff, most of it located on the main river in the upper reaches. No appreciable number reservoirs exist in the tributaries. Because of serious sedimentation, it seems impractical to store water in the middle reaches during the flood season. The storage volume in reserviors available for use during nonflood seasons is rather limited and hardly meets the huge demands of agriculture and industry. Waste of water is not uncommon in some areas. Accompanying the development of industry and urbanization, water pollution in the river system has increasingly drawn the attention of government agencies.

Prospects for Water Resources Utilization

ESTIMATE OF AVAILABLE WATER SUPPLY

Runoff in the Yellow River basin is characterized by its uneven spatial and temporal distribution. It would be difficult to satisfy the various demands of different regions and units in the basin. Accordingly multi-purpose engineering measures should be constructed and put into use in order to redistribute the runoff, including projects for storing, diverting, pumping and transferring water across watersheds.

Considering the current policies and conditions of the national economy, our main tasks in the short run are to reinforce, modify, and integrate the existing structures for the purpose of promoting the reliability and capacity of water storage, diversion and pumping. At present, Longyangxia reservior on the upper reach of the Yellow River has commenced its impoundment. By the year 2000, two reservoirs, at Wanjazhai and Xiaolangdi on the middle reach of the main river, and three reservoirs on tributaries will be built and put into operation. In addition, construction of some medium and small-sized reservoirs on tributaries is being planned. Their purpose is comprehensive water and soil conservation, and the enhancement of runoff regulation. The utilization of water resources may be further promoted by additional water supplied by constructing necessary pumping plants for diversion projects.

According to the analysis of runoff processes in various typical years and the relevant water supply projects, in the upper and lower reaches maximum usable runoff in the upper reach is 23-25 billion cu.m, 22 billion cu.m in moderately dry years and 20 billion cu.m in severely dry years. With further deduction for necessary water consumption for industry and agriculture below Hekouzhen, the above figures are 15-16 billion cu.m on average, and 14 billion cu.m and 12 billion cu.m for moderately and severely dry years, respectively.

After analysis of many years' runoff variation at Huayuankou, the maxmum water volume usable in the lower reaches, neglecting the portion used to sluice the sediment, is about 38-40 billion cu.m, 37 billion cu.m in moderately dry years and 30 billion cu.m in severely dry years, respectively. The lower reach of the Yellow River is marked by aggradation, so that a certain amount of water is needed to maintain sediment transport capacity. To maintain the present status of sedimentation, about 20-24 billion cu.m of water must be allowed to flow through the lower reaches. As a result, the annual maximum volume of runoff available to be used out of the river channel is about 34-38 billion cu.m.

THE PROSPECT OF WATER DEMAND

With respect to irrigation conditions in the years remaining before 2000, efforts should be concentrated on promoting economic benefits instead of enlarging the irrigation area. New projects should be considered only as a necessity. It has been estimated that the total water consumption by agriculture at the end of this century may increase by 20%.

Energy and chemical industrial bases occur in Shanxi province and two large-scale oil fields are located in the Yellow River basin. The necessity of rapid development of industry have stressed energy demands and the rate of industrial development is rapidly spreading in the whole basin. With the development of industry, the booming growth of numerous new cities and expanding population combined with a higher living standard, water consumption per capita is rising at a rapid rate. Total water consumption by the year 2000 will increase by almost three times. As a whole, by the end of this century, the total demand by industries and agriculture in the river basin will soar to 35-36 billion cu.m, a 30% increase in comparision with the demand in 1980.

Water supply and demand are basically balanced with respect to the entire river basin, but the regional discordance of supply and demand will result in a 2.0 billion cu.m water shortage by the year 2000. The main areas facing water deficiencies are the Fenhe and Weihe basins between Longmen and Sanmenxia, centers of economic development for Shanxi and Shaanxi provinces. Due to difficulties involved in the transfer of water, deficiencies there in water supply are more pronounced.

In the long term, the Yellow River basin is of strategic importance to China's economic shift from east to west. With the development of the west in the next century, there must be a great leap forward in the industries and agriculture of the Yellow River basin. It is estimated that in the next fifty years, the energy industries in the river basin will have a high development, especially hydropower stations and energy bases on the upper and middle reaches. Besides the large-scale urbanization accompanied by development in industry, substantial increases in municipal consumption of water are expected.

Industry will have to be developed, as well as irrigation, and the standard of living has to be improved. Since the limited water resources of the Yellow River valley just meet the demands for the year 2000, demands of the coming century can hardly be satisfied. To surmount the serious water shortage, sparing use of water is of primary importance. Given this limited base, water transfer from the Yangtze River should also be taken into account.

SEDIMENTATION IN THE LOWER REACHES OF THE YELLOW RIVER AND ITS BASIC LAWS

Zhao Yean, Pan Xiandi, Fan Zuoying and Han Shaofa
Institute of Hydraulic Research
Yellow River Conservancy Commission
People's Republic of China

Preface

The lower reaches of the Yellow River from Mengjin to the estuary, a distance of over 800 km, traverse the North China Plain. The river is an alluvial river with an intense accumulation of sediment. The large amount of deposition has caused the river bed to rise three to five meters on average, with a maximum of 10 m, higher than the adjacent ground surface behind the levees. Therefore, the river is well known over the world as a 'suspended river.' Because of the elevation of the river bed, flood prevention on the lower reaches is highly important. In fact, aggradation of the river bed is the cause of flood disasters on the lower reaches. Hence, all the problems caused by sediment deposition must be solved properly for flood prevention. The total length of dikes along both banks is 1,400 km. As most dikes were built gradually over hundreds of years, the bases of the dikes are complex and of different materials. This leads to the existence of latent dangers for failure. The dangers are more severe as a result of raising the dikes. The reaches below Gaocun are especially in danger because the river course in this stretch is relatively narrow, flood conveyance capacity is quite low and stages during medium or high floods are relatively high. Hazards of dike breaching by bursting may occur easily in these reaches. The reaches above Gaocun have wider channels. The river frequently changes its course and wanders between the dikes. Dangerous situations can occur because the dikes can be impinged upon by cross currents or by shifting of the main current. The dikes are vulnerable to the main current and different types of structural measures are needed according to the local situation when flood prevention works are built.

This paper deals with channel accretion on the lower reaches of the Yellow River and the basic laws of sedimentation. It would be helpful in reforming the river if the laws of fluvial

L. M. Brush et al. (eds.), Taming the Yellow River: Silt and Floods, 477–516.

processes could be fully understood and the trends of its future development properly predicted.

General Introduction Regarding Erosion and Sedimentation of the Lower Reaches of the Yellow River

The courses of the lower reaches were formed during different historical periods. The course from Mengjin to the confluence of the Qinhe River has been in existence for thousands of years. The course from the river outlet of Qinhe River to Dongbarou, Lankao county, Henan province, is 500 years old. From Dongbatou to Taochengpu, the course was formed on the floodplain caused by a dike breach at Tongwaxiang in 1855. The course from Taochengpu to the estuary of the Yellow River has been formed by river capture of the original Daqing River after a dike breach and has flowed in the current course for more than 130 years.

It is difficult to precisely describe the former erosion and sedimentation process of the river course. The reasons are that there were no systematic surveys of the processes and no detailed records of the past, and only rough estimations of the process from ancient deposits. Moreover, the rising speed of the river bed was usually slowed with respect to frequent dike breaches and the shifting of courses in the past, as great amounts of sediment were carried out of the dikes. However, precise knowledge of the fluvial processes over the past several decades is available.

SEDIMENTATION SINCE 1855

Since 1855, the characteristics of erosion and deposition have been different in different reaches of the lower Yellow River.

The Reach From Mengjin to Dongbatou. At the initial stages of the bank breach at Tongwaxiang, there was a drop of 6 meters or so around the outlet at the breaching point of the dike. This resulted in a retrogressive erosion up the channel, which extended to the outlet of the Qinhe River, a length of about 100 km. Subsequently, the river bed was deeply cut in the reach from the outlet of the Qinhe River to Dongbatou leading to a deep channel with high floodplains. From 1875 to 1905, retrogressive deposition as well as floodplain bank caving took place in this reach in response to the construction of dikes below Dongbatou. The thickness of deposits in the main channel was generally 1 to 2 m. From 1905 to 1985, the thickness increased to 2 to 3 m. The floodplain level around Dongbatou is still 2.5 to 3 m lower than the old floodplain formed in 1855. In comparison only a slight trace of the old floodplain around Huayuankou remains, about 1 m higher than the new one. Generally speaking, the thickness deposition in

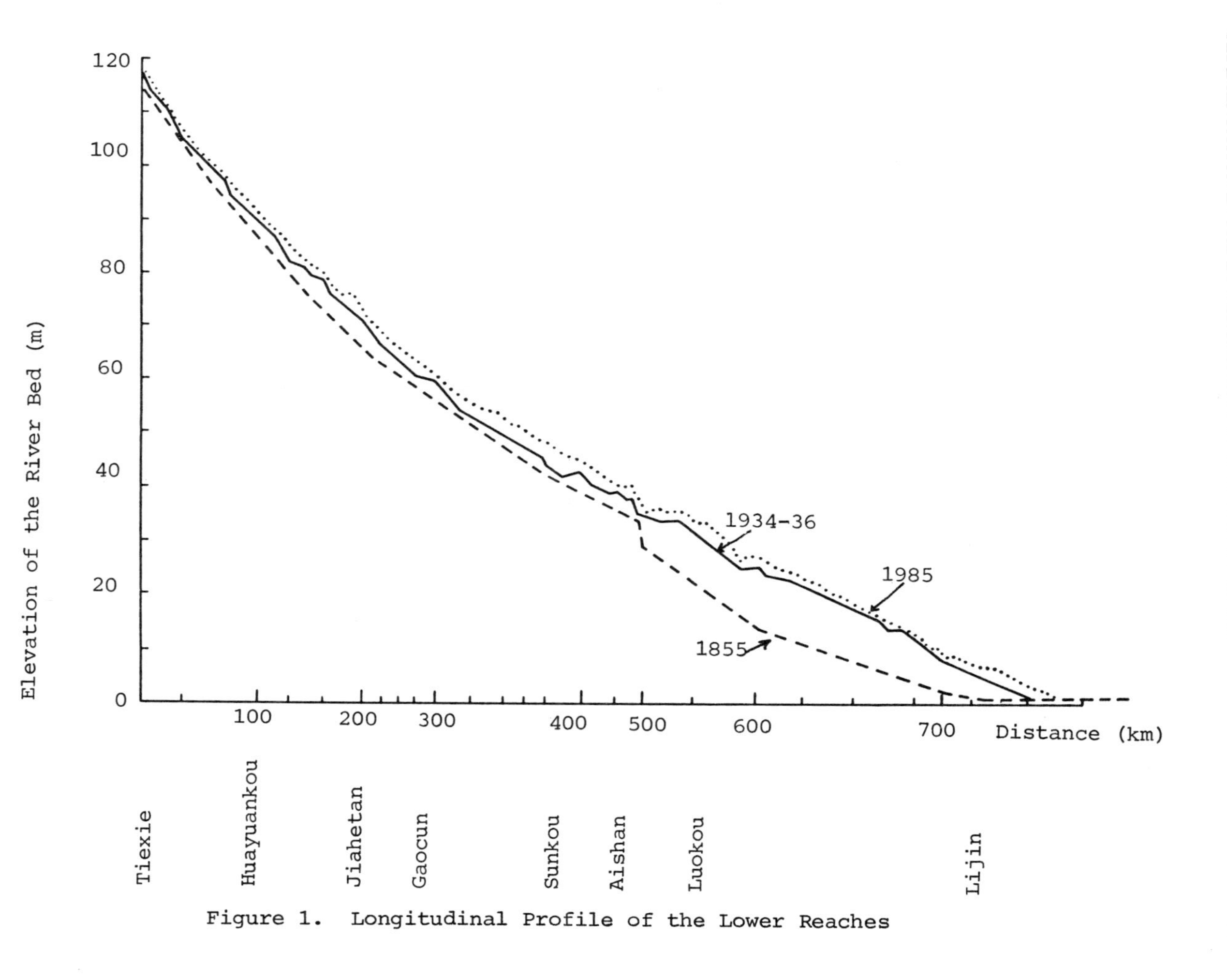

Figure 1. Longitudinal Profile of the Lower Reaches

the reach from Mengjin to the outlet of Qinhe River was 1 to 2 m between 1855 and 1934. After 1935, no substantial amount of deposition has been observed due to the effect of retrogressive erosion resulting from the dike breach at Huayuankou in 1938 and later on, from the construction of the Sanmenxia Reservoir.

The Reach From Dongbatou to Taochengpu. From 1855 to 1875, just after the bank was breached, the area north of the bank of the Yellow River was defended by the Beijindi Dike and there was no dike south of the river. The flow, therefore, flooded freely to an extent of 200 km in length and 100 km in width. This led to an average thickness of deposition of 1 to 2 m on the area.

The Reach Below Taochengpu. This reach was formerly the Daqing River which was an 'unsuspended river' before 1855, but its channel was small, only 30 m more or less in width and less than 3 m in depth. At the start of the avulsion of the Yellow River to the Daqing River, a huge amount of sediment was deposited on floodplain of the upper reach, while the clear water sluiced into the Daqing River scouring the channel both in width and in depth. After the dikes were built on the floodplain in 1875, the channel began to deposit sediment. By 1893, the current dike lines had formed beginning with the formal construction of dikes along both sides of the Daqing River in 1885. Later on, the deposition speeded up. According to the topographic analysis deposition was about 1.5 m from 1891 to 1901, and 1 to 2 m from 1902 to 1972. It can be seen that the deposition in this reach mainly occured in the period of radical regulation of the river bed due to the change of runoff and sediment load conditions after the construction of dikes on the upper reaches. Fig.2 shows the longitudinal profile of the lower reaches.

SEDIMENTATION DURING THE LAST FIVE DECADES

In 1933 an extreme flood originating in the loess area above Sanmenxia led to a flood peak discharge of 22,000 cms at Huayuankou. Severe deposition on the lower reaches took place. From 1933 through 1938, the bank of the high floodplain formed in the 1933 flood gradually caved in year by year and the sediment was deposited in the main channel. The river bed tended to be more shallow and wide. Nevertheless, the quantity of deposition was not noticeable. In June. 1938, the dike breach at Huayuankou caused a complete breaking off of the reach below and retrogressive erosion on the reach above. Hence, a deep channel with a high floodplain was formed extending upstream from the breaching point. In March, 1947,

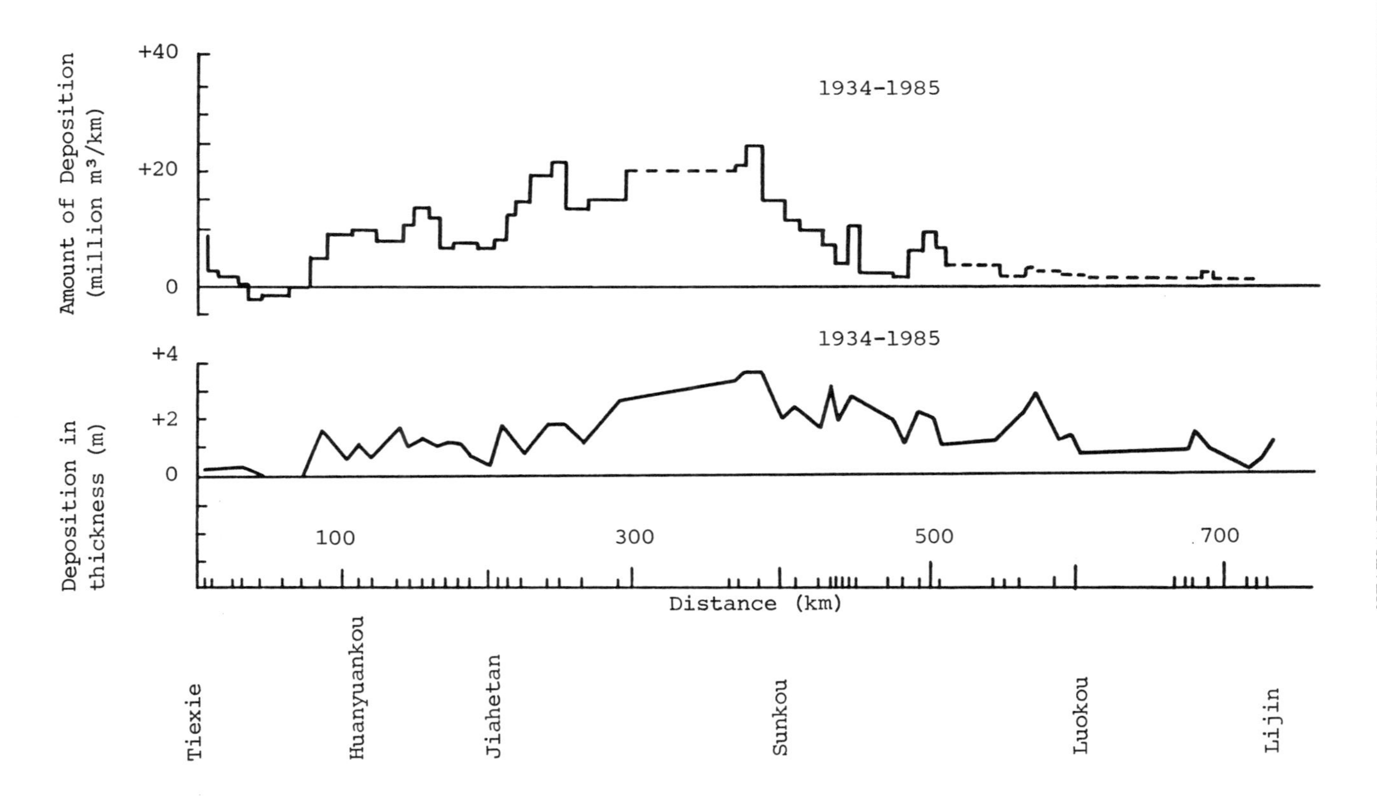

Figure 2. Variation of Deposition Along the River.

the Yellow River flowed once again in the original or present course after the closure of the dike at Huayuankou. Since then no dikes have been breached, but huge quantities of deposition have taken place.

In September 1960, Sanmenxia Reservoir was constructed. It controls 92% of the river-basin, 89% of the runoff and 98% of the sediment load for the lower reaches. Therefore the operation mode of the reservoir decisively affects the fluvial processes of the lower reaches. During the five decades from 1934 to 1985, some 9 billion tons of sediment have been deposited on the reach from Tiexie to Lijin. This was estimated by topograghic maps and converted to weight by using 1.4 tons per cu.m. If the 9 years during which the Yellow River changed its course at Huayuankou by manmade effects is not taken into account, the average annual deposition is 220 million tons. The distribution of the deposition is presented in Fig.3. A tendency can be seen for more sediment deposition on the middle reaches than on the two end reaches. In fact, deposition is 1 to 1.5 m on the upper reach, 2.5 to 3.5 m on the middle and 0.5 to 1 m on the lower. Fig.4 shows the transverse distribution of the deposition.

The general aspects of erosion and deposition on the lower reaches are principally determined by the oncoming runoff and sediment load conditions. In the long run, the annual amount of deposition fluctuates naturally in a periodic way in conformity with the variation of hydrological conditions. In a wet year with a large sediment load, deposition is great. For example, the flood peak discharge at Shaanxian (i.e. present Sanmenxia) was up to 22,000 cms and the maximum measured sediment concentration was 519 k per cu.m. In the reach from Mengjin to Gaocun, the level of the floodplain rose 1 to 1.5 m due to the deposition of 2,200 million tons of sediment although the main channel was scoured by 500 million tons and the whole reach had deposits of 1,700 million tons. During this time the river experienced the most severe deposition for decades. In a dry year with a small sediment load, the river may erode. Because of frequent dike breaches from 1919 to 1949, the average annual deposition was only 200 million tons or so as a large amount of sediment was washed out of the dikes. From 1950 to 1985 the deposition averaged 6,850 million tons, about 200 million tons for each year(*).

* This value equals the difference between the amount of sediment load at the entrance and at the exit of the lower reaches subtracting the amount of sediment withdrawn by irrigation. The amount of sediment at the entrance equals the sum of the amount of sediment load at Sanmenxia, that at Heishiguan on the tributary Yilue River and that at Xiaodung on the tributary Qinhe River. The amount of sediment load at the exit equals the sediment at Lijin.

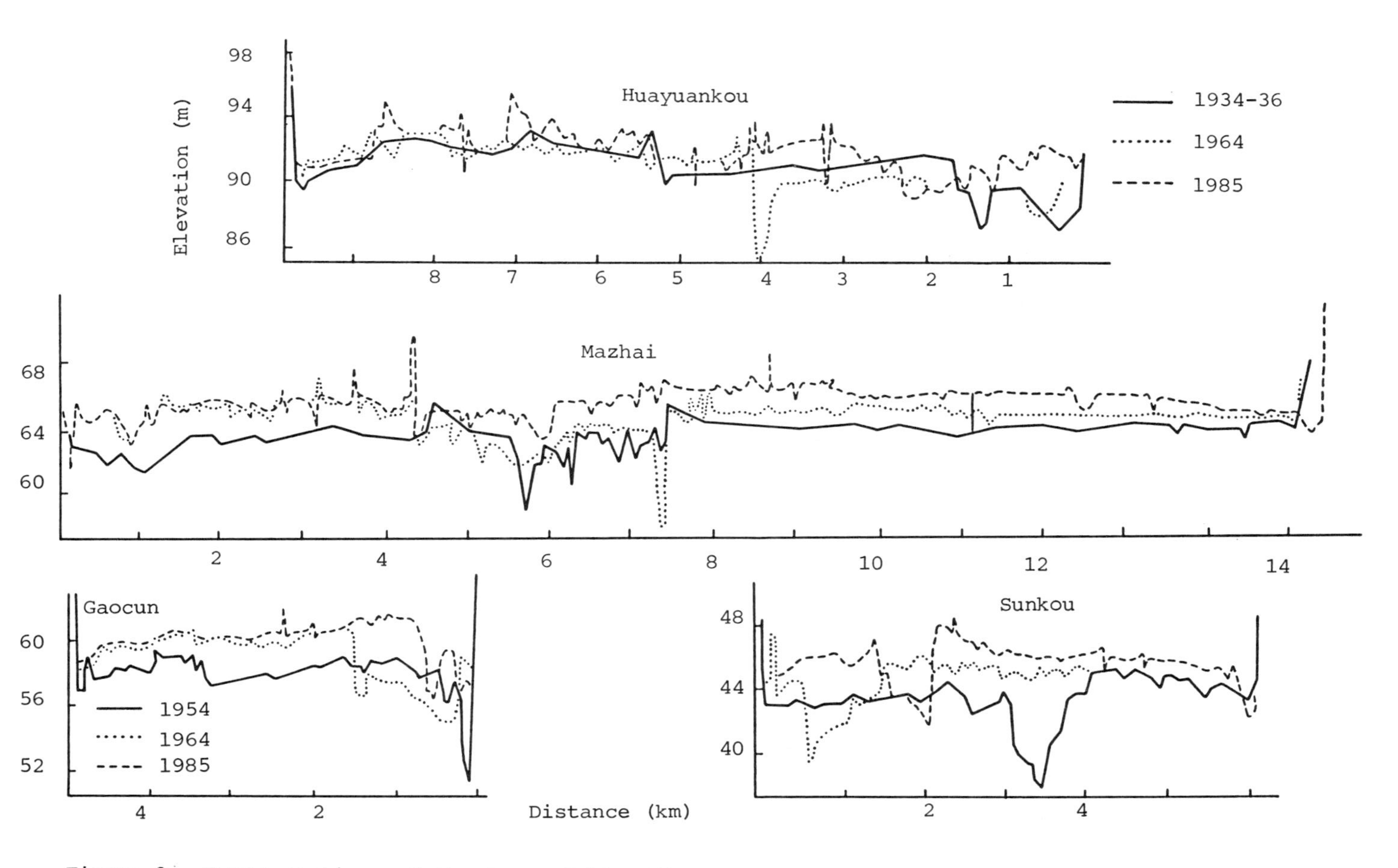

Figure 3. Cross-sections of the Lower Yellow River

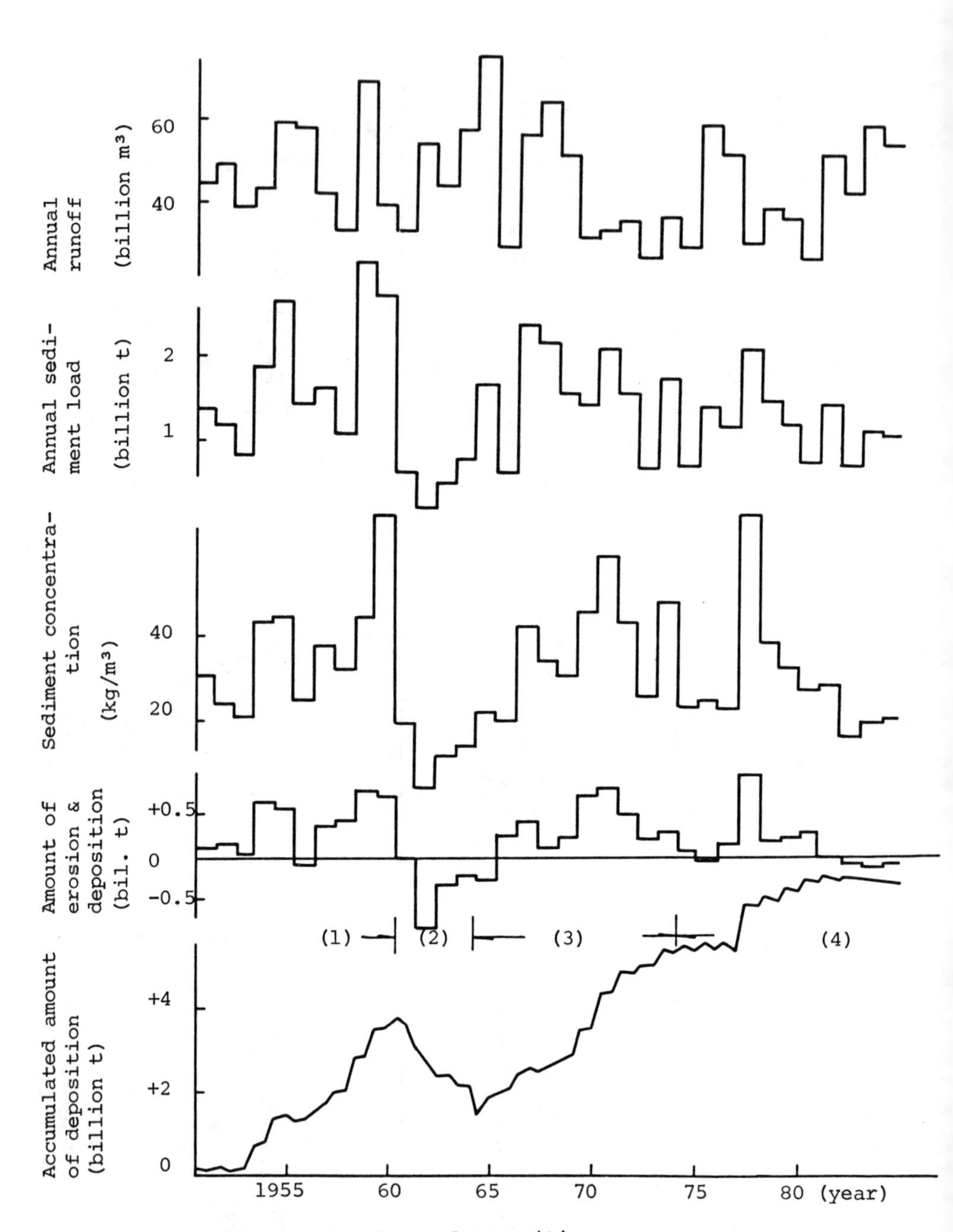

Figure 4. Variation of Erosion and Deposition

Key: (1) natural condition; (2) Sanmenxia Reservoir impounds floods; (3) it retards floods and sluices sediment; (4) it regulates both water and sediment

Taking the statistics accumulatively, deposition was 13.1 billion tons on the reach above Lijin and (or) about 200 million tons annually from 1919 to 1985. It can be estimated that the deposition would be 19.5 billion tons for the same period, nearly 300 million tons annually, if there were no dike breaches, no avulsions and no controls of sediment by the Sanmenxia reservior.

SEDIMENTATION SINCE 1950

Fig. 5 shows the variation of erosion and deposition on the lower reaches since 1950. From 1959 to 1960, an annual average of 361 million tons of sediment was deposited on the lower reaches before construction of the Sanmenxia Reservior (excluding the deposition of 190 million tons on the Dongping Lake), 77% of which was deposited on the floodplains. Fig.6-1 shows the longitudinal distribution of the deposits. It can be seen that more sediment was deposited on the middle reach. In fact, the intensity on the reach from Jiahetan to Gaocun was up to 8 million tons per kilometer. The stage for the discharge of 3,000 cms also reflects the amount of deposition in the main channel (see Fig. 7 and Tab. 1). It is obvious that the stages for the same discharge rise universally along the reaches, in which the maximum rise of stage near Sunkou amounted up to 1.72 m.

The Sanmenxia Reservior commenced impoundment in Sept. 1960, and afterwards, changed its mode of operation to detention only. Prior to Oct. 1964, water and sediment load released from the reservior had the following characteristics: peak discharges of floods were reduced to a great extent, and for medium flows the duration was lengthened.

Runoff processes for low flows tended to be even and the outflow was mainly sediment-free water, except for some fine sediment disposed of by turbidity density currents from the reservior. The lower reaches were scoured by a total amount of 2.31 billion tons. It is shown in Fig. 6-1 that the scour decreases progressively along the reaches. The amount of the erosion on the reach above Gaocun amounted to 73% of the total. In the process of scouring, the floodplains of the wandering reach and the transient reach caved in great quantities. The sediment due to the bank erosion was about 700 million tons, or 30% of the total amount of erosion in the lower reaches. The scour due to the bank caving on the reach above Gaocun was 95% of the total, or 670 million tons. Stages for the same discharge were also reduced progressively along the river, in a range from 2 or more to 0.2 m. At Lijin near the estuary, the river bed rose only slightly.

From Nov. 1964 to Oct. 1973, the Sanmenxia Reservoir was operated in a flood detention mode. The outlet structures were reconstructed in two stages to enlarge the outlet capacity. After 1966, two new outlet facilities were put into

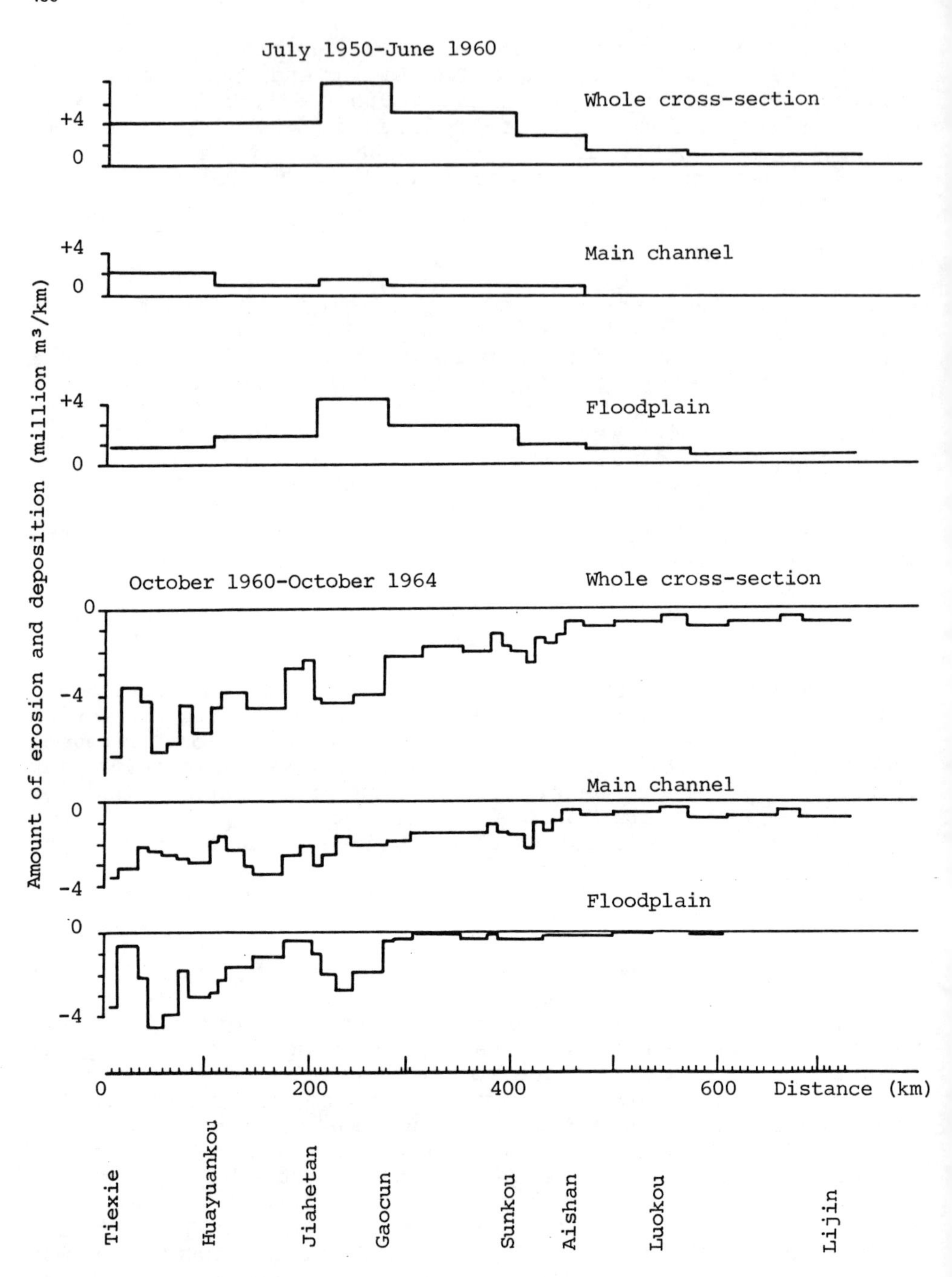

Figure 5. Longitudinal Distribution of Erosion and Deposition

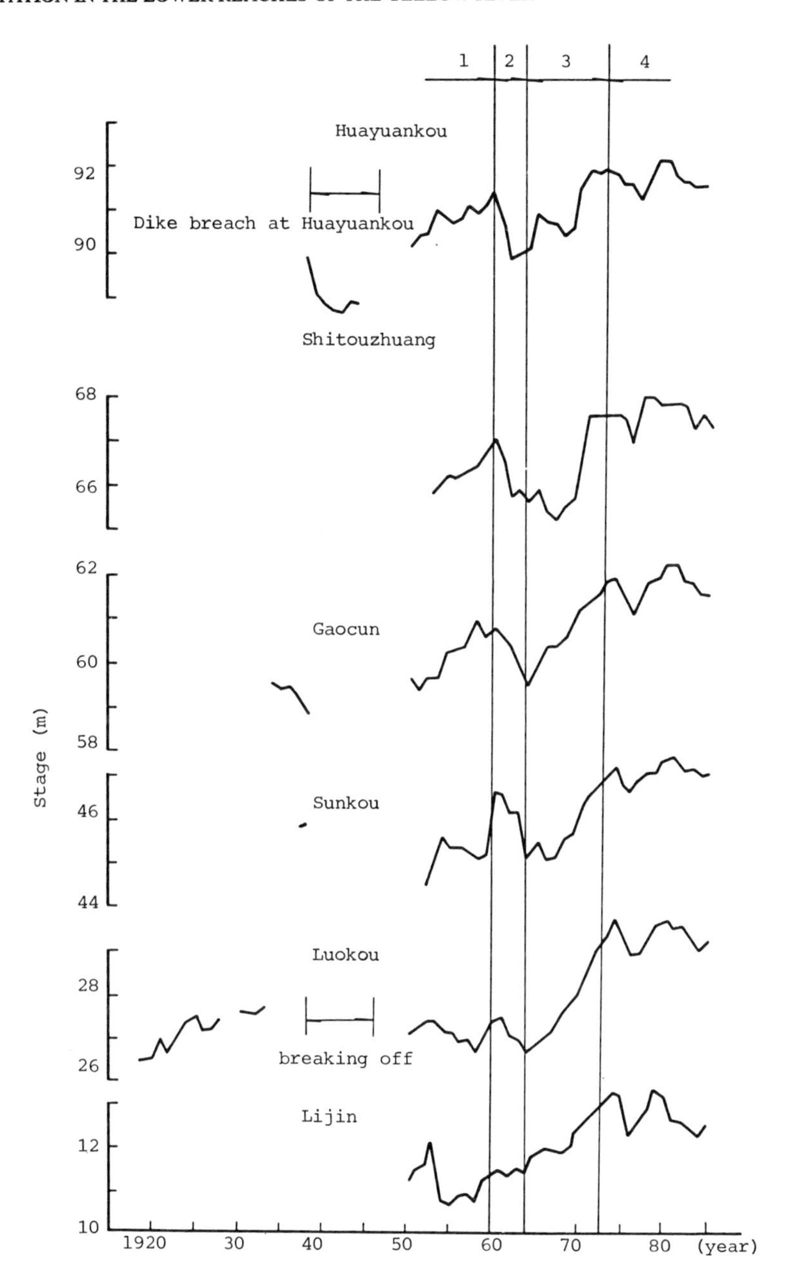

Figure 6. Variation of Stages

Key: 1. natural condition; 2. Sanmenxia Reservoir impounds floods, 3. retards floods and sluices sediment, 4. regulates both water and sediment

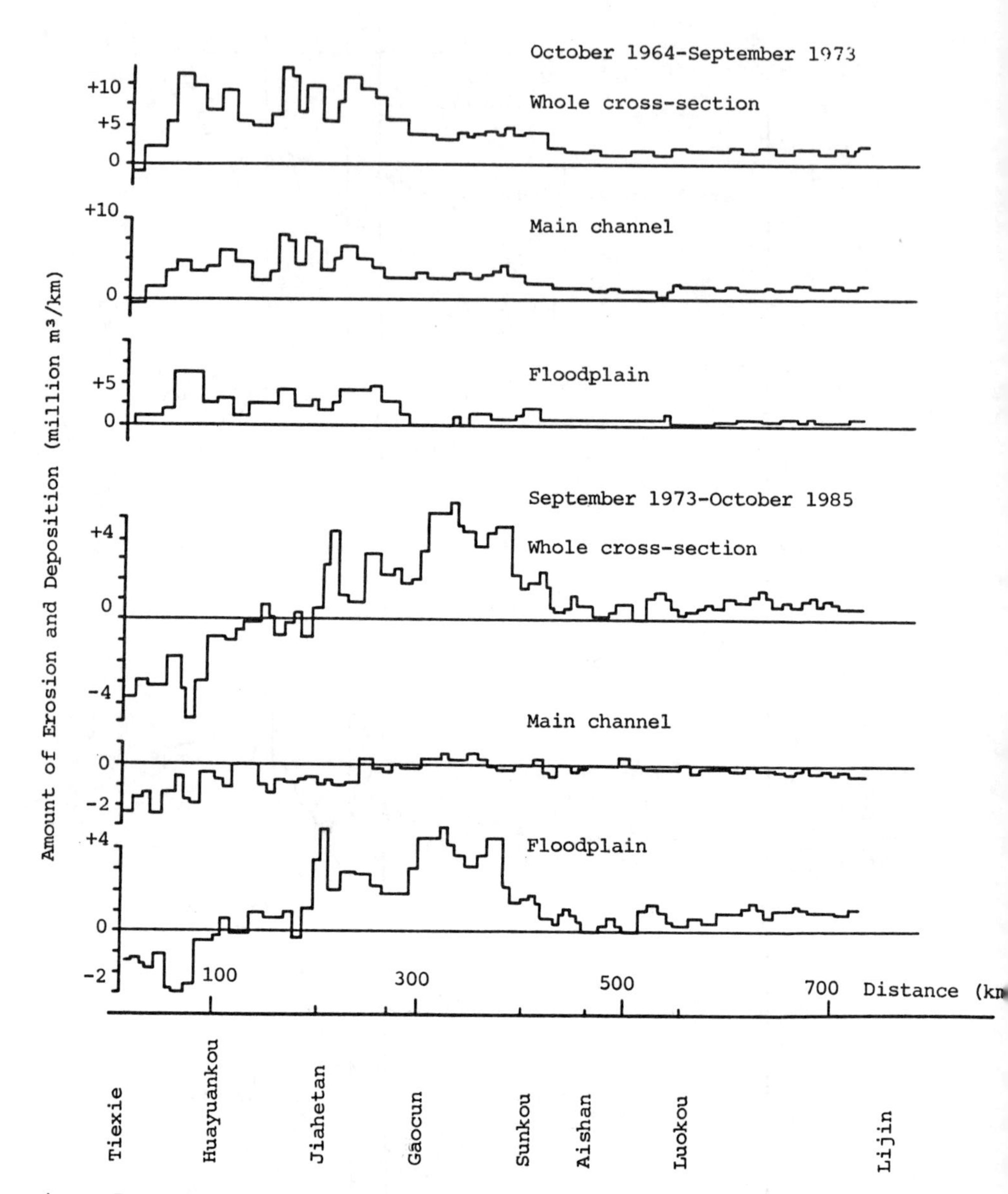

Figure 7. Longitudinal Distribution of Erosion and Deposition.

Table 1 The value of rise/fall of stage for Q = 3,000 cms in different periods

Station	Value of rise/fall (m)					Annual average (m)				
	1950-1960	1960-1964	1964-1973	1973-1985	1950-1985	1950-1960	1960-1964	1964-1973	1973-1985	1950-1985
Tiexie		-2.81	0.64				-0.70	0.07		
Peiyu		-2.16	1.54				-0.54	0.17		
Guan-zhuangyu		-2.07	2.02				-0.52	0.22		
Huayuankou	1.2	-1.30	1.85	-0.43	1.32	0.12	-0.33	0.21	-0.04	0.04
Jiahetan	1.13*	-1.32	1.94	-0.53	1.22*	0.14	-0.33	0.22	-0.04	0.04
Shitou-zhuang		-1.44	2.07	-0.23			-0.36	0.23	-0.02	
Gongcun	1.17	-1.33	2.37	-0.26	1.95	0.12	-0.33	0.26	-0.02	0.06
Liuzhuang	1.05	-1.30	2.35			0.11	-0.33	0.26		
Susizhuang		-1.35	2.20	0.05			-0.34	0.24	0.004	
Xingmiao		-1.78	2.94	0.19			-0.45	0.33	0.016	
Yangji		-1.85	2.24	0.04			-0.46	0.25	0.003	
Sunkou	1.72*	-1.56	1.86	0.04	2.06*	0.22	-0.39	0.21	0.003	0.06
Nanqiao		-0.65	2.21	0.08			-0.16	0.25	0.006	
Aishan	0.56	-0.75	2.25	-0.05	2.01	0.056	-0.19	0.25	-0.004	0.06
Guanzhuang		-0.45	2.35	0.05			-0.11	0.26	0.004	
Beidianzi	0.35	-1.10	2.90	-0.28	1.87	0.035	-0.28	0.32	-0.023	0.05
Luokou	0.26	-0.69	2.63	-0.10	2.10	0.026	-0.17	0.29	-0.008	0.06
Liujiayuan		-0.17	2.17	-0.11			-0.043	0.24	-0.009	
Zhang-xiaotang	0.22	-0.22	1.94	-0.39	1.55	0.022	-0.055	0.22	-0.03	0.04
Daoxu	0.23	-0.30	1.95	-0.47	1.41	0.023	-0.075	0.22	-0.04	0.04
Mawang		-0.4	2.12	-0.67			-0.10	0.24	-0.056	
Lijin	0.20	0.01	1.64	-0.56	1.29	0.02	0.002	0.18	-0.05	0.04

Note: * refers to the data of 1952

use. After reconstruction, the reservoir still retarded floods and sediment load to some extent. On the other hand, a part of the sediment formerly deposited within the reservoir was washed out when it operated at low stages. Therefore, flows with high sediment concentration often occurred when the outlet discharges were small. This led to an incompatability of water flow and sediment load, and in addition, led to serious deposition on the lower reaches. In fact, the total amount of deposition was 3.95 billion tons, and the annual average was 439 million tons in that period, which is even more than that in the period from 1950 to 1960. Fig. 6-2 shows the longitudinal distribution of deposition along the lower reaches. Meanwhile, the transverse distribution changed a lot. 61% of the deposition of the whole section was in the main channel, and 33% was on the floodplain. Almost all of the sediment was deposited on the main channel in the reach below Aishan. Stages for the same discharge generally rose 2 m or so. In some stretches in the reach from Dongbatou to Taochengpu, another "suspended river" was formed within the already "suspended river" affected by the "production dikes."

Since 1974, the operation mode of the Sanmenxia Reservoir has been changed into regulation of both water and sediment, and by impounding clear water and sluicing sediment-laden water. That means the reservoir impounds flows during the nonflood season (from Nov. to next June), and sluices sediment during the flood season (from July to Oct.). The runoff since then approximates the long term average, while the sediment load is much smaller. The deposition on the lower reaches, thus, is relatively small, only 100 million tons as an annual average. Because the reservoir releases clear water during nonflood seasons and the annual sediment load is entirely disposed of in flood seasons, erosion and deposition processes during a single year change quite a lot in the lower reaches. Scour in the channel of an annual average of 100 million tons occurred instead of deposition. In flood seasons the river still undergoes deposition. Longitudinally, scour takes places in the upper reach and deposition in the lower reach (see Fig. 6-2). Deposition on the reach from Gaocun to Sunkou is 3/4 of that on the entire lower reaches. The intensity of deposition below Taochengpu is relatively small. The main channel in both the upper and lower stretches is eroded, while that in middle stretch has deposition, as far as the transverse distribution is concerned. For the entire lower Yellow River the main channels are scoured, while the floodplains undergo deposition. The amount of deposition on the floodplains is greater than the scour on the main channels. The scour on the reach above Huayuankou is mainly in the form of bank erosion of the floodplains. In the rest of the reaches deposition on the floodplains occurs. The stages for the same discharge on the lower reaches generally decrease except in the reach from Susizhuang to Nanqiao. The extent of the fall in the reach above Susizhuang gradually decreases progressively along the river. This appears to be the characteristic erosion along the river. Inversely, in the reach below Luokou, the extent of the fall increases along the river. The shifting of river course in the estuary area to Qingshuigou in 1976 also affects the stage.

In summary it has been shown that both the amount of erosion and deposition and their distribution along the lower reaches vary depending on the operation of the Sanmenxia Reservoir. The erosion and deposition, therefore, are closely related to variations in runoff and sediment load and to operation of the reservoir which modifies the runoff and sediment conditions for the lower reaches.

Basic Law of Channel Accretion in the Lower Reaches

Determining the laws of erosion and sedimentation to find the factors that influence sedimentation is the key to solving the serious problems of aggradation in lower reaches.

ORIGINS OF RUNOFF AND SEDIMENT LOAD AND THEIR EFFECTS

The Yellow River drainage basin is vast in size with complex geographic conditions. 54% of total runoff and only 10% of the sediment originates in the area above Hekouzhen. Major sources of sediment are tributaries in the stretch between Hekouzhen and Longmen, contributing 56% of total sediment load and only 15% of the total runoff. There are three types of sediment sources:

* Sources with high sediment yield and coarse particles: The stretch between Hekouzhen and Longmen; Mailian and Beiluo River.
* Sources with high sediment yield and fine particles: The trunk stream and tributaries of Jinghe River, excpt Malian River; the upper reaches of Weihe River and Fenhe River.
* Sources with low sediment yields: The reaches above Hekouzhen; The tributaries originating from Qinling Mountain in Weihe valley; Yiluohe River and Qinhe River.

The runoff can be divided into six groups, according to the analysis of 130 floods from 1952 to 1983 (see Tab.2). It can be seen from Tab. 2 that floods forming in the source area of coarse sediment may cause serious deposition on the lower reaches, although their frequency of occurrence is only 10%. The amount of deposition due to these floods is usually 40% to 60% of the total. When floods from different origins meet, overbank flow often occurs in lower reaches. The intensity of deposition is relatively great and the main channel is scoured while the floodplain undergoes deposition. For floods originating in the source area with low yield of sediment, the

Table 2 Effect of floods of different groups on deposition in lower reaches in different periods

Flood group	No. of incidence	% of incidence (%)	Intensity of deposition (million t/day)	Percentage of deposition (%)
From 1952 to 1961				
(1)	4	6.9	3.86	4.5
(2)	7	12.1	24.80	38.9
(3)	10	17.3	5.92	10.2
(4)	11	19.0	19.36	55.9
(5)	24	41.3	-1.68	-12.0
(6)	2	3.4	6.12	2.5
Total	58	100.		100.
From 1969 to 1973				
(1)	2	8.7	3.20	3.7
(2)	4	17.4	3.17	76.5
(3)	8	34.8	5.45	28.0
(4)	0	0	0	0
(5)	9	39.1	-1.48	-8.2
(6)	0	0	0	0
Total	23	100.		100.
From 1974 to 1983				
(1)	0	0	0	0
(2)	5	10.2	26.16	61.7
(3)	9	18.4	4.18	15.2
(4)	2	4.1	9.68	11.3
(5)	26	53.1	-0.53	-8.8
(6)	7	14.2	7.16	20.6
Total	49	100.		100.

river courses in the lower reaches are scoured. Sources and features of floods may be different in different hydrologic series. However, the composition of each individual flood is still in the six groups listed in Tab.2.

In Tab.2, the six flood groups are named as following:
(1) A widespread rainfall with low intensity falling over the whole of the middle reaches.
(2) Comparatively large floods from the coarse sediment source area with high sediment concentration.
(3) Floods of medium intensity from the coarse sediment source areas with high concentration, and some supply from areas with small sediment concentration.
(4) A junction of floods from the coarse- and fine-sediment source areas with high concentration and from areas with small sediment concentration.
(5) Floods mainly from areas with small sediment concentration.
(6) Floods from the fine sediment source areas with high concentration.

Fig.8 shows that the average sediment concentrations are generally greater than 150 kg per cu.m from floods originating in source areas of coarse sediment. This leads to serious deposition in the lower reaches. In contrast, if the sediment concentration is less than 50 kg per cu.m during floods with origins with a low yield of sediment, it causes the river to scour or to experience small deposition. If the flood is composed of runoff from all sources, the sediment concentration will vary in a range from 50 to 150 kg per cu.m according to the proportion of runoff from the different sources. With regard to the sources, floods from two particular source areas will lead to serious deposition in lower reaches. These are floods on the tributaries of the Yellow River between Hekouzhen and the confluence point of Wuding River, with the source in the Baiyu Mountains. The modulus of coarse sediment yield in these areas is up to 6,000 to 10,000 tons per sq. km. Therefore, intensive harnessing works in these areas are of vast importance in minimizing deposition in the lower Yellow River.

EFFECTS OF FLOW AND SEDIMENT REGIME ON EROSION AND SEDIMENTATION

In the long run, the lower reaches are zones of deposition. Nevertheless, the deposits are not developed in a monotonic way and erosion may take place intermittently between periods of deposition. For years with abundant runoff and small sediment load, the river erodes or undergoes slight deposition (examples are the years 1952,1955,1961 and 1981 to 1985). The river will deposit, however, in other years with a small amount of runoff and great amount of sediment load (examples

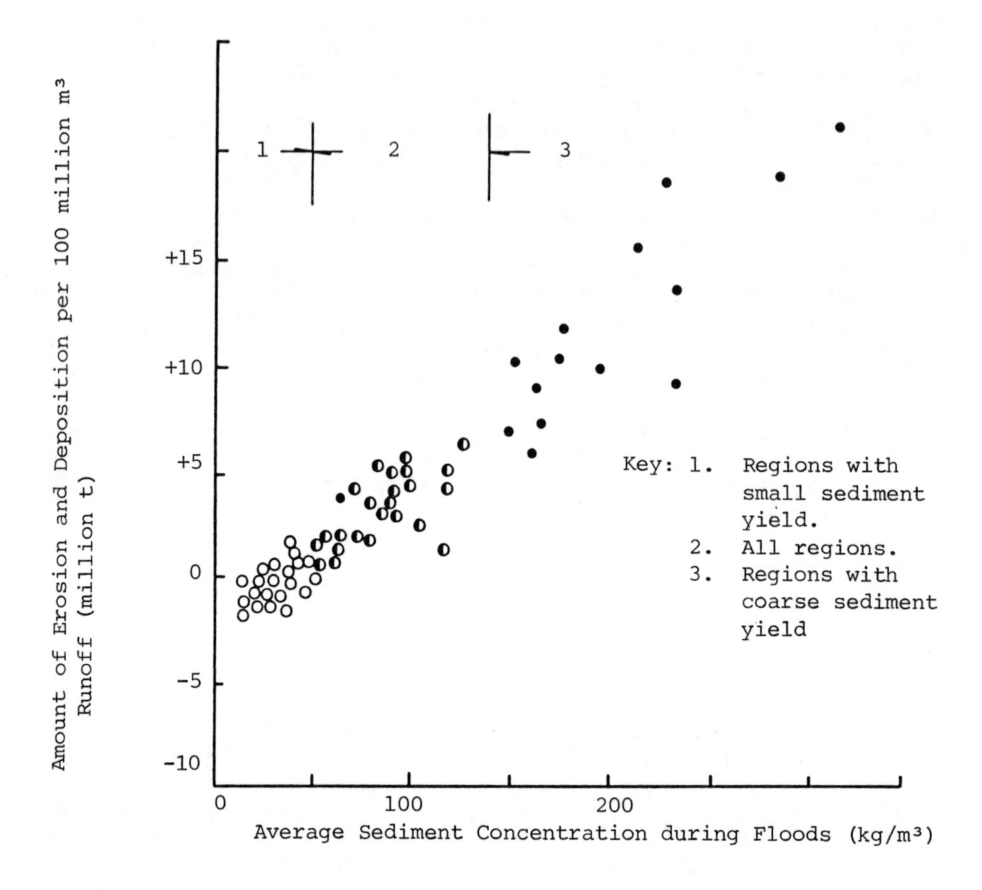

Figure 8. Relationship Between the Average Sediment Concentration and the Amount of Erosion and Deposition During Floods.

are the years 1969, and 1977). Annual deposition during these years usually was up to 700 to 1,000 million tons (see Fig.5). The amount of deposition is closely related to the oncoming runoff and sediment load. It is obvious that the greater the runoff, the less the deposition, sometimes even leading to scour. According to data analyses, on average, a reduction of 3,000 to 4,000 million cubic meters of runoff in the flood season may lead to a 100-million ton increase of deposition in the lower reaches. On the other hand, 100 million tons of reduced sediment load may lead to a decrease of 50 million tons in deposition on the lower reaches in the flood season. At the same time, 100 million tons of decrease in sediment load may lead to 100 million tons of decrease in deposition in a nonflood season.

The amount of deposition per 100 million cu.m of runoff is directly proportional to the annaul average sediment concentration in the lower reaches (see Fig.9). It can be seen from Fig.9 that if the sediment concentration is small, the deposition in the lower reaches is also small, sometimes leading to scour. Inversely, the greater the sediment concentration, the more the deposition will be. There is a critical value of about 20 to 27 kg/cu.m. If the sediment concentration is greater than this, the river course will undergo deposition, otherwise there will be scour. Apparently, the deposition on the lower reaches can be effectively decreased by increasing the amount of runoff and reducing sediment load.

CAPACITY OF SEDIMENT TRANSPORT

As the origins of sediment for different floods are not the same, sediment loads carried by similar flood vary quite a lot. Corresponding to the rapid change of runoff and sediment load, the alluvial bed in the lower reaches regulates itself quite sensitively and intensively, and the transport capacity is adjusted very quickly. As a result, the phrase "the more the sediment load, the more the deposition will be and the more sediment to be disposed" or "the less the sediment load, the less the deposition will be (or scour) and the less sediment to be disposed" has been observed on reaches more or less a thousand kilometers in length. When the sediment load is relatively high, sediment discharge is a function of both the flow discharge and the sediment concentration (see Fig. 10):

$$Gs = KQ^a S_o^b$$

in which Gs---discharge of bed material load, in tons/per s; Q---flow discharge, in cms; S_o---concentration of bed material load of the oncoming flow, in Kg per cu.m; K---a coefficient related to boundary conditions, and size composition of the

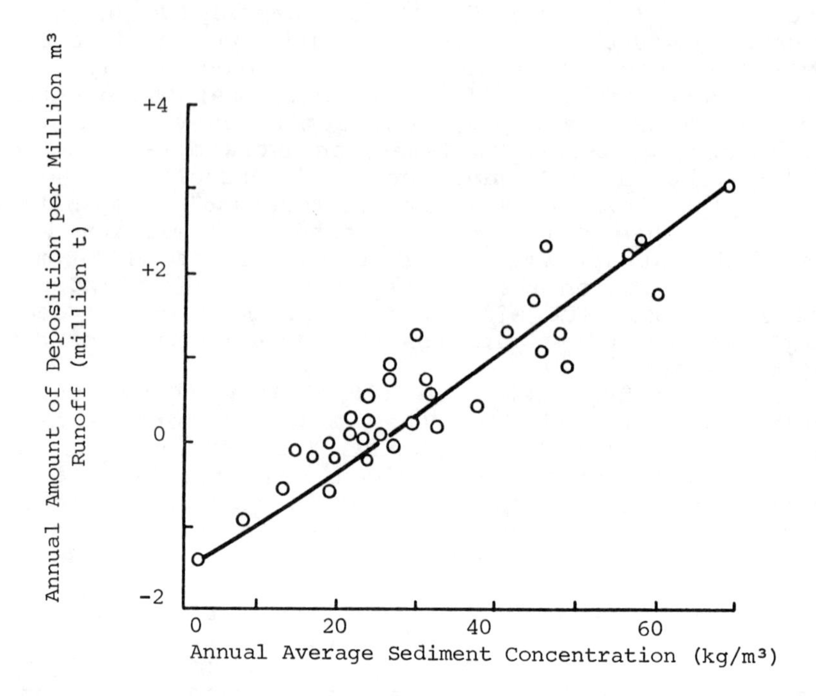

Figure 9. Relationship Between the Deposition per 100 Million m³ Runoff and the Annual Average Sediment Concentration.

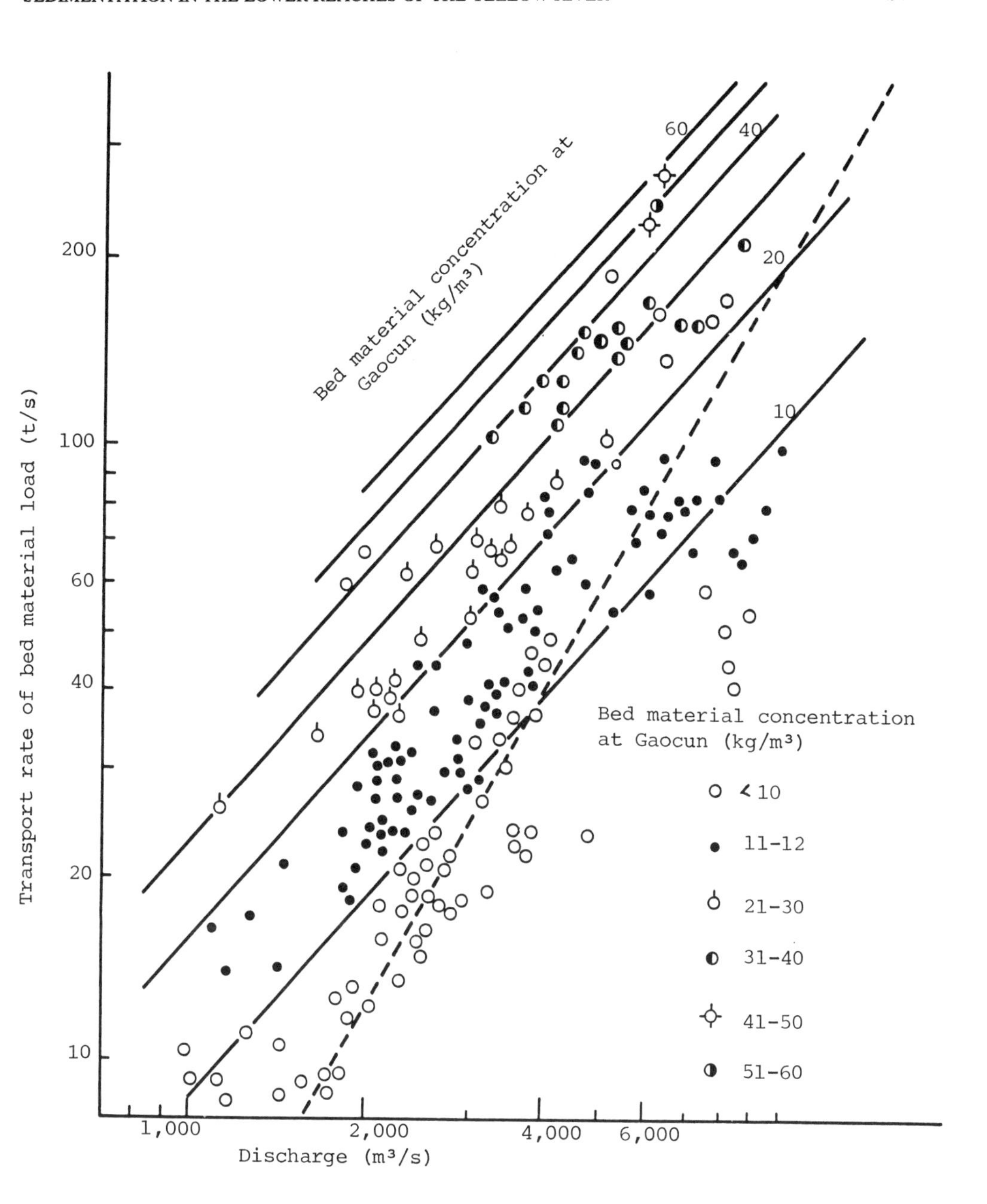

Figure 10. Discharge vs. Bed Material Transport Rate at Sunkou

oncoming sediment, etc.; a and b---indices, with values of 1.1-1.3 and 0.7-0.9, respectively.

The amplitude of fluctuations in the bed load is very large during floods. When runoff and sediment load are relatively large, the bed material will be finer because of deposition. The sediment transport capacity, therefore, will be enhanced. When the runoff and the sediment load are small, the capacity will be reduced due to the armoring caused by scour. Hence, the sediment transport capacity in the lower reaches can also be related to the amount of previous deposition in the reaches.

It is clear according to the above analysis that by increasing the sediment transport capacity under a certain boundary condition, not only water runoff but also sediment should be regulated to make the sediment compatible to the water flow so that heavy sediment loads can be carried by large flows and sediment can be transported more efficiently. This is the basis for regulating water and sediment runoff for the purpose of enlarging the sediment transport capacity in sediment laden rivers. In this way, the location of deposition may be improved and the amount of deposition may be reduced creating a favourable situation for the lower reaches.

When clear water is released from Sanmenxia Reservoir or the oncoming flow has a relatively low sediment concentration, sediment discharge in the lower reaches is in direct proportion to a power of discharge with an exponent of nearly 2. If the same amount of flow were regulated through operation to form an artificial flood instead of being released at an uniform rate, it would have a higher capacity for sediment transport than it naturally does. To study the effects of scour by manmade floods, analyses have been made for floods with low sediment concentrations and clear floodwaters released from Sanmenxia Reservoir. Average discharge of these floods was generally 4,000 to 5,000 cms and sediment concentration was less than 25 kg per cu.m. The flood volume was about 3 to 5 billion cu.m, and usually no overbank flow occurred during these floods. The results of analyses show that the lower reaches scour during these floods, and the sediment concentration at Lijin was 20 to 30 kg per cu.m. It is also known that the amount of water required to transport 0.1 billion tons of sediment is related to the average discharge. The value decreases with the increase of discharge. When the discharge is 4,000 to 5,000 cms, the minimum consumption is approximately 3 billion cu.m for transporting 0.1 billion tons of sediment.

SEDIMENTATION DURING FLOODS WITH HYPERCONCENTRATIONS OF SEDIMENT

Heavy sediment-laden floods with sediment concentrations greater than 400 kg per cu.m occur frequently in the lower reaches (see Tab.3). These floods greatly influence scour and deposition, and flood prevention measures. Their characteristics are as follows:

(A) These floods are formed by storms from the middle drainage basins and are characterized by high, sharp peaks with short durations. After joining the main stream, the floods usually have a peak discharge of 4,000 to 8,000 cms, or an average discharge of 2,000 to 3,000 cms, after the peak flow is reduced by the Sanmenxia Reservoir or detained by channel storage. The average coefficient of sediment load (which equals sediment concentration divided by flow discharge), is greater than 0.04 (with unit kg. sec. per m).

(B) These floods mainly originate from coarse sediment source areas. Usually, runoff from this area is 40% of the total runoff, while the sediment load is more than 90% of the total. Besides, the particles of sediment are quite coarse. In fact, the percentage coarser than 0.05 mm is more than 40% of the total.

(C) Heavy deposition in the lower reaches takes place during these floods. From 1950 to 1983, there are 11 heavy sediment laden floods with a total accumulated duration of only 104 days, contributing 2% of the total runoff and 14% of the total sediment load. However, 54% of the total deposition took place during these floods. Furthermore, the intensity of deposition was great, amounting to 18.8 to 61 million tons a day. The distance covered by deposition was relatively short. Fig. 11 shows the variation of maximum instantaneous sediment concentration along the river for each flood. It can be seen that the value is from 412 to 911 kg per cu.m at Sanmenxia.

In gorge sections from Sanmenxia to Xiaolangdi, sediment concentration remains constant because of the steep slope and the rapid flow. In alluvial reaches below Xiaolangdi, the maximum sediment concentration reduces rapidly, due to the deposition caused by the slowing of the flow. For instance, the concentration falls to 200 to 550 at Huyuankou and to 100 to 250 at Aishan. It changes only slightly in the reach from Aishan to Lijin. In summary, rapid adjustments along the river caused by floods with hyperconcentrations of sediment are noticeable, especially in the reaches above Gaocun.

(D) After the deposition of these floods, quite deep and narrow courses are often formed leading to a great reduction of cross sectional area for the flow. The rate of rise of stage is large and the stage itself is also high. In August 1933, for example, the maximum flood stage in the reaches below Huayuankou were all higher than those in July 1958, in some places by 2-3 meters, although the discharge for the former was less than the latter, and the river bed level of the former was lower also. The appearance of a high stage during the heavily sediment-laden floods may be related to the previous deposition in the channel, and still more, to the rate of rise of stage. Under the same conditions of previous

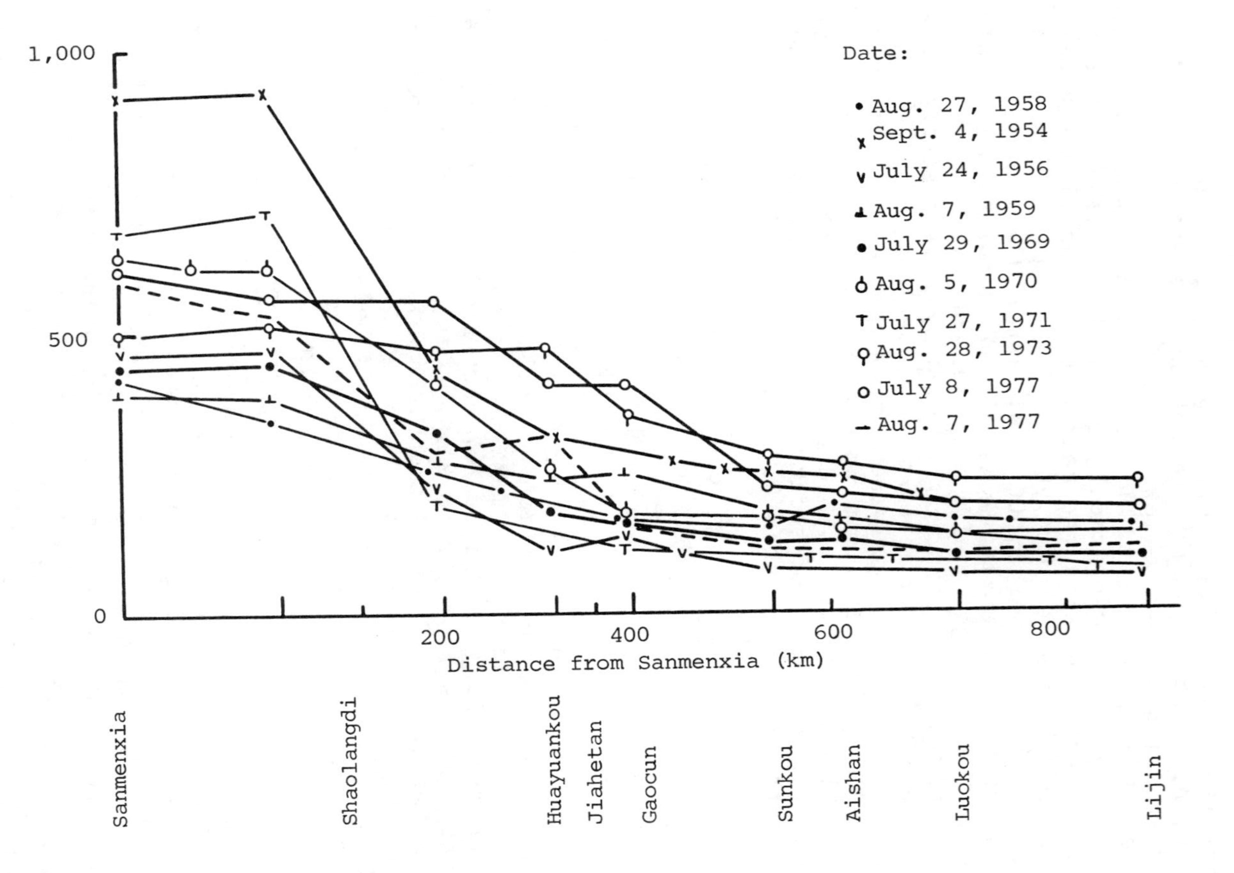

Figure 11. Variation of Maximum Instantaneous Sediment Concentration

deposition and comparable magnitude of floods for floods originating in tributaries between Hekouzhen and Longmen, which is in general hyperconcentrated with coarse particles, the flood stages will be high, because of the intensive deposition and the large rate of rise of stage. For floods originating below Sanmenxia, the stages will be low and quite often accompanied by scour of the main channel.

In Tab.3, Smax--maximum sediment concentration at Sanmenxia; Qmax, S, Q--maximum discharge, average sediment concentration and average at Huayuankou.

Since the cross section becomes deep and narrow, discharge per unit width may reach a suprisingly high value of up to 50 cms per m at some locations, usually in reaches above Huayuankou. This leads to drastic scouring called "river bed turning up." At times the stages rise and fall extremely rapidly. During the propogation of the flood wave, the peak discharge may even increase instead of decreasing due to channel storage effects. This endangers the flood prevention measures at high stages.

Based on the above analysis, a possible method of operating the reservoirs can be proposed. That is; the reservoirs store only heavy sediment-laden floods, which happen only infrequently, but are most hazardous to the lower reaches, while other floods which take place more frenquently and with only small effects on the lower reaches are allowed to be released from the reservoirs. Furthermore, floods which are somewhat favorable for the lower reaches should be conveyed through the reservoirs with minimum detention. This would be good for preserving the live storage capacity of the reservoirs and reducing deposition on the lower reaches as well.

CHARACTERISTICS OF EROSION AND DEPOSITION IN MAIN CHANNEL AND FLOODPLAIN

A planimetric view of the lower reaches looks something like a lotus root. That is to say that the wide sections alternate with narrow ones in a stretch of the river. The area of the floodplain in the lower Yellow River is 80% of the total area between the dikes. Floodplains serve the principal function of flood retardation and desilting. Because of the difference in resistance of the main channel and the floodplain, an exchange of flow as well as a transverse exchange of sediment may occur after the floodplain is inundated. Generally, deposition occurs on the floodplain while the stream flows from a narrow stretch into wide ones. The main channel in the narrow stretch will be scoured as a result of flowing of relatively clear water released from the floodplain. Deposition on the floodplain and erosion in the main channel may be observed along the river for a distance of several hundred kilometers. For example, 1,070 million tons of sediment were deposited on

Table 3 Deposition on lower reaches during floods with hyperconcentration of sediment

Data	Smax kg/ cu.m	Qmax cms	S/Q kg.s/ sq.mi	Runoff billion cu.m	Sediment $t \times 10^9$	Deposi- tion $t \times 10^9$	Intensity of deposi- tion mil.t/day
Aug.18-Aug.25 1953	716	6790	0.0452	1.98	0.35	0.231	28.80
Aug.26-Sept.2 1953	412	8410	0.0436	2.10	0.351	0.150	18.80
Sept.2-Sept.9 1954	590	12300	0.0179	4.75	0.838	0.488	61.00
Jul.23-Jul.29 1956	444	6500	0.0338	2.06	0.313	0.210	30.00
Aug.6-Aug.12 1959	397	7680	0.0439	2.70	0.531	0.265	33.20
Jul.25-Aug.5 1969	435	4500	0.0709	2.16	0.463	0.335	27.86
Aug.4-Aug.17 1970	620	4040	0.1140	2.63	0.830	0.554	39.70
Jul.25-Jul.30 1971	666	5040	0.0732	1.08	0.247	0.200	33.30
Aug.28-Sept.7 1973	477	5890	0.0588	3.17	0.740	0.302	26.26
Jul.4-Jul.15 1977	589	8100	0.0524	3.45	0.802	0.454	37.80
Aug.3-Aug.12 1977	911	10800	0.0586	3.09	0.887	0.582	50.20
Total				29.20	6.35	3.770	
Total amount in period July 1950 to June 1983				10482	46.20	6.99	
Amount produced by flood listed caused in the long period %				2.0	13.7	54	

the floodplains and 860 million tons of sediment were eroded from the main channels during the flood in July, 1958 (see Tab.4). But this is not the whole story. Deposition may occur in the main channel if the sediment load is excessive. The coefficient of sediment load is usually used to determine qualitatively whether there is scour or deposition on the floodplain and the main channel. If the coefficient is greater than 0.015, both receive deposition. If it is smaller than 0.015, the floodplain will be in deposition while the main channel will be scoured. The phenomenon that takes place during a large overbank flood flow is helpful to the stabilization of the main channel, as it is accompanied with high deposition on the floodplains and a deeply scoured main channel. The river may, therefore, benefit after the passage of an extreme flood, although the task of safe guarding the dike system during flooding is not an easy job. It has been widely spread among people living along the river that "A good river course is made by an extreme flood."

A dialectical relationship among floodplains, the main channel and dikes has been summarized based on the long term experience as follows: "Deposition on the floodplain may be helpful to the erosion in main channel, which will be stabilized with the existence of a high floodplain. Floodplains may be kept on site in case of a stabilized channel and it is also beneficial in protecting the dike with a high floodplain serving as a frontier." Also, it has been said as an experience that "a great capacity of flow and sediment discharge, a lower stage during floods and an infrequent shifting of the main current all result from the existence of a main channel. In flood fighting, it would be better to protect the floodplain than the dike, while the key to flood prevention on the floodplain is to stabilize the main channel." In reaches of the Yellow River within Henan province, the river is of a wandering nature. The floodplain and the main channel can only exist for a relatively a short period. Training of the river course for flood prevention purposes is mainly to keep relatively stable the relationship between the floodplain and main channel.

Assuming that sediment of the Yellow River has not been controlled and deposition in lower reaches is unavoidable, overbank flow during relatively large floods should be allowed provided that it will not endanger the flood prevention works. At the same time, deposition in the main channel during low flows and nonflood seasons as well as in floods with peak flow less than bankfull discharge should be minimized. When a new reservoir is to be built, enough outlet discharge should be designed so that retardation of flood peak discharge can be avoided.

PARTICLE DIAMETER OF THE DEPOSITS

Table 4 Amount of erosion and deposition during overbank floods, in 100 million tons

Date	Huayuankou		Huayuankou-Lijin		
	Q cms	Q/S kg.s/sq.mi	Main channel	Flood plain	Whole section
July 6-Aug.1 1953	10700	0.0112	-3.00	3.03	0.03
Aug.15-Sept.1 1953	11700	0.0376	1.49	1.03	2.52
Aug.2-Aug.25 1954	15000	0.0097	-2.08	4.90	-2.82
Aug.28-Sept.9 1954	12300	0.0170			
July 12-Aug.4 1957	13000	0.0119	-4.33	5.27	0.94
July 13-July 23 1958	22300	0.0095	-8.60	10.69	2.09
Total			-16.50	24.9	8.40

According to analyses of samples of the deposits, the particles are coarser in the upper reaches than in lower reaches, coarser in deeper layers than in surface layers, and coarser in the main channel than on the floodplain. The percentage of particles greater than 0.05 mm in diameter in deposits is 80% in surface layers, and 90% or more in deep layers. If the diameter of 0.025 mm is considered as the demarcation for particles of bed material load and wash load, the bed material load was 50% of the total sediment load from 1950 to 1960 before Sanmenxia Reservoir was built. From 1974 on, the reservoir has been operated in a mode of storing water in the nonflood season and operating at a low head during the flood season. However, 65% to 85% of the total deposition was in the size range of bed material load during the same periods (see Tab. 5). Coarse sediment greater than 0.05 mm occupied 42% to 49% of the total deposits, although it was only 20% of total sediment in transport. Furthermore, almost all the sediment with diameters greater then 0.1 mm was deposited in the lower reaches. Coarse sediment is the main composition of deposits in the main channel. By comparision, wash load sediment with a particle diameter smaller than 0.025 mm never deposits in the main channel. Only part of these are deposited on the floodplain during overbank floods, leading to 50% of the total deposition on the floodplain.

For effectively reducing deposition in lower Yellow River, the coarse sediment should be controlled in the middle reaches of the Yellow River. Hence, reservoirs built in this area should be used mainly to detain coarse sediment with particle diameters greater than 0.05 mm, and fine sediment and medium sized sediment (0.025-0.05) should not be detained in order to minimize the amount of deposition. In the nonflood season, Sanmenxia Reservoir is operated to store water and retain all of the coarse sediment. Deposition in the reservoir approximately equals the reduction of deposition in the entire lower reaches.

WATER REQUIRED FOR SEDIMENT TRANSPORTATION

The amount of water required to transport 100 million tons of sediment out of Lijin in the lower reaches is related to the oncoming sediment load. Figs. 12 and 13 indicate that if runoff and sediment load equal the long term average, about 3 billion cu.m of water are required to transport 100 million tons of sediment in the flood season, and 9 billion cubic meters in the nonflood season. That is to say that more water is needed for sediment tranport in the nonflood season than in the flood season. It also indicates that the amount of water required for the transportation is relative to the sediment load as well as the allowable amount of deposition in the river. That means that different amounts of flow are needed to

Table 5 Deposition for different particle diameter of sediment in different periods

Period	Sediment size (mm)	Average July 1950–June 1960 Runoff 10^8 m^3	Average July 1950–June 1960 Sediment 10^8 t	Average July 1950–June 1960 Deposition 10^8 t	Average Nov.1964–Oct.1973 Runoff 10^8 m^3	Average Nov.1964–Oct.1973 Sediment 10^8 t	Average Nov.1964–Oct.1973 Deposition 10^8 t	Average Nov.1973–Oct.1982 Runoff 10^8 m^3	Average Nov.1973–Oct.1982 Sediment 10^8 t	Average Nov.1973–Oct.1982 Deposition 10^8 t
Flood season	<0.025		8.76	0.865		6.48	0.64		5.77	0.98
	0.025–0.05	295.6	3.83	1.068	226	3.23	0.91	239	3.16	0.78
	>0.05		2.75	1.217		3.11	1.72		2.62	1.08
	All sizes		15.34	3.15		12.82	3.27		11.55	2.84
Non-flood season	<0.025		0.94	-0.166		0.97	-0.25		0.15	-0.33
	0.025–0.05	184.0	0.77	0.206	200	0.94	0.29	163	0.06	-0.36
	>0.05		0.90	0.68		1.57	1.08		0.07	-0.31
	All sizes		2.61	0.72		3.48	1.16		0.28	-1.00
Whole year	<0.025		9.7	0.69		7.45	0.43		5.92	0.65
	0.025–0.05	479.6	4.61	1.278	476	4.17	1.20	402	3.22	0.42
	>0.05		3.64	1.90		4.68	2.80		2.69	0.77
	All sizes		17.95	3.87		16.3	4.43		11.83	1.84

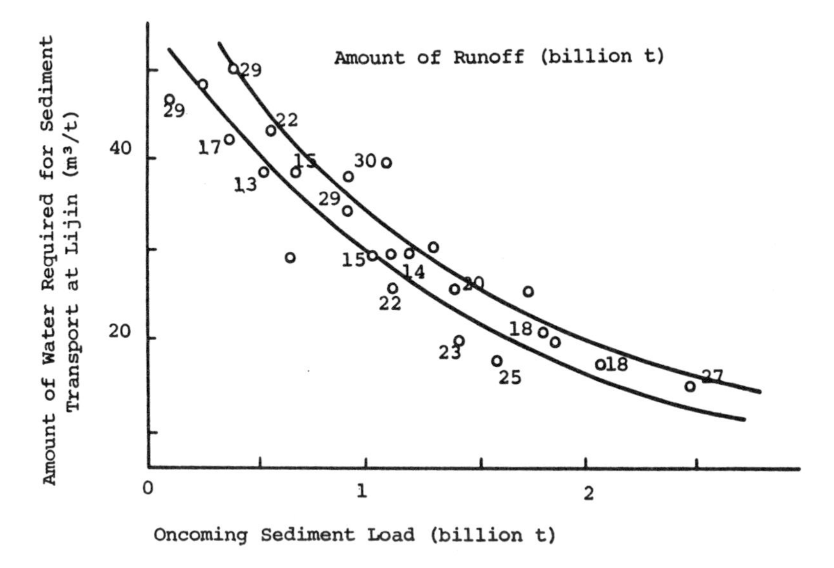

Figure 12. Amount of Water Required for Sediment Transport vs. Oncoming Sediment Load in Flood Season.

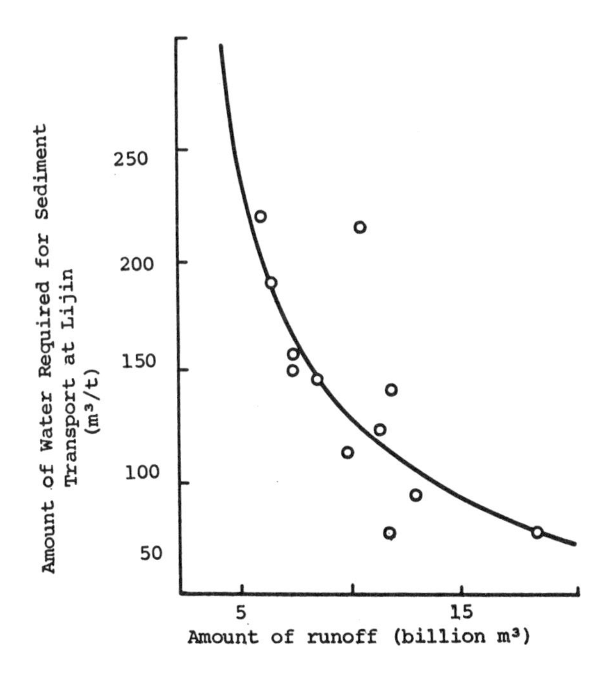

Figure 13. Amount of Water Required for Sediment Transport vs. Oncoming Runoff in Nonflood Season.

cope with variable tolerable limits under specific oncoming sediment loads. To keep the deposition in the lower reaches at current rates, the amount of runoff at Lijin should not be less than 20 billion cu.m.

LAW OF SEDIMENTATION IN REACHES BELOW AISHAN

The lower Yellow River travels a long distance, and its channel is wide in the upper reaches and narrow in the lower. The extent of rise of the river bed in narrow reaches below Aishan, therefore, would be much greater than that in reaches above it for the same amount of deposition. Considering flood prevention, deposition in narrow reaches is very important. The status of deposition in the reaches below Aishan is not only decided by the runoff and sediment conditions from the watershed but also by the adjustment of upstream alluvial reaches. Runoff and sediment load, which have been already adjusted in the upper wide reach, are usually favorable for conveyance in the reach below. Therefore, the rate of deposition in the reach is smaller than that in the upper reach. According to the measured data, this reach is scoured during a flood and has deposition during low flow. The deposition occurring in low flow periods in the flood season or in the nonflood season is most harmful. When sediment-laden flow is released from the upper reach, the reach will be scoured if the discharge is greater than 4,000 cms. The river can be scoured to a great extent during large floods when overbank flow takes place, because a large amount of sediment carried by the flood has already been deposited in the upper reaches resulting in low sediment concentration in the reach. The river bed scoured during one such large flood usually persists for years before it is again silted up. For instance, during the flood in July, 1958, the sediment concentration was 126 kg per cu.m when the flood peaked at Jiahetan. Nevertheless, it was only 20 at Aishan. The amount of deposition on the floodplains above Aishan was 900 million tons, while the amount of scour in the reach from Aishan to Lijin was 150 million tons. Thus, large floods may play an important role in repressing deposition in narrow reaches below Aishan. When clear water is released from the upper reach, the channel scours. The distance of scour varies in a very complex way. Among the major factors are the discharge and the duration of flood. Fig. 14 shows the relationship between discharge and distance of scour. For small discharges, the sediment load supplemented by erosion along the river before entering into the reach may exceed the capacity of sediment transport of the reach and deposition may take place. For the whole reach, erosion takes place only when the discharge is greater than 2,500 cms and the volume of flow is greater than 3 billion cms. For narrow reaches below Aishan, maximum deposition takes place at discharges of 1,000 to 2,000

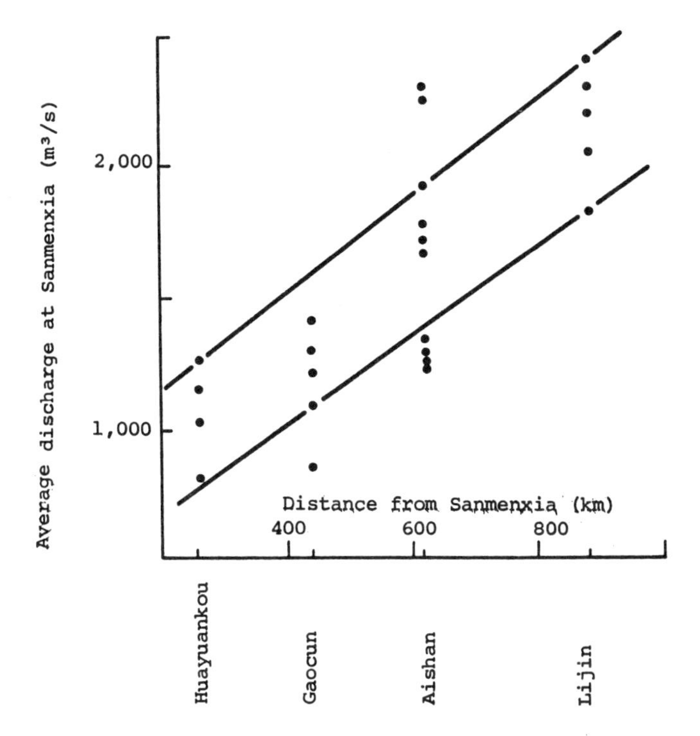

Figure 14. Average Discharge vs. the Distance of Deposition

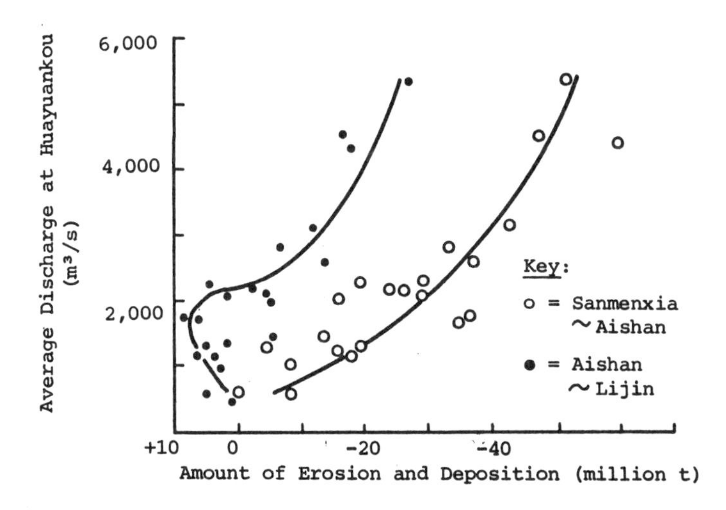

Figure 15. Average Discharge vs. the Erosion and Deposition When Clear Water is Released from Sanmenxia

cms. Deposition also takes place when the discharge is smaller than 1,000 cms, however the absolute quantity of deposition is small, as shown on Fig.15.

It should be pointed out that deposition, the extension of the river course and avulsion occur in the estuary area and change the relative base level of the river. This in turn affects erosion and deposition in the lower reaches. In the 1950s, the main channel had deposition in the reach from Aishan to Daoxu, and scoured below Daoxu. From 1960 to 1964, the main channel above Mawan scoured and had deposition below. All these phenomena were affected by the change in estuary and the stage at Lijin rose about 1 m in the 1950s.

According to the characteristics of the river below Aishan, the transport capacity during large floods may be used as an advantage to minimize the aggradation. By proper operation of the reserviors, the outflow may be regulated to increase the change of occurrence for large flows and to minimize the low flow releases.

Effects of Comprehensive Watershed Management Throughout the River-Basin on Sedimentation in Lower Reaches

A river is a component of a river-basin, naturally, characteristics of the river are determined by features of the river-basin. Erosion and deposition in the river depend on conditions imposed on the river by the basin. The conditions are the following: 1) Runoff conditions---Runoff quantity and regime. 2) Sediment conditions--- Sediment load, its regime and the size distribution of the sediment. 3) Boundary conditions --- The geometric shape of the river course, the relative mobility of the substances composing the river bed and the bank, and manmade structures on the river. Moreover, the estuary conditions affect the river by controlling its base level.

A comprehensive watershed management plan throughout the basin of the Yellow River has been carried on for more than 30 years since liberation. It includes a great number of projects for water and soil conservation and utilization of water resources. There are now 7 projects and hydropower stations on the course of the river and another is under construction. Many large or medium sized reservoirs have been built on tributaries with a total storage capacity of more than 30 billion cu.m. More than 60 million mu of land have been irrigated and adequate water supply provided for cities and industries along the river. In the lower reaches, dikes have been raised and strenthened, and bank protection works have been built or rebuilt at vulnerable spots. In the estuary area, some dikes have been constructed. As a result, the apex point of the delta moves downstream, the coast line becomes shorter and the area of the delta becomes smaller. Meanwhile, the population within the river-basin has doubled over that in the early fifties. As a consquence, the Yellow River is now intensively affected by human activities including comprehensive management measures. Various engineering measures modify the water and sediment flow regime affects the fluvial processes of the lower Yellow River.

BASIC CHARACTERISTICS OF THE CHANGE OF RUNOFF AND SEDIMENT LOAD

Decreases in the Yearly Amount and Distribution of Runoff. With the increase of water usage by irrigation, industries and cities, the amount of runoff in the Yellow River has decreased a great deal. For example, the amount of measured runoff is only 70% of the total natural runoff at Hekouzhen located at the exit of upper reaches. At Lijin, which is the controlling station at the exit of the lower reaches, the measured water is 50% of the total natural runoff. In a dry year, the utilization of water often breaks off the flow in May, June, or early July. Runoff has been regulated by reserviors on the main course and its tributaries, among which regulation by the Lujiaxia reservior is most prominent. Since commencement of operation of the reservoir at the end of 1968, the runoff during flood seasons has been decreased and that in nonflood seasons has been increased. Therefore, the allotment of runoff within a year has been changed. Moreover, ice-flood prevention in lower reaches has been made more difficult since as much as 1,000 million cu.m of water in excess of the normal flow are released from the reservior from November through March.

Peak Discharges of Flood Have Been Decreased and the Regimes of Runoff and Sediment Have Been Changed. Liujiaxia Reservoir and Sanmenxia Reservoir play a major role in the regulation of floods. Outflow processes from the Liujiaxia Reservior after adjustments for the long river reach have become relatively steady which may be considered as the base flow of the lower reaches. According to analysis of 35 floods, the storage of the former corresponds to 40-100% of the runoff measured at Longmen for 9 floods. In the other five floods, the percentage sometimes exceeds 100%. The reduction of flow due to storage in Liujiaxia Reservior would bring about an increase of sediment concentration at Longmen. Before reconstruction of Sanmenxia reservior, peak discharges had been reduced significantly, sometimes by a ratio of 60%. Because of the detention effect, large amounts of sediment are to be sluiced off the reservior, greatly modifying the natural regime. This, in turn, leads to incompatability of water and sediment flow. The situation has been greatly improved after outlet facilities of the reservior were reconstructed.

The Sediment Load Has Been Reduced a Small Amount, But Its Regime Has Been Changed. Heavily Sediment-laden Floods Occur More Frequently. Based on analysis, it can be roughly affirmed that storage of sediment by all types of engineering measures on the main course and tributaries above Sanmenxia since 1970 was 500 million tons as an annual average. However, it does not mean that the same amount of sediment has been reduced that enters the lower reaches. Preliminary analysis, in fact, indicates that the annual average sediment load entering Sanmenxia Reservoir has been reduced by only 150 to 300 million tons. Some 5 billion cu.m of sediment have been deposited in the reservoir since it was put into use. Change of the allotment is mainly a result of a different operational mode of Sanmenxia Reservior. Since it began operating in a mode of impounding clear water during the nonflood season and sluicing sediment-laden flows, sediment brought down from the watershed in an entire year is disposed of in the flood season.

CHANGE OF BOUNDARY CONDITIONS OF THE RIVER BED

Since liberation, a great number of river training works have been built. This leads to changes in some reaches in their patterns, as mobilities of river banks have been changed. The reach below Taochengpu has become a meandering course, as banks have been put under strict control, and the main current cannot flow freely. The reach from Gaocun to Taochengpu, which was originally variable in nature, has evolved to a relatively stable course after the implementation of training works (see Fig.16). In the reach from Mengjin to Gaocun, the extent of silting of the main channel has decreased. On the whole, the river training works control the main current and stabilize the river pattern. It also has some impact on the transverse distribution of sediment.

CHANGE OF THE ESTUARY CONDITIONS

The change of river course in the estuary area of Yellow River follows the law of depositing---extending---course shifting, as the river frequently changes its course to the sea. This, of course, will affect fluvial processes in the reach of the estuary area and on the reaches just above. Since 1855, avulsions have occurred as many as 10 times, among which there were three major shifts of course in the past 40 years (see Fig.17). According to investigation and analysis of relevant materials, the rate of coastline extension was 0.18 km per yr on average from 1954 to 1982. Effects of the extension of the river course on the stage just above the estuary area were also analysed with the result that 15 km of extension may lead to a 1 m in stage at Luojiawuzhi.

Many reservoirs have been built on the upper and middle

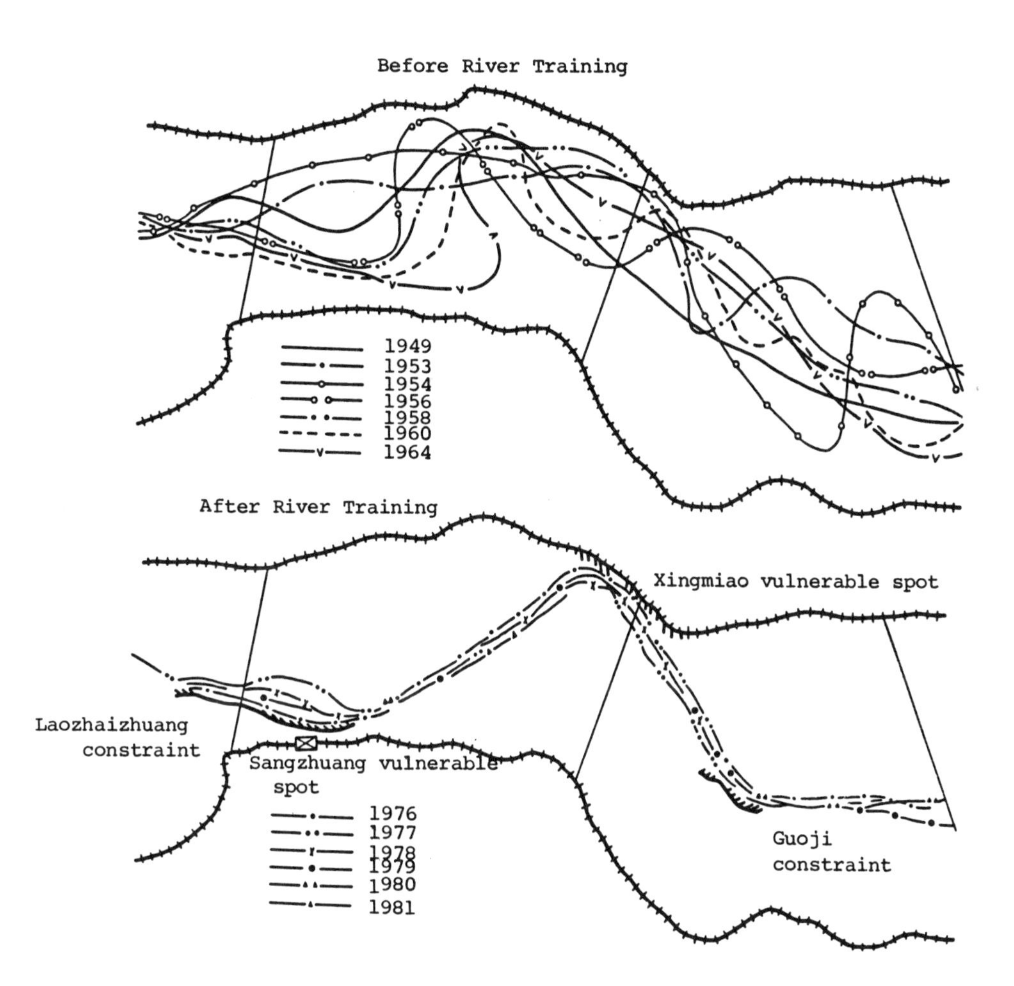

Figure 16. Extent of Shifting of the Main Channel

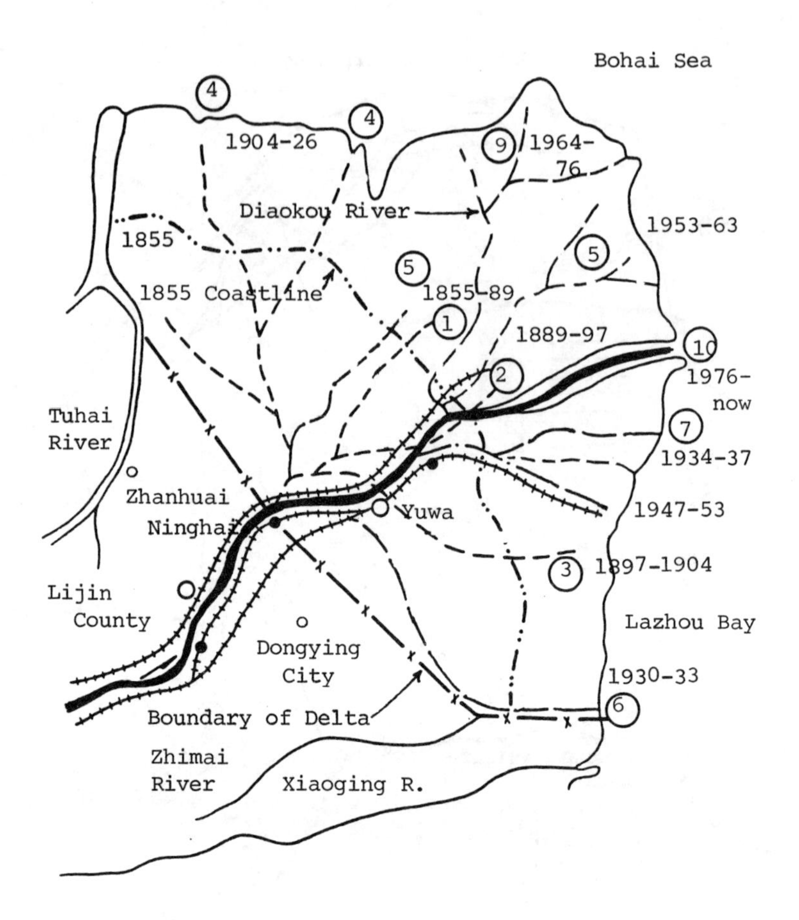

Figure 17. Change of Yellow River Estuary in Recent History

Legend: ◯ Order
o City
● County
-+++++ Dike
----- River course

reaches of the Yellow River as well as on tributaries. A large reservoir, Longyangxia, now under construction with ability of overyear storage, has begun to impound flow. Thus, the reservoirs will further regulate the runoff. As a result, the runoff allotment within a year will tend to be more even. Water reservoirs of the Yellow River will be further utilized, because of water shortage in the river-basin and in the North China Plain. New key projects for multi-purpose development will be set up in the future. Therefore, runoff and sediment load will be further regulated and modified. Water and soil conservation will continue to be carried out. All these will cause a further change of the features of the river-basin. Therefore, it is important and urgent to research these works to evaluate the impacts brought about by these changes in a comprehensive and thorough way.

References

Mai Qiaowei, Zhao Yean, Pan Xiandi and Fan Zuoying, Characteristics of Runoff and Sediment Load and Law of Channel Accretion on the Lower Yellow River, Inst. of Hydraulic Res., YRCC, 1978

Mai Qiaowei, Zhao Yean, and Pan Xiandi, Sediment Problems of the Lower Yellow River, Proceeding of the International Symposium on River Sedimentation, Guanghua Press, 1980

Li Baoru, Hua Zhengben, Fan Zuoying and Chen Shangqun, Changes of the River Channel Downstream of Sanmenxia Reservoir During Silt Detention Period, Proceeding of the International Symposium on River Sedimentation, Guanghua Press, 1980

Circumstance of the Channel of the Lower Yellow River, Design Inst. of Hydroelectric Engineering, YRCC and Inst. of Hydraul. Res., YRCC, 1978

Zhao Yean and Pan Xiandi, Effects of Mankind Activities on the Change of the Circumstance of the Yellow River and the Fluvial Process of the Lower Yellow River, Inst. of Hydraul. Res., YRCC, 1980

Yellow River Conservancy Commission, Channel Accretion on the Lower Yellow River and its Tendency in the Period 1984 Through 1995, Yellow River, No. 3-No.4, 1985

Zhao Yean, Pan Xiandi, Fan Zuoying and Han Shaofa, Fluvial Processes after the Construction of Sanmenxia Reservoir, Inst. of Hydraulic Res., YRCC, 1985

Pang Jiazhen and Si Shuheng, Fluvial Processes of the Yellow

River Estuary, Proceeding of the International Symposium on River Sedimentation, Guanghua Press, 1980

HARNESSING AND DEVELOPMENT OF THE LOESS PLATEAU IN THE MIDDLE REACHES OF THE YELLOW RIVER

Hua Shaozu and Mou Jinze
Division of Soil and Water Conservation
Yellow River Conservancy Commission
People's Republic of China

The Natural Geographical Environment of the Loess Plateau

The Loess Plateau within the Yellow River basin is bounded on the north by Yinshen Mountain and on the south by Qinling Mountain; it stretches to the west of Riyue Mountain and east of Taihang Mountain. Land overlain by loess covers an area of 598,000 square km. populated by nearly 72.7 million people. The administrative divisions within this basin are seven provinces, 46 prefectures and 306 counties. The area subject to severe soil and water losses is 430,000 sq.km, and modulus of the mean annual sediment load reaches 3,700 t/sq.km.

As the Yellow River passes through the Western Region Plates and the North China Plates, the Loess Plateau in the middle portion of the basin is located in the border area of the east and west regional tectonic stress field, marked in general by the deep fractures to the west of Liupan Mountain. Differences in geological tectonic activities during the recent period may be clearly distinguished. In the west part (the upper portion of the Yellow River) the earth surface rises intensively and there exist many canyons; but in the east regions (middle and lower portion of the basin) the earth surface is subsiding or ascending and the alluvial plains or canyons are also alternatively distributed. During the later period of the Himalayas movement, the earth surface of this region had been raised slowly, except that the basins of Yinchuan, Hetao, Taiyuan, Fenhe, Weihe and Luoyang were still in successive subsidence. In early ages, there were stream systems formed by some interior lakes and basins. The development of the Yellow River was greatly influenced by the diminution and evolution of the stream system.

The middle reaches of the Yellow River are the center of loess accumulation in China, where loess with great thickness and complete stratigraphy is widely distributed. The sequence for the loess strata are as follows: (1) early pleistocene

L. M. Brush et al. (eds.), Taming the Yellow River: Silt and Floods, 517–528.

loess (Q1), being named the Wucheng Loess or the paleoloess, also, it is distributed on the "Yuan" (a high table-like plain surface with abruptly descending edges) and part of "Taiyuan" (a terrace overlain by loess) with a thickness of 10-40m. From paleo-earth magnetic measurement, the age was determined approximately as 2.43 million years; (2) Middle pleistocene loess (Q2), being named Lishi Loess or the old loess also, includes multi-layers of 4-8 fossil soil intercalated in the loess; the total thickness reaches 26-28 m and the age is 1.2-0.1 million years; (3) Late pleistocene Loess (Q3), named the Nalan Loess, or the new loess, is scattered throughout the whole region, the thickness reaches 10-30 m in general, intercalated with 1-2 layers of fossil soil, having an age about 0.1 million years; (4) Holocene Loess, named recent loess, with a thickness of one to several meters, dating back about 8,550 or so years.

Since the Pleistocene, the Qinghai-Xizang Plateau has been elevated, thus the warm wet air mass which originates from the Indian Ocean and the Arabian Sea and moves successively forward to the northeast after "landing" at the Bay of Bengal is obstructed by mountains. Meanwhile mountain ranges of the Qinling and Taihang raised intensely, causing the west Pacific Ocean subtropical, high-pressure to subside when it moves to the west and the continental cold air mass constituted mainly by the Mongolia high pressure has gradually controlled the climate of this region. As a consequence, the climate has turned cooler and drier since the Quaternary period. Finally, it has became a region of semi-humid, semi-arid and arid climate, in which the arid and semi-arid areas occupy approximately 80% of the total area.

Precipitation in this region is unevenly distributed and fairly low. In most parts of the region it is only 400-500 mm, decreasing from the southeast toward the northwest. Secondly the precipitation is concentrated in June through September and amounts to 60-70% of the annual total and differs by 3-4 times from one year to the next. Rainstorms with high intensity and duration often take place in the area between Hekouzhen and Longmen, such as the heavy rainstorm on August 1, 1977, which was centered at Wushen county, Nei Mongol Autonomous Region, and amounted to 1,400 mm within 10 hours. Another example was the heavy rain in July 17, 1978 at Majiawaze village, Jinbian county, Shaanxi province, which amounted to 476 mm in 4 hours.

In most parts of the Loess Plateau, solar radiation is very strong, the time of sunshine is long, atmospheric temperature varies greatly between night and day and the effectiveness of the accumulated temperature is high. The mean annual solar radiation amounts to 110-150 kilo calorie per square centimeter; the mean annual sunshine time is about 2,000-3,000 hours; the annual atmospheric temperature varies in the range of 5-14 degrees centigrade.

Most vegetative cover in the middle portion of the Yellow River is secondary in nature. In accord with the climate belt there exists apparently a regular pattern of spatial distribution of plants. Along the direction from the southeast to the northwest its sequences are: forest, forest grasslands, typical grasslands, desolate grasslands, extending to the edge of desert.

On basis of the characteristics of climate, soil, geology, topography, vegetation cover and soil and water losses, in the comprehensive planning conducted in 1955, the Loess Plateau in upper and middle reaches of the Yellow River was classified into nine districts as follows: gullied hilly loess region which may be subdivided into five sub-districts: gullied Loess Plateau, loess terrace, dry grasslands area, wind drifted sandy area, rocky mountains, high grassland, alluvial plain and forest area as shown in Fig.1.

Conditions of Soil and Water Loss and Characteristics of Sediment Yield in the Loess Plateau

The dynamic forces causing erosion in the Loess Plateau may be classified into three categories depending on the macroscopic land form. In the wind drifted sandy area and arid grasslands, wind erosion is predominant. In the gullied hilly loess region, water and gravity erosion are the major forces. In the rocky mountains the erosive patterns are denudation by weathering and water erosion. Studies of the laws of erosion as well as sediment delivery and deposition on sloping ground, first and second order gullies and tributaries have become the foundation work for the development of Loess Plateau. Observations of runoff and sediment yield from experimental plots and small experimental watersheds are carried out by many institutes or experiment stations for soil and water conservation and are distributed all over the middle portion of drainage basin. Observations of flood and sediment discharge are carried out at hundreds of hydrometric stations that are dispersed in the stream net work. Also, soil erosion has been analyzed by means of space or aerial remote-sensing techniques. The status of soil erosion can be monitored and studied by the aforementioned networks.

From a macroscopic point of view, the erosion intensity has the regularity of regional distribution as in the east side of the Liupan Mountain, the erosive intensity decreases from north to south; while in the west side it decreases from east to west. Within a small watershed the erosive intensity varies also in a vertical direction. For example, data observed by the Zizhou Runoff experiment station in Chaba gully show that on the hill top, near the divide, erosion is slight with an erosion modulus of only 274 t/sq.km. However, the middle and lower parts of the slope are in an intensive

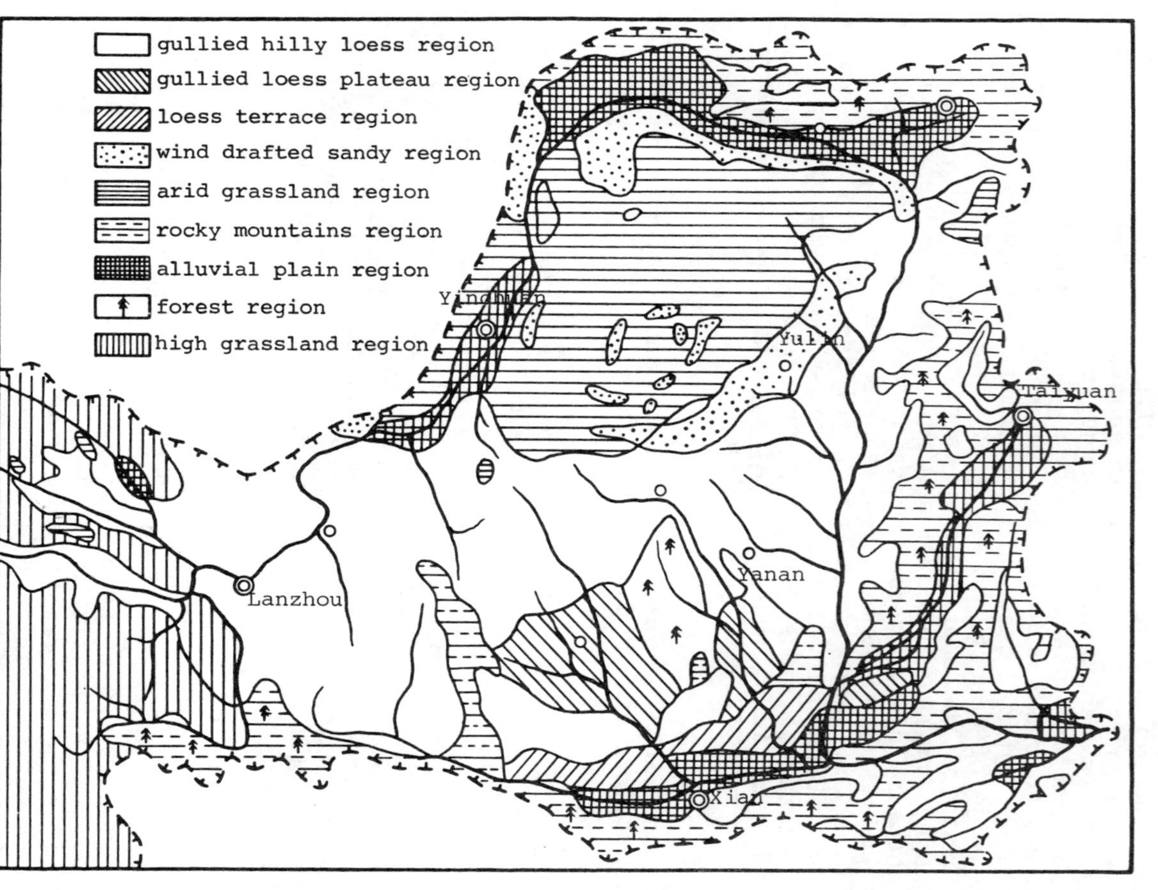

Figure 1. The Types of Soil Erosion in the Middle Reaches of the Yellow River.

erosive belt, where the modulus of erosion reaches 13,800 t/sq.km. Finally, the ravine or gully bank below the gully rim line, due to the mass movement of collapse and landslide occurring in the overhanging cliff, is eroded very severely. The erosion modulus reaches as high as 21,100 t/sq.km or even more. The basis of argument is mainly then the relevant deposits from erosion in ancient times. The maximum thickness of strata in the Weihe River basin during the Tertiary period was 6,000 m, the maximum thickness for the Quaternary period is 1,300m, the thickness for the Holocene strata is 50m, all of these amount to 7,350 m. The thickness of Quaternary strata in Hetao is 700m and 300-400m in the North China Plain.

The typical topography which reflects the erosion history is the stratified topography demonstrated by the first order ridges in the gullied hilly loess region where a complete stratigraphic record exists containing the Wucheng, Lishi Loess and Malan Loess representing the Early, Middle and Later Pleistocene, respectively. Among the loess of each era are intercalated layers of fossil soil that reflect the prevalence of paleoclimate belonging to a sub-tropic environment in that area, according to analysis of mammalian fossils and plant spore pollen. In the early stages of the late Pleistocene the climate appeared relatively warm and wet, but in the later period it turned dry and cold, a condition of desolate grasslands. In the early stages of the Holocene, the post glacial stage, the climate became warm but fluctuated many times causing many alternations of erosion and deposition. It may be concluded that erosion in the Loess Plateau is certainly a geological process, not just a recent event.

The Comprehensive Harnessing of Soil and Water Conservation on the Loess Plateau

Since the founding of the People's Republic of China, the Chinese goverment has attached great importance to Loess Plateau management. In the 1950s, experts from Academia Sinica and other relevant agencies were organized twice to explore over 370,000 sq.km area in the mid Yellow River for natural resources and soil and water conservation. In July of 1955, the second conference of the first National People's Congress issued the "Decision on Integrated Planning of Yellow River to Obviate Inundation and to Develop Water Conservancy." Soil conservation was listed among the chief components of Yellow River management.

Since the 1950s, the Yellow River Conservancy Commission has established the Northwest Project Administrative Bureau of the Yellow River and seven soil and water conservation experimental or extension stations at Tianshui, Shuide, Xifeng, etc. to put into effect the guiding principles of soil and water conservation and to assist relevant provinces or autonomous regions. In 1963, the State Council approved the establishment of the Soil and Water Conservation Commission of the Mid Yellow River to manage this area. Since then, the methodology used to protect and harness the soil and water losses has turned from a pilot demonstration to widespread application within this district.

As the reformation in the countryside vigorously forged ahead during the 1980s, soil conservation was renewed and further encouraged. In June of 1982, Huangpuchuan River, Sanchuanhe River, Wuding River, the three tributaries confluence to the Yellow River, and also Dinxi county which is located in the semi-arid area of central Gansu, were listed among the national conservation pilot projects aided with special funds.

The Soil and Water Conservation Commission of the Mid Yellow River and its administrative body--the Mid Yellow River Soil Conservation and Management Bureau and the Yellow River Conservancy Commission--have been strengthened and intensified. Basin planning on soil and water conservation is now underway and will eventually open up new prospects of conservation works on the Loess Plateau.

The Loess Plateau in the middle portion of the basin joining east to west of interior China occupies a strategic position. Development of the Loess Plateau has to be gradually speeded up to meet the urgent demand of vitalizing the national economy.

For thirty odd years, 130 large and medium reservoirs, more than 30,000 dams for small reservoirs or for land forming by trapping silt have been constructed; 43 million mu of terraces, 3.69 million mu of new arable land were formed through siltation in gullies and also 10 million mu of land were irrigated on a small scale or by warping; 68.7 million mu of trees, 18.1 million mu of grass have been planted as well as 18 million mu of hillside area closed to facilitate aforestation. Over 2,000 small pilot watersheds are harnessed and managed under the sponsorship of the Yellow River Conservancy Commission or provincial goverments.

Besides these important aspects of the soil conservation, other works for harnessing and developing the Loess Plateau consist of large-scale resource surveys on erosion and resource information mapping through different methods of remote sensing, data banks for territory resources and information systems, etc. The interflow and intersection of multiple disciplines, coordination and combination of relevant agencies in management, research and education will give strong technical support to Loess Plateau development.

Conservation Measures on the Loess Plateau

Soil erosion is a major obstacle to a thriving economy on

the Loess Plateau, so that major countermeasures undertaken since the founding of New China are to implement conservation programs consistently and extensively, relying on the active participation of the masses. Works have focused on three aspects: experiment, research and extension of three major conservation measures; comprehensive improvement and management of small watersheds and pilot tributaries.

WATER AND SOIL CONSERVATION MEASURES

All the control works fit generally into three types: cultivation practices, vegetation and engineering measures.

Cultivation Practices. These include deep plowing, contour tillage, furrow cultivation strip planting, crop rotation, and so on. The purpose of treatments is to ensure high productivity as a consequence of conserving water, soil and fertility by means of changing the microrelief, increasing roughness of the ground surface and water infiltration. Experiments indicate that erosion could be reduced by 83.6-100% if comprehensive treatment of deep plowing furrows and increasing application of organic fertilizer were to be implemented. Consequently production could be raised by 33.0-53.8% generally. In Yenan Prefecture, Shaanxi Province, conservation cultivation practices were extended to an area of 2.64 million mu, increasing the grain yield by 65 million kg, constituting 12% of the total production.

Vegetation Measures. There are lots of species suitable for conservation work on the Loess Plateau, such as Sweet Clover, Alfalfa, Astragulus, Caragana korshinskii, Common Seabuchthron, False Indigo, Hedysarum Scoprium as well Locust and Poplar.

Sweet clover, a wild plant in Northwest China tamed by Tianshui conservation experiment station in early times, is honored as "treasure grass" by local inhabitants. Abundance of root nodules renders it resistant to critical conditions and is functional in soil improvement, so that crop production may be promoted by an increase of 73-300%. It was introduced throughout the country during the 1950s, and the planted area reached as high as 10 million mu. Stands of Caragana korshinskii affording animal feed in Shaanxi, Shanxi and Nei Mongol Provinces or Autonomous Region, have exceeded several million mu in recent years.

Aerial seeding experiments were undertaken on a large scale in sparsely populated areas. At the western edge of Naowesu desert in Yulin, Shaanxi, for example, 26,000 mu of sand dunes were air sown with Hedysarum scoparium and Hedysarum mongolicum in 1974. The seedling survival rate ranged from 35.7 to 45.3%. During 1980-1982, 168,000 mu of Caragana korshinskii and Astragalus adsurgens were planted by airplane in Kelan and Pianguan counties in Shanxi, and more than half of the sowing area succeeded. A Y-5 plane may spread 1,700 mu hourly, and the cost is only 3 yuan, 10-20% cheaper than manual sodding.

Upland orchard is a popular measure because local inhabitants may gain high benefits. The Xifeng conservation experiment station demonstrated and extended the planting method during the 1950s, and now 4,700 orchards with a total area of 105,000 mu have been set up in Qingyang prefecture of Gansualone. Annual output of apples exceeds 20 million kg.

Engineering Measures. All the types of terraces are popular: level terraces, narrow or broad base terraces, conservation bench terraces on sloping cropland. Other engineering treatments include contour ditchs, banquettes or gradins on the barren hills, water storages along roads or in villages; check dams and reservoirs in ravines, warping projects and the withdrawal of sediment-laden water during floods, etc.

Some techinques for rapid construction such as sluicing earth materials from the banks and building dams by hydraulic fill, and terracing by machine had been conducted in the late 1960s. The efficiency of dam construction by sluicing siltation is 5-10 times faster than by rolled earth dam, and the cost reduced by 60% or more. During the last decade, in Shaanxi, Shanxi and Nei Mongol provinces (region), 9,700 dams have been built by sluicing siltation. Dam construction has been greatly encouraged. Machines employed in terrace construction bring higher efficiency. An E-75 tractor drawing a plow can be used for cutting terraces with 20-30m wide bench on less than 5 degree slopes. The daily work rate averages 5-8 mu. A land leveller turning a slope of 6-12 degrees into terraces 10 m wide benches may produce 4-6 mu every work day. Efficiency of a bulldozer is one mu of 5-8m wide bench terrace on relatively steep slopes of more than 20 degrees. One machine replaces the work of 100-200 laborers. During the last two decades more than 200,000 mu terraces have been completed by machines in Shanxi province.

Making use of flood water and sediment is an ecomomic and effective method on the middle reaches of the Yellow River and releases less flood and sediment to lower reaches. Varieties of forms of utilizing floodwater as well as sediment are introduced as follows:

(1) Collecting surface runoff from barren hills, villages or roads for warping is extensively used for small agricultural operations. It profits both the sediment retention and crop production. In Zhaizimao village in Fuxian county, Shaanxi for example, 45mu of land being warped had trapped 23,000 cu.m of sediment during a five year period. Crop yield per mu has increased to 300 kg.

(2) Use of flood water with high sediment content for irrigation. There are three large irrigation areas with a

total area of 5 million mu or so where water is diverted from Jinghe River, Luohe River and Weihe River. In the past, irrigation had to stop when the sediment content at the intake exceeded 15% by weight. Since a new regulation was put forward after extensive research, the maximum sediment concentration for irrigation has now reached 41.9-61.6%. The normal content exceeded 25%. 80-100 million tons of sediment have been used for irrigation in the summer of each year. During the period from 1967 to 1986, output of raw cotton increased by 1.1 million kg, in addition to 55 million kg of grain production. The net economic income was more than 20 million yuan.

(3) Use of stream floods for warping. This kind of practice goes back 2,300 years in Zhaolaoyu, Zuping county, Shaanxi, where 2 million cubic meters of flood water with 300-600 thousands tons of sediment are used annually in a drainage area of 200 sq.km. Land for warping has gradually extended up to 34,000 mu, compared with 8,000-12,000 mu before 1949. Flood waters through a 40 km long course of the Chunyang River has been used for warping, which lies on a plain from Zhaolaoyu Gully outlet to the confluence of Shichuan River. Pingliang prefecture in Gansu has 5,000 projects with 277,000 mu of warpland, holding 9.7 million cu.m of sediment per year on average.

(4) Disposal of sediment from reservoir for warping. Several types will be illustrated below:

(a) Impounding clear water and disposing of sediment-laden water for irrigation. Heisongling reservoir in Jingyang, Shaanxi, has a total capacity of 14.3 million cu.m, and the drainage area is 370 sq.km. Since 1962, 8.55 million tons of sediment have been sluiced off the reservoir and 52 million cu.m of flood water containing 11 million cu.m of sediment was directly supplied for irrigaton. The life of the reservoir has been prolonged to 86 years instead of 16 years; irrigation area extended from 80,000-110,000mu, and grain production increased from 100 kg/mu to 300 kg/mu.

(b) Siphoning off sediment from a reservoir. The difference in head available above and below the dam of a reservoir may be used to siphon off sediment through a pressure pipeline laid in the reservoir and joined to the outlet conduit. In Tianjiawan reservoir in Yuci, Shanxi province, with a storage capacity of 9.42 million cu.m, the highest head is 10.28 m, discharge through the pipeline 0.3 cms, the maximum sediment concentration is 814 kg/cu.m, averaging 343 kg/c.m. and disposal of one cubic meter of sediment costs only 0.02 yuan.

(5) Use of flood waters and sediment through dam systems in gullies. A series of desilting dams in gullies has obvious effects on sediment reduction and production increases, due to retention of floods and sediment. In Jingbian county located at the south edge of Maowusu Desert, 1,300 dams with a storage capacity of 1.0 billion cu.m were built in the past three

decades. New arable land formed by damming sediment amounts to 40,000 mu, 30,000 mu of desert has been converted into a valuable land by means of channeling water to sluice sand dunes, and 31,400 mu formed by warping during floods. Sediment yield over the county has decreased fom 56.6 million tons to 6.16 million tons, and grain production increased from 12,180 tons to 60,000 tons. This county has become one of the famous green areas in a desert and a granary and forest area comparable to the area south of Yangtze River where vegetation covers 40% of the land.

In Yulin prefecture in Shaanxi, there are 15,000 dams for desilting and land-forming. Until 1983, available area for sedimentation was one-half million mu, including 324,000 mu of planted area in which 11.3 million cubic m of sediment was trapped, averaging 3500 cu.m/mu. The land area formed behind dams reaches a proportion of 2.1% to the total cropland in the prefecture and contributes 16% of the total production.

SMALL WATERSHED AND PILOT TRIBUTARY MANAGEMENT

Integrated planning and comprehensive improvement with small watersheds as fundamental units is an important approach not only for soil conservation but also for rational utilization and preservation of water and soil resources. Various treatments are laid out and implemented in small watersheds to form a protection system at multiple levels. Treatment on slopes and in gullies, engineering and vegetation measures will be tightly coordinated. In Nianzhuangguo watershed in the vicinity of Yenan city, Shaanxi province, with an area of 54 sq.km, 5,592 mu of terrace, 1,771 mu of damming land, 334 mu of irrigated area, 33,279 mu of tree plantation and 8,534 mu of grass were completed through 20 years of effort. In 1984 the output crop reached 2.06 million kg, averaging 575 kg per capita. Beneficial returns were 41.3 yuan per mu, twice as much as that before management. Since 1970, all the floods and sediment for the past 15 years have essentially been retained in the watershed.

On the basis of small watershed management, pilot tributary programs where sediment load with coarse particles is excessive have been implemented. Wuding River is an example. It drains an area of 30,260 sq km, among which 23,137 sq.km are susceptible to erosion. A long term-management program was put into effect with notable achievement. Up to the end of 1981, 984 thousands mu of terraces, 31 large and medium sized reservoirs and 230 small reservoirs with a total capacity of 1.68 billion cu.m were established. Seven hundred and seventy-one irrigation canals and 4,631 pump stations have been completed with an irrigation area of 827,000 mu. 9,096 dams are available for sediment deposition creating arable land of 300,000 mu of which two-thirds have already been planted. Reforestation has occurred on 5.43 million mu and

1,077 mu of grass has been planted. Farm income totaled 271 million yuan in 1981, 1.65 times higher than that in 1957. Total grain production is 399 million kg, 3.2 times higher than that in 1949. The annual average sediment delivered to the Yellow River was 94.8 million tons during the 1970s instead of 204 million tons produced in the 1960s, a reduction of 53.5%.

Basic Experience, Understanding and Problems of Loess Plateau Management

In thirty odd years, a great deal of work has been done and obvious achievements have been obtained, however many problems remain to be solved. (1) Development in different areas is not in balance; there are many advanced models set up for demonmstration, but they have not been extended to other areas; the rate of development is not rapid enough to meet the demand of the national economy. (2) In existing programs there are many cases, such as salinization and alkalization of the land formed behind desilting dams, low grassland productivity and weak transfer efficiency, as well as low seedling survival rate etc so that beneficial returns either in the economy or ecology cannot be obtained. (3) Cultivaton on sloping land still occupies a major constituent for agricultural production and prime cropland is still insufficient, and this precludes the ability to withstand natural calamities. (4) Sometimes population growth exceeds land productive capacity, so that it is difficult to prohibit reclamation on steep slopes and striping of vegetative cover, meanwhile manmade erosion has been further aggravated due to discarding soil, stone and residue resulting from openpit mining and road construction.

Summarizing the conservation management over the past 30 years, our recognition and experience are as follows: The Loess Plateau in the mid Yellow River is one of national poverty-striken regions with a very weak basis of economic development owing to severe erosion and critical ecologic circumstances. The harnessing and development of the mid Yellow River Plateau Region with a priority on soil and water loss control will have a significant influence on making further advances on the national economy and safety of lower Yellow River and alluvial plain in the drainage basin of the Yellow, Huaihe and Haihe Rivers. Viewed as a whole, if the task is to be accomplished in the long run the initiative of the masses must be brought into play in an effort to control the numerous hills and gullies. The work has to be carried out with guidelines that the harnessing work and its management should be better coordinated and the duties, rights and benefits of the people engaged in the work should be jointly taken into consideration. Attention should be given to the overall consideration of immediate medium and long-term benefits relating to soil and water conservation to lift the people out of poverty and backwardness. It is only on the basis of gaining economic benefits successively that the broad masses of people can be expected to participate in the work of improving the mode of life and the environment before unified benefits to the economy, ecology and society can be realized.

The task of Loess Plateau management is, generally speaking, to improve and create a favorable artificial ecological system, gradually change the primary conditions for local agricultural production, promote rural and regional commodity economy, maintain and improve environment quality for development of energy resources, communication and mine industries. The conflict of supply and demand of growth of population and availability of natural resources should be properly dealt with so that men, resources and environment will be compatible with each other. Soil and water conservation is a large system which is constituted by three sub-systems namely, science and technology, eco-economy and society. Application of optimization techniques by systems analysis and adoption of management methods of systems engineering are needed to solve the problems. Additionally, macroscopic control guided by soft science, multi-disciplinary and departmental coordination, making decisions at a high conprehensive level and scientific basis are urgently needed for further development.

Based on the experience accumulated from long-term practices, we constantly sum up and intensify our understanding, and in turn direct future action. We have increasingly recognized that improvement in procedures of decisionmaking at multiple levels and organizations are so important as to have to be reformed. These contain strategy, principles and direction concerning watershed management, law and regulation, policy on technology and economy, conservation zoning and planning etc. as well as administrative structure and responsibility systems. The former implies the establishment of effective institutions which could make adequate decisions, intercede for coordination, and provide consultation and extension services with reliable information and a feedback system. The latter implies household responsibility on a contract basis to harness small watersheds; contracts or bids for backbone projects undertaken by special construction teams and a duty-bound labor system to build prime agricultural land.

Loess Plateau management should be implemented appropriate to the local conditons and economic strength. Stress should be laid on key areas where large amounts of coarse sediment originate in combination with small watershed management and pilot tributary projects in order to make the most of the integrated system and to open up broad prospects for sediment reduction in the lower reaches and for agricultural and industrial development.

ASSESSMENT OF SEDIMENT REDUCTION DUE TO UPLAND WATER AND SOIL CONSERVATION WORKS IN THE YELLOW RIVER

Xiong Guishu, Xu Jianhua, Gu Bisheng, and Dong Xuena
Yellow River Water Resources Protection Institute
Yellow River Conservancy Commission
People's Republic of China

Introduction

Practicing water and soil conservation in the Yellow River would reduce the sediment carried into the Yellow River and slow the deposition in the lower reaches of the Yellow River. Investigation and study of the sediment retention effect of water conservancy and water and soil conservation has been carried out for some time. By analyzing hydrologic data, it has been verified that during the late 50's the Jiuyuangou gully (with a drainage area of 70.1 sq. km), Nanxiaohe gully (30.6 sq. km), and Luergou gully (12.0 sq. km) retained 50-80% of sediment a short time after water and soil conservation practices were put into effect. Hydrologic data in the early 70's shows that sediment discharge has been reduced 50% or more in rivers such as the Wuding River (30,127 sq. km), Fenhe River (38,928 sq. km.), Qinshui River (14,480 sq. km), and Dahei River (6,835 sq. km.) after initiating the practice of water conservancy and water and soil conservation. Up to the early 80's, through the analysis of the relationship between rainfall and runoff and the comparison between processes of runoff and sediment among similar river basins, it has been proven that the sediment carried into the Yellow River from 1970 to 1984 has been reduced by 300 million tons annually on average. It can be seen that reservoirs and water and soil conservation works as methods of sediment control in the upper and middle drainage basins are very helpful in reducing deposition in the lower reaches. Hydrologic data also make it clear that the state of both runoff and sedimentation in the lower reaches of the river has changed a lot following installation of major projects on the main stem and tributaries. Table 1 illustrates the state of runoff and sediment discharge at Huayuankou station as well as the sediment deposition in the lower reaches during different

L. M. Brush et al. (eds.), Taming the Yellow River: Silt and Floods, 529–543.

Table 1. The amount of flow and sediment discharge and deposition on the lower reach at Huayuankou station in different periods

periods		1950.7 --- 1960.6	1960.7 --- 1970.6	1970.7 --- 1980.6	1980.7 --- 1985.6	1919 --- 1960
runoff amount at Huayuankou	100mil cu.m.	480	503	380	460	472.4
sediment discharge at Huayuankou	100mil tons	17.9	11.7	13.8	9.39	14.8
deposition on lower reach	100mil tons	3.61	0.03	3.20	0.64	

periods. As shown in the Table, sediment entering into the lower reaches has decreased gradually decade by decade and deposition has also decreased.

In order to predict the future development in the lower reaches of the Yellow River it is necessary to predict the changes of runoff and sediment in future decades. Most important is to predict the probable quantity of sediment retained by reservoirs and by soil conservation works in future decades. To solve this problem it is necessary to study the relationship between development of water and soil conservation and natural factors, and also the output of cereal production from which to estimate the probable development of water and soil conservation works and the trend of reduction of sediment yield.

Relationship Between Water and Soil Conservation and Natural Factors

Sediment of the Yellow River is mainly produced during storms and floods, but there is no monotonic relationship between soil erosion and rainfall in the middle drainage basins. Fig.1 shows that the envelope curve of the relation between annual precipitation and annual modulus of sediment transport looks like an outline of a bell. The maximum modulus of sediment transport corresponds to a rainfall of 450 mm in the loess gullied hilly area, while in the loess gullied plateau and earth and rocky mountain areas the maximum modulus corresponds to 550 mm rainfall. Similar relationships can also be found in other countries. The figure reveals an inner relationship among precipitation, soil character and vegetation, though only in rough outline.

In recent years some scientists and soil conservationists have studied the spatial pattern of rainfall and radiation conditions in the Yellow River basin, and defined boundaries which divided belts suitable for growth of arbors, bushes and grasses (see Figure 2).

The arbor belt is distributed among areas where precipitation is 550 mm or more; the bush belt is in areas susceptible to severe soil loss, e.g., in northern Shaanxi, western Shanxi, Beiluohe basin, Malianhe basin and Jinhe's upper reaches. Bush with a poor canopy density has only a small effect in controlling loss of water and soil. Limited by available rainfall and radiation conditions it would be difficult to control water and soil loss by reforestation in those areas where most of the sediment originates.

Regarding sediment transport, the lower river course of the Yellow River reveals characteristics of deposition of coarse and fine sediments. In the deposits from Xiaolangdi to Lijin the coarse sediment, of which the grain diameter is over 0.05mm, accounts for 69% of the total deposition, but the

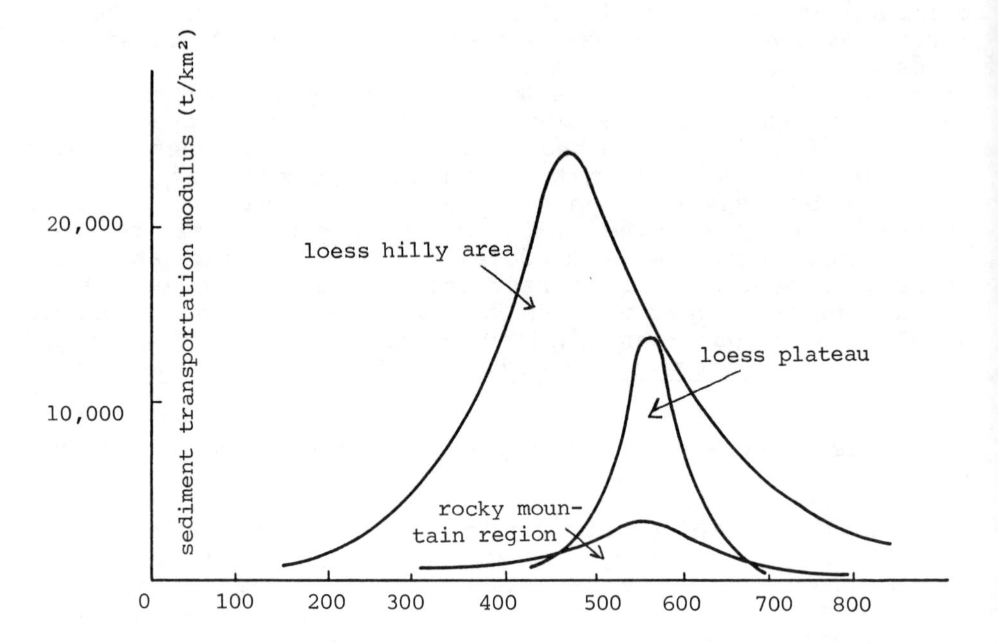

Figure 1. Modulus of Annual Rainfall and Sediment Transport in the Middle Reaches of the Yellow River.

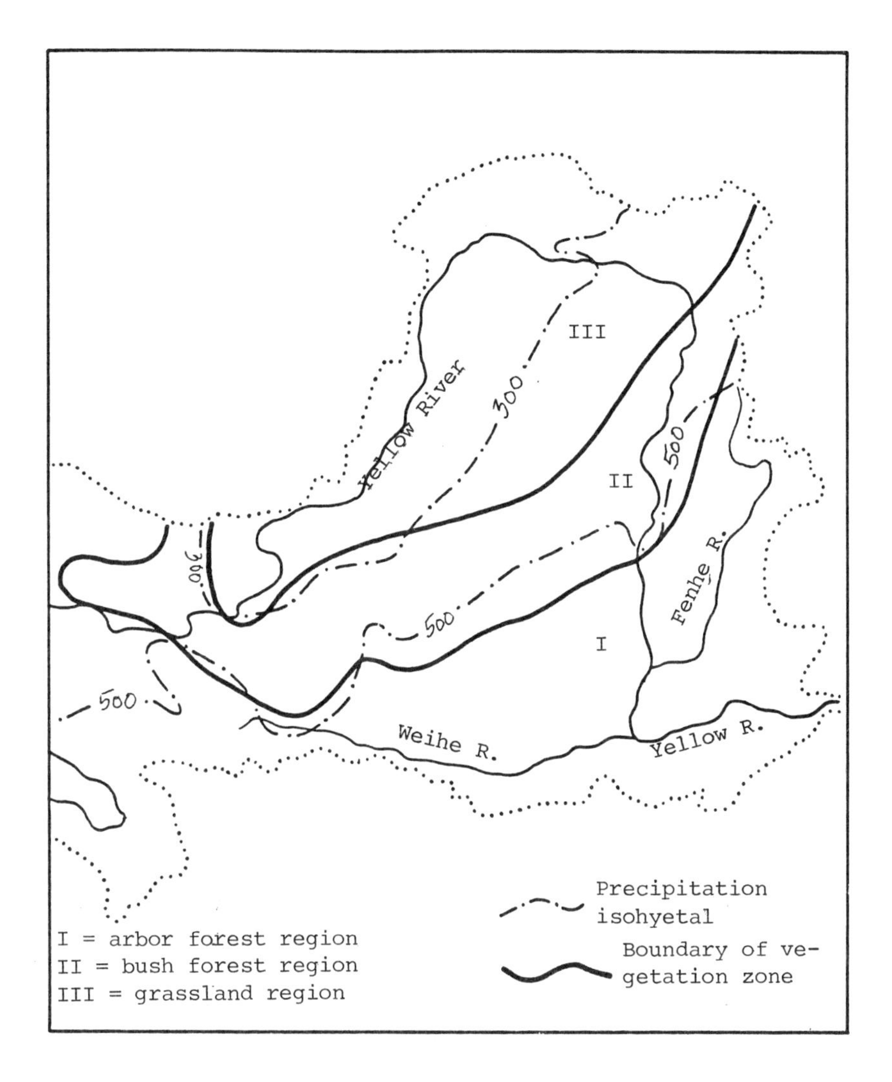

Figure 2. Map of Precipitation Isobath and Vegetation Covering Line

coarse sediment being transported in the lower reaches is only 20% of the total sediment discharge. A great part of the coarse sediment is deposited on the river bed. It has been estimated that 30% of the coarse sediment comes from an area of 100 thousand sq.km but that 50% of this is from an area of only 38 thousand sq.km. These areas are in the region between Hekouzhen and Sanmenxia. It has also been discovered that the source area of coarse sediment is also the area where large amounts of sediment originate. This fact is of significance for controlling the sediment yield from main source areas and for slowing the aggradation in the lower reaches. In the past 20 years the source area where most coarse sediment originates has been considered a focal point for practicing water and soil conservation.

Relationship Between Water and Soil Conservation and Cereal Production

According to the comprehensive plans for harnessing the Yellow River (1955), a decision was made to turn 16% of the cultivated land into grassland and forest within 12 years. After 30 years of practice, however, the area of cultivated land has increased rather than decreased. Though the data are not very accurate, it is certain that the phenomena of reclaiming wasteland has been in existence for a long time. The area of cultivated land correlates with population and cereal yield. Between 1985 and 1949, for instance, the population of Shaanxi province increased by 1.28 times, the cereal yield increased by 1.87 times, the average yield per mu increased from 60.8 kg to 119 kg and the yield per capita increased from 252 kg to 316.5 kg. The data show that in last 37 years in Shaanxi, the population has increased greatly as has cereal production. The increased area of irrigation and the silting behind dams and terraces account for the large increase of cereal production. In 1949, the cereal yield of irrigated land in Shaanxi was 21% of the total yield, while in 1985 the cereal yield from irrigated land, land behind dams and terraces amounted to 70% of the total yield. Water conservancy and water and soil conservation works had played a most important role in solving the agricultural production problem in Shaanxi province. By data analysis the correlation of the development of water resources and soil conservation works with the population and cereal yield in other provinces in the Yellow River basin is generally quite similar to that found in Shaanxi province.

It is possible to turn some cultivated land back to woodland and grassland in the Yellow River basin. The approach is to extend or enlarge the area of basic cultivated land in which production could be assured and to abandon cultivation on dry sloping lands. The prerequisite for not using sloping land for cultivation is to solve the problem of cereal production. If the production could not meet the demand for living, cultivation on sloping land would not be turned into forest. Water and soil conservation can surely promote agricutural production, however, it does not certainly conform to reality to ask for too fast a rate to turn too much cultivated land into forest by implementation of water and soil conservation.

Mechanism of Sediment Retention by Water Conservancy and Water and Soil Conservation Works

Sediment in the river originates from soil erosion by rainstorms and the accompanying runoff. Because rainstorm vary year by year, the annual amount of soil erosion is quite different. If this fact were not thoroughly recognized, and several wet years were to take place consecutively, it would give people a false impression that no effects of sediment retention could be observed by practicing water and soil conservation and building dams. On the contrary, when dry seasons are encountered, the effect of water and soil conservation and reservoirs as well as check dams would be over-estimated. It is thus necessary to discuss the sediment retention mechanism of water and soil conservation for evaluation of sediment retention effects in the past and a prediction of the effects in the future.

To build a terrace is to change the steep slope susceptible to erosion into strips of flat ground, tier upon tier, promoting an increase in permeability during rainstorms and consequently reducing the loss of water and soil. In this way the erosion rate can be slowed down. But when catastrophic rainstorms take place the terraces will be damaged and erosion will still continue. The effect of forest and grass on water and soil conservation is to protect the soil surface from rainstorm splash and runoff erosion by means of a layer of dead leaves, branches and sod. The plant leaves also have a function by increasing permeability and decreasing runoff through rainfall interception. But if the vegetation cover is not dense enough and dead leaves and branches are not thick enough to cover the ground surface, the sediment retention effect is relatively small. As mentioned above, measures on the slopes such as terraces, forests and grasslands etc. now existing in the basin can only restrain to some extent the soil erosion but can not completely control it.

Another type of measure is check dams, dams to form land by siltation in gullies and watersheds. The projects obstruct the sediment flowing from catchment areas and provide storage capacity. Their effect in detaining sediment will disappear after the storage capacity for sediment retention is depleted. These projects can also play a role in controlling gully collapse under the range covered by the deposits. Because there are about ten thousand projects of this kind in the middle Yellow River drainage basin, it is rather difficult and complicated to count the sediment retention effects one by one. Previous studies by the first writer have shown that the accumulative sediment retention in reservoirs and check dam groups in a basin may be described by the theory of storage in depressions given by G.L.Kalinin. The accumulative sediment retention may be calculated by

$$V_t = V_m \left(1-e^{\frac{-FotWs}{F\,V_m}}\right) \tag{1}$$

where,

V_t: accumulative sediment retention after t years of reservoir operation
V_m: total storage volume for sediment retention
t: years of reservoir operation
Ws: annual sediment yield produced in a basin, in volume unit
Fo: catchment area of reservoirs and check dams in a basin, in area unit
F: basin area, in area unit,
e: base of natural logarithm.

As the formula shows, when the accumulative sediment yield in catchment area gets large enough, the exponent in equation approaches 0. The total sediment retention at this time approaches the total storage capacity of the group of reservoirs. The sediment produced in the basin would still move downward through the reservoirs in steady streams. It is obvious that the reservoirs and check dams would only play a role in slowing down the rate of sediment yield from the basin, but in the long run, they cannot totally stop the water and soil loss.

The erosion that rainstorms and runoff exert on soil is a natural force. Earth's surface is moulded into mountains and rivers under the long-term effect of natural forces. Compared to the natural forces, man's influence seems relatively insignificant. Though the total sediment can be blocked temporarily by practicing water and soil conservation, it should be only regarded as a temporary measure. For instance, after building a large number of check dams in the mid-50's in the Jiuyuangou basin, no sediment came from the gully in normal years. During the extreme flood in 1958, the earth dams burst one by one and part of the sediment was still discharged downstream. The sediment retention effect decreased somewhat. Generally speaking, water and soil conservation can only restrain and slow down the loss of water and soil but they cannot put it completely under control. In addition, soil loss may be aggravated by human activities.

Analysis of the Effects of Water and Soil Conservation Projects in Reducing Sediment from 1970 to 1984

The status of water and soil conservation in the middle basin has been described eleswhere in detail. Here no more descriptions will be given.

Up to now, two kinds of evaluation methods have been used to compute the effects of water and soil conservation on the reduction of sediment as well as the effects of water conservancy works. This approach will be taken into account for work done on all kinds of projects and then the sediment retention evaluated on all projects. Its precision depends on statistical information obtained through observations.

Another approach is the hydrologic method which involves analyzing the variation of sediment discharge in rivers, before and after the projects were carried out in a basin, by means of data from hydrometric stations at all levels. Its precision depends on the reliability of the computing method.

The arguments and data used in both methods are not the same, and the meaning of computed results is also somewhat different. Though there are some differences in both methods, the data can be compared. The differences are as follows:

(1) The method of water and soil conservation gives the result of the amounts of sediment retention in projects, not the amount of sediment reduction in rivers. As long as there are some projects, the relevent sediment retention can be computed. The hydrologic method gives the amount of decrease in sediment discharge because of the effect of the projects. Restrained by the laws of sediment transport and accuracy of sediment measurement, the variations in sediment discharge in the river can be found in the measured data only when a sufficient number of projects exist.

(2) It would be difficult to consider the influence of damages or projects under repair and the influence of uneven runoff and sediment yield by the method of water and soil conservation, but for the hydrologic method, these influences will be reflected, at least in theory.

(3) Functions of the self adjustment of sediment transport in alluvial rivers can not be reflected in results given by the method of water and soil conservation. However, the hydrologic method gives computed results after the sediment discharge has been adjusted by the river.

(4) Both the hydrologic method and method of water and soil conservation can be used in analyzing the past state of sediment retention. For prediction purposes, only the method

of water and soil conservation can be used. The hydrologic method, however, can be used as a reference for the method of water and soil conservation.

Computations by the method of water and soil conservation, statistics on the amount of water conservancy and soil conservation works in the Yellow River have been used, as well as data on the areas of reforestation, grasslands, terraces and new land formed behind dams, irrigation areas as well as the deposition in large, middle and small reservoirs etc. Because the survival rates of trees and grasses planted in areas subject to severe soil loss are quite low, their sediment retention effect is too small to be taken into account in the computations.

The computation of hydrologic method adopted at present is mainly based on comparison of results with and without engineering measures in similar watersheds. The difference of sediment discharge under the same precipitation conditions is taken as the sediment retention effect of the projects. By summing up the computational results by both methods, the effects of sediment reduction and retention in the Yellow River basin are listed in Table 2. As the Table shows, the calculated amount of sediment reduction and retention in the basin of the Yellow River as a whole is not the same as that of the total divisions. The reason for this difference is mainly computational error and, secondly, the influence of adjustment in sediment load in the process of transportation. It can be concluded that from 1970 to 1984, because of the projects, sediment discharge in the Yellow River has been decreased by about 250 to 330 million tons per year, which is about 15% of 1.6 billion tons of average annual sediment load at the Sanmenxia gaging station. In order to keep a 15% reduction of sediment load per year from now on, more projects should be built, especially in areas where large amounts of coarse sediment occur.

Estimation of Sediment Retention in Middle Reaches of the Yellow River in the Next Few Decades

An estimate of sediment retention in the next decades can be done only under some assumed conditions, which are:

(1) It is noted that, by reviewing the data obtained in the past thirty years, effective measures of sediment retention in the Yellow River basin were mainly the terraces, check dams and reservoirs. Only these projects are considered in the computation of sediment retention for the future.

(2) Planting of trees and grasses is also a kind of soil conservation measure to be used in the Yellow River basin, but it is difficult to ascertain the quantity of work done or the sediment retention by these measures. Terraces, check dams and reservoirs have a great effect on sediment retention, but once

Table 2. The effects of sediment reduction and retention

zones	periods	sediment retention calculated by the method of water & soil conservation (10,000tons/year)					
		terrace	land behind dams	large & middle reser-voirs	small reser-voirs	irrigation	total
above He-kouzhen on Yellow River	1950-1969			1350	1600	3870	6820
	1970-1984			10216	1883	3293	15392
Hekouzhen to Longmen	1950-1969			1555	87		1642
	1970-1984	2119	14054	4631	1348		22152
above He-jin on FenheRiver	1950-1969			1870			1870
	1970-1984	84	2604	1234	207		4129
above Huaxian on WeiheRiver	1950-1969			1263	165	1950	3378
	1970-1984	1554	22020	2954	1049	3174	10751
above Zhangtou on Beiluo River	1950-1969			12	4	275	291
	1970-1984	88	886	139	9	763	1885
Hekouzhen to area of Huaxian, Hejin and Zhuangtou	1950-1969			4739	256	2225	7220
	1970-1984	3844	19564	9065	2706	3937	39116
above the area of Longmen, Huaxian, Hejin and Zhuangtou	1950-1969			6089	1856	6095	14040
	1970-1984	(3844)	(19564)	19281	4589	7230	54508

Table 2 - continued

zones	periods	sediment retention calculated by the hydrology method (10,000tons/year)	
		by comparison of results with sediment similarity	by comparison of results with precipitation similarity, over 30 mm
above Hekouzhen on Yellow River	1950-1969 1970-1984	4600	(4600)
Hekouzhen to Longmen	1950-1969 1970-1984	18383	20000
above Hejin on Fenhe River	1950-1969 1970-1984	3741	3600
above Huaxian on Weihe River	1950-1969 1970-1984	-1438	2930
above Zhangtou on Beiluo River	1950-1969 1970-1984	883	2000
Hekouzhen to area of Huaxian,Hejin and Zhuangtou	1950-1969 1970-1984	21300	20000
above the area of Longmen,Huaxian, Hejin & Zhuangtou	1950-1969 1970-1984	25900	33130

a heavy rainstorm takes place failure and damage of some of the projects would be inevitable, and according to past analysis, about 10% of original sediment deposits behind the dams could be sluiced off from the reservoirs. How to properly predict this on a large scale and for a long period remains a difficult problem. Comprehensively considering those different effects, positive or negative mentioned above, in our computation of sediment retention neither the effect of trees and grass nor the probable damages were taken into account.

(3) Warping by withdrawal of floodwater is also a measure which takes a great deal of sediment from the river. Some measure of this has been made. Because it is restrained by topographic conditions and flood sources, the prospect of its future development is difficult to say. Hence, warping is not considered.

The program of computation is as follows:

(1) Sustained effects in the next decades of only those existing projects are to be considered in the first step. That is, assuming no new projects are built, predictions will be made of the effects from existing projects. This data can be used as a basis for comparison with other methods.

The computation is done in four zones, including the area above Huaxian in Weihe River, above Hejin in Fenhe River, above Zhuangtou in Beiluo River and the area between Hekouzhen and Longmen. In these four areas there are 15.84 million mu of terraces, 1.52 million mu of land formed behind check dams (roughly estimating that there are about 6.03 billion cu.m of total storage capacity behind check dams), and 6.4 billion cubic meters of total storage capacity in reservoirs. These projects blocked sediment for an average of 99.1 million tons per year from 1960 to 1969, 366.5 million tons from 1970 to 1979, and 286 million tons from 1980 to 1984.

In computing sediment retention in the future the amount of sediment intercepted by terraces is calculated by using a local modulus of sediment yield on slopes multiplied by the area of terraces. For sediment retention of check dams and large, middle and small reservoirs, retention is computed by the formula introduced in this paper according to depression storage theory.

The sediment retention in watersheds between Hekouzhen and Shanxian in various periods is as follows:

Periods	Sediment retention (million tons per year)
1960-1969	99.1
1970-1979	366.5
1980-1984	285.0
2000	154.8
2030	100.5

The results show that the level of sediment retention at

present would go below 100 million tons per year rapidly if the middle Yellow River basins are not harnessed and no more check dams are built in the next 60 years. This situation needs attention.

(2) The second step of computation is to assume that new projects will occur at the average rate of progress seen during the past 30 years. By this step the estimation for future sediment retention can only be made roughly. To implement this program requires the building of terraces on 5.3 million mu, check dams with 2 billion cubic meters of storage capacity, and reservoirs with 2.1 billion cubic meters of storage capacity for sediment retention per decade. The method of computation is the same as the former. The result is as follows:

Periods	Sediment Retention (million tons each year)
2000	350
2030	484

The result shows that present level of sediment retention or slightly more could be obtained if the basin were harnessed in the next 60 years according to same rate of progress as during the past 30 years. The sediment carried into the Yellow River might be reduced by about 400 million tons.

Because the system of responsibility in agricultural production and agricultural economic reform has changed a lot compared to the past, there is little possibility of carrying out water and soil conservation works and building reservoirs in past ways. Now a certain proportion of the rural labor force in quite a few districts is engaged in business, light industry, mining industry and other sidelines. There is also a tendency for growth of the urban population. The departments concerned have predicted that decrease of the rural labor force is an inevitable trend. But the beneficial situation to water and soil conservation is that the economy becomes more flexible and both the government's financial capability and peasants' income are much more than that in the past. The projects which were built primarily by manual labor in the past will now be built more by mechanical power. It is difficult to estimate the effect of this change accurately. But the standard is surely not very low requiring a rate of progress to be kept the same as in the past 30 years. Pressure must be maintained for building check dams and reservoirs for sediment retention and terraced fields if an average reduction of 400 million tons of sediment is to be expected each year for future decades.

Our government has now attached importance to fully harnessing the Loess Plateau, and many scientific research studies have been organized. It is necessary to study how control measures in the middle drainage basin can be made compatible with the development of our national economy. Especially, studies of reforestation and grass planting should be encouraged for the purposes of minimizing, on a permanent basis, sediment yield from the basin.

References

(1) Xiong Guishu, Influence of water and soil conservation work on flood runoff. Yellow River, 1965

(2) Hua Shaozu, Mou Jinze, Exploitation and harnessing of Loess Plateau in the middle reaches of Yellow River. (paper from this symposium)

(3) Xiong Guishu, Effect of water and soil conservation in upper and middle reaches of Yellow River upon sediment reduction. Yellow River, 14th, 1986

(4) Xu Jianhua, Analysis of effect of sediment retention of water and soil conservation work in middle reaches of Yellow River. Soil and Water Conservation in China, 12th, 1986

(5) Gu Bisheng, Proof of arbor plantation in large area of Loess Plateau. (paper from Yellow River Water Resources Protection Institute), Feb. 1987

(6) Dong Xuena, Analysis of sediment retention by water and soil conservation projects by utilizing precipitation data. (paper from Yellow River Water Resources Protection Institute), Feb. 1987

EVOLUTION OF THE COURSE OF THE LOWER YELLOW RIVER IN PAST HISTORY

Xu Fuling
Yellow River Conservancy Commission
People's Republic of China

The Yellow River is a well-known heavily sediment laden river with a long history. Its sediment content ranks first in the world. The flood waters of the Yellow River used to run wild far back into ancient times. It was said that Yu the Great diverted the flood waters into the Bohai Sea through "nine rivers" to control floods. The so-called "nine rivers" means many rivers, and does not mean exactly nine. It is quite possible that the Yellow River formed many forks and emptied into the sea due to shifting of its main course by natural fluvial processes of sedimentation on the great plain. The dikes on both banks of the lower Yellow River below Mengjin in Henan province were built gradually before 400 B.C. (the period of the Warring States) and the disasters caused by wild floods have been reduced ever after. The channel within the dikes rose because of continuous sedimentation and a suspended river was formed. If there were a breach, an avulsion might take place. For thousands of years, the lower Yellow River changed course and shifted many times, flowing northeastwards to the Bohai Sea or southeastwards into the Huanghai Sea, involving a total area of 250,000 sq.km from Tianjin in the north to Changjiang and Huaihe River in the south.

Factors Influencing Dike Breaching

According to incomplete statistics, there were more than 1500 dike breaches during the 2000 years or more before 1949 in the lower reaches of the river. There were many reasons for this. When flood waters exceeded the conveyance capacity of the channel during large and major floods, dike breaches might occur due to overflowing. For instance, in 1761 A.D. (26th year in the reign of Emperor Qianlong, in Qing dynasty), a flood of 32,000 cms occurred at Huayuankou and 27 breaches occurred due to overtopping of the dikes in the reach between Wuzhi, in Henan and Caoxian county, in Shandong. During the

L. M. Brush et al. (eds.), Taming the Yellow River: Silt and Floods, 545–555.

1933 flood, with a peak discharge of 22,000 cms at Shaanzhou, levees were breached at more than 30 places near Changyuan and about 0.3 m of sediment was deposited on top of the embankments at that time. All of these are examples of dike breaches due to overtopping. When the river impinges on the dikes causing cave-ins and slides, if rush repairing could not be done in time, a dike breach might take place. Such breaches occurred at Dagong, Fengqiu, in Henan in 1803 (8th year of the reign of Jiaqing, Qing dynasty), at Zhangjiawan, Kaifeng in 1841 (21st year of Daoguang, Qing dynasty), Jiubao, Zhongmu in 1843 (23rd year of the reign of Daoguang, Qing dynasty), Tongwaxiang in 1855 (5th year in the reign of Xianfeng, Qing dynasty) and Guangtai, Fengqiu in 1934. When the flow approaches the embankment, serious piping and seepage may take place due to hidden defects in the dikes. The embankment may slump and fail if remedial measures have not been taken. The phenomenon often referred to as dike breaching due to bursting occurred at Shiqiao, Zhengzhou in 1837 (13th year in the reign of Guangxu, Qing dynasty) and Dongzhuang, Juancheng, in Shandong in 1935. In addition, there were also man-made breaches, such as the breach at Huayankou in 1938. In history, more breaches took place at ordinary sections than at vulnerable spots and more breaches occurred due to impingement on and bursts of the dike than that due to overflowing.

The lower Yellow River is a typical aggrading river and siltation, dike breaching and shifting of its main course easily take place. The river may easily shift its course if the dike breaches frequently and excessive aggradation gives more opportunities for dike breaching. There is a cause and effect relationship among the three. In the past people have talked about the possible reduction of the amount of deposition by releases of part of the sediment load through the embankment by the frequent dike breaches, but the severe result caused by siltation downstream of the breaching point has been overlooked. For instance, in 1676 (15th year in the reign of Kangxi, Qing dynasty) water levels of the Yellow River and Huaihe River raised simultaneously and the dike system at many places burst. It was recorded in the "Outline of River Improvement" that for 150 km or more downstream of Qingjianpu, Huaian, in Jiangsu to the mouth of the Yellow River, "the river originally was 0.5-1 km to 2-2.5 km wide, now it has narrowed to 30-60 m and is 6-9 m or even more than 15 m deep, and shallows to barely over one meter." It was dredged later on a large scale during the closure of the breached dike.

In 1781 (46th year in the reign of Qianlong, Qing dynasty), the dike breached at Qinglonggang, Yifeng (now Lankao). A new levee had to be built beyond the south dike due to serious deposition below the breach, and it could be blocked only by building a new diversion channel 74.5 km long from Sanbao, Lanyang (now Lankao) to Qibao, Shangqiu and the old channel dredged from Shangqiu to Xuzhou to divert the water into the old channel of Shangqiu. In 1819 (24th year in the reign of Jiaqing), a wet year with high sediment yield, the dikes on both banks were breached at 8 places in a 50 km reach from Kaifeng to Yiyang within 4 days during a flood in July. A 25 km long diversion channel of 3-9 m in depth had been dug to facilitate the closure. During a flood in August, another dike breach took place at Jiubao, Wuzhi. On the second day after the closure on March 12, 1820, the dike breached again at Sanbao, Yifeng. It showed that it was easy to breach continuously due to serious deposition in the original channel. Similar cases have been witnessed many times in history. Thus, the initial channel would undergo severe deposition subsequent to each breach, and provid conditions for the next breach, such as a sudden change of the channel pattern during reformation of the channel. It became a vicious cycle that the more breaches the more siltation, and vice versa, this resulted in reducing the capability of flood conveyance in the channel. Once a large flood occurred it would breach and turn to a new channel, because the conveyance capacity of the old channel was far from enough. As Pan Jixun of the Ming dynasty put it: "the capture of a new course does not take place immediately after a dike breach, but if nothing is done, the flow in the main channel becomes sluggish and channel aggradation intensifies, leading to more breaches, until finally avulsion takes place." Of course, the influence of headward deposition due to the river mouth extension was also one of the primary elements accounting for the aggradation of the lower reaches of the river.

Major Avulsions

Avulsion refers to major changes of the river course wherein the course will not return to its original position subsequent to a dike breach and empty into the sea at another location. For an avulsion, the area of influence is very large and finally a definite course with completed dike system may be formed. One of the following cases is not considered as an avulsion.

AVULSION CONDITONS

The river changes its course after a dike breaches at certain location and returns to the original course again after blocking. For instance, in 132 B.C. (3rd year in the reign of Yangguang, under Emperor Wu of the Han dynasty), the Yellow River dike breached at Piaozi, in Puyang. The flood waters moved southeastward and inundated the area for 23 years. In 1938, the man-made breach at Huayuankou, Zhengzhou, remained for 9 years. These two breaches were finally closed and the river returned to its original course.

A new channel separated from the old one and flowed back to the main course after some distance. For example, in 109 B.C. (2nd year in the reign of Yuanfeng, under Emperor Wu of the Han dynasty), the dike breached to the north of Guantao, and water was diverted into the Tunshi River, but returned to the main channel at Xiuxian county (now Wuqiao county).

A bifurcated river channel can run parallel to the sea. In 1060 (5th year of the reign of Jiayou of the Song dynasty) such an event took place, wherein the dike breached at fascine works No.6 at Daming, and a separate branch channel called Forking flowed to the sea separately.

The first case mentioned above shows that although the duration of flooding was quite long, a fixed course had not formed creating only a large area under inundation and the old course was resumed at last after closure. The second and third cases are those where the main channel remains in place while a separate branch is formed, which should not be considered an avulsion. Many authors mentioned elsewhere state that there were 26 major changes of course or avulsion of the Yellow River in past history. If the situation was analyzed based on the above criteria, the author thinks that only 5 major changes have taken place in history. The others are merely dike breaches causing serious disasters and blocked later. The following are the 5 major ones (see map attached):

SPECIFIC MAJOR AVULSIONS

In 602 B.C. (the 5th Year in the Reign of Emporor Ding of the Zhou Dynasty) the Dike Breached at Suxukou. According to "Yugong" the channel of Yu "flows through Luorui (now the confluence area of Luohe River, in Gongxian) to Dapi (one staying in Xingyang, another in the Mount Dapi, in Junxian county) in the east, passes through Jianghe River (now Zhanghe River) to Dalu (Daluze) to the north and affects the Jiuhe River, before emptying into the sea by the Nihe River." The flow path was now roughly through Mengjin, Xingyang, Wuzhi, Yangwu, Huaxian, Neihuang in Henan province and Guangzong and north of Julu (so called Daluze) in Hebei province and reached the Jiuhe River separately before emptying to the Bohai Sea at Jinghai county. The dike breached at ancient Suxukou in Junxian county in 602 B.C., according to the conclusion drawn by Mr. Huwei of the Qing dynasty (there is still contention among historians on that). Since then a new channel has been formed below Suxukou to Puyang in the east and Neihuang, Qingfeng, Nanle, Daming and Guantao in the northeast and empties into the Bohai Sea at Huanghua in the east. In 132 B.C. (3rd year in the reign of Yuanguang, under Emperor Wu of the Han dynasty) the Yellow River dike breached at Piaozi, in Puyang. The floodwaters flowed southeastward to merge with Sihe River before capturing the Huaihe River. The gap formed

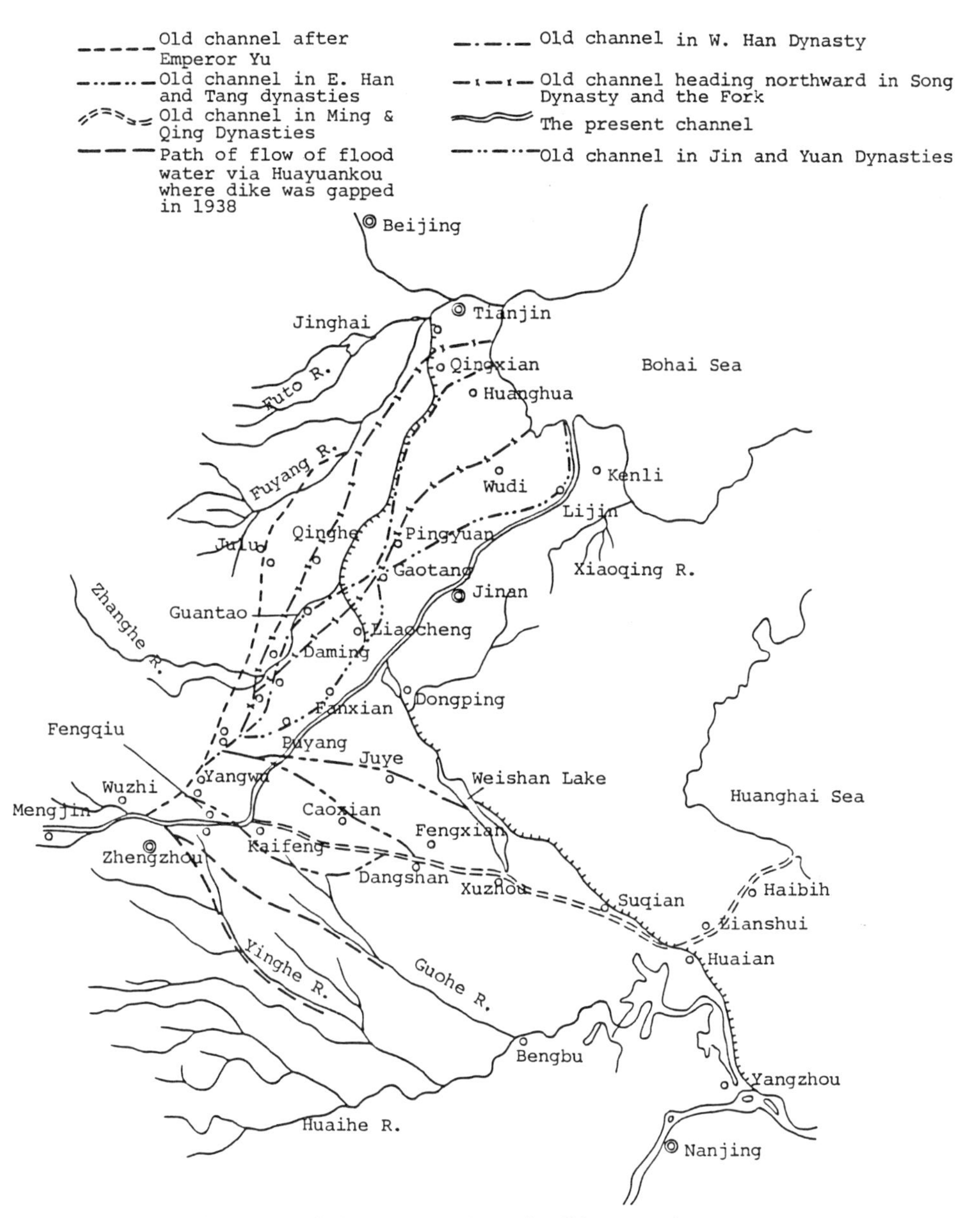

Map of Changes of Course of the Yellow River in All Dynasties

by the dike breach was closed in 109 B.C. In 17 B.C. (4th year in the reign of Hongjia of the Han dynasty), the flood waters overflowed the dike at Qinghe. For more than two decades the gaps were not closed, and the flooded area below Guantao was criss-crossed by streams. As described in Han Annals in the chapter on Drainage Network, "the river overflowed and breached to the north of Weijun (in Puyang), the boundaries hard to distinguish," signifying that the course was out of control and in danger of shifting and changing its course.

The Dike Breach at Weijun in 11 A.D. (3rd Year in the Reign of Wang Mang). In that year the dike breached at Weijun (now west of Puyang) when Wang Mang was in power. He considered that "the water flowed eastward after breaching, no need to worry about floods any more for Yuancheng, thus, the gap need not be closed," (Biography of Wang Mang of Later Han Annals). Yuancheng (now Daming in Henan) was the place of his ancestor's grave, when the water changed its direction to the east, his ancestor's grave would not be flooded, that was why he prefered not to close the gap. As a consequence a major avulsion took place. The old channel in West Han dynasty was called "ancient water of the Great River" or "River of Wang Mang" by historians. The basic flow path of the Yellow River subsequent to the change of course was to the east of Liaocheng and north to the Daqing River. Upstream of Puyang, the river maintained its old channel since the West Han dynasty.

Downstream of Puyang, the Yellow River did not possess a fixed bed for nearly 60 years and wandered all over the inundated area. When Wang Jing, a water commissioner, was appointed to direct the work of harnessing the Yellow River in 69 A.D. (12th year in the reign of Yong Ping, under Emperor Ming of the Han dynasty), dikes were built along the course of the channel causing the water to pass through Puyang, Fanxian, in Henan and Gaotang, Pingyuan, in Shandong and emptying into the sea at Wuli. It had taken place in the Wei, Jin, Northern and Southern dynasties, Tang and Song dynasties. Dike breaches became more frequent in the reach between Puyang and Huaxian by the end of Tang and early Song dynasties.

A Major Change of Course Occurred at Shanghu, in Pyuang in 1048 A.D. (8th Year in the Reign of Qingli of Song Dynasty). In 1034 A.D., the dike breached at the fascine works of Henglong in Chanzhi. The flood waters flowed through Gaotang, Pingyuan and emptied into the sea by joining three tributaries of Chi, Jin and You. In 1048 A.D. a major change of course northward from Shanghu occurred at Puyang. The river took its course mainly at the west of Liaocheng and flowed through Guantao, Qinghe, Jixian and Qianningjun (in Qingxian) before emptying into the sea. This was referred to as the "northern course." In 1060 A.D., a dike breached at fascine works No.6 in Weijun (now west of Nanle), and the river forked out by way of Gaotang, Pingyuan and east of Wuli to reach the sea. This was referred to as the "eastern course." At first, both reached the sea, later the river was forced three times to flow back to the eastern course for military purposes against Liao troops. But they all failed until the end of the Northern Song dynasty.

In 1128 A.D. (2nd Year of the Reign of Jiangyuan of Song Dynasty) the Dike was Breached By Du Chong. During the reign of Zhao Gou of the Southern Song dynasty, Du Chong, the commanding officer in Kaifeng, breached the dike of the Yellow River at the west of Huanxian, in Henan with the intention of stopping the southward movement of Jin troops. Much harm resulted from the dike breach. Subsequent to this breaching, as described in the chapter on Water Course in the History of the Jin dynasty, "the river overflowed or breached the dike so frequently during the past decades that the course changed unceasingly," and the river reached the Huaihe River in several forks. In 1194 (the 5th year of the reign of Mingchang of the Jin dynasty), a dike breached at Yangwu (now Yuanyang), and the flow path was generally along Yuanyang, Fengqiu, Changyuan, Dangshan and Xuzhou, merging with the Huaihe River via the Sihe River.

Maintenance for navigation of grain transported along the Great Canal was the chief aim of harnessing the Yellow River in the Yuan and Ming dynasties, and stress was laid on reinforcing the northern dikes. In the early 15th century, the Yellow River captured the Huaihe River via four routes on the south bank below Zhengzhou. In 1495 A.D. (8th year of the reign of Hongzhi in Ming dynasty), a dike was built on the northern bank at a distance from the channel, called the Taihang dike, which led from Yanjin, of Henan to Fengxian, of Jiangsu to maintain transport of grain along the Great Canal to the north of the Yellow River. This compelled the river to trend toward the south, and to pour the whole Yellow River into the Huaihe River which had only a limited capacity for flood conveyance. Subsequent to 1546 A.D. (25th year of the reign of Jiajing of the Ming dynasty), dikes were built on both banks in the reach between Kaifeng and Dangshang, giving the river a fixed channel from Lankao to Shangqiu, Dangshan, Xuzhou and Suqian, and merging with the Lianshui River before reaching the Huanghai Sea. This river course is known as the "old course of the Ming and Qing dynasty." During the 700 years of capturing of the Huaihe River by the Yellow River, severe aggradation was brought about on the lower reaches of the former (below Huaihe of Jiangsu), and dike breaches became even more frequent.

In 1855 (5th Year of the Reign of Xianfeng of the Qing

Dynasty) a Major Change of Course Occurred at Tongwaxiang. In the flood season of 1855, a dike breached at Tongwaxiang in Lanyang (now Lankao county) because the flow pattern changed and the main current impinged on the dike downstream of the vulnerable spots where no bank protection works were in existence. Rush repair work could not be done quicky enough. This happened during the peasant uprisings of the Taiping Heavenly Kingdom, and Emperor Xianfeng gave an imperial edict to suspend closing. Since then the Yellow River took the course of the Daqing River to empty itself into the Bohai Sea below Jijin county in Shandong. Subsequent to the major change of course, the river wandered over a vast area of the country for more than two decades. The dike to the north served as a guide wall, whereas there were no defenses on the vast south area in the counties of Dingtao, Shanxian, Caoxian, Chengwu, and Jinxiang in Shandong was inundated. Dikes of both banks downstream of Tongwaxiang were gradually built by 1877 and joined those existing above Tongwaxiang to form the present channel.

Tendency of Evolution

For thousands of years, five major changes of course took place in the lower Yellow River. It can be seen from the past history of the evolution of the river that:

1. During 1730 years from 602 B.C. to 1128 A.D., the course of the lower Yellow River flowed north of the present channel, shifting was characterized by an apex located somewhere between Junxian and Huaxian of Henan province. The Zhanghe River was the western boundary for the change of course, and Daqing River the eastern. The river emptied into the Bohai Sea by any of the courses by Huanghua, Wuli and Qingxian. The duration was different for the river subsequent to each change of course as Table 1 shows.

2. From 1128 to 1855, the lower Yellow River flowed to the south of the present channel due to the breach made by Du Chong and captured the Huaihe River before emptying into the Huanghai Sea. The northern boundaries of changes of river course were the Daqing River, and the extreme southern limit was the Yinghe and Huaihe River. From Wuzhi, in Henan, the apex, the flow path gradually shifted southward. The turning point on the southern bank was Guangwu Mt. (now Mangshan Mt.) and Aushan Mt. to the east, both eastern hills. The terrain constrained somewhat the trend of streamflow toward northeast. Subsequent to breaching of the dike at Yangwu in 1194, the river did not flow in Xinxiang and Jixian county any more. Aushan Mt. was gradually scoured away owing to the southward shifts of the river. In 1356 A.D., the county of Heyin to the north of Guangwu Mt. was submerged and the river ran out of control. The river took the path southward of Yangwu after the

Table 1.

Name of old channel	River reach	Flowing time (year)	Remarks
Old channel in W. Han dynasty	Wuzhi-Junxian	1700	The reach was too aged to be researched, but the time that the river no longer ran through dated later than 1194. If it was counted from the existence of the dikes in Warring States, the river flowed about 1,700 years.
	Junxian-Puyang	1700	It flowed in W. Han, Tang and Song dynasty, the time of no water flow about in 1170s and the river existed about 1700 years.
	Puyang-Guantao	580	It was formed in 602 B.C., subsequent to the change of course at Suxukou, the actual flowing time was 580 years except 23 years breach at Piaozi in W. Han dynasty.
Old channel in E. Han dynasty	East of Puyang	1038	It flowed eastward subsequent to the breach at Weijin in 11 A.D. and the flow stopped in 1048, flowing time 1037 years. The dike buildings existed after 69 A.D. and about 900 years history.
Old channel in N. Song dynasty	North of Puyang	80	It flowed about 80 years from 1048 to 1128.

dike breached at Sunjiadu, Xingze in 1448. Thence, the river gradually evolved to take the course below Zhengzhou to merge with the Huaihe River in Jiangsu (old channel in Ming and Qing dynasty), and flowed there about 727 years with the dike building on both banks for 400 to 500 years.

3. From 1855 to 1987, the lower Yellow River flowed in its present course which is the boundary of the southward and northward course and the shortest path to the sea. From east of Lankao to the river mouth of Lijin, in Shandong the river has run for 123 years, excluding 9 years after the man-made breach at Huayuankou. The reach above Lankao is still the old channel from Ming and Qing dynasties, with the dike system on both banks about 500 years old. Subsequent to the change of course at Tongwaxiang, high floodplains have been left on both banks due to headward erosion, and the situation has not recovered to the pre- avulsion days even though the river bed has severely aggraded. Compared to the old channels of 500-1000 years ago, the present course is still quite young. The Yellow River flooded frequently after the change of course at Tongwaxiang prior to 1949, and became a suspended river as well as the watershed of Huaihe and Haihe River due to aggradation. Since the founding of new China, the embankments on both banks have been heightened and strengthened three times, Sanmenxia Reservoir on the mainstem has been built and certain areas along the side of the river have been reserved for detention basins. Large-scale training works have been constructed to limit shifting of the thalweg, which is advantageous to water and sediment conveyance, and allowing the river to be used to its design capacity for flood control. There have been no dike breaches during the summer and autumn floods for more than 40 years, an exceptional case rarely seen in past history. Nevertheless, major floods might happen and the river bed would still be aggraded before floods and sediment could be effectively put under control. For the sake of flood prevention in a long run, firstly, Xiaolangdi project, a key project for flood control in the flood prevention system on the lower Yellow River, should be built. This construction would bring about benefits of flood prevention over a long period of time, raise the ability to safeguard the lower Yellow River during floods with a frequency of once in one thousand years.

Sediment deposition on the lower course could be reduced appreciably though reservoir regulation and management. Soil and water conservation works on the loess plateau area should be continued and intensified. This would include reforestation and constuction of terraced fields in a planned way, and building key projects for controlling gullies cooperating with reservoirs on the main river and its tributaries to further control floods and sediment. The dikes should be heightened and strengthened and the hidden defects be remedied by engineering measures such as warping, which will play an important role in maintaining the designed capacity of flood conveyance, preventing, bursting and overtopping as well. River training works and floodplain management should be done continuously to further stabilize the channel for dominant flow to prevent dike breaches due to impingement. Withdrawal of water from the Yellow River for warping by means of existing intakes along both banks of the river to raise the ground level behind the levee should be continued. Measures available today to harness the river are more advanced than during the dynasties in history. The present course will be kept for a long time to benefit the people on both banks in the wake of development of science and comprehensive management of the entire river.

REVIEW OF MEASURES FOR REDUCING SEDIMENTATION OF THE LOWER YELLOW RIVER

Chen Zhilin
Reconnaissance Planning & Design Institute
Yellow River Conservancy Commission
People's Republic of China

Introduction

The lower Yellow River is a heavily sediment-laden river with severe aggradation. At the present time, the river bed is commonly 3-5 meters higher than the adjoining ground, in some stretches even 10 meters or so higher. Hence the Yellow River has been famously called the "suspended river". The continuously rising river bed poses a serious threat to flood prevention on the Lower Yellow River. Harnessing and development of the Yellow River as a whole requires that special attention be given to decreasing aggradation.

Basic Estimate of Variation of Water and Sediment in the Basin

The variation of river channel erosion and aggradation is the result of the interaction of water inflow, silt factors, boundaries of the river and conditions of the river mouth. Therefore, an estimate of the trends in variation of water and sediment in the upper and middle reaches of the Yellow River must be made initially.

Hydrological observations include a series of 56 water years, from July 1919 to June 1975. These data indicate a mean annual runoff of some 42 billion cu.m. at the site of Sanmenxia and some 47 billion cu.m at Huayuankou. Adjusting the series by taking into consideration each year's water consumption for irrigation and storage by large reservoirs, the corresponding natural runoff at Sanmenxia and Huayuankou is 50 billion cu.m and 56 billion cu.m per year, respectively. The results of investigations in 1979 indicate that the total water consumption for industry and agriculture in the entire Yellow River basin reaches 27.1 billion cu.m, of which 17.3 billion is in the catchment above Huayuankou and 9.8 billion cu.m below Huayuankou. The total consumption for industry and

L. M. Brush et al. (eds.), Taming the Yellow River: Silt and Floods, 557–584.

agriculture is as much as 48.4% of the river's mean annual runoff. Detention of sediment in reservoirs and diversion of water and sediment from the river for irrigation have caused the inflow of sediment to the Yellow River to decrease, and water and soil conservation measures and hydraulic works on the middle reaches also play a positive role. The mean annual sediment retained by all measures since 1970 is some 0.51 billion tons (see table 1). Different analyses show that all the measures reduced the sediment load annually by 0.15-0.3 billion tons on average for the stations at Longmen, Huaxian, Hejin, and Zhuangtou. These stations control the sediment inflow to the Sanmenxia Reservoir.

Since 1960 many water conservation works have been built on the upper and middle reaches as well as on their tributaries. The two large reservoirs at Liujiaxia and Sanmenxia influence the water and silt carried by the middle reaches. Sanmenxia had an original storage volume of 16.2 billion cu.m at an elevation of 340 m. Storage began in 1960. Liujiaxia Reservoir on the main river stem of the upper reach has a storage capacity of 5.7 billion cu.m and storage began in October 1968. The total number of small and medium sized reservoirs exceeding one million cu.m in capacity built on tributaries of upper and middle reaches is 739 with a total storage capacity of 6.59 billion cu.m. In addition, water use by industry and agriculture is increasing steadily. All these factors cause the quantities of water and sediment observed in the Yellow River basin to be influenced by human activities to varying degrees. Therefore, inflow of water and sediment can be computed (see Table 2) under the present conditions by taking into account the hydraulic projects mentioned above, the effects of water and soil conservation measures and water consumption for industry and agriculture on the upper and middle reaches. From Table 2 it can be seen that based on two series, the shorter one from July 1919 to June 1960 and the longer from July 1919 to June 1984, the summation of the mean annual inflow sediment observed at the stations of Sanmenxia, Heishiguan and Xiaodong is 1.57-1.67 billion tons. Although a series of low sediment discharges has appeared since 1974, the mean annual incoming sediment load is still nearly 1.6 billion tons. Using a conservative estimate of the effect of sediment detention by various engineering measures, the mean annual incoming sediment load under the present conditions is some 1.44 billion tons, 10% less than the annual observed mean.

Depending upon projections of the social and economic development within the basin, the total water consumption for industry and agriculture will be some 40 cu.m, or as much as 71.4% of the Yellow River's natural mean annual runoff. This water consumption is the limiting factor on the water and sediment conditions along the whole river. If consumption is further increased, aggradation in the lower reaches would speed up and the task of flood prevention on these lower

Table 1. The annual mean sediment retained by various measures

Item	Reservoirs on Main River	Reservoirs on Tributaries	Sediment Diverted by Irrigation	Land Formed behind the Desilting Dam	Terrace Field	Total
Annual Mean Silt Retained and Diverted	0.1 billion tons	0.12	0.06	0.2	0.03	0.51
%	20	24	12	39	6	100

Table 2. Incoming water and sediment observed and under present conditions per year on an average

Items / Series	Summation of Stations of Longmen +Huayuankou+Hejin+Zhuangtou (*1)				Summation of stations Sanmenxia +Heishiguan+Xiaodong (*2)			
	Annual mean Inflow Water (billion cu.m)		Annual mean Inflow sediment (billion tons)		Annual mean Inflow water (billion cu.m)		Annual mean Inflow sediment (billion tons)	
	Observed	Computed	Observed	Computed	Observed	Computed	Observed	Computed
Jul 1919-Jun 1960	42.666	33.997	1.622	1.483	47.496	37.585	1.636	1.439
Jul 1960-Jun 1984	41.298	40.235	1.394	1.394	44.313	43.032	(1.445)	1.440
Jul 1919-Jun 1984	42.136	36.300	1.538	1.456	46.321	39.600	1.566	1.439

(*1) Inflow to the Sanmenxia Reservoir.
(*2) Inflow to the Lower Yellow River.
( ) Sanmenxia Reservoir retained sediment returned to original

reaches would be much more difficult.

Since more conservation facilities and soil conservation measures will be implemented in the future, the benefits of reducing sediment from entering the river can be expected to remain. It is feasible to take into consideration the condition of the river basin since 1970 in predicting the incoming sediment in the year 2000. The predicted value of inflow water and sediment in the year 2000 will be taken as the basis for analyzing the influence of engineering measures on erosion and deposition in the lower reaches. Water and sediment data from July 1950 to June 1975 were used for the predictions. After estimating properly the possible amount of reduction, the total annual mean runoff of the three stations, Sanmenxia, Heishiguan and Xiaodong, on the main river and tributaries would be 34.24 billion cu.m, a decrease of 11 billion cu.m, or 24.1% of the present runoff. The decrease in the flood season amounted to 64.5% of the reduction for the year. The predicted mean annual incoming sediment is 1.37 billion tons reduced by 76 million tons, or 5.2% from the observed sediment load. Sanmenxia Reservoir is operated to store water and silt in the nonflood season (November through June of the next year) and to dispose of sediment through lowering the level during the flood season. The incoming sediment load for a year will be, basically, that which is delivered into lower reaches during the flood season only. As a consequence, the load in the flood season will be increased, rather than decreased, by 0.166 billion tons.

In addition to Sanmenxia and Liujiaxia, the regulation of Longyangxia Reservoir must also be taken into account in order to predict incoming water and sediment in the year 2000. Longyangxia Reservoir. It is located on the main river upstream from Liujiaxia Reservoir and is a huge reservoir with a storage capacity of 26.8 cu.m. The reservoir which controls 36% of the annual runoff produces further adjustment in the natural runoff processes through its operation that commenced impoundment in October 1986. Compared with other existing projects, inflow to the lower reaches will be reduced by some 4.0 billion cu.m per year on an average in the flood season. At present, Sanmenxia reservoir is operated in such a way that almost all the annual sediment load will be deposited in the lower reaches during the flood season. If inflow were reduced in the flood season, aggradation would become more serious.

Present Status and Prediction of Erosion and Deposition in the Lower Yellow River

Erosion and deposition in the Lower Yellow River depends mainly on the inflow of water and sediment. Before Sanmenxia Reservoir was completed, the river was very strongly aggrading. During the period from 1919 to 1949, the annual incoming water volume was 46.8 billion cu.m and sediment 1.56 billion tons. The dikes broke many times and large amounts of sediment were flushed out of the banks reducing the intensity of deposition on the lower reaches. By estimation, if there had been no breaches or avulsions, the sediment deposited on river bed of the Lower Yellow River would have been 0.32 billion tons per year. During the 33 years from July 1950 to June 1983 the mean annual water and sediment inflow were respectively 44.9 billion cu.m and 1.4 billion tons. Since Sanmenxia reservoir began operation in September 196 annual deposition in the lower reaches averaged some 0.211 billion tons. By the year 2000, the predicted mean annual sediment inflow is 1.373 billion tons. As noted earlier, through the operation of Sanmenxia Reservoir, sediment will be released only in flood seasons. From calculations based on the present conditions of the lower course, it is estimated that the mean annual amount of sediment deposited on the river bed will be 0.379 billion tons (see table 3).

Effects on Reduction of Sediment in the Lower Reaches by Construction of Key Projects on the Middle Yellow River

PRACTICAL EXPERIENCES IN REDUCING DEPOSITION ON THE LOWER REACHES BY THE SANMENXIA RESERVOIR

During the period of impoundment from September 1960 to March 1962, reservoir deposition was serious. Although in March 1962 the mode of operation was changed into flood-detention only during the flood season, the reservoir was still undergoing deposition due to insufficient outlet capacity. In the 10-year period from July 1960 to June 1970, the total sediment retained in Sanmenxia Reservoir amounted to some 5.641 billion tons while the accumulated amount of sediment in the lower reaches was only 3 million tons and the river had been eroded by 2.31 billion tons. This implies that the retention of sediment in Sanmenxia delayed the development of aggradation in the Lower Yellow River by almost 10 years. If the number of years from 1960 to the year when water stages corresponding to a discharge of 3000 cms, the period during which the stage was lowered and then raised to the same level as in 1960, as a result of erosion and sedimentation, are taken as the years that reduction of deposition was effective at a given locality, then the number of years was more than 10 years in the upper reach to Huayuankou, 10 years on the reach between Huayuankou and Gaochun, 7-9 years between Aishan and Luokou, and 5-6 years from Yangfang to Lijin. The number of years becoming less as one proceeds downstream (see fig. 1).

Table 3. The annual mean inflow water and sediment, annual mean erosion and deposition on lower reaches from July 1950 to June 1975

Items / period	Annual mean Inflow Water (billion cu.m)		Annual mean Inflow Sediment (billion tons)		Annual mean Erosion & Deposition (billion tons)	
	Observed	In 2000's Level	Observed	In 2000's Level	Observed	In 2000's Level
Flood Season	26.25	19.09	1.192	1.358	0.183	0.481
Nonflood season	19.65	15.15	0.257	0.015	0.039	-0.102
Year	45.90	34.24	1.449	1.373	0.222	0.379

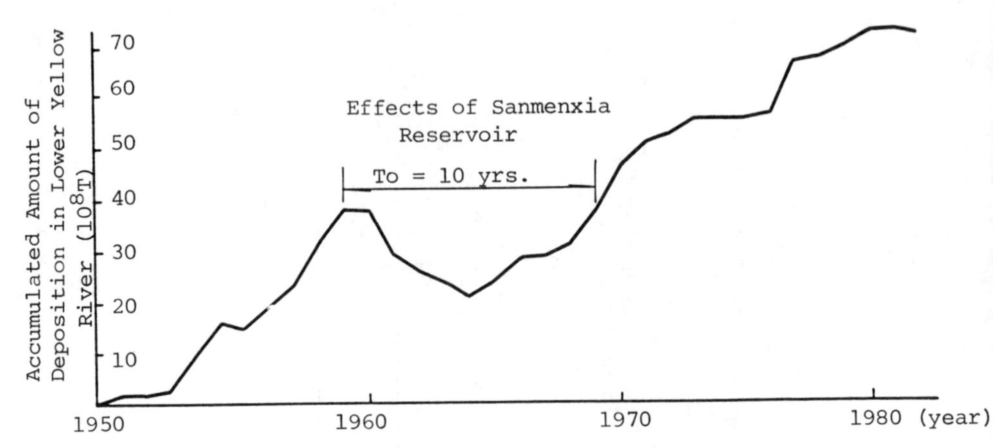

Figure 1a. Course of Accumulated Amount of Deposition in the Lower Course of the Yellow River.

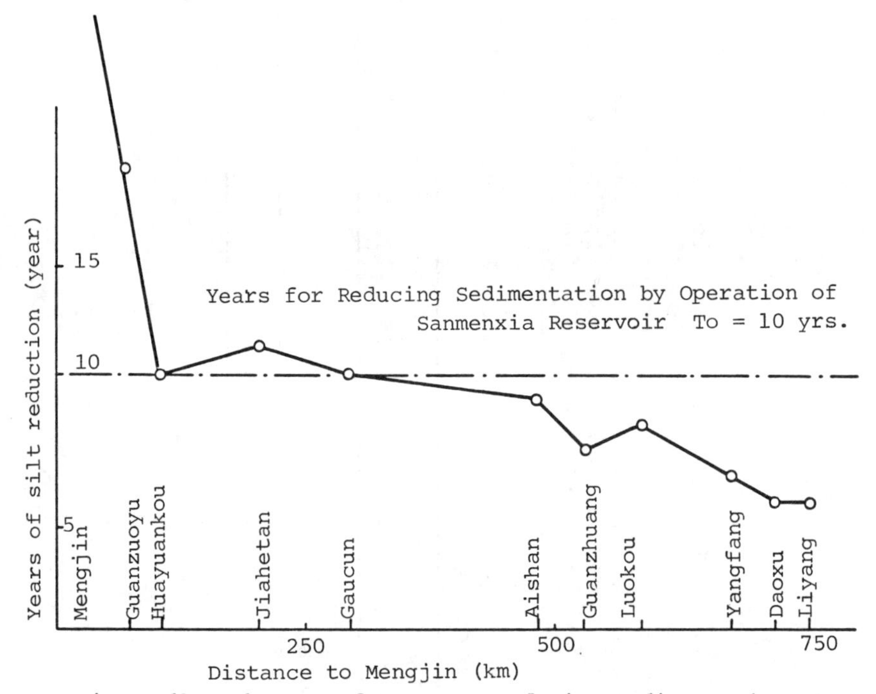

Figure 1b. Changes of Years on Reducing Sedimentation Along River Course by Operation of Sanmenxia Reservoir

Figure 1. Effects of Reduction of Sedimentation in the Lower Reaches by Operation of Sanmenxia Reservoir

The river in the Sanmenxia Reservoir above Tongguan is an alluvial stream. The initial operation raised the stream bed at Tongguan by 4.5 m and deposition from the reservoir retrogressively extended upstream along the Weihe River and the main river. This brought a serious threat to the alluvial plain and the industrial-agricultural base centered at Xian. Meanwhile, reservoir deposition was extraordinarily serious. Operating for 4 years, 4 billion cu.m of storage was deposited below the stage of 335 m and the reservoir lost 41% of its storage capacity. After being twice restructured, the reservoir outlet capacity corresponding to stages of 300 m and 315 m were respectively 3240 cms and 9390 cms, and reservoir deposition has been controlled through lowering the operational water level during floods. Capacity in the main river channel has recovered by 1.05 billion cu.m below Tongguan and the mean bed elevation at Tongguan has fallen by 2 m according to observations since November 1973.

During the period from Nov, 1964 to Dec, 1973, a part of the previous deposits in the main channel were flushed out of the reservoir through bottom outlets which were reopened and through two new tunnels. The total sediment deposited in the lower reaches was up to 3.92 billion tons, or 0.432 billion tons per year, the most serious since 1950.

Since 1984, the operational level in Sanmenxia Reservoir has been lowered in the flood season, generally to 300-305 m in order to scour the silt retained during the previous impoundment period. Impoundment in the nonflood season in which the highest water level is limited to 326 m will not have any prolonged influence on the river bed elevation at Tongguan. The storage capacity available to be used for regulation is some 1.8 billion cu.m. The present operation of Sanmenxia Reservoir not only has minimized the reservoir deposition below Tongguan and steadily maintained a specified effective storage capacity, but also has reduced aggradation in the lower river channel. Based on the difference between inflow and outflow of the sediment discharge of Sanmenxia Reservoir observed at Tongguan and Sanmenxia station from July 1974 to June 1984, annual average storage of sediment in the nonflood season was about 0.132 billion tons. This amount, or perhaps more, was flushed out of the reservoir in flood seasons. Sediment transport capacity is rather high during the flood season and quite low at normal discharges in the nonflood seasons. It has been shown by analysis and computation that a reduction of river deposition by 0.06-0.08 billion tons may be expected if the disposal of sediment is concentrated only in flood season. Since storage of sediment during the low water season will be balanced by ejection of sediment in the flood season, it is reasonable to say that the effect of reducing aggradation in the Lower Yellow River through regulation of water and sediment by Sanmenxia Reservoir will continue to be effective.

Practical experience at Sanmenxia Reservoir has shown that regulation of sediment as well as water in the reservoir can be accomplished and, if the reservoir can maintain a definite effective storage capacity over the long run, a long term impact on the reduction of deposition in the lower reaches can be expected.

XIAOLANGDI RESERVOIR: A PROJECT IN THE PLANNING STAGE

Xiaolangdi Reservoir will be built at the exit of the last gorge of the middle reaches, 131 km downstream from Sanmenxia Reservoir. According to the prediction for the year of 2000,the reservoir's mean annual water inflow will be some 31.49 billion cu.m, or 92% of the mean annual incoming water of the whole Yellow River basin. The mean annual incoming sediment will be some 1.345 billion tons with a suspended sediment median diameter of 0.036 mm, or 98% of the mean annual sediment inflow for the whole basin. Therefore, this project will occupy a position of strategic importance in harnessing the Yellow River. The developing tasks of Xiaolangdi Reservoir are flood prevention and reduction of deposition as well as water supply, irrigation and power generation, which means that the reservoir will be a multi-purpose project.

The proposed Xiaolangdi Reservoir has a total capacity of 12.56 billion cu.m at a stage of 275 m with a maximum height of 173 m and capacity reserved for sediment storage of 7.25 billion cu.m. Through lowering the water level to dispose of sediment during the flood seasons and staging in the nonflood seasons, after the capacity for sediment storage is fully utilized, an effective storage capacity of 5.0 billion cu.m, of which 4 billion cu.m for flood control and another 1 billion cu.m for regulation of sediment, can be expected.

In order to fully utilize Xiaolangdi Reservoir to reduce deposition in the lower reaches, the operational method of gradually raising the water level is recommended so that the reservoir may mainly eject fine sediment and retain the coarser sediment. Observations at Sanmenxia Reservoir show that the percentage of the outflow of suspended sediment coarser than 0.025 mm will increase with an increase in the ratio of sediment disposal, i.e. the ratio of sediment outflow to sediment inflow (see fig. 2). It is quite possible to achieve the purpose of retaining much coarser sediment in the reservoir by controlling the ratio of sediment charged through a policy of raising the operation level in steps provided that enough outlet facilities are installed.

According to analyses of the characteristics of the sediment transported in the Lower Yellow River, there exists almost no

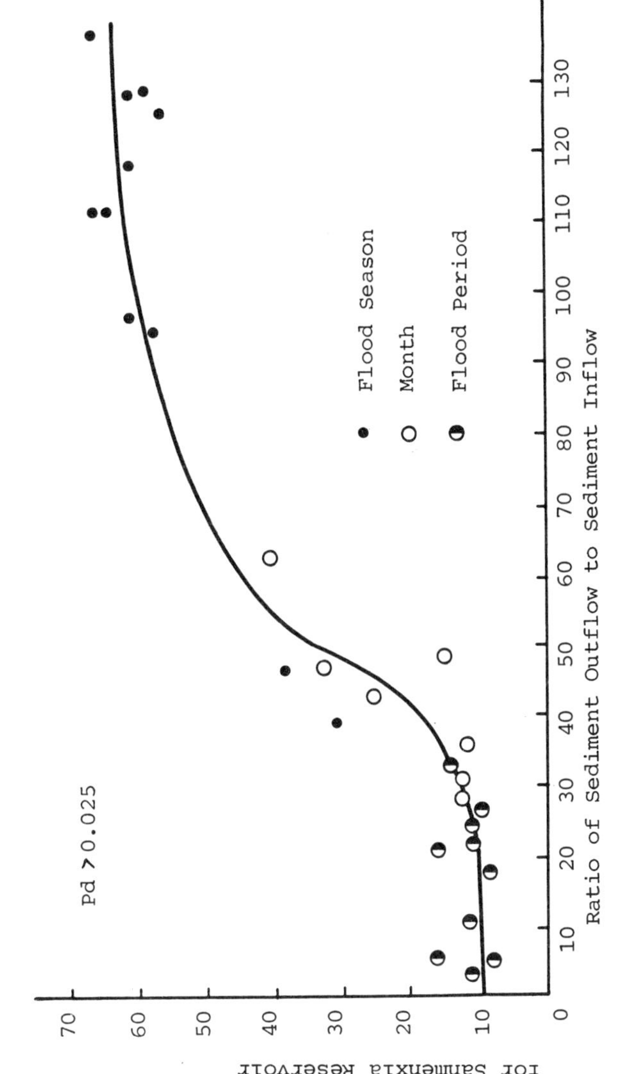

Figure 2. Relationship Between Ratio of Sediment Outflow to Sediment Inflow and Pd >0.025mm for Sanmenxia Reservoir

fine sediment, less than 0.025 mm in diameter, on the main river's bed. This part of the fine sediment can be regarded as wash load. Due to deposition from overbank flow, the capacity for disposal of wash load is only 90% and the transport capacity of the suspended sediment coarser than 0.025 mm is about 70%. Therefore, the effect of the reservoir's retention of coarser sediment in reducing the aggradation in the lower reaches could be brought into full play by taking advantage of the large sediment transport capacity of the river channel.

There are two indices with which to express the effect of reduction of deposition by Xiaolangdi Reservoir. The first is the ratio of the amount of sediment retained in reservoir to the amount of reduction in deposition in the Lower river or, in other words, amount of sediment retained in reservoir per unit of amount of reduction in channel deposition. The reduction in deposition means the difference between the channel deposition before completion of this project and channel deposition influenced by sediment stored in reservoir during the same period. Nevertheless, the smaller the ratio, the greater would be the effect of reduction in deposition. The second index is the number of years during which the reduction in deposition is effective. This again represents the ratio of the total amount of reduction in deposition after completing the project to the annual average amount of channel deposition prior to the project.

Two different modes have been considered for bringing Xiaolangdi Reservoir on line. One would be to raise the reservoir stage up to the normal stage in one step to store water and sediment as is commonly adopted for a storage reservoir in order to fill up the reservoir and generate electricity as soon as possible and to obtain the benefit of high power-generation during the initial stages of reservoir operation. This is based on the premise that the storage capacity required for flood prevention is preserved and that the reservoir will operate in a way so that sediment deposition is minimized in the lower Yellow River channel. At the assigned stage of 250 m, clear water released from the reservoir will persist for 9 years to facilitate the erosion in the lower channel. The second mode would be to raise the reservoir stage gradually so that the reservoir can store coarse sediment. The initial regulation stage during flood season would be 200 m and stage would be raised up to 250 m step by step. After the sediment storage is fully realized, the operation level will gradually be lowered to 230 m when the reservoir turns to normal operation. The effects produced by these two different operational methods mentioned above through storage of water and sediment in the reservoir to reduce deposition can be seen in Table 4.

When the reservoir capacity allocated to sediment storage is fully attained, the storage capacity reserved for regulating sediment can still be utilized for the regulation

Table 4. Effects of Xiaolangdi Reservoir

Operation Methods	Number of Years for the Sediment Storage Capacity Fully deposited (years)	Sediment Retained in the Reservoir billion (cu.m)	Sediment Retained in the Reservoir billion (tons)	Mean Ratio of Sediment Outflow to Inflow (%)	Reduction in Deposition on Lower Reaches (b. tons)	Ratio of Retained to Reduction in Deposition (b. tons)	Number of Years for Reduction in Deposition (year)
Stage Raised in One Step	13	7.25	9.425	46.2	5.71	1.65	15
Stage Raised Gradually	31	7.25	7.425	77.2	7.25	1.30	19.2

of water and sediment in addition to the controlled-operation of Sanmenxia Reservoir, to maintain the effectiveness of the reduction of deposition in lower reaches. The regulation of water and sediment in Xiaolangdi Reservoir can be divided into two parts: regulation of water and sediment in the flood season and an artificial flood in the nonflood season.

Regulation of Water and Sediment in Flood Season. Water and sediment in the Yellow River vary markedly. Floods with high concentrations of sediment usually cause severe deposition in the lower river. In 1977, deposition from two flood events with hyperconcentrations of sediment amounted to 1.0 billion tons or almost 4.7 times the annual mean deposition. According to the 33 years of record before 1983, floods with high sediment concentrations have taken place 11 times. The volume of flood and sediment runoff were respectively 19.7% and 13.7% of the total in the past 33 years, and the deposition in the lower reaches amounted to 54.2% of the total for the 33 years. Therefore, it is possible to use 1 billion cu.m of sediment regulation capacity in Xiaolangdi Reservoir to temporarily store a part of these floods until a normal year occurs with a low sediment load, or a year of abundant water occurs to flush the sediment and recover the sediment regulating capacity. It is estimated that by means of this type of multi-year regulation of water and sediment, deposition in the lower reaches might be further reduced by 0.04 billion tons per year on average.

Artificial Floods in Nonflood Seasons. The Xiaolangdi Reservoir during nonflood seasons can be operated to release a discharge of 5000 cms which corresponds to the bankfull discharge of the lower reaches. If this is done at the right moment after the requirements for flood control and irrigation are met, it can remarkably improve the sediment transport capacity of the river course. Since the demand for water in the lower reaches increases day by day, the artificial flood would be made once every three years on average. The average volume of water for each artificial flood will be about 4 billion cu.m. Although the artificial flood may promote erosion in the lower reaches, deposition may be induced by river bed armoring and widening of the channel cross section. The net effect of reduction in deposition for each artificial flood is estimated to be 0.06 billion tons, equivalent to an annual average of 0.02 billion tons.

To sum up, proper regulation of water and sediment in Xiaolangdi Reservoir will reduce deposition in lower Yellow River by 0.06 billion tons per year on average.

Longmen Reservoir: Project in the Planning Stage. The proposal Longmen project will be located at the exit of gorges 274 km upstream from the Sanmenxia Reservoir. According to projections for the year 2000, the annual inflow will be 24.66 billion cu.m and the annual inflow of sediment will be 0.94 billion tons. The purpose for this project is flood control, reduction of channel deposition, hydropower and irrigation.

The reservoir will have an original capacity of 11.4 billion cu.m, of which 7.5 billion cu.m will be available for sediment storage. When this capacity is filled, through discharge of sediment by lowering the water level in flood season and storage of water in the nonflood season, an effective capacity of 4.34 billion cu.m can be obtained, of which 3.26 billion cu.m will be for flood control and the rest, 1.08 billion cu.m for sediment regulation.

Impacts of the Longmen Reservoir on reduction of deposition in the lower reaches can be divided in to two parts: the first is the stretch from Yumenkou, the exit of the gorges below Longmen to Sanmenxia, and the second is the lower reaches below the exit of Xiaolangdi gorge. At Longmen, the average medium diameter of suspended sediment is 0.045 mm which is coarser than the suspended sediment flowing into Xiaolangdi Reservoir. Eighty-one percent of the coarse sediment (greater than 0.05 mm) will be controlled by the project so that Longmen Reservoir will aid in reducing the aggradation even better than Xiaolangdi Reservoir.

Like Xiaolangdi Reservoir, two different operational methods have been studied for the Longmen Reservoir. The first would raise the water stage ine one step at once to store water and sediment; the second would gradually raise the water level in order to retain coarser sediment as much as possible in the reservoir. When the sediment storage is filled, the operational level will be managed to maintain a large effective storage capacity. Different effects of these two different operation methods are shown in table 5. It can be seen that raising the reservoir stage gradually will be more effective in inducing deposition of coarse sediment. The river course from Longmen to Sanmenxia will be mainly influenced by the operation of Longmen Reservoir.

Qikou Reservoir: Project in the Planning Stage. The Qikou project is to be built upstream from the Longmen project in the middle part of gorges of the middle Yellow River. According to projections for the year 2000, it will have a mean annual inflow of 21.86 billion cu.m and a mean annual incoming sediment load of 0.586 billion tons. It will control 54% of the coarse sediment (greater than 0.05 mm) of the Yellow River. The project will be developed for power-generation first, then for runoff regulation and storage of coarse sediment while taking account of water-supply and irrigation.

Qikou Reservoir is proposed to have a total volume of 12.48 billion cu.m including a sediment storage capacity of 11.05 billion cu.m and an effective storage capacity for long term use of 2.7 billion cu.m. Qikou Reservoir is to be raised to its

Table 5. Effects of Longmen Reservoir

Stage Operation Methods	Number of Years for the Sediment Storage Capacity Fully deposited (years)	Sediment Retained in the Reservoir billion (cu.m)	Sediment Retained in the Reservoir billion (tons)	Mean Ratio of Sediment Outflow to Inflow (%)	Reduction in Deposition LOM/SMX	Reduction in Deposition SMX/LIJ	Ratio of Retained Sed. to Reduction in Deposition (%)	Number of Years for Reduction in Deposition (year)
Raised in One Step	11	7.5	9.75	10	1.976	4.48	1.51	12
Gradually	19	7.5	9.75	50	3.25	4.875	1.20	13

LOM stands for Longmen, SMX for Sanmenxia, LIJ for Lijin

operating level at once. After the sediment storage is fully filled, the water level will be decreased gradually in the flood seasons in order to retain effective storage. In the low water seasons, the water level can reach 785 m at most.

Retaining silt and reducing deposition are to be the same as at Longmen, again reflected in two parts, one from Longmen to Sanmenxia and another below Xiaolangdi. The effects of decreased deposition can be seen in Table 6.

Influences of Flood Plains in the Middle and Lower Reaches.

Along the wandering reaches of the middle Yellow River after it passes out of the gorges there are broad flood plains; hence, diverting water and warping, combined with river regulation, can also slow deposition in the lower reaches. At present, according to our studies, two localities are available for warping, one is the flood plain on both banks between Tongguan and Longmen and the other is Wenmeng plain below Xiaolangdi on the north bank of the river.

THE FLOOD PLAIN BETWEEN LONGMEN AND TONGGUAN

This river course is 130 km in length with an average width of 8.5 km. The total area of the river is about 1,130 sq.km, of which the area of flood plain is about 600 sq.km and is inhabited by more than one hundred thousand people. The warping is to be done in combination with the proposed Longmen Reservoir. According to the plan, a low dam will be built at nearby Yumenkou, exit of the gorge, with intake gates on both sides of the dam to divert water for warping and incorporating hydropower generation. Taking into account the flood control operation of Sanmenxia Reservoir, warping will be carried out above 335 m in elevation in an area of 566 sq.km. Warping works are to be arranged in conformity with the overall planning including river training and bank protection works. The diverting discharge is some 500 cms of which 300 cms is for the west bank and the rest for the east.

Diverting water for warping reduces the amount of sediment. The reduction in deposition in the lower reaches is shown on Table 7.

WENMENG FLOOD PLAIN

Wenmeng flood plain on the north bank of the river below the exit of Xiaolangdi gorge has a length of 91 km and an average width of 3.7 km. The total area of the plain is some 338 sq.km. At present land is cultivated up to 0.4 million mu and 0.1 million people will be influenced by warping on the plain. Two schemes of warping have been studied: the first is to build the Xixiayuan project near the gorge exit to divert

Table 6. Effects of Qikou Reservoir

Operation Methods	Number of Yrs before Sediment Storage Capacity Being fully fill-ed (years)	Sediment Retained in the Reservoir billion (tons)	Sediment Retained in the Reservoir billion (cu.m)	Mean Ratio of Sediment Outflow to Inflow (%)	Reduction in b.tons, Deposition LOM/SMX	Reduction in b.tons, Deposition SMX/LIJ	Ratio of Retained Sed.to Reduc.in Deposition (%)	Number of Years for Reduction in Deposition (years)
Stage Raised in One Step	30	14.49	11.05	15	2.868	7.816	1.36	20

LOM stands for Longmen, SMX for Sanmenxia, LIJ for Lijin

Table 7. Effects of warping on flood plain

Name of Flood Plain under Warping	Location of Key Project	Area Deposited (sq km)	Mean Thickness of Deposits (m)	Total Sediment Deposited (b.tons)	Diverting Discharge (cu.m/s)	Reduction in Deposition (b.tons)	Ratio of Sediment Retained (%)	Number of Years for Reduction in Deposition (year)
Long-Tong Stretch	Yumen-kou	566	24	19.32	500	10.12	1.91	27
Wenmeng Plain	Xixia-yuan	338	17	8.04	1000	4.39	1.83	12
Wenmeng Plain	Xiao-langdi	338	34	16.10	500	11.10	1.45	29

water for warping with a diverting discharge of 1,000 cms; the second is to divert water from outlets provided for sediment ejection on the left bank of Xiaolangdi dam with a diverting discharge of 500 cms. The diverted sediment laden water, after passing through the winding mountain canal, is to be desilted on the Wenmeng plain. It is equivalent to an enlargement of the sediment storage of Xiaolangdi Reservoir for the purpose of reducing deposition in the lower reaches.

Warping operations will be carried out when the discharge is 2000-6000 cms and the ratio of sediment concentration to discharge is greater than 0.03. Channel aggradation in the lower reaches will be reduced as shown in Table 7.

Reduction in Deposition by Water Transfer from the Yangtze to the Yellow River

The distribution of water resources is quite uneven in China. It has abundant water in the south and a shortage of water in the north. In the long run, to meet the increasing demand of water for the national economy and to develop the great north-west of China, a south-to-north water transfer is imperative. Transporting the abundant water of the Yangtze River to supplement the Yellow River and to assist in scouring the lower river course can also contribute to reducing deposition.

Two different routes of south-to-north water transfer may produce remarkable effects on the reduction of deposition. One is called the middle route in which water is conducted from the Danjiangkou Reservoir on the Hanjiang River, a tributary of the middle Yangtze River, into the Yellow River at a place 20 km upstream from the Beijing-Guangzhou railway bridge. The second is the west route which includes three feasible lines proposed in preliminary studies: one involves building a dam on Tongtian River and pumping water by steps into the upper reaches of the Yellow River, the annual transfer would be 10 billion cu.m; the second is to build a dam on the Tadu River drawing water into the Yellow River at an annual rate of 5 billion cu.m; the third is to build a dam on Yalong River conducting an annual volume of 5 billion cu.m of water into the Yellow River. The south-to-north water transfer along the west route, through regulation of Longyangxia and Liujiaxia reservoirs, will increase the power generating capacity of the hydro-power stations on the upper Yellow River and meet the water requirements for development of the north-west region of China. Some of the water can also be used to supplement the inflow of water to the lower reaches to decrease channel aggradation.

The effects of the south-to-north water transfer by the middle or west routes on the reduction of deposition are compared in table 8.

Table 8. Effects of supplying water to the Yellow River in different routes

Routes	Reduction in deposition (billion tons)	Increment of Water for Transporting Sediment in the Yellow River in Flood Season (billion cu.m)				
	Reaches	2.0	5.0	10.0	15.0	20.0
West Route	Longmen --- Sanmenxia	0.022	0.054	0.108	0.162	0.216
	Sanmenxia --- Lijin	0.039	0.098	0.196	0.294	0.392
Middle Route	Sanmenxia --- Lijin	0.050	0.125	0.250	0.375	0.500

Estimate of Long-Term Effects of Reduction in Deposition by Water and Soil Conservation Measures on Tributaries of the Middle Yellow River

Tributaries of the middle Yellow River between Hekouzhen and Sanmenxia are the main sources of sediment that flow into the river, as well as the major region of comprehensive control of the Yellow River including water and soil conservation. Since 1970, the annual mean incoming sediment of the Yellow River has been reduced by 0.2 billion tons on average by all water and soil conservation measures in the upper and middle reaches of Yellow River. From Table 1 it can be seen that sediment detained by reservoirs on the tributaries and by desilting dams is 63% of the sediment detained by all the measures. The key problem is how to steadily maintain the efforts to reduce sediment by water and soil conservation measures in the long run once the existing reservoirs and desilting dams are fully deposited.

Assuming that the engineering works on tributaries of the middle reaches constructed at a rate comparable to that since 1970, it is possible to reduce sediment by 0.4 billion tons per year over the next 50 years. This means that the annual mean incoming sediment of the Yellow River will decrease from 1.6 billion tons to 1.2 billion tons.

The small and medium sized reservoirs and desilting dams built in the region of accelerated erosion will raise the local base level, and, along with increasing sediment detained by the dams and the enlargement of silted land, gravitational erosion to gullies will be somewhat alleviated. Taking the area of Wuding River, a tributary of the middle Yellow River as a source of coarse sediment as well as a region under severe erosion, as an example, the variation of the erosion modulus based on natural conditions and including both existing works (including the earth dams formed by natural landslide) and the complete undestroyed basin (small basins surrounded by mountain ridges) can be used to show the function of control of works on natural erosion. The ratio of the area of deposition in the reservoir to the area of sediment yield is referred to here as an erosion-control parameter which is taken as an index reflecting the erosion-control ability of a reservoir. In the same physiographic regions, erosion in an area of high sediment yield obviously decreases with an increase of erosion parameter (see Table 9).

All of the above information indicates that when reservoirs and desilting dams are filled with sediment, erosion will be weakened under the action of erosion-control of existing works. The impacts of reduction of sediment will be effective in the long run even if sediment were sluiced off the

Table 9. Relationship between erosion-control and erosion module in the source area of the Wuding River

Items	Hydrological Station of Jingbian	Jiucheng Reservoir	Xinqiao Reservoir	Ximawan Basin	Lijiagou Naturale Reservoir
Area of Sediment Yield (km)	445	270	707	104	8.84
Number of Yrs in Statistics	13	17	14	33	130
Area of deposition in Reservoirs (sq.km)	0	5.92	38.48	5.5	0.86
Erosion-Control Parameter (%)	0	2.19	5.4	5.3	9.7
Erosion Modulus (t/sq.km.year)	20800	17360	17200	7550	1600

reservoirs instead of being retained. The effectiveness depends upon the magnitude of the erosion-control parameter. For example, the Lijiagou Dam is a natural dam formed by a bankslide. In the past 130 years neither water nor sediment has flowed out of the gully and floods have never overflowed the dam. This may be related to the fact that the erosion-control parameter has been kept at 0.1 and incoming sediment has decreased 92%, slowing down the reservoir deposition. Therefore, different measures aimed at increasing the erosion-control parameters can be adopted depending upon different situations. These include building silt trapping reservoirs or check dams upstream to increase the erosion-control parameter for a given sediment yielding area and for check dams controlling an area of some 10 sq.km. Raising the dam height on the deposited material behind the dam in stages and reforming the outlet works will also increase the erosion-control parameter. This will function to reduce sedimentation. Since the flood prevention ability of check dams or small desilting dams is quite low, the works should be designed so that the outlet works could still stand even though the dam itself burst. In fact, an earth dam is easy to store at a small cost.

After implementation of all of the aforementioned measures, in addition to strengthening other water and soil conservation measures, it is expected that in next 50 years the average annual incoming sediment of the Yellow River can be kept at a level of 1.2 billion tons. The average annual incoming water, coordinated with the realization of a south-to-north water transfer project, may be maintained at 40 billion cu.m., which is almost the same as the incoming water and sediment observed during July 1983. The sum of the average annual water and sediment observed at Sanmenxia, Heishiguan and Xiaodong are respectively 40.49 billion cu.m. and 1.176 billion tons with a deposition of 0.189 billion tons per year on average. Through water and sediment regulation by large reservoirs on the upper and middle reaches, as well as artificial flooding, deposition in the Lower Yellow River can be kept at 0.1 billion tons per year so that strong aggradation in the course of the Lower Yellow River will be changed markedly.

Comprehensive Review of all the Measures

The fundamental problem of the Yellow River is sediment. It is predicted, based on the knowledge, already obtained, that the middle and lower reaches of the Yellow River will remain a heavily sediment-laden and that the river bed of the lower reaches will continue to rise for a long time to come. Therefore, the ideal plan to harness the Yellow River should assure that all construction and operation of major projects on the main stream of the Yellow River must favor flood prevention and reduction of deposition in the lower reaches (see Table 10.). Solving the problem of sedimentation on the Yellow River is a very hard task and it is the key factor in making decisions on major projects on the middle reaches. The design of high-dams and reservoirs with large capacities is needed to retain more sediment.

Looking into all sedimentation-reduction measures, the first thing to build is Xiaolangdi Reservoir as soon as possible. Since Xiaolongdi Reservoir is located at a key position for controlling inflowing water and sediment to the Lower Yellow River, it will play a very important role in flood prevention and reduction of deposition. It will also cause favorable conditions for the application of other measures, such as warping on the Wenmeng flood plain, regulation of water and sediment and further renovation of the river course. None of these measures would achieve the expected effects without Xiaolongdi Reservoir. Benefits to be obtained by operation of Xiaolongdi Reservoir in reducing deposition in the lower reaches may last for 20 years. Through regulation of water and sediment it may be used continuously to reduce deposition to some extent.

It would be practical to build Qikou hydropower station after completion of the Xiaolongdi Reservoir. At present, the power system in the north of China is seriously short of peak-regulating capacity. This problem can be resolved by building the Qikou project to form an electricity system relying mainly on thermal power while providing the assistance of hydropower to regulate the peak. Qikou project also has a larger sediment storage capacity than Xiaolongdi Reservoir and the effect of retaining sediment and reducing deposition is comparable to that of Xiaolongdi Reservoir. The effective decrease of deposition in the lower reaches will be 20 years. Operation of the Qikou project in coordination with the Xiaolongdi Reservoir will increase the number of years of benefit to 40. This will provide valuable time and experience in the work of harnessing the middle reaches of the Yellow River.

Longmen reservoir is a project with multi-purpose benefits. At first, it will play an important role in improving the backwater area of Sanmenxia Reservoir and the Lower Weihe River. In addition, it may also markedly reduce the deposition in the Lower Yellow River. The regulation of water and sediment can be coordinated through combined operations of Longmen, Xiaolongdi and Sanmenxia reservoirs. These three reservoirs will have prolonged effects on the reduction of sedimentation in the lower reaches. They are to be considered as priority projects and constructed.

Warping on Wenmeng flood plain and on the flood plain in Longmen-Tongguan stretch also are major projects intended to reduce deposition. What is needed is the determination of the order of priority for the construction of those projects.

Table 10. Effects of all the measures

Items	Unit	Projects on Main River			Warping Works in Flood Plains			South-to-north Water Transfer	
		Xiao-langdi	Qikou	Longmen	Longmen-Tongguan	Wenmeng Plain: Xiao-langdi	Wenmeng Plain: Xixia-yuan	Middle Route	West Route
Diverting Discharge	cms				500	500	1000		
Annual Wtr Diverted	billion cu.m								200
Install. Generation Capacity	mw	1560	1600	2160	144		144		
Anl. Power Production	kwhr	5.45	5.15	9.04	0.6		0.47		
Amt. of Earth & Rock wks.	bil-lion cu.m	0.048	0.014	0.026	0.444	0.202	0.184		0.2
Concrete	bil. cu.m	2.7	1.297	1.913	1.132	1.6	0.039		11.13
Population Affected	1000	137.6	53.7	3.6	42	100	100		
Area of Deposition	sq.km				566	338	338		
Capac. for sediment storage	bil. cu.m	7.25	11.05	7.5	13.8	11.5	5.74		
	bil. tons	9.425	14.49	9.75	19.32	16.1	8.04		
Mean thick.	m				24	34	12-17		
Reduction in Depo-sition/lwr. reaches		7.27	7.02	4.5	10.12	11.1	4.39		
No.of years		19	20	12	27	29	12		
Anl .Incr. of Water								10	7.0
Anl.Reduc. Deposition								0.25	0.137

This depends on relevant economical analyses. Among these projects, the diversion of floods for warping on Wenmeng flood plain in combination with silt ejection outlets at Xiaolongdi Reservoir is a complex and huge project. Since it has a notable effect and may provide some practical experience for further study on the transportation of water with high sediment content through the man made canals, it is a project worth studying.

South-to-north water transfer is mainly designed to solve problem of water deficiency in the arid north. However, the problem of how much water can be supplemented to the Lower Yellow River for conveyance of sediment has to be further researched. The river bed of the lower reaches is raised by deposition which is mainly caused by a shortage of water and an abundance of sediment. Serious consideration should be given to the study of the feasibility of supplementing water through transfers to facilitate sediment transport. Even construction of south-to-north water transfer project along the middle or west route would be very difficult. However, it must be done in order to provide water for the arid north and to help develop the great north-west region. Sediment in the Yellow River comes mainly from the loess plateau. er and soil conservation works is the only way to change the appearance of loess plateau, but it also is important in solving the problem of sedimentation of the Yellow River.

In the past four decades, large-scale water and soil conservation work was launched on the loess plateau. The principles or guidelines drawn were "harnessing in a comprehensive, concentrative, and continuous way", "combining biological measures with engineering measures", "combining harnessing on slopes with harnessing in gullies", or "comprehensive harnessing of small watersheds".

Through practice it is understood that controlling the loess plateau is an arduous and long term task which depends on local conditions and the economy. The loess plateau is a vast and sparsely populated area, containing a porous and fragmented soil, in an arid climate, with a small but concentrated rainfall, experiencing intense water and soil erosion, and with a low yield from plants and grasses. In addition, as the water and soil conservation measures are crude, the control standard will not be very high and great damages to check dams and small reservoirs can hardly be avoided. When storm-floods occur, the deposition is rapid and dams may fail requiring a large amount of work to restore the structures. All of these factors make the rate of increase of effect of water and soil conservation measures far lower than the rate of increase in the amount of control works. Therefore, unremitting efforts are required to ensure that the water and soil conservation measures are effective.

To sum up, to solve the problems of sedimentation of the Yellow River, a policy involving a variety of measures for comprehensive control needs to be adopted. Besides the measures discussed above, it is also necessary to do river training and control structures at the estuary to increase the silt-transporting capacity of the river in transporting more sediment into the sea. The rate of aggradation eventually can be slowed down and safe flood prevention can be ensured with the present dikes.

ESTIMATE OF REDUCTION IN DEPOSITION IN LOWER YELLOW RIVER BY WARPING ON FLOODPLAINS

Hong Shangchi and Chen Zhilin
Reconnaissance, Planning and Design Institute
Yellow River Conservancy Commission
People's Republic of China

The Necessity for Large Scale Warping on Floodplains

(1). Large scale warping along the middle and lower Yellow River is one of the important ways to retard deposition on the downstream riverbed. Water and sediment conditions are the basic factors that decide whether aggradation or degradation takes place. Deposition will take place if runoff is small and sediment excessive, and the riverbed will be scoured or the rate of deposition will be slowed if the sediment load is low. The sediment detained in Sanmenxia reservoir amounted to 5.6 billion tons between September 1960 and June 1970. The cumulative amount of deposition in the lower Yellow River was very small and the river was approximately in balance, slowing the trend of aggradation by almost 10 years. However, there is a limit to the capacity of the reservoir to detain sediment. For this reason, using large-scale warping on the floodplains along the river will help the situation.

(2). Many places are available for warping along both sides of the main stem downstream from Longmen to the mouth of the Yellow River. Some of the wide floodplains in the middle reaches below Longmen may be used for warping. There is a natural capability for deposition by means of flood detention because the difference in height between the floodplain and channel is small, and the plain is easily flooded. At present the floodplain is overlain by an abundant sandy soil or saline-alkali land for which the efficiency of utilization is very low. The population is also small. Combined with the realignment of the river channel, warping by way of flood diversion may retard deposition but may also help develop agriculture in this area. These kinds of floodplains include the broad floodplain between Longmen and Tongguan and that in Wen-Meng counties.

Other choice sites for warping are the lowlands behind the

L. M. Brush et al. (eds.), Taming the Yellow River: Silt and Floods, 585–596.

levee. Because the riverbed rises year by year, drainage is difficult and waterlogging, salinity and alkalinity are very serious problems. Agricultural production is very low in the waterlogged lowlands. Warping behind the river, strengthening dikes and renovating saline-alkali soil in these areas will not only benefit flood protection and development of agriculture, but will also retard deposition areas in Yuanyang, Fongqiu, Dongming, and Taiqian counties.

The third possible area for warping is the wide floodplain area nearly 3,000 km, between the dikes in the lower Yellow River. Warping may be implemented due to the existence of fairly large transverse slopes over the floodplain. In addition, due to the unceasing fluctuation of the river, many waterlogged lowlands are formed in that area, where the groundwater table is high, and development of agriculture is not suitable. Thus it is a good place for renovating the soil by warping.

(3). Experimental information indicates that the larger the factor of incoming load (rate of sediment concentration to discharge) entering the downstream reach, the more serious the deposition. Statistics show that there were a total of 11 floods where the maximum sediment concentration exceeded 400 kg/cu.m at Sanmenxia from 1950 to 1983. The total amounts of water and sediment of the 11 floods with hyperconcentrations of sediment were only 2% of the 34-year total incoming water and 14% of total incoming sediment, respectively. But the amount of deposition was 54% of total sedimentation and 70% of athe sediment was deposited in the channel. These problems illustrate that deposition in the lower course is mainly concentrated during several events with heavy sediment loads. If water and sediment exceed the sediment transport capacity of the lower course, the flood waters can be drawn off into the warping areas. In addition the waters may be detained in the reservoir temporarily and sluiced off during lower stages. Deposition in the main channel of the lower reaches may be reduced to some extent.

Feasibility of Large Scale Warping in the Floodplain

In the past, warping was for different purposes in the middlestream and downstream of Yellow River and some successful results have occurred:

(1). Shibahu is located in Kenli county, Shandong province, 30 km from the Yellow River to the northwest of the Bohai Sea. This area is 13 km wide from north to south with 400 sq.km of total area, of which cultivated land is only 10%. Salinization of the area is serious and agricultural production is very low and unstable. From July 28 to Aug. 15,

1975, a total of 370 million cubic meters of water and 25 million tons of sediment, 10% of the sediment load at Lijing, were drawn into this area during 16 days and 8 hours. The discharge of the Yellow River was 3,000 cms and the factor of incoming sediment was less than 0.02. The water withdrawn from the side sluice gate no longer returned to the Yellow River.

(2). Three irrigation areas in Jinghe, Luohe and Weihe river are situated at Guanzhong and are important regions for wheat and cotton production. The irrigated areas in these regions are 1.32, 0.74 and 3 million mu respectively. Details of these areas are shown in Table 1. Before the 1970s the rule, "Shut down sluice gate as long as the sediment concentration exceeds 15% by weight," had always been used. This rule limited the days of irrigation because of the high sediment content in the Jinghe, Luohe and Weihe rivers. Investigation of the characteristics of flow with high sediment concentration and renovation of the canal system, no noticeable deposition took place although the concentration allowed into the canal was up to 35-60% (450--950 kg/cu.m). This example indicates that as long as appropriate conditions exist, great quantities of sediment can be transported by only a limited amount of water.

(3). Siltation at historic dike breaching sites can lessen the danger of dike bursts. The total area of the waterlogged pits left over after the dike was blocked back at the breaching point of Huayuankou, Zhengzhou in 1947 was 2,530 mu and the depth of the largest pit was 13 meters. The danger from piping was very serious. The waterlogged pits were filled later by warping and the danger was alleviated. After that, a new idea for strengthening dikes by means of warping was tried out in 1970. Since 1974, it has been adopted as a routine measure for strengthening the dikes, cumulative total of 25 million cubic meters on the back side was placed on the dike by the end of 1985. The main measures taken are removal through intake gates and siphonage. The waterlogged pits can be filled by gravity at times when the sediment concentration is large. Where warping by gravity is impossible because insufficient head is available, pumping may be used. Also, home-made suction dredges are used to advantage for warping areas in the vicinity of vulnerable spots.

(4). Storing clear water and draining muddy water from medium and small reservoirs in the loess areas facilitates the transport of sediment directly to the irrigation area such as at Hesonglin, Hunlinjin and Hengshan reservoirs. For example, by means of proper operation of the outlet gates, water with concentrations of 550 kg/cu.m frequently passes through the west main canal of Hengshan resrvoir in Shanxi province. Sediment used for the irrigation area is closely incorporated

Table. 1 General situation of three irrigation areas

name of irrigation area	Renming yinjing	Baojixia yinwei		Renming yinluo
		Yuanshang	Yuanxia	
Location of head of canal	Zhangjia kou	Lingjia cun	Weijia bao	
diversion capacity (cms)	50	50	50	18.5
slope	1/2000 - 1/2800	1/2000 - 1/4000	1/2000 - 1/3000	1/2000 - 1/3000
length of main canal (km)	21.2	170.2	17.0	21.5
mean silt load in flood season (kg/cu.m)	312	189		345
rate of silt load in flood season to annual silt load (%)	92.8	76.3		90
amount of silt diversion (million tons)	816	700	327	769

with that drained from the reservoir. The sediment which was used for irrigation between 1975 and 1979 was 2.2 million tons, or 79.3% of the total volume of the outgoing load. This greatly reduced the sediment transported to the lower reach of the river.

Summary of Warping and Project Planning in the Middle and Downstream Parts of the Yellow River

Fig.1 shows the layout of a large-scale warping area in the middle and downstream parts of the Yellow River. The planning is described as follows:

CENTRALIZED WARPING ON FLOODPLAINS

Warping areas between Longmen and Tongguan. The area is a wide floodplain from Longmen to Tongguan along the Yellow River. It is 125 km long, 8.5km wide on average and has a net area of 600 sq.km. The inside of the floodplain contains loess hills which are from 50 to 200 meters higher than the plain. From Yumenkou to Chaoyi and Hanyang the warping area of the plain is about 556 sq.km. The elevation is as high as 335 meters and the area has a population of more than 40,000 people. This project is planned in coordination with the proposed Longmen reservoir project. A diversion dam is proposed at Yumenkou below Longmen, from which sediment-laden water can be withdrawn for warping. Sluices will be set up at both sides with an outlet discharge capacity of 500 cms--300 cms at the west side, and 200 cms at the east side. The warping area is divided into nine parts and enclosing dikes for flood control will be constructed along the river. Narrow strip canals will be set up in each part, and each part will be warped. The longitudinal slope is 1/5,000 and the expected height of aggradation is on the average, 24 m. The total volume available for sediment deposition is 19.3 billion tons. Since the two main canals will be built gradually, intakes of the existing pump stations may be shifted to the canal in stages in order to keep the site of the pumping station stable and the water diversion reliable. As the project is undertaken, appropriate arrangements, as well as some protection works, should be made for the outlet of some relatively large rivers such as the Fenhe and Sushui. At the same time, in order to keep floodplains stable it will be important to plan and harness the main course of the Yellow River to protect the floodplain from bank erosion and to stablize the main channel.

Wenmengtan warping area. Wenmengtan is situated at the north

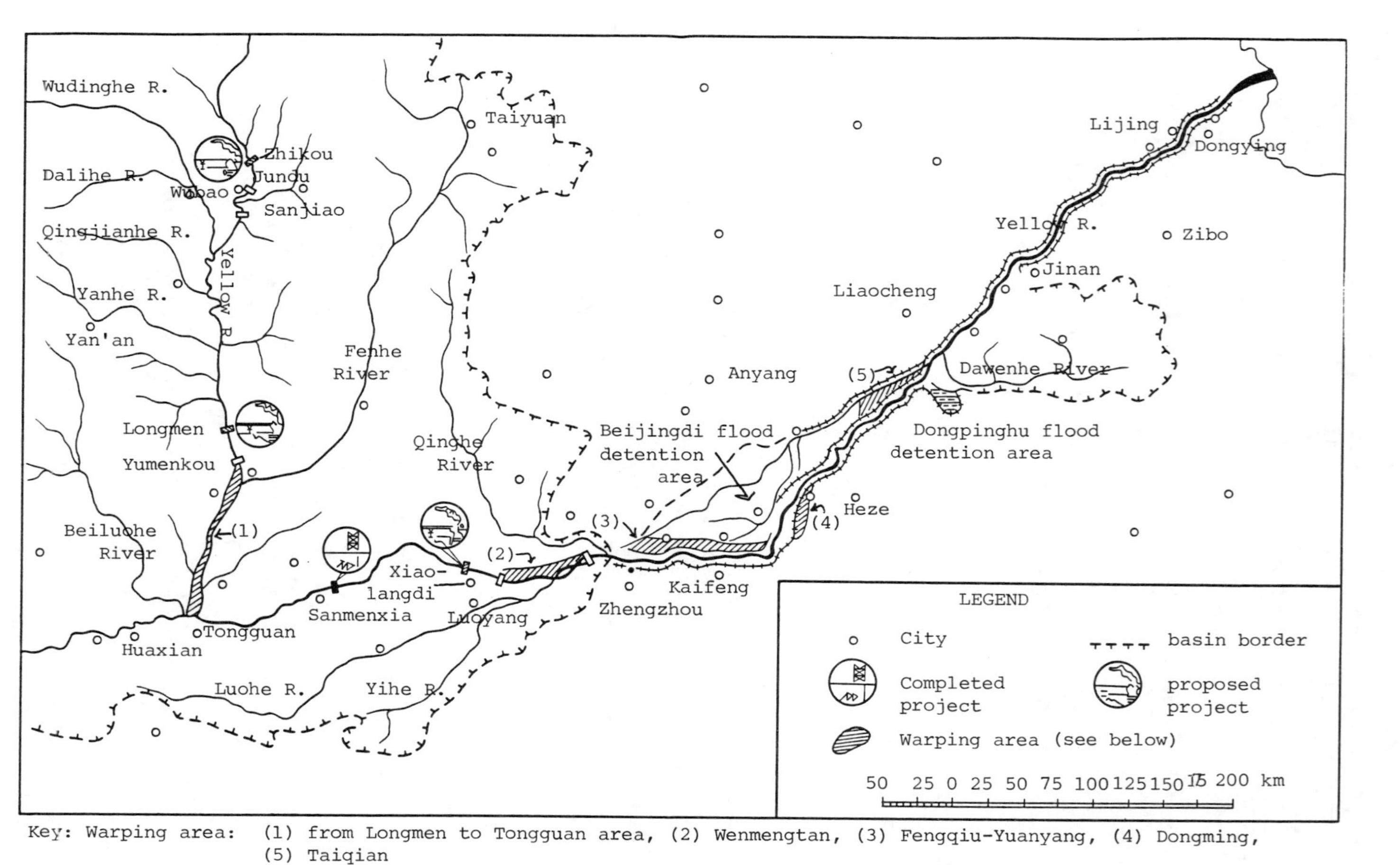

Key: Warping area: (1) from Longmen to Tongguan area, (2) Wenmengtan, (3) Fengqiu-Yuanyang, (4) Dongming, (5) Taiqian

Figure 1. Plan of Various Measures for Reducing Deposition in the Lower Yellow River.

side of the river at the exit of the canyon in the middle reach of the Yellow River. It is 91 km long between Baipo on the west and the mouth of Qinghe River on the east. It is 3.7 km wide on average from the Yellow River to the Qingfengling Mountains, and has 338 sq. km. of warping area. Another river, the Manghe, flowing from west to east, empties into the Yellow River at the mouth of the Qinghe. There are 400,000 mu of cultivated land inside the floodplain and 100,000 residents. This project coordinated with the proposed Xiaolangdi Reservoir, is to withdraw water by means of a sluice tunnel on the left bank and to make the water flow into the warping area at the east of Baipuo village through a canal with a discharge capacity of 500 cms. The south boundary of the project is the proposed boundary line of the river training works of the Yellow River. The north boundary is the old Manghe River, the dikes of the Yellow River in Mengxian county, and the course-changing engineering works of the new Menghe River. The warping area is divided into three parts: the Mengxian, Wenxian and Wuzhixian county boundaries. Gates are to be installed between two of the parts and are to be used for either drawing in, or draining out, water between each of the two parts with a designed discharge of 500 cms. The expected average depth of sediment aggradation is 34 m and the total deposition would be 16.1 billion tons. In order to consolidate the engineering work, realignment and regulation of the course of the Yellow River are needed because of the braided channel. Whether or not it will affect the salinization of the adjacent areas when the sediment aggradation in the warping area is as high as 30 meters is unknown. The conveyance capacity of the Manghe River may be limited by the water level of Qinghe River after the Manghe River course is changed to join the Qinghe River. These problems should be studied in the future.

CENTRALIZED WARPING OUTSIDE OF DIKES

Yuanyang-Fengqiu warping area. The Yuanyang-Fengqiu warping area is 102 km long from the dikes in Yuanyang county to Chanfang in Fengqiu. It is 5.2 km wide on average, has an area of 532 sq. km, and has 167,000 residents. The difference in elevation in front of and behind the dike in this part of the Yellow River is 6-7 m, and the maximum, 11 m. This area is located at the back side of the dike and is saline-alkali, waterlogged lowland within 2-3 km from the dikes. Agricultural productivity has always been very low at this site. The natural Wenyian canal is the main drainage channel in this area. Because of sediment accretion in the river bed, and the backing up of water during floods at the lower end, the canal has difficulty draining and becomes waterlogged. According to the plan, this area is to be divided into three parts. Three sluice gates are to be constructed at Jiadi dike, Yueshi and Hongqi sluice, respectively, and a sluice for draining out the water will be built at the Chanfang. The design discharge is 600 cms. The mean depth of aggradation in the warping area is expected to be 5.2 m on a slope of 1/6,900, and the total sediment deposition would be 3.89 billion tons. Due to the dense population in the area, the problem of relocating local inhabitants has to be solved. In order to prevent salinization in the area adjacent to the warping area, a ditch for the interception of seepage along the outside boundary of the area will be built to enclose the dikes so that the effect of the side-seepage may be minimized.

Dongming warping area. The Dongming warping area is 54 km long reaching from Simingtang in Lankao county, Henan province on the south to Luzhuang village in Heze county, Shandong province on the north. It is 12.6 km wide from the dikes of the Yellow River on the west to Lankon highway on the east. The total warping area is about 644 sq.km and the average slope of the land surface is 1/5,000. There are 275,000 people living there, and 787,000 mu of cultivated land. Of this, 180,000 mu is a saline-alkali sand desert. Warping will occur by taking in water at Sanyizhai and draining it out at Luzhuang village, with designed discharge capacity of 1,000 cms. The whole area may be divided into 3 or 4 narrow regions, and warped in turn from west to east. The mean aggradation depth is 2 m, and the total sediment deposition will be 1.8 billion tons. Because of the large population it is necessary to make appropriate arrangements for relocation. In addition, some questions also need to be considered regarding flood control and transportation.

Taiqian warping area . The Taiqian warping area is located downstream of the Jingdihe River and 62.6 km from Menglou River on the west to Zhangzhuang sluice on the east. It is 5.9 km wide from the dikes of the Yellow River on the south to Jingdihe River on the north. The whole area is 410 sq. km. and the slope of the land surface is 1/6,000. 260,000 people live there and cultivate 420,000 mu. Waterlogging has been very serious in the past because the surface becomes lower toward the east. Due to sediment accretion of the riverbed, drainage is made more difficult because of the Jingdi canal. The plan is to take water in at Xingmiao and drain it out to the Yellow River via Zhangzhuang village with a discharge under 1,200 cms. The whole area will be divided into two parts. The lower part is to be warped first, followed by the upper part. The expected mean depth of aggradation is to be 2 m, and the total sediment deposition would be 1.26 billion tons. After warping, the capacity in the flood detention region north of Jingdi will be affected. Again, relocation of the inhabitants in this area needs to be carefully arranged.

PROPOSAL OF WARPING ALONG THE DIKES

The floodplain of the lower Yellow River from Mengjing to the mouth is about 3,000 sq.km in area, three-fourths of the total area of the river course. These broad areas have the function of detaining floods and sediment during large floods. According to surveying data from 1965 to 1985 unfavorable situations such as a high channel bed, a low floodplain and a low-lying area near the dikes exist below Dongbatou. In the reach from Chanfang at Fengqiu to Changhuan village, for example, the channel is 0.4-0.8 meters higher than one plain and 2-4 meters higher than the low-lying area near the dikes. This is called the "suspended river" in the "suspended river", because the Yellow River itself is usually called the "suspended river." The transverse slope of the plain is 1/2,000-1/3,000, or three times as much as the longitudinal slope in the reach from Dongbatou to Gaocun. The transverse slope in other floodplains is up to 1/4,000, so it is poor for flood conveyance in the downstream part of the Yellow River and there is danger of an abrupt change of the river course. In order to change the situation, deposition by artificial flood diversion should be undertaken in addition to natural deposition. Statistics show that the river channel along the dikes is 790 km long, generally 100-300 m wide and 1-2 m deep. There are 46 larger interlacing channels on the floodplain with a total length of 314 km. These channels should be filled in by artificial measures. The amount of silt which is warped by way of artificial flood diversion is less than that of natural deposition, but it helps to strengthen the dikes and stabilize the water channels.

Analysis of Retarding Downstream Deposition by Large Scale Warping

(1). The data show that for constant inflow, if the sediment is reduced by 250 million tons at 4 stations (Longmen, Huaxian, Hejing and Zhuangtou), the sediment load entering the downstream reaches might decrease by 200 million tons, and deposition downstream in the Yellow River would be reduced by 100 million tons. The effect of retarding sediment deposition would be greater if floods with large sediment concentrations no longer took place. Comparing the situation in 1971 with 1977, the effect of retarding sediment deposition may be seen in Table 2. As far as warping in the Longmen-Tongguan reach and the Wenmengtan area is concerned, if the deposition process is coordinated with the operation of Longmen and Xiaolangdi reservoirs, the proper time of warping could be better controlled and more rational than that which takes place naturally, and the retardation of deposition in the lower reaches would be greater. Considering the fact that

Table.2 Relationship between decreasing sediment and decreasing deposition in different parts of the Yellow River

time of measurement		73 - 74 75 - 76 76 - 77 77 - 78	80 - 81 81 - 82 82 - 83 83 - 84	1977	1971
Longmen, Huaxian, Hejing (average number in flood season)	amount of water (10^8 cu.m)	413.4	408.8	132	127.5
	among of sediment (10^8 tons)	15.35	7.4	15.5	6.56
	amount of decre. silt (10^8 tons)	7.95		8.94	
Sanmenxia, Heishiguan and Xiaodong (mean number in flood season)	amount of water (10^8 cu.m)	442.3	445.6	175.9	154.0
	amount of sediment	15.77	9.10	20.64	12.07
	amount of decre. silt (10^8 tons)	6.67		8.57	
course between Sanmenxia and Lijing (mean number in flood season)	amount of scouring deposition	3.37	0.32	10.06	5.38
	amount of decreasing deposition	3.05		4.68	
rate of decreing deposition	second iterm	2.61		1.91	
	third iterm	2.19		1.83	

deposition in the Longmen-Tongguan reach will be the first to be reduced by warping in this reach, the reducing deposition ratio in the lower reaches further downstream is taken as 1.91 for safety reasons. The reducing deposition ratio is defined as the ratio of the amount of reduction of incoming sediment load to the amount of reduction of deposition in the lower Yellow River. Because Xiaolangdi reservoir controls the diversion of water and sediment for deposition in the Wenmengtan warping area, the retaining capacity in the warping area is actually an enlargement of the capacity allocated to sediment retention in the reservoir. Therefore, a value of 1.45 is used which is similar to the effect of deposition in Xiaolangdi reservoir during the period of sediment detention. For the three warping areas--Yuanyang-Fenqiu, Dongming and Taiqian--where no reservoir control is available, and for the areas far from the river course, the ratios selected are 2.0. 2.14, and 3.3 respectively. Details are shown in Table 3.

It should be pointed out that each warping area is not likely to detain more than 500 million tons of sediment. Silt deposition in the warping areas can only retard the deposition in the lower reaches, but cannot free it entirely from deposition. The so-called "number of years retarding deposition" is a relative index only, and means that the river course might undergo deposition eventually.

According to the analysis mentioned above, 35.4 billion tons of sediment might be deposited at the Longmen-Tongguan reach and Wenmengtan, thereby reducing deposition by 21.2 billion tons in the lower reaches. This is equivalent to extending for 56 years the present status of the lower Yellow River.

PROBLEMS

Large scale warping on the floodplains and behind dikes is a possible measure for retarding deposition in lower Yellow River. This paper describes briefly the possible sites for large scale warping, discusses the major engineering works and estimates their probable effects. However, there are also some problems to be considered:

a. Large numbers of people need to be relocated in each warping area. Appropriate arrangments are required.
b. The land after warping may be like a desert if the sizes of the deposits are not controlled properly.
c. Industry and agriculture in the warping area will be affected temporarily.
d. The feasibility of long-distance warping by means of heavily silt-laden flows with a relatively large discharges and the formation of such flows through reservoir outlets by reservoir operation should be further studied.

Table 3. Decreasing deposition in lower reach affected by various warping areas

name of warping area	course between Longmen and Tongguan	Wenmengtan	Yuanyang-Fengqiu	Dongming	Taiqian
plan of engineering	Yumenkou	Xiaolangdi	diversion without dike	diversion without dike	
area of warping area (sq.km)	566	338	532	644	410
total silt deposition (10^8 tons)	193.2	161.0	38.9	18.0	12.6
mean thickness of deposition (m)	24	34	5.2	2.0	2.0
discharge of diversion (cms)	500	500	600	1000	1200
rate of decreasing deposition	1.91	1.45	2.0	2.14	3.3
amount of decreasing deposition in lower reach (10^8 tons)	101.2	111.0	19.5	8.4	3.8
amount of deposition in lower reach in present (10^8 tons)	3.79	3.79	3.79	3.79	3.79
number of year of decreasing deposition	26.7	29.3	5.1	2.2	1.0

EFFECTS AND SIGNIFICANCE OF WATER AND SEDIMENT REGULATION BY RESERVOIRS IN THE MIDDLE YELLOW RIVER

Qian Yiying
Institute of Hydraulic Research
Yellow River Conservancy Commission
People's Republic of China

ABSTRACT: This paper describes chiefly water and sediment regulation with different operational modes subsequent to the re-construction of Sanmenxia reservoir, and analyses sedimentation in the reservoir and the lower river course. Regulation of sediment must be taken into account together with regulation of runoff. A certain storage capacity has been maintained for long-term use for multi-purpose developments of flood control, irrigation and power generation and for sediment conveyance in the lower river course as well through rational regulation of water and sediment in reservoirs built on middle reaches of the Yellow River. Reservoir operational modes for water and sediment regulation are also discussed in the paper.

The total annual water runoff and sediment load at Sanmenxia is 42 billion cu.m and 1.6 billion tons respectively. More than 90% of the total sediment comes from the loess area of the middle Yellow River basin. This load causes severe deposition in the lower river course and brings about many problems in the development of the water resources of Yellow River. Since the founding of new China, a number of large and medium sized reservoirs have been built on the main stem and tributaries of the Yellow River for flood control, irrigation and power generation. Great benefits have been obtained and much experience gained in dealing with sedimentation problems in the reservoirs. Sanmenxia Reservoir was put into operation and commenced impoundment in Sept. 1960. The operational method of detaining floods and sluicing sediment was adopted in March 1962 due to serious siltation in the reservoir during its initial storage period. In 1964, a decision was made to reconstruct and enlarge the outlet discharge capacity to facilitate the disposal of sediment. By the end of 1973, reconstruction was complete and releases from the reservoir were controlled in such a way that relatively clear water was stored during the non-flood season and sediment retained during this season was flushed out of the reservoir during the flood season. This paper discusses the conditions of water and

L. M. Brush et al. (eds.), Taming the Yellow River: Silt and Floods, 597–616.

sediment regulation by reservoir operation to maintain effective storage capacity for long-term use. The influence of sedimentation on the lower Yellow River and the possible approaches to development of multi-purpose benefits and the control of hazards with reservoirs are also discussed.

Conditions Required to Maintain Storage Capacity for Long-Term Use

FEATURES OF INCOMING WATER AND SEDIMENT FROM THE MIDDLE YELLOW RIVER

It is well known that large amounts of sediment and low runoff are produced from the area of the middle Yellow River. Floods are characterized by sharp peaks and abrupt rises and falls of the hydrograph. Temporal distributions of water and sediment are extremely uneven. Usually the flood season (from June to October) makes up 60-80% of the total annual runoff and about 90% of the total sediment. The amount of sediment often concentrates during a few floods. For instance, in 1933, the annual sediment load at Shaanxian Hydrometric Station was 3.91 billion tons, 2.44 times as much as the average annual sediment load. The sediment load from Aug. 9 to 13 amounted to 2.12 billion tons, making up 54.3% of the total in that year and 1.33 times of the average annual sediment load. This was the main reason for severe siltation in reservoirs and in the river course. The outlet facilities of Sanmenxia Reservoir were rebuilt in recognition of the aforementioned characteristics of water and sediment. Guidelines for the reconstruction included making full use of the capacity of flood and sediment conveyance of the existing river course. The reservoir was not designed to detain medium or small floods but to discharge water and sediment as much as possible, while keeping a certain reservoir capacity for purposes of controlling major floods and impounding clear water for irrigation and municipal use. For this objective, the operational mode of the reservoir was changed to store clear water and release muddy water, i.e. impounding water during non-flood seasons and releasing water for regulating ice floods and irrigation. Usually the operation level for ice flood control is lower than 326 m. The stored water supplements spring irrigation and industrial use of water through reservoir regulation. The Sanmenxia Reservoir has begun to develop benefits from its multipurpose use subsequent to re-construction.

FEATURES OF SEDIMENTATION IN RESERVOIRS

Sedimentation in a reservoir depends on the conditions of the incoming water and sediment, the discharge capacity, the operational mode of outlets as well as the topography of the reservoir. Fig.1 shows the distribution of deposits in the reservoir under different operating conditions. It can be seen from the figure that during the impounding period up to Oct. 1961, i.e. the first year of operation, siltation was in the form of a delta. The apex of the delta was near section 31, and the cross section was raised evenly in the transverse direction, making no distinction between the main channel and the floodplain. The reservoir has been managed to retard only floods since March 1962. Due to the fact that the dead storage of the reservoir was not fully used, deposition took place in the reservoir continuously but the efficiency of retaining sediment was reduced. During the wet year of 1964, the annual water and sediment inflow were 69.7 billion cu.m and 3.06 billion tons, respectively. The reservoir was severely silted because the outlet capacity was too small and the sluice holes were located too high. About 1.95 billion tons of sediment was deposited in the reservoir during the flood season, which constituted 70% of the total sediment entering the reservoir, and represented 41% of the total deposition in the reservoir. This was the most serious year of siltation. Longitudinal deposition was in the form of a cone, and channel and floodplains were formed. This can be seen from the transverse section, wherein the floodplain rose simultaneously with the main channel. The width of siltation on the floodplain changed with the width of the valley while the siltation in the main channel varied. In 1973, deposition occurred with the main channel being eroded because the reservoir was used for flood detention only during the flood season and the outlet discharge was larger after reconstruction. It can be seen on the figure that the main channel has lowered by erosion with basically no changes in the floodplains. "A high floodplain and deep main channel" had been formed, about one billion cu.m of channel storage below Tongguan (cs,41) was recovered and the elevation of the river bed at Tongguan dropped 2 m. After the flood season of 1973. During the nonflood season with lower sediment concentration, the reservoir impounded water and retained sediment for comprehensive use for ice flood control, power generation and irrigation. During flood seasons with higher sediment concentration, the operation level was controlled to release both water and sediment basically keeping a balance between erosion and sedimentation. In Fig.1 the longitudinal profile of October 1983 shows the sedimentation subsequent to the operational mode of impounding clear water and releasing muddy water. It can be seen from the figure that, there was no appreciable change in the longitudinal profile, except a slight modification of the width and depth of the channel. The floodplain surface did not change while the reservoir storage changed by about 100-200 million cu.m in 10 years and the river bed elevation at Tongguan changed within 1 m. Generally speaking, deposition

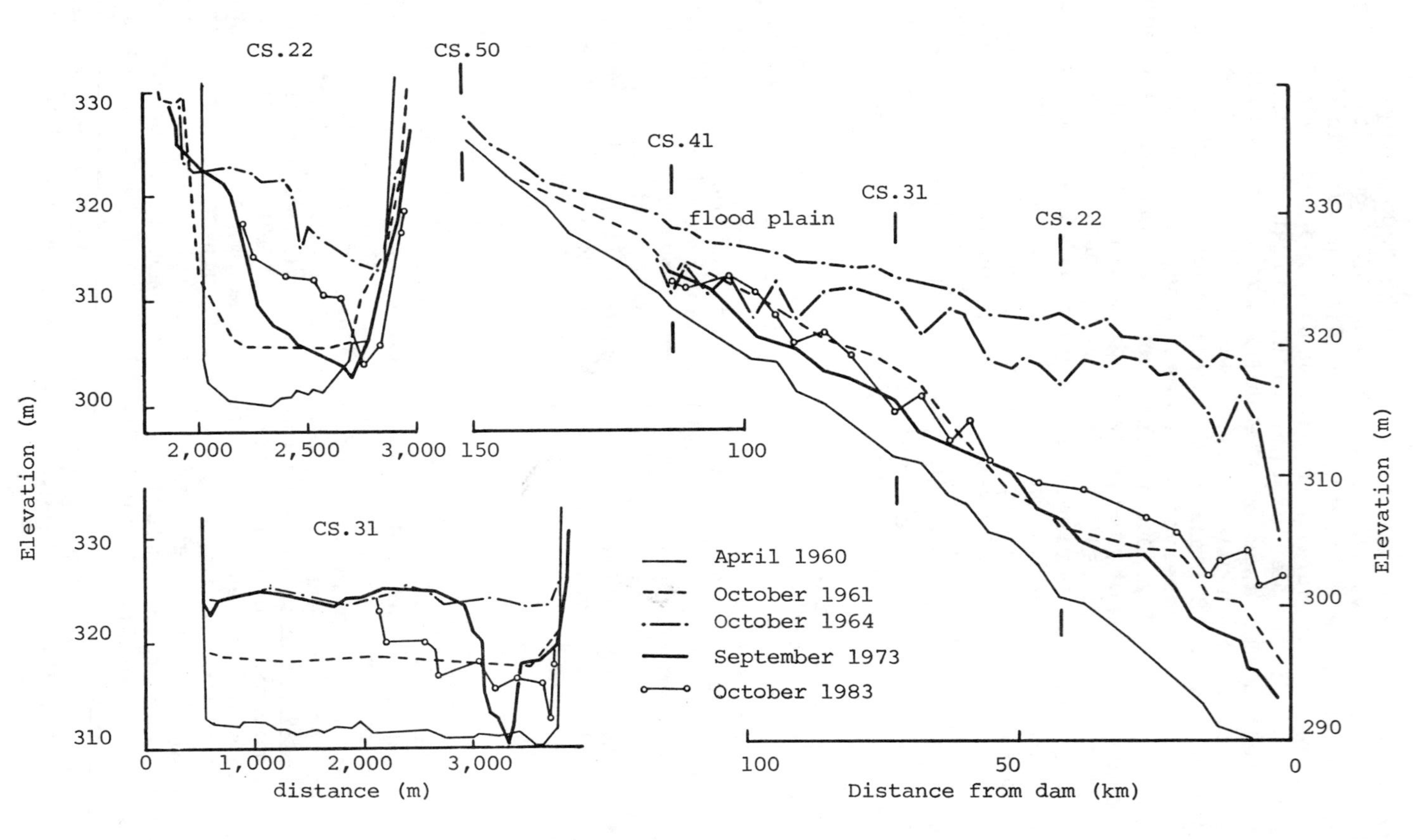

Figure 1. Longitudinal and Transverse Profile of Sanmenxia Reservoir.

occurred during the non-flood period and scour during the flood season. About 3 billion cu.m of reservoir capacity below elevation 330 m below Tongguan has been kept for long term use and the reservoir siltation and upstream extension of backwater deposits had been put under control as planned.

CONDITIONS REQUIRED TO KEEP THE RESERVOIR FOR LONG-TERM USE

To sum up, sediment problems should be properly dealt with when building a reservoir on a sediment-laden river. If siltation could be made to change to erosion and deposition intermittently to achieve the balance of sedimentation in a certain period, a certain reservoir storage capacity could be kept for the long term. The conditions required to keep the reservoir available for long term use depend on the outlet discharge capacity and the operational mode of the reservoir.

Broadly speaking, the outlet discharge capacity of a reservoir means the discharge capacity at different reservoir stages. The benefits of reservoir operations are closely related to the discharge capacity. From the point of view of keeping the reservoir available for long term use, the discharge capacity at elevation of dead storage is the most decisive factor in determining the formation of a relative stable morphology of the river including the slope of river bed and channel morphology in the reservoir. For instance, prior to reconstruction, an attempt was made to change the operational mode to reduce the amount of deposition. Siltation in the reservoir, however, was still very severe and sedimentation moved upstream because the discharge capacity of the reservoir was too small. The elevation set for dead storage had lost its meaning. On other hand, if the discharge capacity is too great, it would make the layout of sluicing structures difficult and increase the initial cost of key projects. In the design of reservoir projects on a sediment laden river, the discharge capacity at an elevation relevant to dead storage should not only satisfy the demands similar to the reservoir project built on a river with less sediment load, but also fulfill the requirements for keeping an effective reservoir capacity for the long term. Data for Sanmenxia and Naodehai Reservoir indicate that the discharge capacity should be made equal to or somewhat greater than bankfull discharge of the natural river course. Thus, it is possible through the reservoir operation to regulate water and sediment in a reasonable way. Channel morphology in the reservoir proper will be formed in a way which is compatible with the bankfull discharge to allow some reservoir storage for long term use. In the process of reservoir operation, emphasis should be paid to the regulation of sediment while regulating the water runoff for fully enjoying the comprehensive benefits of the reservoir, and preventing and controlling unfavorable upstream and downstream effects. There are many approaches to regulating water and sediment and the influence on the lower river course will vary accordingly.

Impacts on Sedimentation in the Lower Yellow River

OPERATION FOR STORAGE OF WATER

During the early stages of impoundment in Sanmenxia Reservoir, runoff was controlled and managed and the sediment discharged from the reservoir was only 7.0% of that which entered the reservoir. The sediment leaving the reservoir consisting entirely of fine particles was released by density currents. The operational mode was changed to flood detention in March 1962, although deposition still took place in the reservoir as the dead storage had not yet been filled up and only small amount of fine sediment was released from the reservoir. Owing to the changed conditions of water and sediment released, the lower reaches changed fundamentally and the lower Yellow River changed from an aggrading river to a degrading one. The length of erosion of the river course was fairly long and extended continuously downstream with time. For instance, at the initial period of impoundment in Sanmenxia Reservoir in April 1961, the mean discharge was 1,370 cms and length of the river reach in which sediment concentration had recovered was 207 km. Under the same conditions of discharge, the recovered length of sediment concentration reached 483 km in Jan. 1963. The length of erosion of the river course is also related to the discharge, and the duration and volume of runoff. When discharges are greater than 2,500 cms, a duration of 14 days and volume of runoff of about 3 billion cu.m, erosion of the river could extend to Lijin, about 800 km downstream. From Sept. 1960 to Oct. 1964, a great amount of sediment was retained in the reservoir and 2.31 billion tons of sediment were eroded in the lower river. 73.5% of the erosion took place in the reach above Gaocun. Siltation in this reach prior to construction of the reservoir was 55% of the total amount of deposition in the lower reaches. This indicates that severe erosion would take place in reaches where aggradation in pre-dam days was the strongest.

During periods of scouring by clear water released from the reservoir, the extent of erosion was different at different reaches of the lower Yellow River and changes of river morphology were different as well. In the upper part of the lower river course, i.e. the reach between Tiexie and Peiyu changed from a wandering braided reach into smooth flowing reach with a single main channel as shown in Fig.2. The channel became deep and wide simultaneously in the reach from Huayuankou to Gaocun. River bends developed with scouring on the concave sides and the banks of the high floodplain collapsed. The river pattern did not change and remained braided. The slope of the river from Tiexie to Huayuankou changed a little from the original 0.000255 to 0.00024. The river bed from Huayuankou to Aishan was deepened almost parallel, with basically no change in the slope. In the stretch below Aishan, the river bed at Lijin not was not lowered, but raised 0.14 m during this period due to the influence of backwater from the estuary, hence, the slope decreased slightly.

During the period of erosion by clear water, the sediment discharge corresponding to the same discharge at each hydrometric station of the lower Yellow River decreased greatly in a short time, see Fig.3. The main reason was that a great amount of sediment was retained by the reservoir. Sediment was scoured from the river bed which was coarser and the supply of wash load was greatly reduced. In addition, the river bed coarsened and the roughness coefficient increased, reducing appreciably the capacity for sediment transport. When the river course was scoured, the river bed would be lowered and the flood conveyance capacity increased. Consider the water level at a discharge of 1,000 cms. prior to construction of the reservoir as basis for comparison. In 1964 the discharge capacity in the reach between Huayuankou and Gaocun under the same water level increased by 6,300-8,000 cms and 1,300 cms in the reach between Aishan and Luokou. Reaches in the delta had silted continuously and the flood conveyance capacity at Lijin Hydrometric Station was reduced slightly.

OPERATION FOR FLOOD DETENTION AND SEDIMENT SLUICING

During the period of operation for flood detention and sediment sluicing, all the outlet gates of the reservoir are opened. Limited by the available outlet capacity, the peak flow was reduced and during the recession sediment would be sluiced from the reservoir and the duration of medium flow lengthened. For example, on August 13, 1964, the peak discharge at Tongguan was 12,400 cu.m per sec. and the outlet discharge was 4,870 cms. Figure 4 shows the relationship between the ratio of peak discharge out of the reservoir and the peak discharge of inflow into the reservoir. It can be seen from the figure that as the outlet discharge capacity of the reservoir increases, attenuation of peak discharge decreases.

Figure 5 shows that the relation between water and sediment

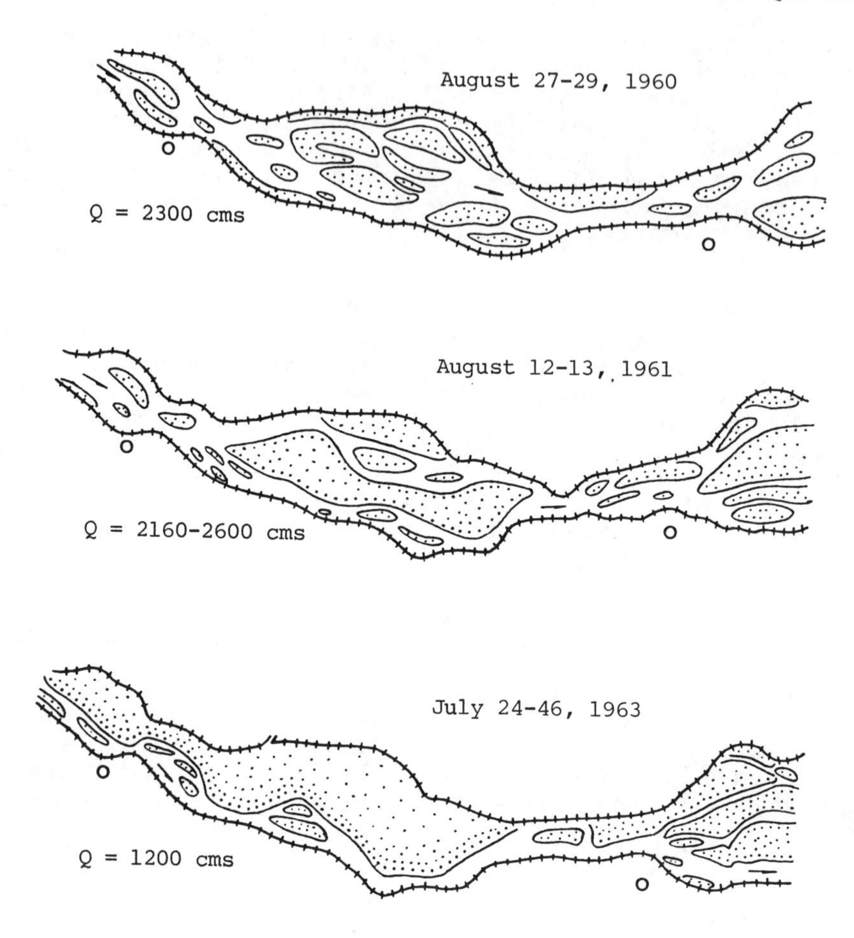

Figure 2. Fluviomorphological Processes in Stretch Tiexie to Peiyu

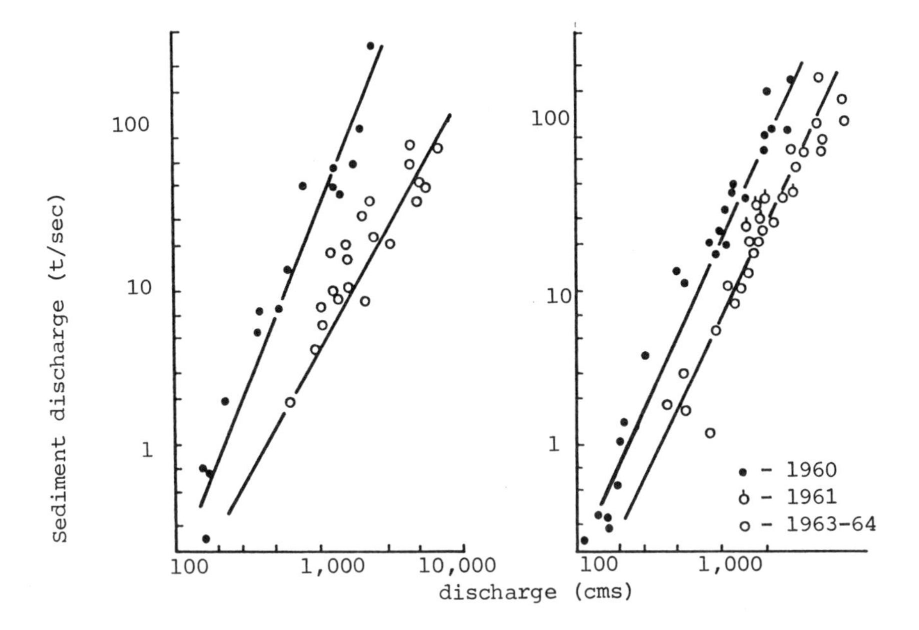

Figure 3. Sediment Discharge versus Water Discharge at Huayuankou (left) and Gaocun (right)

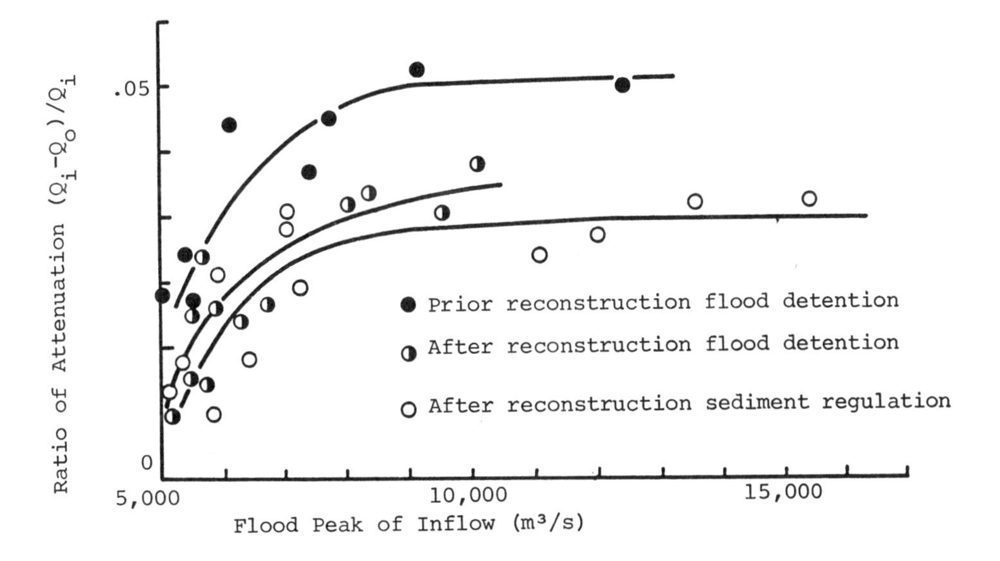

Figure 4. Attenuation of Flood Peak in Sanmenxia Reservoir

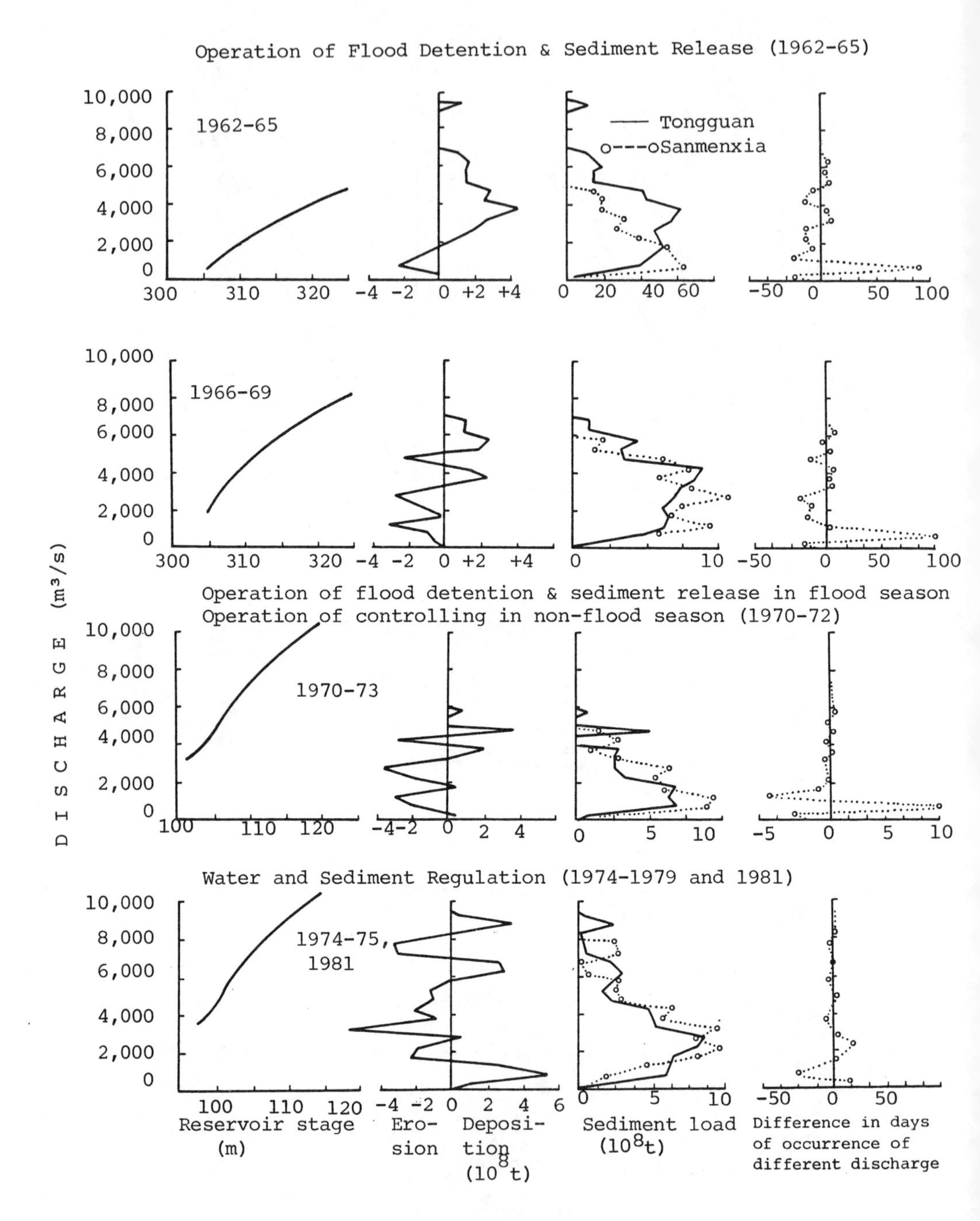

Figure 5. Sediment Transportation Under Different Operational Modes and Sluicing Scale in Sanmenxia Reservoir

tons in the predam days (July 1950 to June 1960). The deposition in the reach above Gaocun was 68% of the total, with 15% in the reach between Aishan and Lijin. Both these figures are greater than the proportion in predam days (deposition in these two reaches were 55.1% and 12.5% of the total, respectively). Deposition in the reach between Gaocun and Aishan was 17% which is smaller than the proportion in predam days. With the operation of flood detention and sediment sluicing, changes in the average elevation of the main channel expressed in terms of water level difference at a discharge of 3,000 cms are shown on Fig.6. The reach between Huayuankou and Lijin was raised about 2 m in parallel and the difference of water level in the reach above Huayuankou was gradually reduced. Table 1 indicates the status of sedimentation in the lower reaches under different stages of peak discharge during different operational periods of Sanmenxia Reservoir. It can be seen from this table that the temporal distribution of erosion and deposition in the lower Yellow River was changed due to operation of the scheme of flood detention and sediment sluicing within a year. The amount of deposition during the flood season was reduced from 80% prior to construction to 74%. In the non-flood season, deposition increased from 20% prior to construction to 26% cms. There were no flood discharges exceeding 10,000 cms. in the 9 years due to the reservoir. Erosion has taken place in the lower river course by dominant floods of 6,000-10,000 cms. Siltation in the lower river course was caused mainly by medium and small sized floods with discharges of 2,000-6,000 cms. This is an increase from 7.7% prior to reservoir construction to 72.7%, almost 10 times. During operation of the flood detention and sediment sluicing, deposition occurred in all medium and small floods in the lower river course and took place mainly in the main channel. This makes the difference in elevation between the main channel and floodplain smaller and causes the bankfull discharge to be reduced from 6,000-7,000 cms. to 2,000-4,000 cms. Thus, on the one hand the river became wide, shallow and scattered. When a medium sized flood occurred, the river course would be scoured and river pattern would change greatly. The main flow shifted frequently, forming cross-currents constantly, causing a dangerous situation for flood protection works. On the other hand, the bankfull discharge decreased, causing small floods to overflow floodplains which increased the resistance of flow and siltation in the river course by decreasing the sediment transport capacity.

Subsequent to the operation of flood detention and sediment sluicing at Sanmenxia Reservoir, the bed material of the lower river course became finer and sediment transport capacity

Table 1. Erosion and Deposition Under Different Peak Floods in the Lower Yellow River

Item	Prior to Construction Jly 1950-Jun 1960		Operation of flood detention & sediment sluicing Nov.1964-Oct.1973		Operation impounding clear water and releasing muddy wtr Nov.1973-Oct.1982	
	Amt. of sedimentation (10^8 tns)	% of annual total	Amt. of sedimentation (10^8 tns)	% of annual total	Amt. of Sedimentation (10^8 tns)	% of annual total
Annual total	3.61	100	3.99	100	1.66	100
I. Flood season	2.90	80.3	2.95	73.9	2.56	154.2
1.flood peak period	2.72	75.3	2.58	64.7	2.20	132.5
Maximum peak discharge (cms) >10000	0.94	26	0	0	0.60	36.2
6000-10000	1.50	41.6	-0.32	-8.0	0.56	33.7
4000-6000	0.25	6.9	1.17	29.3	0.09	5.4
2000-4000	0.03	0.8	1.73	43.4	0.95	57.2
2. Period of medium & low flow	0.18	5.0	0.37	9.2	0.36	21.7
II. Non-flood season	0.71	19.7	1.04	26.1	-0.90	-54.2

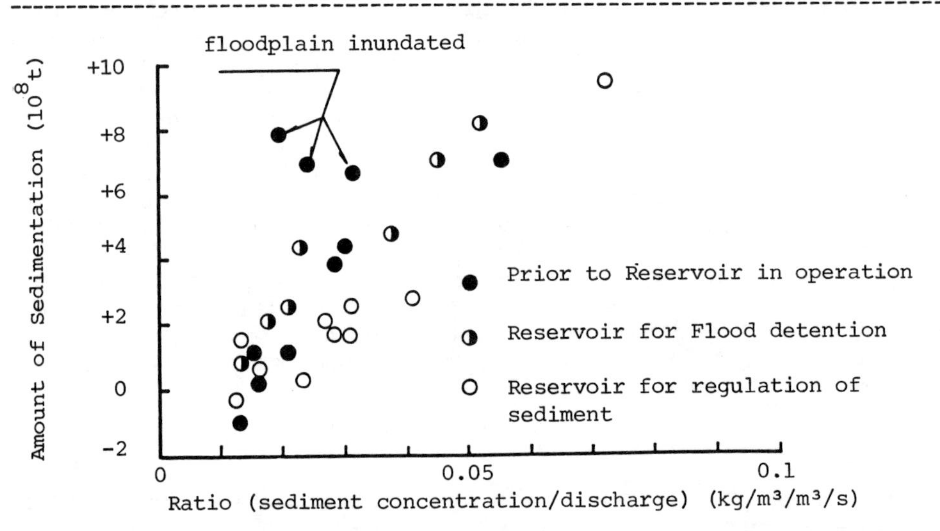

Figure 6. Amount of Sediment Vs. Ratio of Sediment Concentration to Discharge

had been changed greatly owing to the modification of incoming water and sediment of lower Yellow River through regulation by the reservoir. Prior to reconstruction from 1962-1965, the mean daily discharge of outflow was smaller than 2,000 cms, erosion took place in the reservoir and the sediment concentration of outflow was greater than that of inflow. At discharges greater than 2,000 cms, the reservoir retains flood waters and holds sediment making the sediment concentration of outflow smaller than that of inflow. These conditions changed drastically the relation of water and sediment in the natural river. Greater discharges with relatively higher sediment concentrations and vice versa changed into smaller discharges with relatively higher sediment concentration and greater discharges with relatively small concentrations. After reconstruction of Sanmenxia Reservoir, the outlet discharge capacity was increased and its function of reducing flood peaks was reduced. The maximum mean daily discharge of outflow increased to 6,000 cms., and the discharge under which the reservoir scoured was increased to 3,500 cms. The relation of water and sediment in the natural river course had changed similarly. Therefore, the sediment disposed in non-flood seasons was increased from 211 million tons before building the reservoir to 348 million tons afterwards. Mean sediment concentrations during the non-flood seasons increased from 11.4 kg per cu.m to 17.4 kg per cu.m which caused aggradation in the lower Yellow River.

The sediment transport capacity in the lower Yellow River is not only a function of discharge but is also related to the sediment concentration of the inflow, i.e.

$$G_s = K\, Q^{\alpha}\, S_o^{\beta}$$

in which,

Gs-----sediment discharge of bed material load t/s;
Q -----discharge cms;
S_o-----sediment concentration of bed material load at upstream station representing inflow to the section in kg/cu.m;
k -----coefficient relevant to the boundary condition of the river bed;
α and β -----exponents, equal to 1.1-1.3 and 0.7-0.9, respectively;

Thus it can be seen that changes in the incoming water and sediment would inevitably cause variation in sedimentation in the lower Yellow River.

After the flood season of 1964, the lower river course began to backfill. The average annual deposition to October 1973 amounted to 439 million tons, greater than the 361 million

increased during the process of backfill. The relation between discharge and sediment discharge at Huayuankou at the end of flood season in 1964 had recovered to the level of the predam period. However, conveyance capacity greatly decreased because of severe siltation in the main channel. For instance, the peak discharge at Huayuankou in August 1973 was only 5,000 cms. The 270 km long reach from Mengjin to Gaocun underwent heavy deposition during the flood and the water level in the 160 km long reach downstream of Huayuankou was 0.2-0.4 m higher than that of peak discharge 22,300 of cms. which occurred in 1958.

OPERATING BY IMPOUNDING CLEAR WATER AND RELEASING MUDDY WATER

The influence on water and sediment regulation where operating by impounding clear water and releasing muddy water in Sanmenxia Reservoir has been explained. It can be seen also from Figs. 4 and 5 that cutting down peak discharges during the period of operation of impounding clear water and releasing muddy water was smaller than that in the operation of flood detention under the same discharge. Mean daily outflow during floods could be increased to 8,000 cms., and discharge under which the reservoir would be scoured was essentially in the range of 2,000-6,000 cms. For discharges smaller than 2,000 cms., sediment would be retained in the reservoir and the sediment concentration of the outflow reduced tremendously.

The annual mean inflow of lower Yellow River from Nov. 1973 to Oct. 1983 was 41.7 billion cu.m and the mean annual sediment load was 1.16 billion tons, amounting to 74.1% of the average value. During this period a series of low water years with small amounts of sediment occurred. The total deposition in the lower river course was 1.57 billion tons, with an average annual value of only 157 million tons. This was the period of smallest deposition in recent years in the lower Yellow River, except for the period when the waters were first impounded in Sanmenxia Reservoir. The low incoming water and sediment was the main reason for the low rate of aggradation. In addition, the operation of impounding clear water and releasing muddy water also had certain effects. Fig.7 shows the influence on sedimentation of the lower river course by the ratio of sediment concentration to discharge, a parameter representative of the incoming flow. It can be seen from the figure that during the period of operation of flood detention in Sanmenxia Reservoir, siltation was most serious in the lower river course and during the operation of impounding clear water and releasing muddy water, deposition was the least. In the pre-dam period all the points were located between the above two cases except three events during which flood water overflowed the floodplains resulting in heavy deposition. Table 2 shows basically the same condition of

Table 2. Contrast of Water and Sediment Condition and Amount of Sedimentation in flood Seasons of 1974 and 1980

Item			1974	1980
W.S. elevation at damsite at Shijiantau	(m)	mean	303.56	301.87
		Maximum	308.28	311.22
		Minimum	293.19	289.51
Sediment load at Tongguan (million tons)		Total load	552	466
		Coarse sediment	72.9	52.5
Summation of Sanmenxia + Tributaries	Vol. runoff	(billion cu.m)	13.4	14.6
	Sediment load (million tons)	Total load	654	692
		Coarse sediment	148	180
Maximum peak discharge at Huayuankou (cms)			4150	4440
Amount of erosion and/or deposition (million tons)	Tongguan-Sanmenxia	Total load	-97	-220
		Coarse sediment	-75	-128
	Sanmenxia-Lijin	Total load	185	287
		Coarse sediment	79	143
Percent of sediment conveyed to sea		Total load	71.4	41.5
		Coarse sediment	46.6	20.6
Amount of erosion and deposition in the reach of Tongguan and Sanmenxia under different discharge (million tons)		<1000	-10	-36
		~2000	-9	-63
		~3000	-19	-122
		~4000	-60	

Note : coarse sediment mentioned in the table refers to sediment grain sizes greater than 0.05 mm.

incoming water and sediment during flood seasons in 1980 and 1974. However, the operation level in front of the dam was set lower than 300 m in elevation for a long period (end of July to late of Sept.) for sluicing in the flood season of 1980. The reservoir was controlled and better operated during the flood season of 1974, with its operation level varying between 302-305 m from July to the middle of October, except from 23-27, of October, when the water level was lowered to less than 300 m for maintenance work. Owing to the different operational modes of the reservoir, erosion under different discharges was also different in the reservoir proper. The erosion mainly occurred when the discharge was 3,000-4,000 cms. in 1974, and 2,000-3,000 cms. in 1980, 100 million tons more sediment were deposited in lower course of the Yellow River in 1980 than in 1974, among this 63 million tons of coarse sediment were deposited. The ratio of sediment disposal in the lower river course decreased from 71.7% to 41.5% within which the ratio of sediment disposal for coarse sediment decreased from 46.6% to 20.6%. Deposition in the lower river course under the operation of impounding clear water and releasing muddy water was less than that under the operation of flood detention. The main reason was that the amount of reservoir erosion has been reduced, and greater discharges prevailed.

From Table 1 it can be seen that since impounding clear water and releasing muddy water, basically clear water is released in nonflood season. The status of aggradation in predam years and in the period of operation for flood detention has been changed into one of erosion for different discharges. In a flood season the proportion of siltation at various discharges has changed greatly. For instance, erosion in periods of operation for flood detention has caused siltation for discharges greater than 6,000 cms. The proportion of the amount of siltation at this discharge to the total amount is basically the same as that prior to construction of the reservoir. For discharges smaller than 4,000 cms., the proportion of siltation has increased greatly. If the amount of sedimentation in non-flood season is added to that in flood season at discharges smaller than 4,000 cms., the proportion of the amount of sedimentation at this discharge to the total amount would be basically the same as that prior to construction of the reservoir. Thus, the proportion of the amount of deposition at different discharges within a year is the same as that prior to construction of the reservoir. During the 10 years, on average, the annual deposition in the lower river course was 145 million tons with erosion of 48 million tons in the main channel and deposition in the floodplain of 203 million tons. It can be seen from

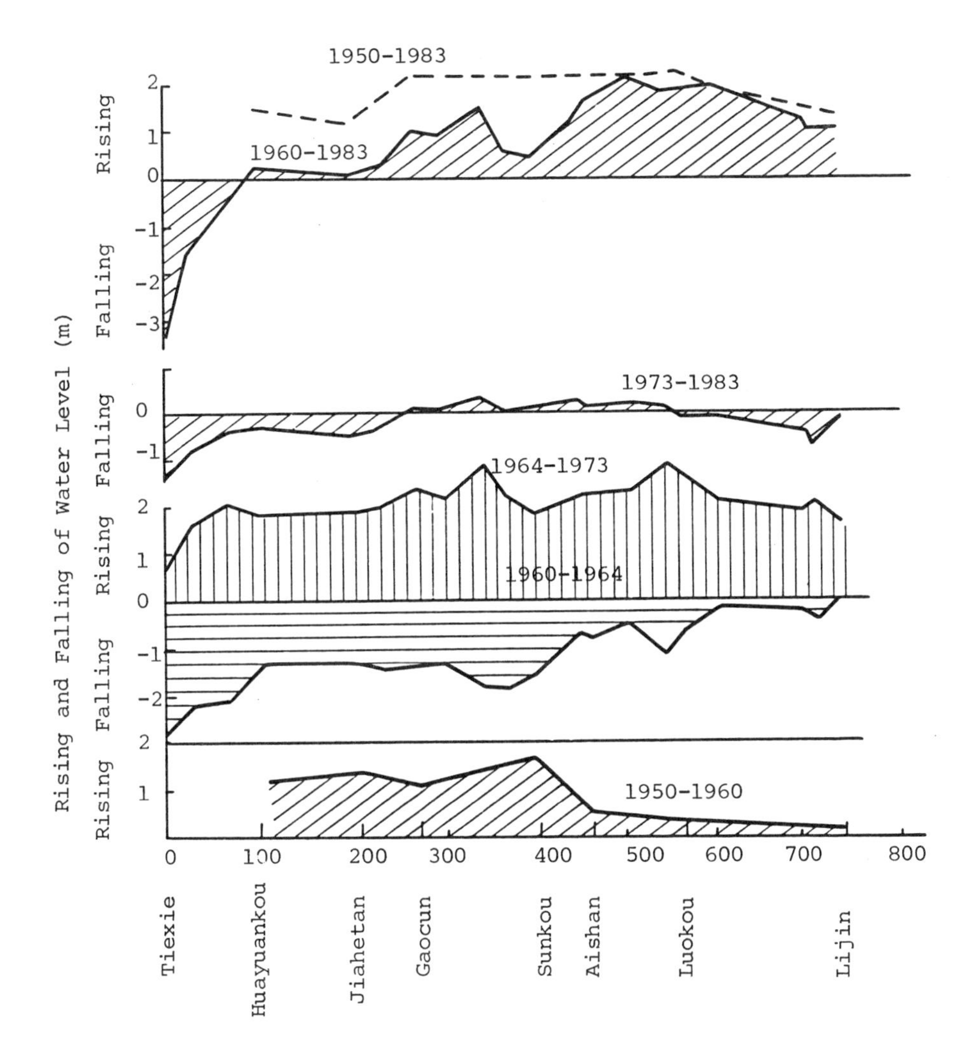

Figure 7. Changes of Water Level in Different Periods Along the Course of the Lower Yellow River.

Fig.7 that the water level under the same discharge in reaches above Gaocun and downstream of Luokou has lowered, while the water level under same discharge in the reach from Gaocun to Luokou rose. This indicates that the main channel in most parts of the lower Yellow River was eroded, although some reaches were raised. Erosion takes the form of down cutting as well as widening. Some reaches were impinged upon and scoured during medium and low flows which lasted a long time, causing great changes in the river pattern except in reaches below Gaocun. Here, subsequent to river training works, the river pattern was basically under control and was stabilized. The floodplain was raised by siltation and the difference in elevation of the main channel and floodplain increased. Bankfull discharge increased from the period of flood detention from 2,000-4,000 cms to 5,000-6,000 cms. at present which is close to the bankfull discharge prior to construction of the reservoir.

To sum up, the effects of reducing siltation in the lower river course by different operational modes may be grouped into two cases, one is the effect of reducing siltation in the lower river course by sediment retention in the reservoir. This effect may be expressed by the ratio of sediment retention in the reservoir to the amount of reduction of deposition in lower reaches, i.e. amount of sediment retained in the reservoir relevant to a reduction of deposition in lower Yellow River of 100 million tons. According to the data from Sanmenxia Reservoir, during the period of operation for flood detention the ratio was 1.63-1.72. In design of the proposed Xiaolangdi Reservoir, sediment will be retained totally in non-flood season. In flood seasons, when the ratio of sediment concentration to discharge is greater than 0.035, the ratio of sediment disposal from the reservoir will be controlled below 44%. Ratio of sediment retaining in the reservoir to the reduction of deposition in the lower reaches is 1.1. The second case regulates water and sediment rationally by improving the compatibility of water and sediment flows and increasing bankfull discharge and sediment transport capacity and, consequently, decreasing deposition in the river course.

Necessity and Possibility of Regulating Water and Sediment Properly by Reservoir Operation

The Yellow River is deficient in water and abundant in sediment. Sediment mainly comes from the area between Hekouzhen and Sanmenxia. Large differences in the relation between water and sediment may be observed for floods originating or joining from different source areas. For instance, in a year when small runoff and excessive sediment yields from the middle basin occur, serious deposition will take place in the lower reaches. In contrast, in a year with abundant water runoff and less sediment, little siltation will take place. When the natural water and sediment flow are not compatible, artificial regulation is needed to optimize the process. It is anticipated that due to storage in flood seasons by large reservoirs in the upper basin and increased amounts of water withdrawn for irrigation and industrial use, the occurrence of less runoff with excessive amounts of sediment may become more frequent in the flood season. Deposition in the lower river course would be more serious and bankfull discharge would be decreased. As a consequence, the main channel would wither and the capacity for flood conveyance would decrease making flood prevention more difficult. Besides, due to an increase in discharge in non-flood season attention should be paid to the requirement of better control during periods of ice flood.

Experience with the reconstruction and operation of Sanmenxia Reservoir is valuable and has provided a better model of building reservoirs on a river laden with fine sediment like the Yellow River. However, the existing capacity for regulation of sediment in Sanmenxia Reservoir is limited. It may be operated in a way to impound clear water and release muddy water but has only limited effects on the regulation of water and sediment. With respect to the reduction of deposition in the lower river course, the annual reduction of deposition was estimated to be about 30-50 million tons in the last 10 years, on average. In the design of Xiaolangdi Reservoir, it is proposed that the surplus water in the non-flood season be used to make artificial floods. Peak discharge would be made equal to the bankfull discharge of the lower river course, and the volume for each flood would be about 4 billion cu.m. On average, reduction of deposition of about 1.2-1.3 million tons in the lower river could be expected for every 100 million cu.m of water released in the artificial floods. Thus, the effect of sediment regulation in the reduction of sedimentation in the lower river by a single reservoir is not very large. Meanwhile, the incompatible relationship between water and sediment originating from different sources is also rather difficult to modify by operation of a single reservoir alone.

In order to modify the relationship of incompatible water and sediment originating from different sources, and also to change the situation of the shrinking channel, a place should be selected where water and sediment can be regulated according to the geographical position of the source area of water and sediment runoff. Reservoirs regulated jointly would optimize the process of water and sediment flow to prevent shrinkage of the channel as well as for increasing the bankfull discharge and capability of sediment transport.

The best localities for sediment regulation are Xiaolangdi and Longmen Reservoirs which are situated close to the lower

Yellow River where the reduction of deposition in the lower reaches would be best achieved. The ideal place for water regulation is located upstream from reaches where the majority of the sediment enters the main river. If reservoirs are built to regulate downstream reaches, they might be operated jointly for water and sediment control. According to the design of the outlet discharge capacity of the proposed Xiaolangdi project, bankfull discharge in the lower river course would be increased from 5,000-6,000 cms to 6,000-7,000 cms. The sediment transport capacity could be further increased and channel morphology could be improved by increasing the difference in elevation between the main channel and floodplain.

References

Li Baoru, et al., "Change of the River Channel Downstream of Sanmenxia Reservoir During Period of Reservoir Impoundment," Proc. Intern. Symposium on River Sedimentation, Beijing Guanhua Press, 1980.

Long Yuqian, "Yellow River Flood and Sediment Problems," Bilateral Seminar on River Engineering Experiences and River Hydraulics, Tokyo, 1984.

Long Yuqian, Zhang Qishun, "Sediment Regulation Problems in Sanmenxia Reservoir," Water Supply and Management, Vol 5, No.4/5, 1981.

Pu Neida, et al., "Problem of Sedimentation in Liujiaxia and Yanggouxia Reservoir," Proc. Intern. Symposium on River Sedimentation, Beijing Guanhua Press, 1980.

Qian Yiying, Han Shaofa, "Effect of Operation in Sanmenxia Reservoir on the Adjustment of Lower River Reaches," Journal People's Yellow River, Vol 6., 1984.

Zhang Qishun, Long Yuqian, "Sedimentation Problems of the Sanmenxia Reservoir," Proc. Intern. Symposium on River Sedimentation, Beijing Guanhua Press, 1980.

Wang Jidi, Zhang Jiajin, "The Fluvial Process of the Lower Liaohe River after the Construction of Reservoirs," Journal of Sediment Research, Beijing, No.4, 1982.

EFFECTS OF RIVER TRAINING WORKS ON FLOOD CONTROL

Hu Yisan and Xu Fuling
Yellow River Conservancy Commission
People's Republic of China

ABSTRACT: Owing to sedimentation, the lower course of the Yellow River has become a "suspended" river. The dikes are threatened by upward or downward shifts of the main current. Bifurucated channels may take up the main flow and attack parts of the dikes where no protection works are placed. Flow may be directed at angles which put flood prevention works in jeopardy. At times dike breaches have caused very grave disasters to people living along the Yellow River. Since 1950, river training works have been carried out reach by reach on the lower Yellow River. The completed training structures have stood severe tests from various floods for years. Improvements have been to pattern the cross-section of the river course and to avoid unfavorable wandering of the main currents. River training works in the lower reaches have played a remarkable role in ensuring the safety of people from floods.

As the Yellow River flows out of the gorges at Mengjin county in Henan, the channel broadens and the slope decreases so that sediment is deposited and aggradation occurs. The length of the channel from Mengjin to the river mouth is 878 km. Most of the Yellow River in this area is confined by dikes on both banks. Because of this, the channel becomes narrower and the slope gentler from upstream to downstream (see Fig.1). According to the features of the channel, it can be divided into four parts with different characteristics (see Fig.2).

The channel on the lower reaches of the Yellow River primarily has compound sections often with two terraces below Dongbatou. Owing to headward erosion after the avulsion took place in 1855, one more terrace was added to the channel so that on the reach upstream of Dongbatou there are three terraces (floodplains). Usually the area between the second terrace of both banks is called the channel for medium floods. The total area of the channel river bed between the dikes on both banks below Mengjin is 4,240 sq.km, of which the channel occupies 713 sq.km and floodplain (including the second and third terrace) 3,527 sq.km. Large floodplains occur on the wider reaches above Taochengbu. Most of the floodplain areas

L. M. Brush et al. (eds.), Taming the Yellow River: Silt and Floods, 617–636.

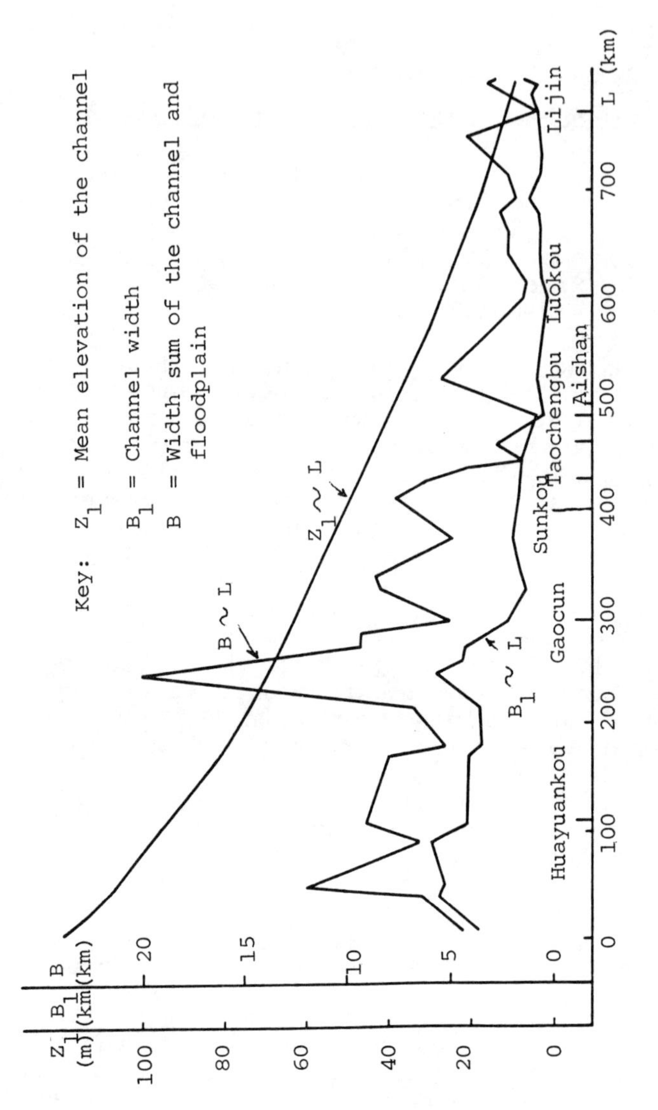

Figure 1. Changes of River Width Along the Course

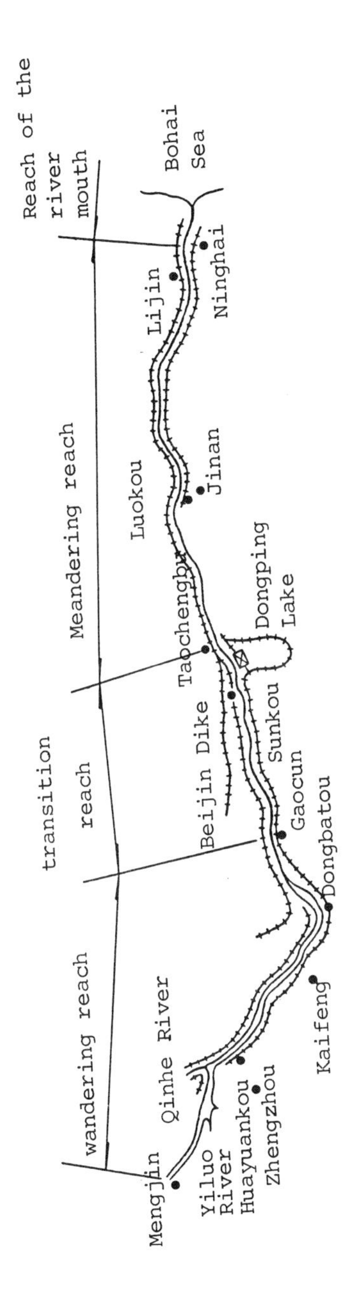

Figure 2. Plane Drawing of Lower Course of the Yellow River.

are fertile land because of sedimentation. At present, there are 2,002 villages populated with 1.268 million people in the area. The channel formed by dominant medium floods on the lower reaches of the Yellow River is the main path for floods and bankfull discharge changes with time and space all the way along the lower course of the river. These flows are generally about 5,000 cms if a relative equilibrium value is selected.

Owing to the deposition of sediment, the river bed rises. The surface of the floodplain is generally 3-5 m, higher than the adjoining land behind the levees on both banks. At some places the maximum difference in elevation is 10 m, which is why the Yellow River is called the "suspended river" (see Fig.3).

Since 1958, the width of the river course for conveying floods has been narrowed and the probability of overflowing the floodplains has been reduced by production dikes on the floodplains by local inhabitants. Most of the sediment is deposited in the river channel between production dikes on both banks and some is deposited on the floodplains between the production dikes and the main levees. At some sections, the floodplain on the front side of production dikes is higher than that on the back side. The mean elevation of the main channel is even higher than that of the floodplain, and a second stage suspended river has formed which is detrimental to flood prevention. A typical cross section is shown in Fig.4.

Safety of the Embankments is Threatened by Changes in Channel Pattern

One of the main characteristics of the channel evolution is the changes which have occurred by scouring and deposition in the channel causing frequent shifting of the main current. The main fluctuations in the regime and pattern of flow are caused by back and forth movements of the bends, shifting of the main current and natural cutoffs. The intensity of the change lessens from upstream to downstream. In meandering reaches, the changes in channel patterns occur because of the development of bends. The maxium distance for shifts within 24 hours is several tens of meters while the flow rolls seriously at wandering sections. During a past flood, the thalweg shifted more than 100 m daily on average, and the maximum was thousands of meters. In the 1954 flood, the main current near Liuyuankou, Kaifeng city, shifted from the north bank to the south, then back to the north bank again. The main channel wandered north and south for 6 km within 24 hrs. These rapid changes are unimaginable.

Sometimes the changes of flow patterns directly threaten the dikes. This makes flood prevention difficult. Some of the main forms are as follows.

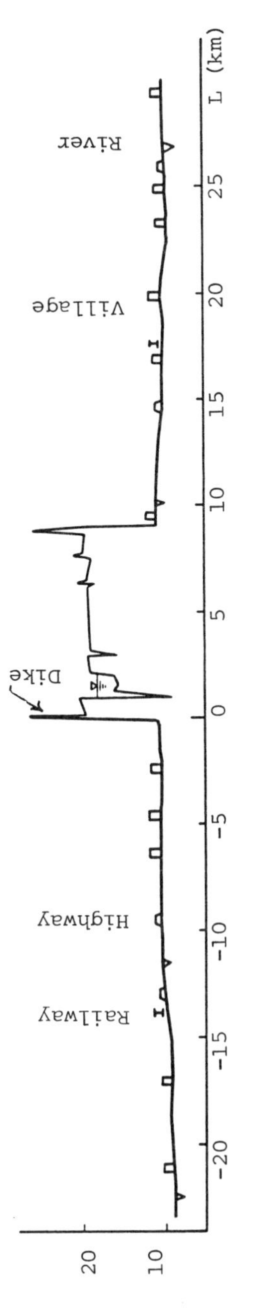

Figure 3. A Suspended River of the Lower Yellow River

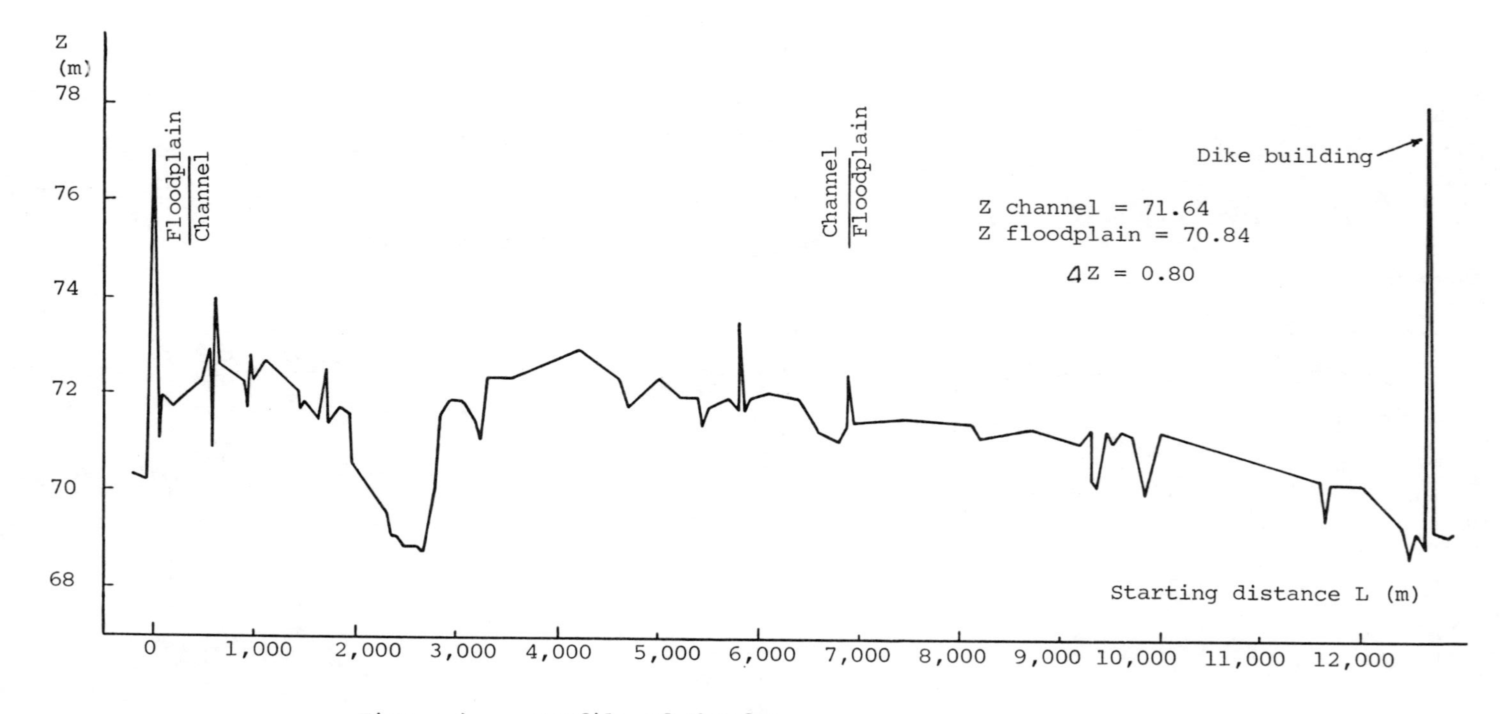

Figure 4. Profile of Chanfang

GREAT CHANGES OF FLOW PATTERN

Before construction of the training works, the distance between the dikes was small, only 0.5-2 km in the meandering reaches, yet dangerous situations constantly occurred at ordinary sections where no bank protection works existed. Such was the case during the flood season of 1949. There were 9 vulnerable spots on the river because the channel pattern moved and the main current extended downstream for 2 km. The floodplain slid and caved in at many locations. The rehabilitation work took more than 40 days and put the flood prevention works in jeopardy.

FORKS TAKING UP THE MAIN FLOW

At reaches where the distance between the two dikes is wider, more sediment is deposited at the edge of the floodplain. The nearer the flood is to the dike, the less the deposition of sediment. Transverse slopes formed on the floodplain, which are steeper than the longitudinal slope of the channel. Because of topography and the spatial variation of the roughness of the floodplain surface, flow over the floodplains is unevenly distributed. Deposits occur in the areas with slow flow and scour occurs where the flow is concentrated causing forks to form on the surface of the floodplain. The bifurcated flows frequently join together at the foot of the dikes forming a river channel along the dike. Fig.5 shows the distribution of forks and channels along dikes below Dongbatou. This occurred before the flow patterns were effectively controlled. Flood waters flowing over the floodplain would flow along forks and reach the river channel at the foot of the dikes, impinging upon the dike at dangerous angles. When the river course is in a bend, the flood water may not overflow, but the main flow may be captured by the shorter forks, also threatening the safety of the dikes. For instance, in the reach from Laojuntang to Baocheng, the main current originally was on the left bank. A fork in front of a vulnerable spot at Huangzhai developed on the right bank (see Fig.6a), and the lower part of the regulation works at Laojuntang was cut off from the current. The main current was out of control and the fork became the main course of the river in June 1978 (see Fig.6b). In order to guarantee the safety of the dikes, a vulnerable spot 1,500 m in length at Wuzhang was built at the ordinary section above the vulnerable spot at Huangzhai. Later the current approached this vulnerable spot once again and attacked the vulnerable spot at Wuzhang. The bend changed its location by moving back and forth increasing the cost of maintenance and also making flood prevention more difficult.

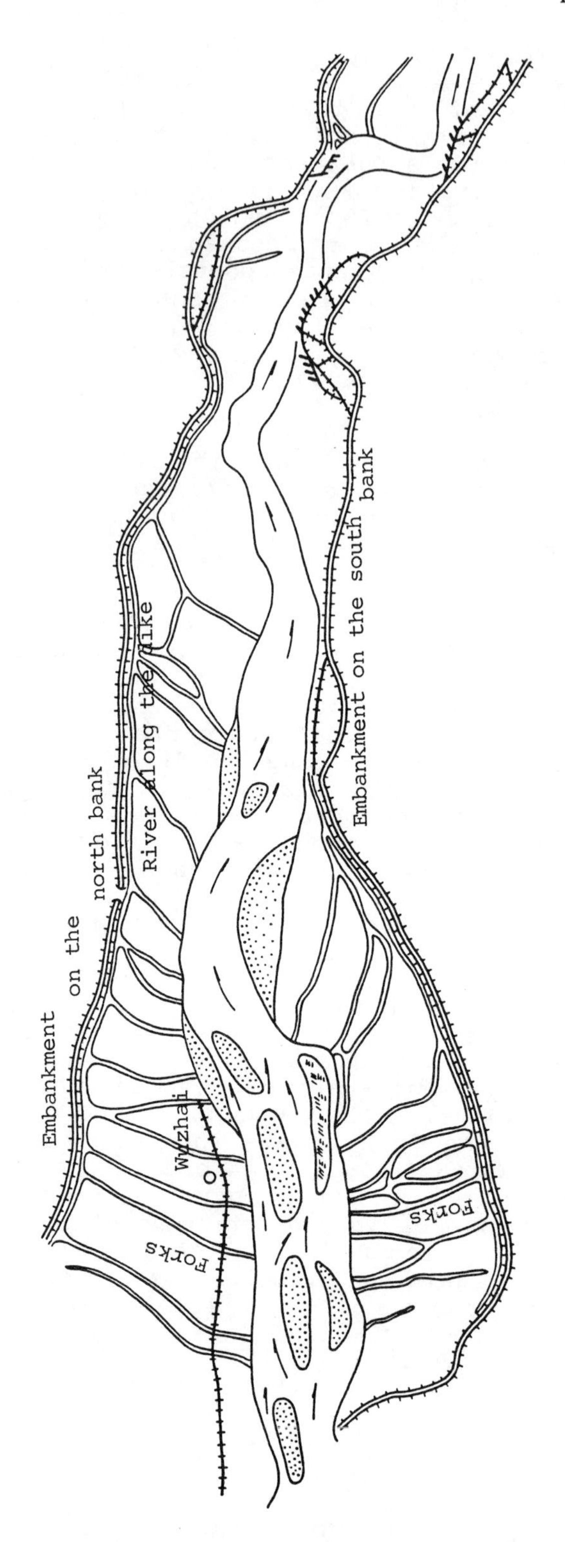

Figure 5. Distribution of Forks and Rivers Along Dikes below Dongbatou

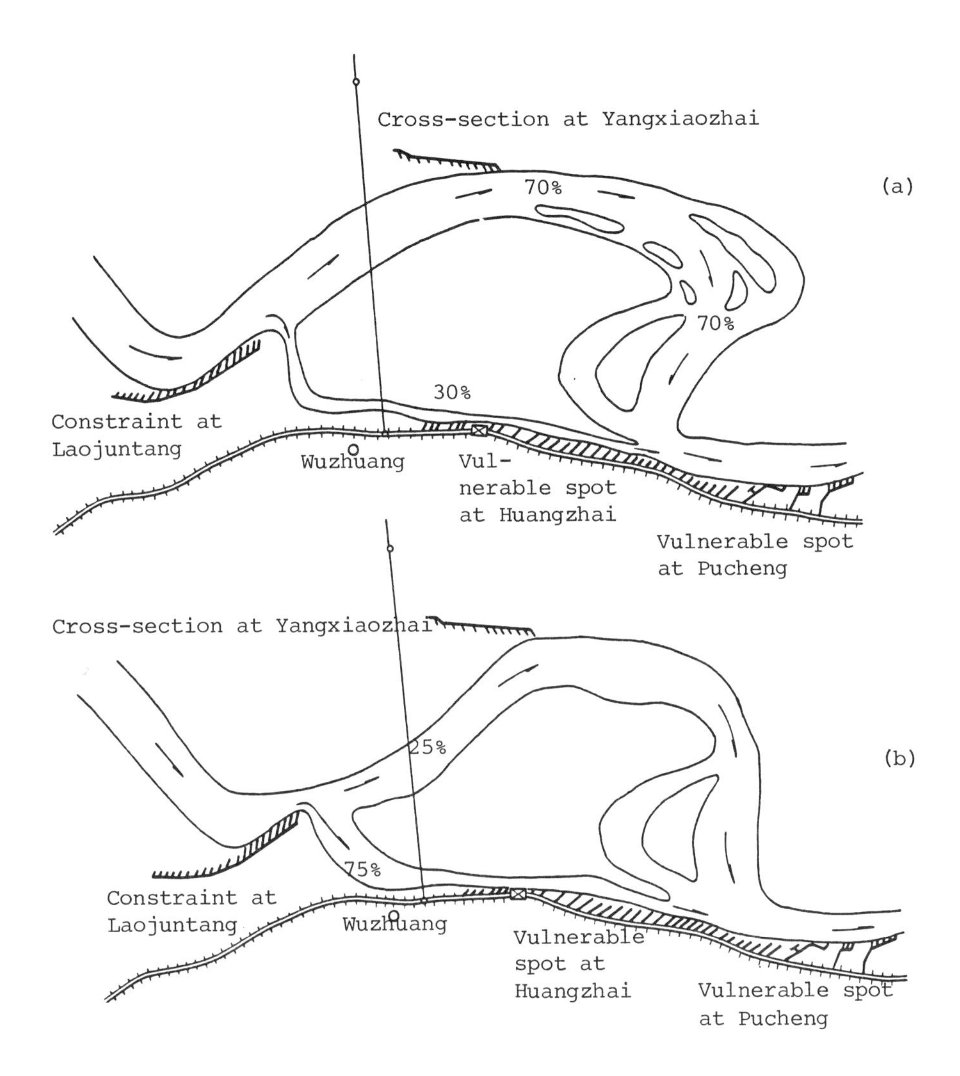

Figure 6. Forks in the Reach of Laojuntang and Pucheng.

DIKES IMPINGED ON BY FLOWS AT A GREAT ANGLE

In the wandering reach, the channel is wide and shallow, has many bars and the flow divides into shallow braids. During the course of changes in the channel pattern, transverse currents often form. Under the action of the cross currents, concave banks may slide and cave in. The tip of the floodplain on the convex banks extends constantly, narrowing the water surface and increasing scour. This leads to rapid erosion of the floodplain against which the main flow impinges. The safety of dikes or weak spots is threatened when directly confronted by the flow. For example, the transverse flow impinged upon the lower part of the vulnerable spot at Huayuankou in early October 1964 (see Fig. 7). The width of the water surface was reduced to about 150 m late in October, leading to concentration of the flow and increasing the risk of scour. Riprap severely slumped and the safety of the dike was maintained through a major effort of rushed repairs. As much as 11,600 cu.m of rock was used for the quick repair at spur dike No. 191 (Dongbatou).

It can be seen from the above description that the dikes are threatened constantly by changes in flow patterns. Breaching may take place if rush repairs and protection works are not done quickly. During the rise of the 1843 flood (June of 23rd year in the reign of Emperor Daoguang in Qing dynasty), the main current suddenly shifted to the ordinary section at Jiubao where no defensive works were in place. The dike finally breached due to difficulties in collecting and transporting materials for rush repairs. During the rising period of the major flood in August 1934, water flowed over the levees at Guanmeng, rushed to the levees at Changyuan through forks and caused several breaches. In the flood of 1868 (the 7th year in the reign of Emperor Tung Chih in Qing dynasty), the flow pattern changed, and sliding and caving-in of blocked bends occurred. The old neglected vulnerable spot at the suburban area of Zhengzhou (Fengzhuang village in Xingyang), was attacked by the floodflow, causing a dike to breach because no materials were available for rush repairs. As for the narrow reaches downstream of Taochengbu, some sharp bends with small radii of curvature were often formed during the process of channel change. Here ice and water are easily jammed up or detained during ice runs, making the water surface elevation rise rapidly and causing dikes to breach. From 1883 to 1936, dikes failed during the ice run in 21 years. Since the river channel is higher than the adjoining land of both banks, it is not easy to block the flow after breaching, therefore, each breach brought extremely serious disasters to the people living along the Yellow River.

Training Works in the Lower Yellow River

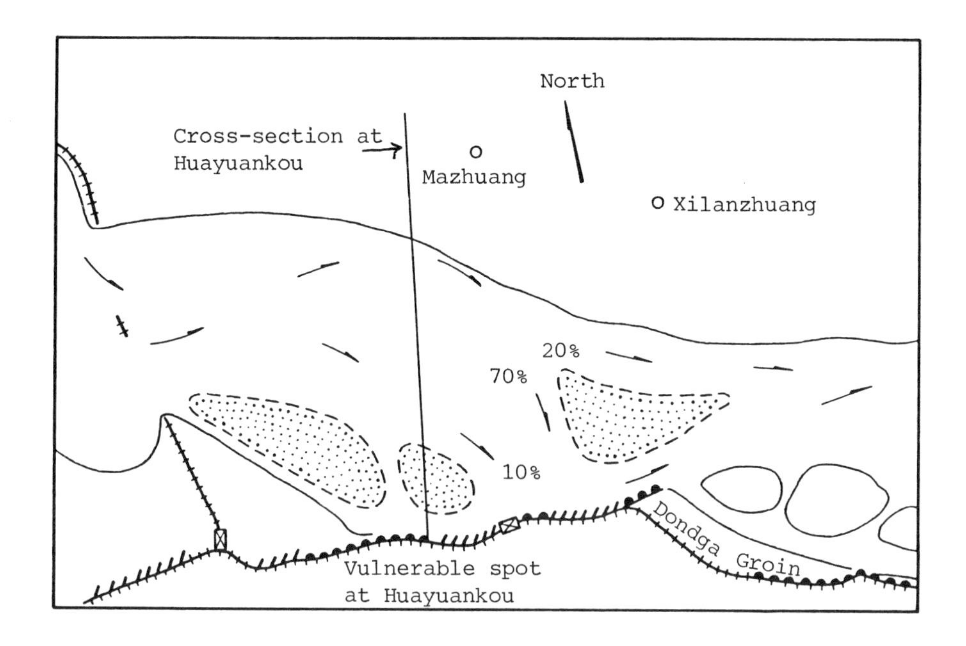

Figure 7. Flow Pattern of Downstream Section of Huayuandou Vulnerable Spots in Early October 1964.

During the 2,000 years before 1949, there were more than 1500 dike breaches on the lower Yellow River. Generally speaking, more breaching happened at ordinary sections than at vulnerable spots and more breaches were due to bursting or impingement of the flow and less by overflows. In order to ensure the safety of the dikes, training works on the lower Yellow River were planned and implemented in the early 1950's to control the channel pattern. This was done to help prevent floods and ice floods. The works were installed to take into account diversion of water through sluices and culverts for irrigation, industry and domestic use, while meeting the requirements of navigation for transporting rocks and materials and the safety of the people living on the floodplain.

The guidelines for river training require unified planning with due consideration for both upstream and downstream reaches as well as for both banks. The channel and floodplain should be controlled comprehensively. Engineering works should be planned and constructed in conformity with the basic laws of fluvial processes and local conditions. According to necessity, regulation works should be placed at primary localities with priority. Local materials should be used and structures should be appropriate to local conditions in order to save money.

MEASURES OF TRAINING WORKS ARE COMPRISED OF BANK PROTECTION AT VULNERABLE SPOTS OF THE DIKES AND CONSTRAINTS

In order to resist scouring, spur dikes, stack dikes (short groins) and revetments were built along the embankments at vulnerable spots. The spur dikes, stack dikes and revetments built on the floodplains to protect the floodplains as well as the embankments and to control and correct the flow are constraints, or bank protection works.

For more than 30 years the downstream reaches were trained and later the upstream reaches. Within a reach, structures were built in sequence from upstream to downstream according to requirements and training principles mentioned above. Training works at different river reaches can be described as follows:

Meandering reach. People realized through the rush work in 1949 that flood fighting, even in the fairly narrow meandering reach, could not be controlled if one relied on the vulnerable spots built along the embankments. The remedial measures were to select bends where experiments were made to build permeable piles and willow dams and later to change the light engineering works of permeable piles and willow dams into willow and rock ones. Within the experimental reach the results of controlling the channel pattern and protecting floodplains through siltation behind the dams were good from the point of view of controlling the dominant flow. From 1952 to 1954, many bank protection works of stack dikes with willow and rock were constructed at the vulnerable spots. By the end of fifties, the length of vulnerable spots and bank protection works was 60% of the total length of the reach. These structures were tested by major floods in 1954, 1957 and 1958. More works were built in the sixties and the meandering reach downstream of Taochengbu was put under control.

Transition Reach. Calling on the experience gained from the training works of the meandering reach, some constraints were constructed and some vulnerable spots were re-adjusted. Continuous building with careful planning was carried out from 1966 to 1974. At present, there are 49 river training works with a total length of 123 km, making up 75% of the length of the river course in the reach. The regulation works withstood the bigger floods of 1976 and 1982. In recent years, there have been no big changes in channel patterns and the river course is, for the most part, controlled.

Wandering Reach. The flow pattern changes rapidly at the wandering reach above Gaocun where constructing training works is more difficult than at the transition reach. The width of the water surface downstream of the constraints at Chanfeng in the reach between Dongbatou and Dawangzhai used to be 3 km or so before the flood season in 1974 (see Fig.8). This flood separated into 2 or 3 forks by shallow bars, and after the flood season an "S"-shaped bend was formed. Subsequent to 1966, constraints were built in the wandering reach between Dongbatou and Gaocun. Works were built in areas where river training works were planned through 1978. Though the length of the existing projects is far from enough to meet the requirements, it has played an important role in limiting the shifting of the channel. An integral system of training works has not been formed and the flow not confined in the wandering reach above Dongbatou. This work should be speeded up in the future.

Main Functions of River Regulation on Flood Control

Through training works, the lower course of the Yellow River has been controlled at different levels. At the controlled meandering reach, the passive situation of rush repairs, as happened in 1949, has been changed completely. The channel pattern changes slightly during flood season and is in good condition for flood control. It has been proved by several floods that the constraints built on the floodplains are capable of controlling the flow pattern. If overflows should occur they would probably experience some damage which should

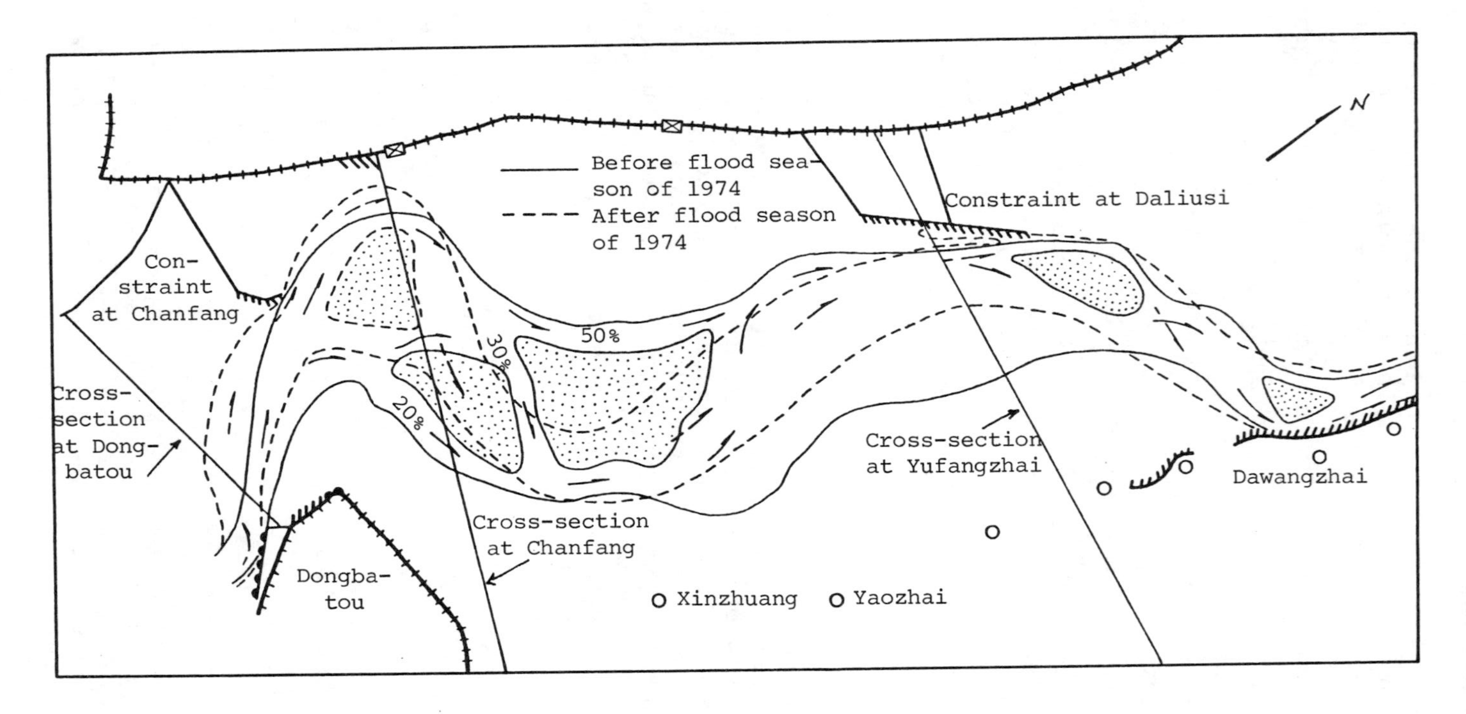

Figure 8. River Pattern Before and After Flood Season of 1974.

be easy to repair. The constraints and river training works also function well in the transition and wandering reaches.

TRANSITION REACH

Preventing the Development of Unfavorable River Pattern. Without the training works, main currents flowing in the transverse direction may occur. For instance, after the flood season of 1949 the main current shifted from the vulnerable spots at Susizhuan to Fangchangzhi where a big bend on the floodplain was formed. This resulted in serious sliding and caving-in on the floodplains of the left bank and the main current impinged directly towards the embankment (see Fig.9.a). Abnormal bends developed easily at some reaches before the training works were established. One of many possible examples of the reach between Susizhuan and Yingfang is shown on Fig. 9 where the ratio of the bend to the neck in 1959 was 2.81. The excessive curvature of the river course is unfavorable to flood conveyance due to the long flow path and great resistance in the channel. Compared with normal bends, the water surface elevation at the upstream end is forced to rise in order to convey major floods at the same discharge. As a consequence, the river may change its course quickly at high water levels and erosion such as slides and cave-ins at low water may take place. This would throw flood control into passivity. During the ice-run period, the ice can easily block the river and form an ice jam because of the small radius of curvature and rapid change of flow direction. Transverse currents have been eliminated for the most part and the formation of abnormal bends limited so that the reach between Susizhuan and Yingfang has become a smooth-going bend and the passive situation of flood control has been changed as shown in Fig.9.b.

Reducing the Amount of Shifting. Shifting of the thalweg was analyzed in order to compare the change of river pattern with and without training works. Mean changes of the thalweg before and after training in the transition reach are shown in Table 1. It can be seen that the mean range and intensity of shifting of the thalweg has been reduced 65% and 60% respectively after training and the maxmium amount of shifting reduced from 5,400 m to 1,850 m.

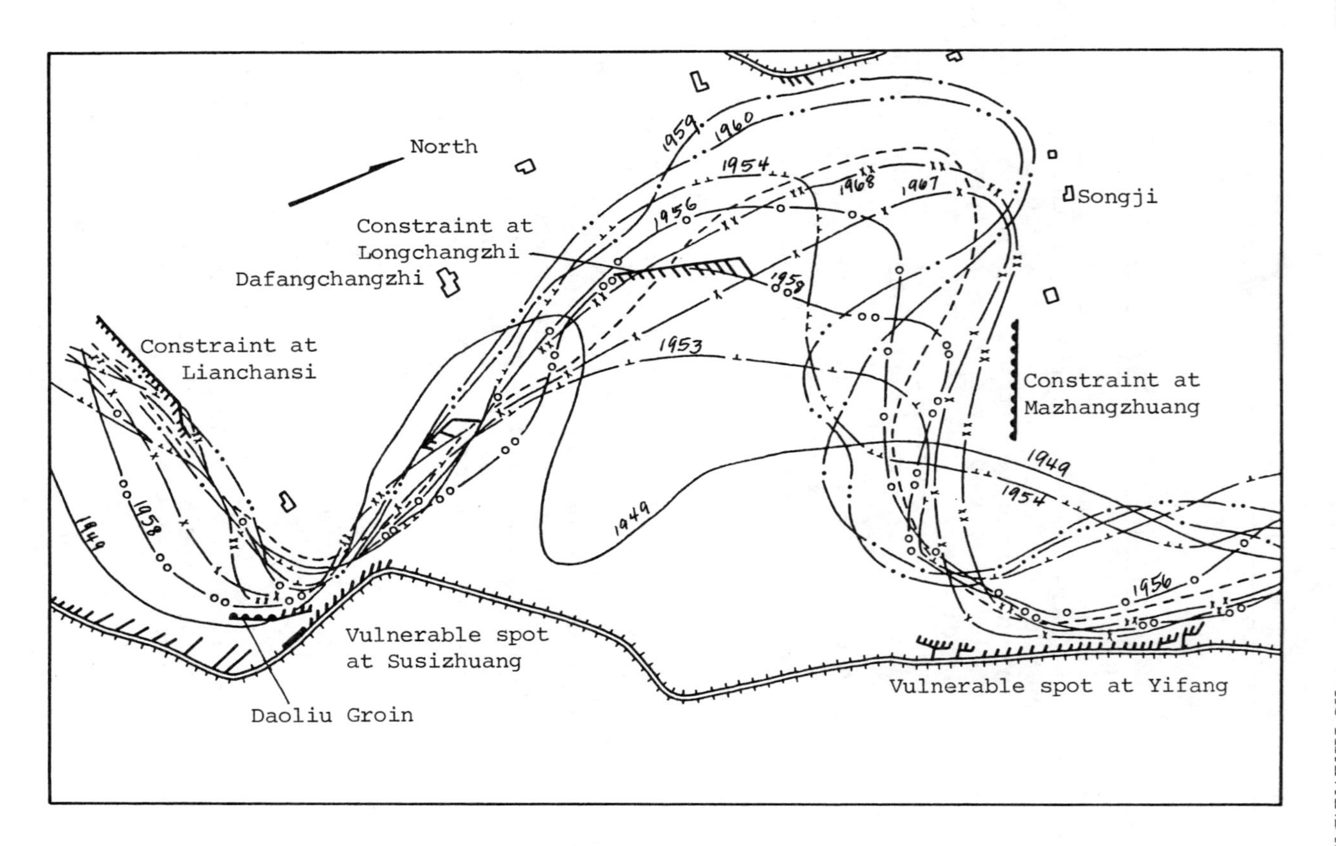

Figure 9(a). Thalweg of the Reach Between Susizhuang and Yingfang After Training.

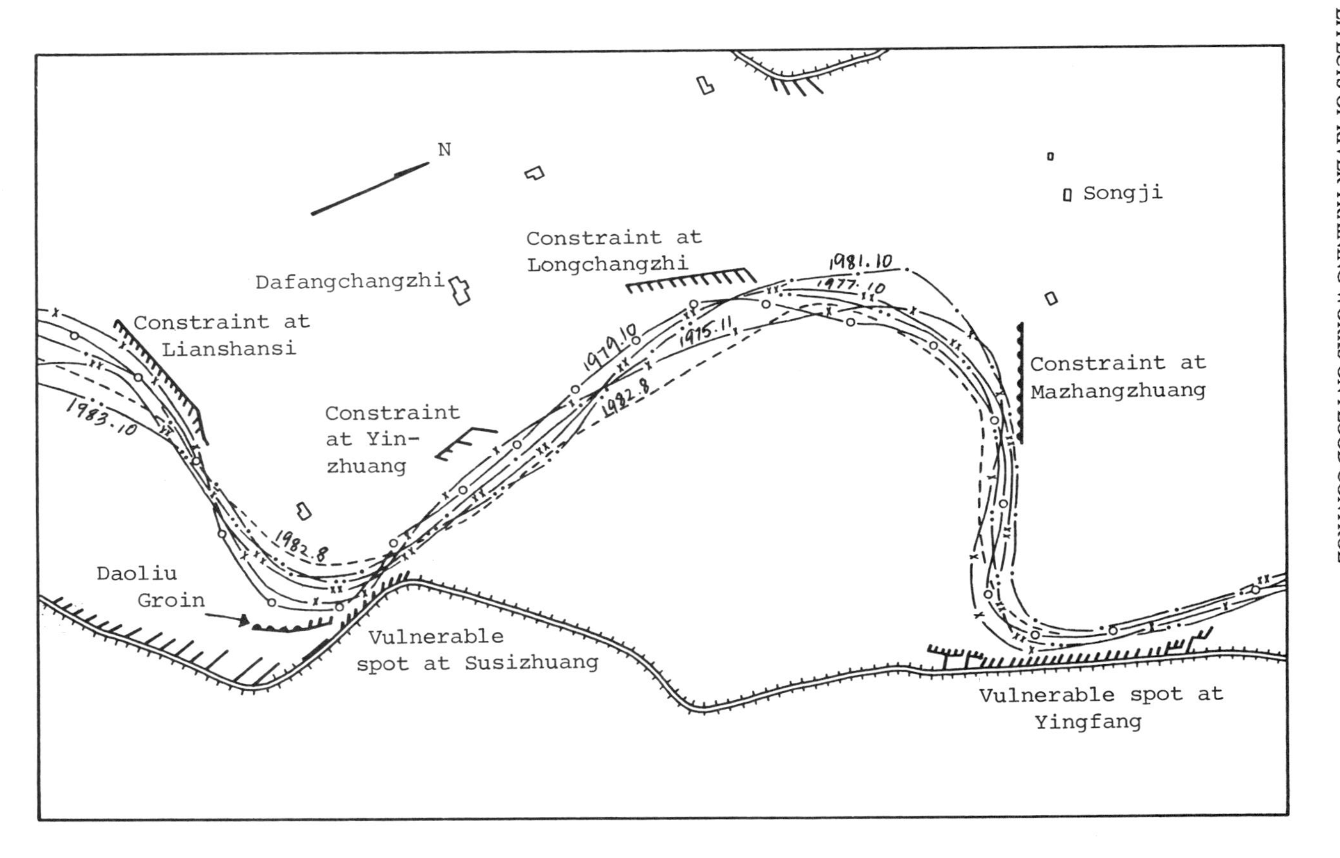

Figure 9 (b). Thalweg of the Reach Between Susizhuang and Yangfang after Training.

Table 1. Changes of the main flow in the reach of Gaocun and Taochengbu

Item	1949-1960 (1)	1975-1984 (2)	(2)/(1) (3)
Shifting scope (m)	1802	631	35%
Shifting intensity (m/year)	425	171	40%

Improvement of Channel Morphology in Transverse Sections. The main channel is the primary passage for flood waters. Therefore, the deeper and narrower the course, the better it is for flood conveyance. Favorable changes of mean depth (H) and width (B) under bankfull discharge of the channel have occurred following training. Comparing the periods of 1975-1984 with 1949-1960, the mean water depth has increased from 1.47-2.77 m to 2.13-4.26 m, and the ratio of width to depth of the cross section reduced relatively from 12-45 to 6-19. Meanwhile the amplitude of annual changes of the coefficient of the river course have become smaller and smaller which shows that the trained river course is tending towards a deeper and narrower channel.

Remarkable Effects on the Safety of Flood Control. It has been proved through practice since 1975, that the river training works have played a significant role in controlling and training the main flow, in protecting the floodplains, and safeguarding the main dikes. There have been no major, abrupt changes in the channel pattern for the past ten years. During the floods of 1976 and 1982, the duration of the dominant medium flood discharges were longer and flood peaks were greater. Some constraints were overtopped, and some spur dikes and stack dikes burst. Some areas between the spur dikes were flooded and newly formed ditches and breaches occurred in the production levees at the head of the project. All this provides the stage for a sudden change of the channel pattern, but this did not happen. Because of the training works, a more stable channel for the flow has been formed. Although some of the contraints were overtopped, and some spur dikes and stack dikes burst in the upper and middle parts of the project, and rock materials from spur or stack dikes slumped during the flood peaks, the section was controlled to a certain degree. The main channel is deeper and its boundaries have a greater ability to control and correct the flow. Therefore, subsequent to floods, the main current comes back to the original path again. Obviously the river training works have played a significant role in flood control.

WANDERING REACH

The wandering reach above Gaocun still undergoes movement, but the scope of wandering is small compared with that in the past. The training works in the reach between Dongbatou and Gaocun play an important role in flood prevention.

Dongbatou is near the breaching point that was breached in the avulsion at Tongwaxing in 1855. The water surface became wider, and most of the coarse sediment was deposited. The original aggraded delta has become the present river course. Shifting and wandering under such a boundary condition may occur quite easily. In more than a hundred years, breaching took place many times in the reach below Dongbatou. After building some river training works, the scope of shifting of the main flow has been reduced and the mean range of shifting in the reach was reduced to 2,435 m from 1949 to 1960, and 1,700 m from 1979 to 1984. The training works are playing a notable role in controlling sudden changes of the channel pattern. For instance, during the major flood of 1933, the peak dischange at Sanmenxia of 22,000 cms (20,400 cms at Huayuankou) overflowed the dike at Guanmeng and directly impinged upon the dike somewhere near Dacheji, Changyuan county. The flood threatened the ordinary sections and caused breaches at several places as the flood passed by Dongbatou. During the flood in 1982, the peak discharge at Huayuankou was 15,300 cms. The flood waters rushed down with a terrifying force and great volume. The high water level provided the conditions for shifting. The river pattern downstream from Dongbatou during the flood peak period was similar to that in 1933, moving in the direction of the levees at Guanmeng. Before coming to the embankment at Guanmeng, the main flow turned to the right bank. A dangerous situation was prevented because the flow was corrected by the constraints at Chanfang on the left bank. Although the river training works in this reach are not perfected and the earth foundation for some spur dikes and stack dikes was damaged, the training works still play a certain role in controlling the channel pattern.

Concluding Remarks

A SUDDEN CHANGE OF THE CHANNEL PATTERN MIGHT LEAD TO VERY GRAVE CONSEQUENCES

Through river training works the main flow can be put under control, and the river patterns stablized. This helps safeguard the floodplains and embankments.

ENSURING FLOOD PREVENTION ON THE LOWER YELLOW RIVER IS A TASK OF TOP PROIORITY

At present, the freeboard of the dikes is 2.1m-3.0m. The dike could be heightened to prevent overflows, and strengthened by measures to remedy hidden defects in the foundation and body of the dikes and by warping to prevent bursts. Major efforts should be made to increase the river training works controlling and correcting the flow as well as those required to reduce chances of abrupt changes of channel patterns. Meanwhile comprehensive control work should be carried out to reduce the speed of siltation in the lower course. The function of sediment and flood water conveyance of the existing channel should be fully used to safeguard the main dikes of the Yellow River.

THE XIAOLANGDI PROJECT

This project, which will be bult in the lower part of middle reaches of the Yellow River, will hold sediment but will also regulate water and sediment discharges. Therefore, an opportunity exists to utilize the training works in a planned way in order to minimize the effects on the lower reaches created by the completion of Xiaolangdi project.

TRAINING WORKS ON THE LOWER REACHES

Through 30-odd years, training works on the lower reaches of the Yellow River have led to the control of flow through the meandering reach. The transition reach is also controlled for the most part. Work on the wandering reach has achieved even better results, but the difficulty of constructing the training works is far greater there than on the other reaches. The requirement of controlling the main flow is far from accomplished. Thus, how effectively to do further training works in the wandering reach is a problem that remains to be solved.

IMPROVEMENT OF THE MOUTH OF THE YELLOW RIVER AND SEDIMENT DISPOSAL

Ding Liuyi
River Engineering Division
Yellow River Conservancy Commission
People's Republic of China

Basic Law of the Evolution of the Estuary of the Yellow River

Evolution of the estuary reflects the comprehensive interaction of two dynamic factors, the river and the sea. Its capricious characteristic is caused by the huge sediment load of the Yellow River, the shallow depth and the weak tidal currents in the gulf area.

CHARACTERISTICS OF ONCOMING WATER AND SEDIMENT

According to statistical data at Lijin from 1950 to 1985, the average annual runoff, sediment load and concentration of sediment entering the estuary of the Yellow River are 41.9 billion cu.m, 1.05 billion tons and 25 kg/cu.m, respectively. The combination of small water discharge and heavy sediment load ranks the river first in the world in sediment transport. Table 1 compares the water and sediment discharge of the Yellow River with several large rivers.

Incoming annual water and sediment discharge are unevenly distributed over the year. During the flood season, from July to October, water runoff amounts to 60% of the total annual runoff. The sediment load is more concentrated, comprising nearly 85% of the total annual value. Incoming sediment is greatly affected by the upstream river course. Table 2 shows that the annual sediment discharge was about 2 billion tons at Lijing Station in 1954, 1958, 1964 and 1967. However, at Huanyuanko the largest value was 1.7 times the lowest during these same years. On the other hand, in 1964 and 1977 the incoming sediment was nearly the same at Huayuankou Station but varied almost twofold at Lijing Station. Even though clear water was released from Sanmenxia Reservoir in 1961, the annual sediment discharge was still 0.9 billion tons at Lijing Station.

L. M. Brush et al. (eds.), Taming the Yellow River: Silt and Floods, 637–655.

Table. 1 Statistics of water and sediment between China and several countries in the world

country	river	station	mean annual water runoff 100 million cubic m	mean annual sediment load 100 million tons	av. sediment concentration kg/cu.m.
Bangladesh	Brahmaputra	estuary	3850	8.0	2.1
	Ganges	estuary	3710	16.0	4.3
India	Kosi	Chutra	570	1.9	3.3
Pakistan	Indus	Coterli	1740	4.8	2.8
U.S.A.	Mississippi	estuary	5640	3.44	0.6
Brazil	Amazon	estuary	57200	4.0	0.07
Egypt	Nile		895	1.22	1.40
China	Changjiang	Datong	9480	4.83	0.51
	Haihe River		90	0.096	1.07
	Pearl River	Wuzhou	2940	0.834	0.28
	Qiantanjian	Lucifu	312	0.05	0.16
	Yellow	Lijin	419	10.5	25.1

Table. 2 Comparison of the amount of sediment load between Huayuankou and Lijing in typical years (100 million tons)

station	year 1954	1958	1964	1967	1977
Huayuaukou	21.3	27.8	16.4	20.5	17.5
Lijin	19.8	21.0	20.3	20.9	9.45

FEATURES OF THE DELTA COASTAL REGION

The coastal region of the Yellow River delta has a width of 20 km outside the low tide line. Generally speaking, its depth is quite shallow and the slope gentle. Different shore areas are quite different. For example, the middle part of this region near Shenxiangou is the deepest and steepest one, with a maximum depth of up to 20 m, and isobaths of 5, 10, 15 m lying only 3.3, 8 and 14 km away, respectively. The southern part of the Laizhou Gulf is shallow with a maximum depth of only 10 m and isobaths of 5 and 10 m located 8.1 and 22.5 km from shore, respectively (see Fig. 1).

Because of sediment deposition, a sandspit protrudes into the sea at all the old courses which once had been the main stem of the river. Outside the sandspit, especially in the Shenxiangou Sea region, there is a much higher gradient of water depth. With the sandspit extending seaward continuously, the water depth becomes deeper and deeper, and the slope of the bottom of the sea also gets much steeper. A comparison of longitudinal profile before and after Diaokouhe is shown in Fig. 1.

In the coastal region, tide levels vary only about 1 m. Tides affect the river course for about 20 km up the Yellow River. The tide type is a half-day tide, which is mainly controlled by an M2 tidal branch. There is a strong current region out of the mouth of Shenxiangou. The maximum speed of tide could be up to 1.5 m/s, which becomes smaller along the two shores. For example, the current is only 0.4 m/s at the mouth of Xiaoqinghe in Laizhou Gulf. The tidal current in the coastal region of the delta has a small rotating ellipticity. Directions of the flowing and ebbing tidal currents are almost parallel to the coastline. In the coastal region west of Goukouhe, the maximum flowing tidal current goes to the north-west and the ebbing tidal current flows south-east, causing counter-clockwise rotation. But in the coastal region

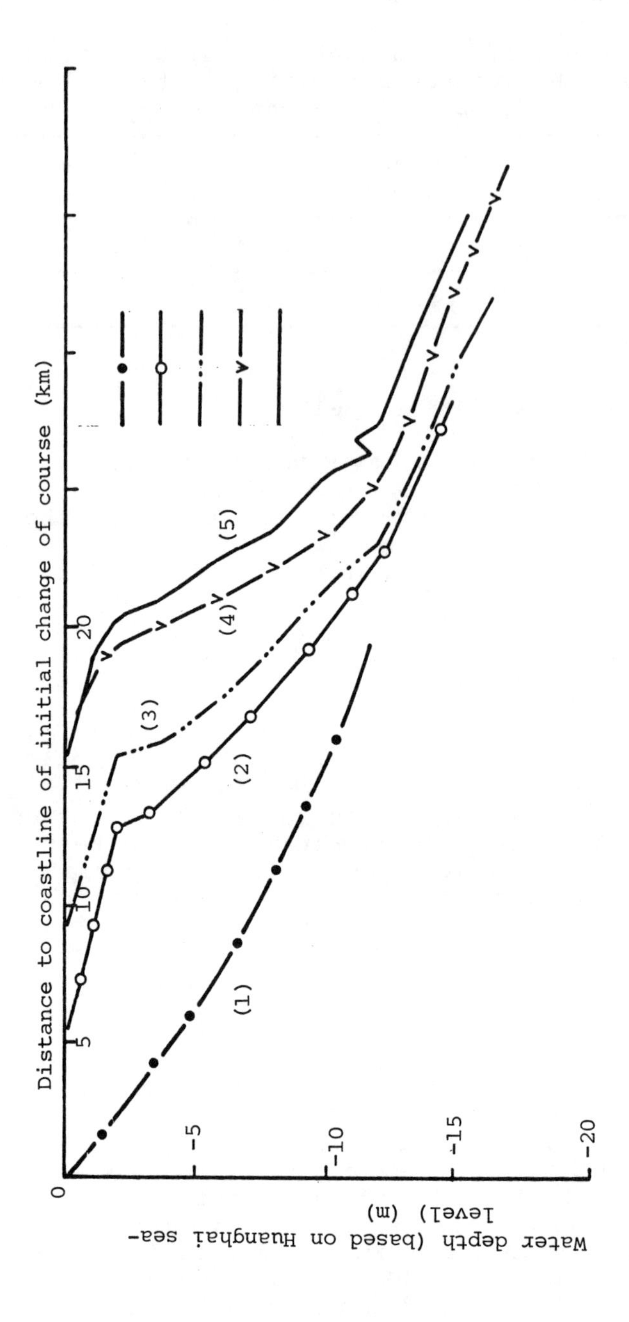

Figure 1. The Comparison of Longitudinal Profile Before and After Outstreaming Through the Section 27 of Diaokouhe River.

to the south of Shenxiangou (in the east of the delta), the maximum flowing tidal current goes to the south and the ebbing tidal current flows north causing a clockwise rotation. Because of shallow depth in the coastal region and the frictional effect on the sea, the former part of the tidal wave becomes steeper, the latter gentler, so that durations of ebb and flow are different. This means that the ebbing tidal current takes a longer time than the flood tidal current. The speed of the latter is more than the former, leading to an unequal exchange of sediment. This is one of the reasons why some of the sediment moved to the sea has been transported to both sides.

After removing the parts of periodic tidal current from the actually measured sea currents, the remaining part is called the residual current (or prevailing current) which is affected by wind, density differences, topography, the Coriolis force, river runoff and so on. Within the Bohai Sea, the direction of the residual current on the surface is the same as that of the wind. The intensity of the sea current is mainly caused by wind which, not counting any other factors, is 3-16 cm/s. When the monsoon comes from the south in the spring and summer, the sea current flows to the north and with the north monsoon in the winter, it flows south. The prevailing current on the bottom of the sea is a residual flow caused by tidal action free from the influence of wind. This is also a non-periodic current change because of topographic effects. Its speed is 1-10 cm/s, and flows toward the northeast during the whole year. Because the prevailing current maintains the same direction over a long period, though not very strong, it can keep sediment in transport in one direction and pass into the deep sea region.

EVOLUTION OF THE YELLOW RIVER DELTA

As many as 1 billion tons of sediment are annually carried into the estuary region of the Yellow River and empty into the sea. Because of shallow water depth and weak tidal energy in the estuary region, the sediment transport capacity affected by the sea is much less than the quantity of sediment moved to the sea. Only a small part of the sediment can be transported into the deep ocean area by waves, tides or the prevailing currents. The majority of sediment is deposited in the Delta and coastal region leading to a rapid extension of the deposits and making the slope of the river course even milder. When deposition and extension build up to a certain point and the river course is unable to drain off floods and sediment, a new entrance to the sea will be established. This kind of branching and shifting of the river course starts from the end of the channel and enlarges its area step by step as it moves upstream. When the uppermost point was near Ninghai, an avulsion occurred in the Delta. Based on this process, another cycle of deposition, extension, wandering and avulsion occurs. This kind of cycle makes the entrance to the sea shift continually, and also extends the coastline seaward.

The vast delta in the estuary of the Yellow River, which extends from Taoerhe to Zhimagou, with the apex at Ninghai, is about 5450 sq.km (see Fig. 2).

This area was formed by 7 avulsions between 1855 and 1953. Three changes have happened since 1953, but the apex of the delta has moved down to Yuwa, 30 km from Ninghai, affecting an area of about 2,200 sq.km within Chezhigou and Nandaiti. Because the area has become smaller, the coastline has pushed seaward more quickly. Throughout the period from 1855 to 1954, there were 64 years when the river actually took this course into the sea. From the changes of coastline, it has been found that a total area of 1510 sq.km of land was gained within the 128 km width, at the rate of 23.6 sq.km per year. The coastline has pushed forward 11.8 km on the average, at a rate of 0.18 km per year. Of course, within certain flow paths the extension of the coastline is much faster. During the period from 1954 to 1982, a total area of 1,000 sq.km of land was gained, or about 34.5 sq.km per year. This small delta has an 80 km coastline which has pushed seaward for 12.5 km, at the rate of 0.43 km per year.

DISTRIBUTION OF DEPOSITION AT RIVER MOUTH AND EXTENSION AND RETREAT OF THE COASTLINE

From cross sections of the river channel and topographic maps of the bottom of the sea in the shore region, it is known that among the total 1.05 billion tons of sediment entering the estuary area, 23% (0.24 bilion tons) is deposited in the channel above the low tide line or the zero line of Dagu Datum. On the delta surface, another 44% (0.462 billion tons) is deposited in the coastal region so that it forms a huge sandspit, only about 1/3 of the total incoming sediment moves seaward under dynamic ocean effects (see Table 3).

Since there are different conditions of seaward extension for each flow path to the sea, the ratio of each part of the sediment deposited is also different. As shown in Table 3, when the channel of the Yellow River passed through Shenxiangou or Diaokouhe, the proportion of outgoing quantity was more than 1/3 of total sediment under strong sea currents, but through Qingshuihe it was only 1/4. If the river were to move further to the south, outgoing sediment would be even less because of the weak sea currents. The ratio of the distribution of sediment to each part also differs with different stages of each flow path. Generally speaking, at the beginning of an avulsion, the entrance to the sea is something

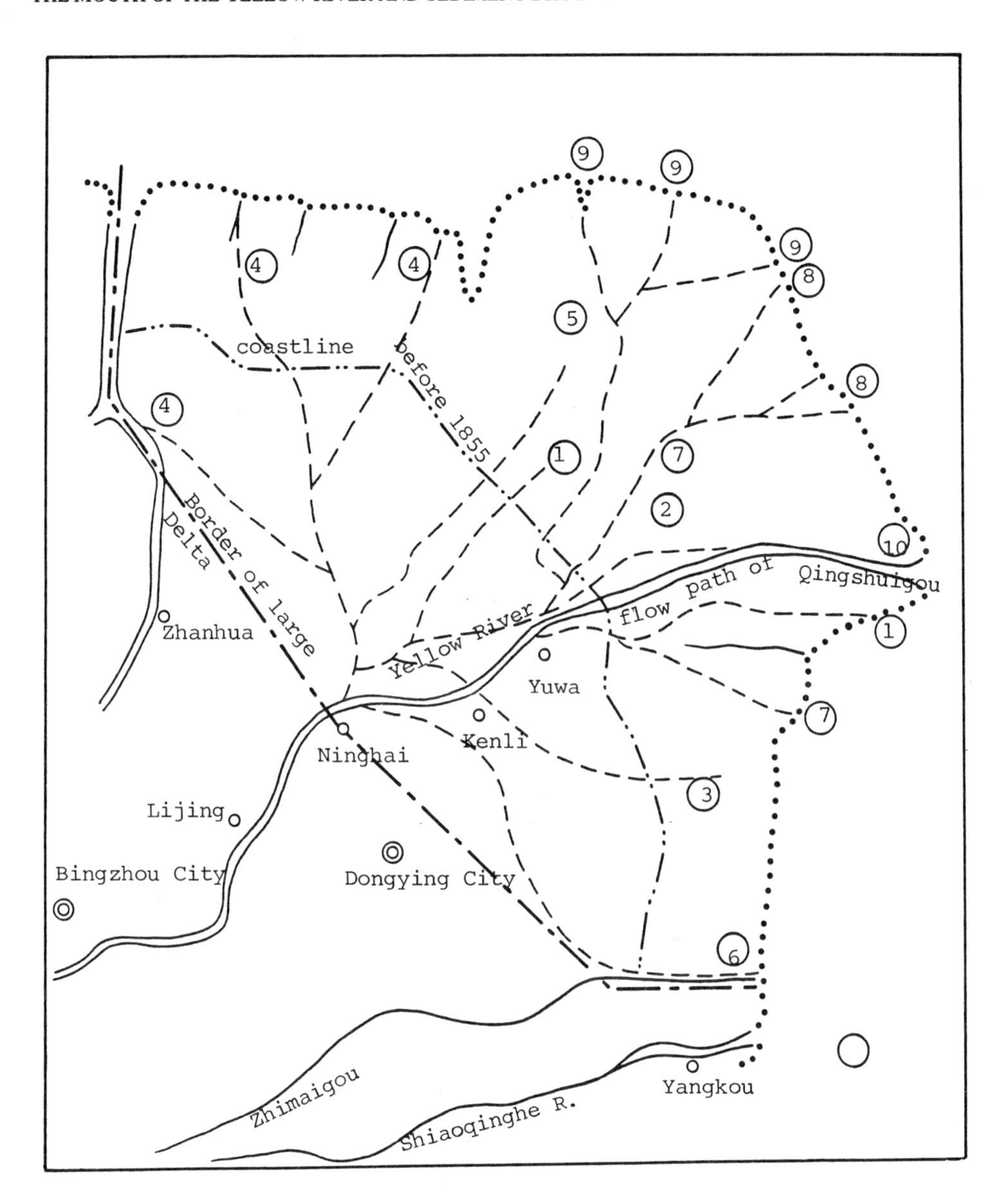

Figure 2. Evolution of Estuary of Yellow River in History

Table. 3 The distribution of deposition at the estuary of the Yellow River (100 million tons)

Time interval	1950-1960 Shenxiangou		1964-1976 Diaokouhe		1976-1983 Qingshuigou		average	
sediment load at Lijin	13.2	%	10.8	%	8.39	%	10.5	%
Location								
:Inland (above zero level in Dagu)	3.5	26.0	2.33	21.6	1.88	22.4	2.42	23
:coastal region of delta	4.7	36.0	4.76	44.1	4.34	51.7	4.62	44
:deep sea	5.0	38.0	3.71	34.3	2.17	25.9	3.46	33

like a concave bay which is sheltered by sandspits along two sides of the old channel. Here more than 90% of the sediment is deposited and forms new land. After the sandspit begins to protrude, more sediment can be transported to deep sea by the ocean dynamics. For instance, more than 50% of the sediment may be transported by the strong sea current in Shenxiangou and Diaokouhe channel. Later on in the development, the ratio becomes smaller again because of the extension of the river mouth and the branching of the channel. Once the huge sediment load is deposited at the entrance to the sea and in the old river mouth where streamflow has been stopped, the sandspits and beaches will be eroded to various degrees by wind-generated waves. It is quite clear that retreats have occurred at the old river mouth. As shown in Table 4, during the period from 1964 to 1976, the river changed its course by passing through Diaokouhe instead of Shenxiangou. Because of the strong sea current outside Shenxiangou, the coastline in this range retreated up to 166 sq.km which was equal to 1/3 of the land formed (500 sq.km) in the Diaokouhe area at the same period. But the longer the elapsed time without streamflow, the wider and milder the coastline becomes, and dynamic effects of the ocean become much weaker and the coastline resists sea currents. So the rate of retreat becomes slow. This phenomenon can be found not only in the coastline area without streamflow but also in the active coastline were extension and retreat are carried out alternatively. Because of the shifting of the entrance to the sea, retreat also can be found in the course originally taken by the river. However, it is not as clear here as it is for the coastline far from the river mouth.

THE LOWER COURSE AFFECTED BY ESTUARY EVOLUTION

Accretion and extension as well as shifting and change of course at the estuary are equivalent in many ways to the change of the base level of the river which results in readjustment of the longitudinal profile of river channels. This will affect aggradation and degradation in the course of the Lower Yellow River during the readjustment process. At the beginning of an avulsion, the retrogressive erosion due to shortening of river length takes place in the course upstream from the shifting point, with the extension of sandspits. The channel slope becomes more gentle so that retrogressive accretion occurs. This kind of effect was found mainly in Loukou below Lijing. Because of continuous shifting, the lower reach will be raised and lowered periodically. When avulsions take place consecutively across the whole area of the delta completing a so-called "large cycle," during which the coastline extends seaward through the delta, ascension of water level as a result of extension can not be eliminated by changing course in the estuary so that the effect can last

Table. 4 Retreat and deposition in Delta coastline of the Yellow River

Flow path	time interval	Deposition of coastline:		Retreat of coastline	
		area sq.km.	velocity : sq.km/yr	area sq.km.	velocity sq.km./yr
Shenxiangou, Tianshuigou	1947-1954	240	34.3	24	3.4
Shenxiangou	1954-1964	286	28.6	42	4.2
Diaokouhe	1964-1976	500	4.0	166	13.3

over a long period of time. Because later avulsions will take place on the new and widespread extended coastline, the original base level is relatively changed. This problem needs attention.

From the length of the flow path from the beginning to the end of the extension of the coastline of the delta and from the water level of the upper reach, it can be shown that the first "large cycle" ended with the sixth avulsion in 1929. At that time, the original river length of the flow path had extended about 15 km making the river at the estuary and the channel near that area rise about 1m. At present, the process is in the later stage of a second large cycle. Since the range of course changes has been limited due to man-made effects, it can be anticipated that the average length and speed of coastline extension will be much greater at the end of the second large cycle than at the first stage.

Review of the History of Improvement of the River Mouth

The unceasing aggradation, extension, shift and change of course brings about two kinds of problems. One is that with aggradation and extension of the river course and coastline in the delta region, the river bed and water level in the estuary and upstream river reaches will rise. This in turn will increase the menace of floods and ice runs. Another problem is that the river course near the entrance to the sea is often unstable. Before 1949, most of the estuary of the Yellow River was undeveloped, desolate and uninhabited. The aforesaid two problems were not important, and demands for improvement of estuary were not urgent. Therefore only a few dikes were built along river reaches upstream from the apex of the delta. This allows the river course to shift freely over the estuary and results in a widespread delta and an enlarged area for sediment deposition. Furthermore, it retards the process of seaward extension of the coastline of the delta and the rate of rise of the water level along the course of the estuary.

From the 1950s, the estuary of the Yellow River has been gradually developed and utilized, and many farms, animal husbandry, forests and some irrigation and drainage projects were established. During the 1960s, Shenli oil field was developed, and the social and economic situation in the estuary has improved greatly. With production developing, it was urgent to prevent floods. Furthermore the freely changing river course in the estuary region could not be allowed. In order to protect development of industry and agriculture in the estuary from flooding and ice runs, a series of rectifications for the estuary were carried out. Three times the embankments along the Yellow River were raised and riprap was used to protect dikes at vulnerable spots so that embankments could bear the rise of water level in the river.

New dikes were constructed along both banks further downstream and connected with the existing dike from the apex of the delta. About 30 km of narrow channel exists between Mawan and Wangzhuang, and is easily blocked during ice runs. This reach was widened to relieve the menace of ice floods to the oil base at Dongyin. Artificial changes of course were made in 1953, 1964 and 1967 respectively. Because of these changes of course, the stream path was shortened, base level has dropped, and the slope is steeper which has caused retrogressive erosion and lowering of flood level of the river above the shifting point. This ensures safety in the Lower Yellow River and is considered an important step in harnessing the estuary.

Although all of these changes were more or less artificially made, the first two were not planned. They were implemented temporarily based on the flood situation at the time. At the beginning of the 1960s, the oil field in the estuary was just being explored, and most of the small delta below Yuwai was not developed. Therefore the changes in course did not need to be restricted. However, later in the 1960's, more and more oil resources were found. Any temporary changes of course without planning in advance could bring about many losses and other effects. Therefore the change of course in 1976 was carefully planned in advance according to the topography of the delta, seaward conditions, the requirements of oil and other industries as well as agricultural development. The alternative course at Qingshuigou was chosen in coordination with other necessary projects, including relocation of inhabitants, construction of dikes on both banks and digging of a pilot channel. When the flood level, due to accretion, rose higher than the level originally set up for safety, the dikes and the oil field faced a menace. The old flow path was blocked and caused the water to flow through a new course to Qingshuigou. Because everything was prepared beforehand, losses and the effects due to an artificial change of course were minimized. This change, of course, also resulted in lowering water stages in the lower reaches upstream from the shifting point due to both retrogressive and along-course erosion.

Improvement of the Estuary and Sediment Disposal

Flood protection is the main task required to improve the estuary of the Yellow River. The problem mainly results from excessive accretion and rapid extension at the estuary due to the enormous sediment load in the Yellow River. This problem may be solved in two ways. One is to reduce sediment load entering the estuary, and the other is to adapt some measures in accordance with laws of estuary evolution to retard the rate of accretion and extension of estuary and delta coastline as much as possible. This will prevent the water stage and river bed from rising too quickly. On the premise that sediment discharge from the upland area cannot be appreciably reduced in the near future, the problem is how to dispose of the huge amount of sediment which enters the estuary.

As mentioned previously, the development and health of the estuary have depended upon the changes in course of the river. First there was free shifting, then restricted shifting, and now planned shifting. In recent years, the oil field development has spread all over the delta and other related installations have been connected in a network. It is now impossible to find a new flow path which is free from effects of the oil field development. Changes of course will bring about some losses and the effects on the oil industry are important. therefore, a stable flow path is eagerly sought.

Because of relatively weak tides and wave energy, the majority of sediment carried into sea has accumulated near the river mouth, forming new land. If only one flow path is available, extension with only a transverse range for sediment deposition of about 30-40 km would proceed at rapid rate. If arrangements could be made in advance for several flow paths to the sea, the problem would be ameliorated. It should be noted that the land formation by sediment has both disadvantages and advantages. The former is that it causes extension of the estuary and raises the water level. The latter is that it can yield land of several hundreds of thousands of sq.km each year which is a benefit for oil exploration. At present some oil fields on the delta are situated on lands formed from sediment deposited in recent decades. So, taking development of the oil field and delta into account, the planned change of course will not only make the coastline of the delta extend seaward equally but also minimize the adverse effects. Furthermore, it would be beneficial to oil prospecting which could change from offshore onshore exploration which is cheaper. Meanwhile, due to the characteristics of the extension and retreat of the coastline from fresh water inflow and with strong ocean dynamic effects, the amount of retreat may be as much as 1/4 of extension, or even up to 1/3. Therefore several flow paths should be found and used alternatively in areas where the coastline has strong dynamic effects so that function of retreat can be fully utilized. In addition, the old waterways should be dredged and obstacles like sand bars cleared away. All of these methods are intended to keep the extension of the flow path within a certain limit, reducing as much as possible the raising of water levels in the upstream reaches.

If the flow path is fixed and the river in the estuary is relatively stable it would be beneficial for development of industry and agriculture. Though warping and soil ameliorating could be used on the delta to dispose of some sediment, the amount would be small. Most of the sediment will be transported into the sea. If the intent is to retard the rate of extension in the estuary, it is necessary to fully utilize the seaboard conditions so that more sediment can be carried forward into the sea.

The seaboard condition means two things, one is how deep or steep the seaboard is; another is how strong dynamic factors are, such as waves, tides and prevailing currents. The latter determine the fraction of sediment moved seaward. When huge amounts of sediment are carried by the Yellow River, most of the sediment, except the smaller sizes finer than, say, 0.007mm, will be deposited in the vicinity of the entrance to the sea. Deposition will occur there because the section widens, there are backwater effects, and some flocculation will occur. A part of the near shore deposits will be transported into the deep sea by waves and currents, the quantity of which depends on the strength of the sea currents. In general, the seaboard out from the delta is mild and shallow, but the situation is difficult at individual locations on the coastline. The coastline in the north Bohai Gulf is much better than that in the south of Laizhou Gulf. The steepest seaboard with the strongest sea current is located around Shenxiangou. In addition, judging from the retreat, while the stream flows through Diaokouhe, the longitudinal profile of the Chahe (near now Qingshuigou) features erosion in the upper part and deposition in the lower part. This means that retreat is found on isobath lines only at 2m and deposition at 5, 10 and 15m. If the isobaths are entirely in retreat this indicates that seaboard condition at the north is much better than that at the south. This is the main reason flow paths at the north take a much longer time to move than those at south. In conclusion, it is very important to choose a suitable seaboard for the entrance to the sea to retard the rate of extension and prolong the time of flow.

Judging from the different evolutional stages of one flow path, at first the rate of extension is very fast. The more it extends, the nearer it approaches the deep seaboard in which there is more space for sediment deposition. The proportion of sediment transported to sea will increase because of ocean dynamic effects, so the rate of extension will be retarded. Therefore within a certain range, it is advantageous to keep the river running as long as permissible along one path. At present , the more realistic methods are reinforcing and appropriately extending the dikes on both banks, and also carrying out some river training works to prevent wandering so that the river's capacity to dispose of flood and sediment will be improved. The relationship between extending the length of sandspits, L, and the rise of water level G may be expressed by: $G=J*L$, J stands for the slope of the river course in the estuary. By river regulation, the channel in the

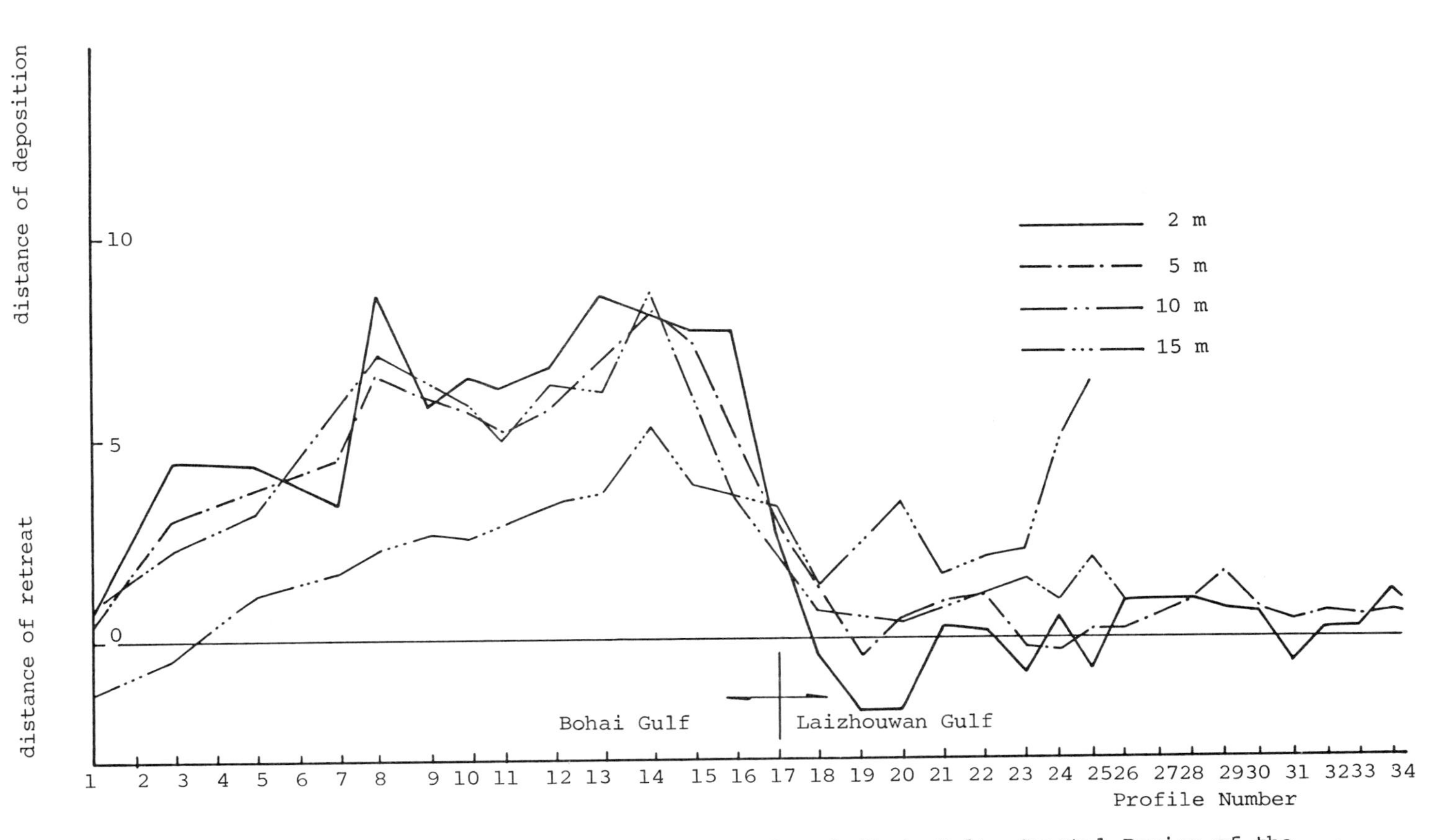

Figure 3-1. Analysis of Aggradation and Degradation of Isobath in Delta Coastal Region of the Yellow River from 1975 to 1980.

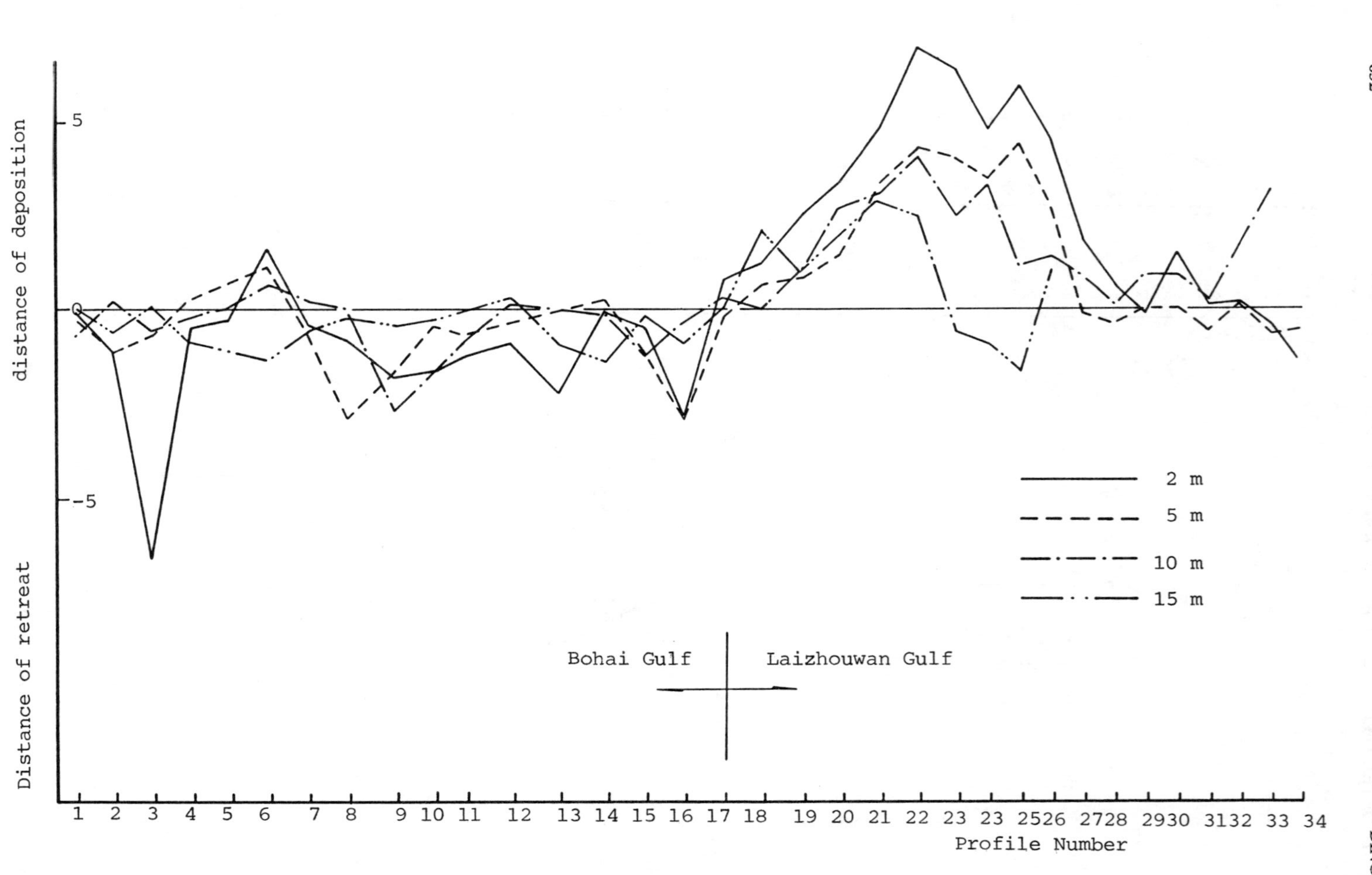

Figure 3-2. Analysis of Aggradation and Degradation of Isobath in Delta Coastal Region of Yellow River from 1975 to 1980.

estuary will become narrow and deep, so J will be reduced accordingly. This means that under a certain extending rate, G will be reduced so that the effect of extension on the rise of water level in upstream reaches can be reduced.

FLOOD DIVERSION AND WARPING TO REDUCE SEDIMENT TRANSPORT TO THE SEA THROUGH STABILIZING THE FLOW PATHS

If during the flood season, flood diversions reduce flood water levels and avoid unnecessary changes of course in the estuary, the safety of oil fields will be assured. If sediment carried by flood flows is distributed as widely as possible on the surface of delta through several flood-ways, the land surface underlain by saline-alkali soil would not only be improved but the amount of sediment transported into the sea could be greatly reduced. But the main problem is that flood diversion works cost too much and the affected area is too large. Relocation of inhabitants would be necessary if diversion were conducted on a large scale. On the other hand, if the scale were small, effects of retarding the extension into the estuary would be not so prominent. This method could be chosen as an auxiliary measure in controlling the estuary.

STABILIZING THE FLOW PATH BY SEPARATING THE FLOOD AND LOW FLOW PATH

The main channel will, in general, be eroded during floods and undergo deposition during low flows. These facts can be used to advantage. The river course might be kept stable and free from excessive deposition by diverting low flows below a certain discharge to another path. These characteristics reflect the basic laws of sedimentation and also hold true for the entire narrow lower reaches in Shandong province. But important factors include the status of previous sedimentation and channel morphology. It would be difficult to determine a definite flow value for distinguishing between scour and deposition. In addition, there is a very strong automatic adjusting action in the alluvial course of the Lower Yellow River. Scour and deposition of the river course and the conveyance for water and sediment of the channel during flood and dry seasons are mutually interrelated and conditioned upon one another. So it is very difficult to achieve the expected results if only one factor is considered.

DREDGING

There are 1.1 billion tons of sediment carried annually into the estuary region, nearly 1/2 of which (about 5 billion cu.m) is deposited in the area resulting in rapid seaward extension of the sandspits at the estuary. Excavation of large amounts of sediment by mechanical dredgers would presumably bring extension of sandspits under control so that a single flow path could be stabilized. It has also been proposed that the excavated sediment could be transported and disposed of on near lowlands to ameliorate the saline and alkali soil. This method looks feasible. Because the incoming sediment may not be reduced appreciably in the near future, it would be too costly to use machines to excavate large amount of sediment each year. The problem of disposing of so much sediment annually is not easy to solve. Therefore, it seems impossible to implement such a plan at present or even in the near future. Nevertheless, dredging may be an efficient auxiliary measure. On the whole, there are some practical problem with these plans. The available capacity on top of the delta is rather limited and cannot accommodate an enormous amount of sediment. The capacity of the seaboard outside the delta is much bigger than inside the delta. So, in harnessing the estuary, proper utilization of the seaboard to adjust and hold sediment still remains a major problem.

Conclusion

(1). Large sediment loads are carried by the Yellow River and relatively weak ocean dynamic factors in Bohai Gulf and Laizhou Gulf lead to a particular evolutionary law which includes cycles of deposition, extension, shifting of channels and changes of the river course. This is the main reason why retrogressive deposition and scouring happen in the estuary.

(2). The main task of improving the estuary is to solve the flood protection problem caused by unceasing deposition, extension and shifting. Before 1949, most parts of estuary were saline-alkali wasteland and a region with a small population. Therefore there was no need for regulation of the river course. The natural change of the entrance to the sea resulted in enlargement of the sediment deposition area and retarded the rate of seaward shifting of the coastline in the delta. Since 1949, agriculture and industry in the estuary area have increased. In addition, the oil resources have been developed and a series of channel rectifications carried out in the estuary region. The river has been changed to a pre-determined course according to a plan instead of being allowed to change naturally. In all probability, the flow path into the sea cannot be maitained stable over a long period of time due to the enormous amount of sediment being delivered to the area.

(3) The huge sediment load carried by the Yellow River into the estuary region may not be reduced appreciably within a short period of time. The key to harnessing the estuary is to dispose of the huge sediment loads so that the rate of rise in

stage and extension of the river mouth can be slowed as much as possible.

(4). There is a very large capacity for sediment disposal in the seaboard of the delta which must be fully utilized. First, a reach should be chosen with good conditions and plans made to provide several flow paths to the sea. This will allow retreat of the protruding sandspits and retard the average rate of seaward extension of the coastline. Then, as the flow path extends seaward, it enters deeper water where the currents are stronger and rate of extension will be slower. Measures should be taken as soon as possible to prolong the instream time of flow paths and to stabilize them.

(5). To prolong the instream time of a flow within a certain extending reach, it would be practicable to strengthen and to extend the dikes on both banks. Rectification of the river course should be made so that the course in the estuary can be kept free from deposition and improved in its capacity to convey flood water and sediment.

References

(1) Xie Jianheeng, Pang Jiazen and Ding Liuyi: 1965, The Basic Law and Situation of the Yellow River Estuary, YRCC Report

(2) Lu Lingyi: 1987, The Evolution and Improvement of the Yellow River estuary, "Research and Pratice of the Yellow River," Water Resource and Electric Power Press

(3) Wang Kaichen: 1985, Improvement of the Mouth of the Yellow River Journal "People's Yellow River," No. 3

(4) Li Zegang: 1984, Change of Deposition and Retreat of the Yellow River Delta, Report, Institute of Hydraulic Research, YRCC

SEDIMENT PROBLEMS IN DIVERTING WATER FOR WATER SUPPLY IN THE LOWER YELLOW RIVER

Tong Erxun
Yellow River Conservancy Commission
People's Republic of China

General Presentation of Diverting Water for Water Supply

INTRODUCTION

After the founding of New China, great changes have taken place in the development of water diversion from the Yellow River for irrigation such as improving soil structure by warping, planting rice, and methods of treating alkaline soils. Before liberation, none of these measures existed. At present, 67 diversion projects and 40 siphon works have been built with a total design diversion capacity of 3,900 cms. There are 86 irrigation projects, which totally control an area of nearly 30 million mu (Fig.1). Since diverting water from the river, the accumulative volume of diverted water is 202 billion cu.m, the volume of sediment diverted, 4.3 billion tons and the accumulative area of warped land 3.23 million mu. In recent years, the amount of water comsumed for industrial and domestic use in the Yellow River basin is 1 billion cu.m per year. Four billion cu.m will be needed in the long term. Some 1.75 billion cu.m has been transferred to the city of Tienjin in North China. Water resources of the Yellow River have been regarded as indispensable for the growth of industry and agriculture in the North China Plain, particularly in the drainage basins of the Huanghe and Haihe rivers as well as in areas in the vicinity of the Yellow River.

Fig.1

DIVERTING WATER AND SEDIMENT FROM THE YELLOW

Sediment contained in the diverted water is closely related to the sediment content of river water, that is to say, silt must be diverted while water is being diverted, and the sediment content will be higher when the river has a heavy sediment load.

L. M. Brush et al. (eds.), Taming the Yellow River: Silt and Floods, 657–676.

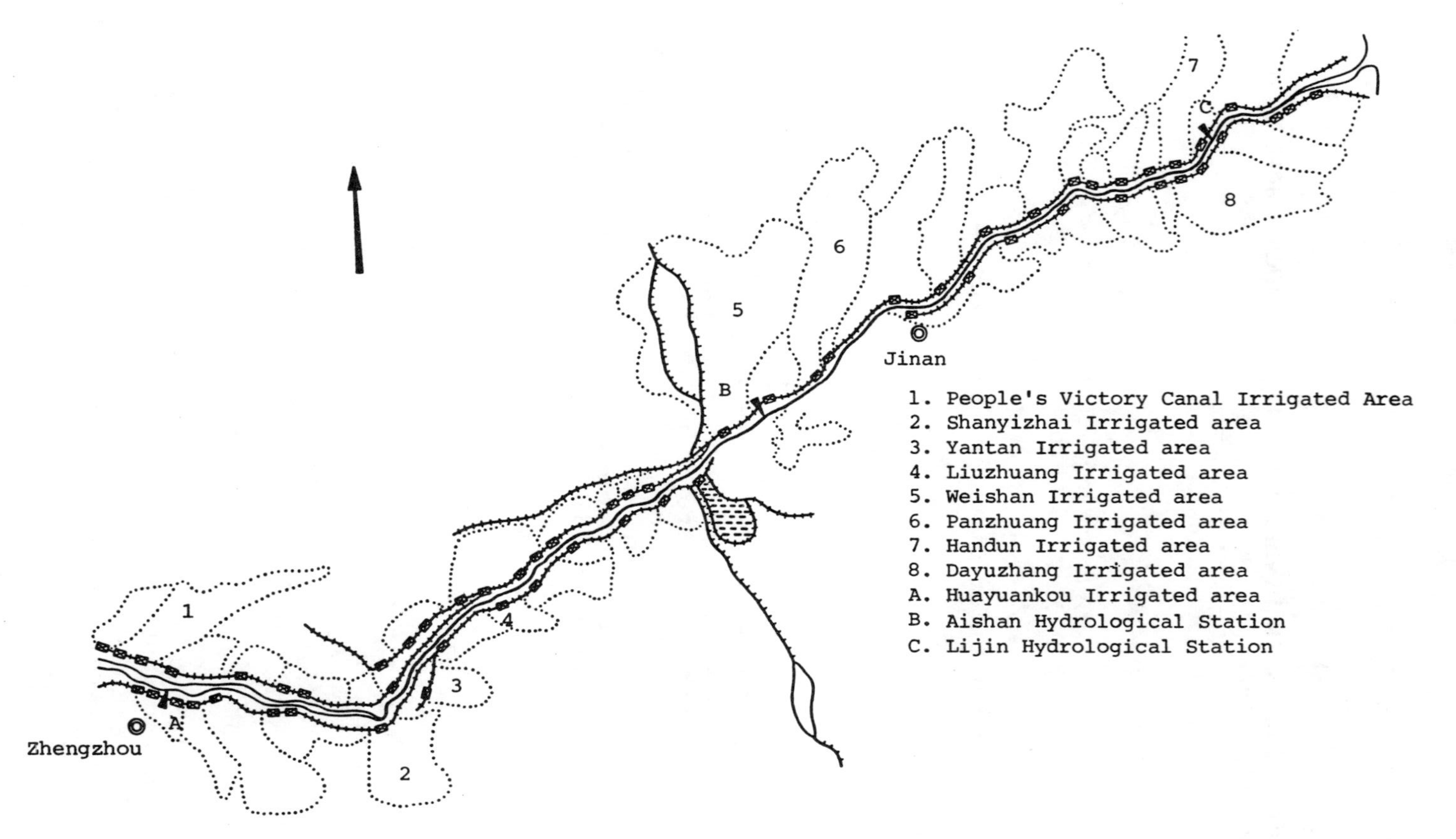

Figure 1. Map of the Irrigated Area in the Lower Yellow River.

In recent decades the average annual water diverted amounts to 9.2 billion cu.m, or 20% of the total runoff and sediment load of the Yellow River. During flood seasons, the sediment diverted is approximately 78% of the total diverted and the amount of diverted water is only 40% of the total for the whole year. The average sediment content is 30 kg/cu.m. During the non-flood season, the diverted sediment amounts to only 22% and the water diverted to 60% for the whole year, with an average sediment content of over 7 kg/cu.m. Moreover, during the period of May and June when the demand for water is the greatest in the low flow season, water diverted is approximately 80% to 90% of the incoming flow of the Yellow River. During the period of storage in Sanmanxia reservoir clear water is released, the river bed is scoured, and the sediment content entering the downstream reach is, on average, 6-9 kg/cu.m and the sediment is relatively coarse. If the timing of the diversions is not well planned, heavy deposition in the canal and in the drainage channels would be unavoidable and would require a great deal of manpower and finance for cleaning.

FEATURES OF DIVERTING WATER FROM THE YELLOW RIVER FOR IRRIGATION

Damless Diverting. The upper part of the downstream reach of the Yellow River has a wandering channel with the river bed higher than the adjacent land, and the main channel is unstable. Therefore many gates rather than dams are used for diversion. However, river training works used for flood control have difficulty in meeting the requirements for diversion. Some diversion works are far away from the main flow and canals must be dug for diversion.

The River Bed is Aggrading. Coarse sediment may easily pass through the diversion gates. In order to meet the diversion demands during the dry season, sometimes the bottom-slots of the gates are designed to be 1-3 m lower than the river bed. Sediment in the low layers may easily go through the diversion works during this time.

The Area to be Irrigated is Flat and the Canal has a Low Ability to Transport Sediment. Bank collapse frequently occurs with unstable side slopes.

Deposition of Sediment due to Diversion in the Irrigated Area

The annual average amount of sediment diverted for one mu of irrigated area is approximately 6.6 cu.m. For each mu of irrigated land, the amount of sediment from the canal system is 1.9 cu.m. So, for diverting one cubic meter of water an additional 0.02 yuan (RMB) is required. A great deal of money is needed if the total amount of water to be withdrawn is appreciable.

In the initial period of diversion, there were many deep pools and low billows with alkaline lands along both banks. These were used for sediment settling and improvement of the soil structure. Now, most of these areas have been turned into good farmland. As a consequence, the possibility of using uncultivated land for desilting has become smaller.

According to some typical surveyed data, the distribution of diverted sediment is as follows: sediment deposited in the settling pool covers 35%-45%; sediment diverted to the farmland covers 15%-25%; sediment deposited in drainage ditches covers 10%-20%; and sediment returned to the drainage system including canals for disposal of waste water covers 5%-15%. In the irrigated area of the People's Victory Canal, some measures have been taken to transport more sediment to the land, as a result, sediment deposited in the settling pools is reduced to 22% of the total diverted sediment, while sediment transported into farmland approaches 35%. In the irrigated area, the distribution of sediment deposition has a direct relation to the layout and operation of the projects. The distribution of sediment deposition in a few important irrigated areas is shown on Table 1.

A large amount of work is needed to clean out deposition in the canals. In some irrigated areas, sediment taken from the canal system is disposed of on nearby farmland, on highways and some sediment is piled up along the canal, which becomes an obstacle to the cleaning operation. On the other hand, due to the sorting of irregularly deposited sediment, some farmland in the vicinity of the headworks of the canals has become sandy and pollutes the environment. This greatly influences the production of agriculture and livelihood of people inhabiting this area.

A great deal of sediment is deposited in the main drainage canal. Based on investigations, 200 million cu.m have been deposited in 11 main canals. Twelve point eight million cu.m, 35% of the total deposited material, was deposited in the main drainage canals due to waste water. The heaviest deposits in the canals are up to 3 m deep. This reduces the drainage ability to only 30-70% of the original design.

Much sediment may also be deposited in the canals in front of the diversion intakes. Taking the People's Victory Canal as an example, the annual deposited sediment in front of the intake amounted to 140,000 cu.m. In the Liu-Zhuang diversion canal, during 1971 sediment was cleaned out 5 times, amounting to 1.2 millions cu.m. At Hanjading gate, in September 1983, the canal was operated to draw off 35 million cu.m of water for 20 days. However, 126,000 cu.m sediment was deposited in the canal, amounting to 53% of the total diverted sediment with an average deposited thickness of 3.1 m. One thing that

Table 1. Distribution of Sediment Deposition in the Canals of Irrigated Areas

Name of Area	Annual volume of water diverted 10^9 m	Annual volume of sediment diverted 10^4 tons	Distribution of deposits in percent			
			Settling basins	Canals	Farmland	Rivers
People's Victory Canal	6.17	693	22.4	27.5	34.7	15.4
Liuzhuang	5.10	653	76.6	4.5	7.4	11.5
Weishan	6.75	610	30.6	46.6	4.9	17.9
Panzhuang	5.40	570	20.6	46.8	4.9	27.7
Dayuzhang	7.00	880	40.0	29.0	26.0	5.0
Yantan	4.00	736	81.0	2.8	0	16.2

must be mentioned is that this place has a very high groundwater table and dredging of quick sand is extremely difficult.

From the above description, it is clearly seen that many sediment problems exist due to diversions from the Yellow River. How to deal with the sediment now is a prerequisite for future withdrawal of water for irrigation and for water supply.

Relation Between Diversion of Water and Sediment at the Head of the Diversion

RELATION BETWEEN RATIO OF DIVIDED FLOW AND SEDIMENT

In diverting some discharge, the ratio of diverted sediment (the ratio of sediment diverted to the sediment load in the river) to diverted flow (the ratio of flow diverted to the river discharge) is more or less closely related to the ratio of the flows. This relationship between the ratio of divided flow and sediment can be found using the observed data from several diversion intakes in the lower Yellow River. When the ratio of divided flow is small, or in other words, the flow of the river is large, the ratio of diverted sediment and ratio of diverted flow gradually becomes large. When flow in the main river is low, then the ratio of diverted sediment is larger than that of diverted flow.

Curves of 5 diversion intakes, People's Victory Canal, Heigangkou, Sanyizhai, Dayuzhang and Weishan, can be seen on Fig.2. Although there are differences among the curves, the ratios of sediment withdrawal are not the same, however, trends in the variations are the same at those points on the curves where ratios of divided flows at each diversion intake are the same as the ratio of diverted sediment. That is, the sediment content in the diverted flow at the intersection points with a slope of 45% is equal to that of the river. The dividing points for People's Victory Canal, Heigangkou, Shangyizhai, and Weishan are as follows respectively: 3%, 3%, 10%, 20%. Points measured in Dayuzhang diversion intakes are all below the slope line with a slope of 45%. That is, the sediment content of diverted flow is less than sediment content of the river. Here, data for the large ratio of divided flow is sufficient. Consequently, as far as reducing sediment of diverted flow is concerned, Dayuzhang is the best among the diversion intakes.

It can also be seen that, when the ratio of divided flow is more than 50%, the ratio of sediment is close to 100%. That is, the ratio of sediment content of diverted flow to that of

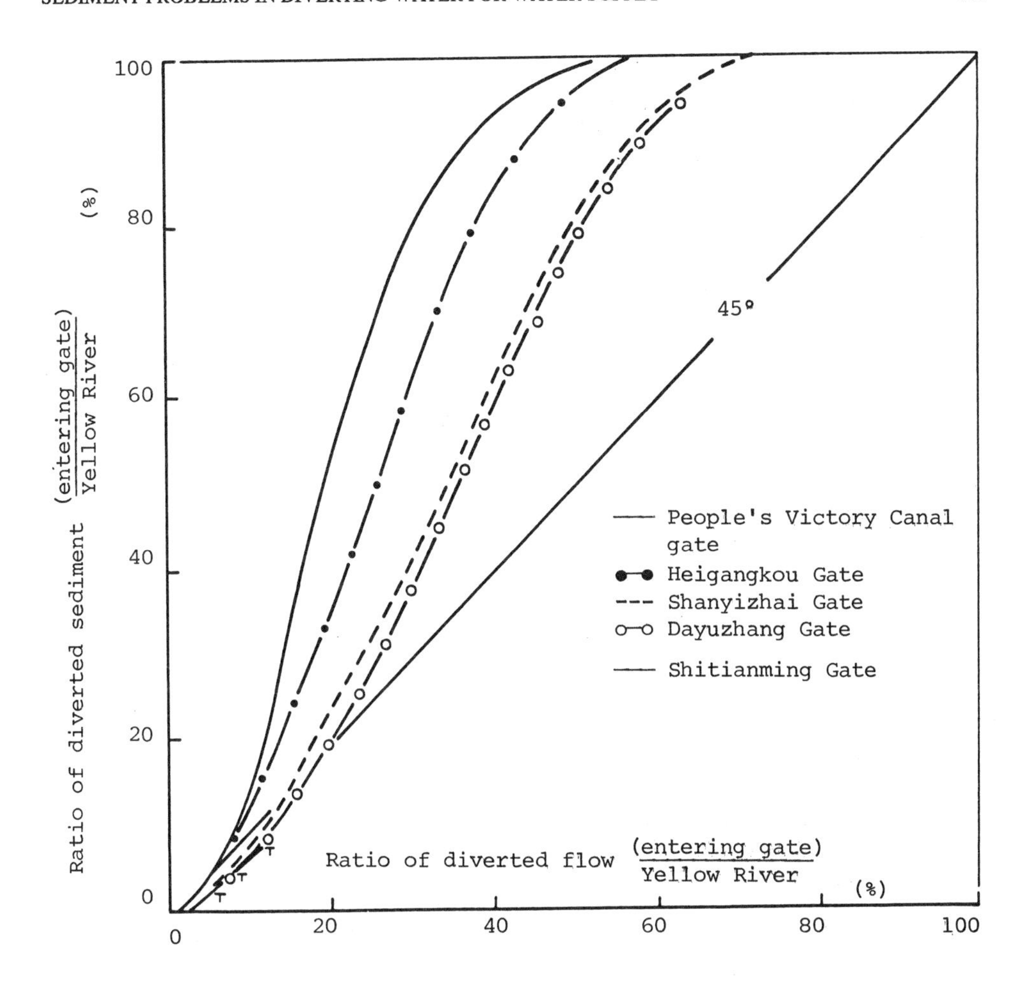

Figure 2. Relation Between Diverted Flow and Sediment of Intake Gate in the Lower Yellow River.

the river is nearly equal to 2.

Actually, scouring and the digging up of canals on floodplains accounts for more sediment being withdrawn into the canals. Not all the sediment is from upstream. In the diversion intake of People's Victory Canal, for example, during the period from August, 1860 to April, 1961, the sediment content of the divided flow was often more then 2-5 times that in the Yellow River. At that time the main current was far from the intake, and water levels on both banks differed by 0.6-2 m. Scouring took place while water passed through the intake canal on the floodplain. Large of amounts of sediment as well as coarse sediment was taken into the canal as shown on Fig.3.

The relation between the ratio of divided flow and ratio of sediment content can be obtained from data for several diversion intakes measured in situ from 1982 to 1983 (Fig. 4). It can be seen that when the ratio of divided flow is more than 5%-10%, the sediment content in the diverted flow is more than the sediment content of the Yellow River.

THE RELATION BETWEEN THE RATIO OF DIVERTED FLOW AND RIVER CHANNEL DEPOSITION

In order to find out whether deposition in the original channel will increase or not after both discharge and sediment load are reduced by the dividing flow, analyses of the observed data have been made.

Suppose that after diversion the sediment content of the diverted flow and that in the river are the same. The empirical relation between the ratio of deposition of the river channel and the ratio of the diverted flow and the ratio of oncoming sediment is as follows:

$$\frac{\Delta W}{W} = \left[1 - \frac{(1-K)^{0.24}}{(\rho_O/\rho_T)^{0.23}}\right][1 - K]$$

where:

$\frac{\Delta W}{W}$ - ratio of amount of deposition to that of oncoming sediment

ρ_O/ρ_T - ratio of sediment concentration of oncoming flow to the transport capacity

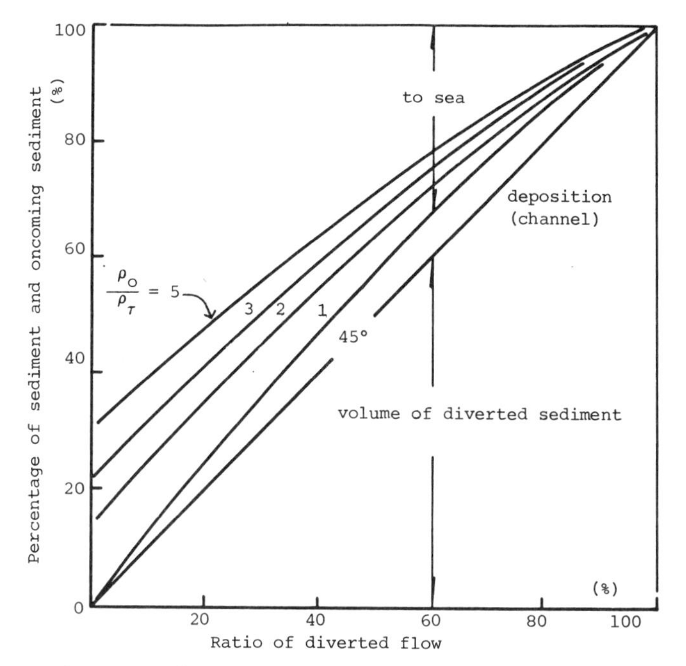

Figure 3. Distribution of Diverted Flow and Sediment in the Lower Yellow River.

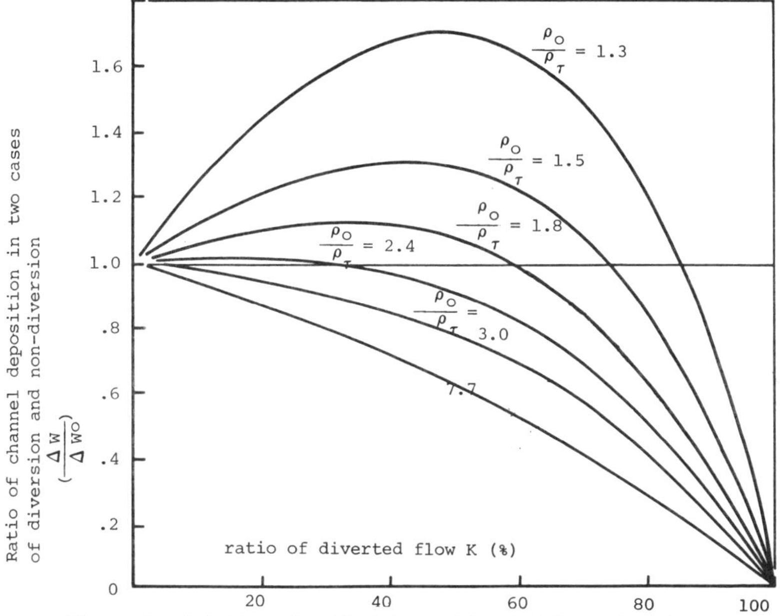

Figure 4. Relation of ratio between Diverted Flow of Kinds of Oncoming Sediment and Channel Deposition.

k - ratio of diverted flow to the river discharge

In the reach where scouring and deposition are in balance, that is, when the ratio of oncoming sediment equals 1 (or $\rho_0 = \rho_T$), the relation between the ratio of deposition and ratio of diverted flow is shown in Fig.5. When the ratio of diverted flow is 60%, the ratio of the amount of deposition to the amount of incoming sediment is a maximum, amounting to approximately 8% of the incoming sediment. It can be seen that the more the flow is diverted, the more sediment is diverted. Although, deposition is slightly increased in the river channel, sediment going to the sea is greatly reduced.

In a river reach under aggradation, in which the ratio of oncoming sediment is greater than 1, whether deposition increases in the river channel after diversion or not depends on the ratio of flows being diverted, see Fig.6 for instance, when the ratio of incoming sediment is 1.8, and ratio of diverted flow is more than 58%, deposition will be somewhat decreased.

The analyses also show that after diversion in an aggradation reach, it is not clear whether the deposition will be increased or not, but the larger the ratio of incoming sediment and the ratio of diverted flow, the less will be the deposition in the river channel.

The relation between ratio of incoming sediment and ratio of flow being diverted at which deposition in river channel will not be increased is shown in Fig.7.

The influence on deposition in the reach of Yellow River in Shandong Province after diversion of water in Henan Province is based on limited available data in lower Yellow River. The relation between the ratio of diverted flow to deposition in the river channel can be seen on Fig.8. Using the coefficient of oncoming sediment (ratio of sediment concentration to discharge) as a parameter, it can be seen that when the coefficient of incoming sediment is 0.01, scouring and deposition in the river in Shandong Province is very small, and when the coefficient of incoming sediment is large, the trend of the curve is almost the same as given by Fig.5.

Prof. Qian Ning previously analyzed the relationship between the ratio of diverted flow and sediment at which deposition would not increase at Huayuankou under different floods. (See Fig.9) It can be seen that no additional deposition takes place when the ratio of diverted sediment is only a little

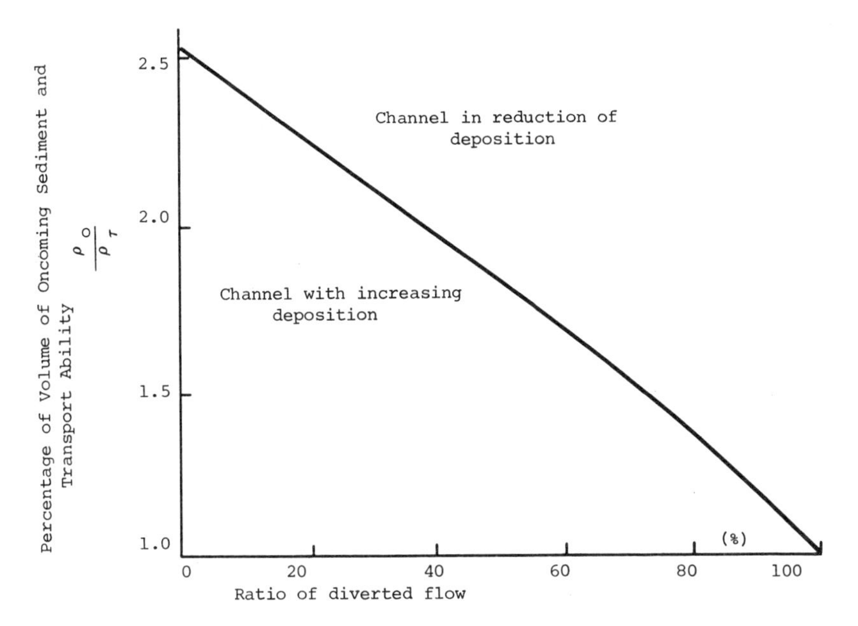

Figure 5. Relation Between Ratio of Diverted Flow of Maximum of Increasing and Reducing Deposition in Channel and Ratio of Oncoming Sediment.

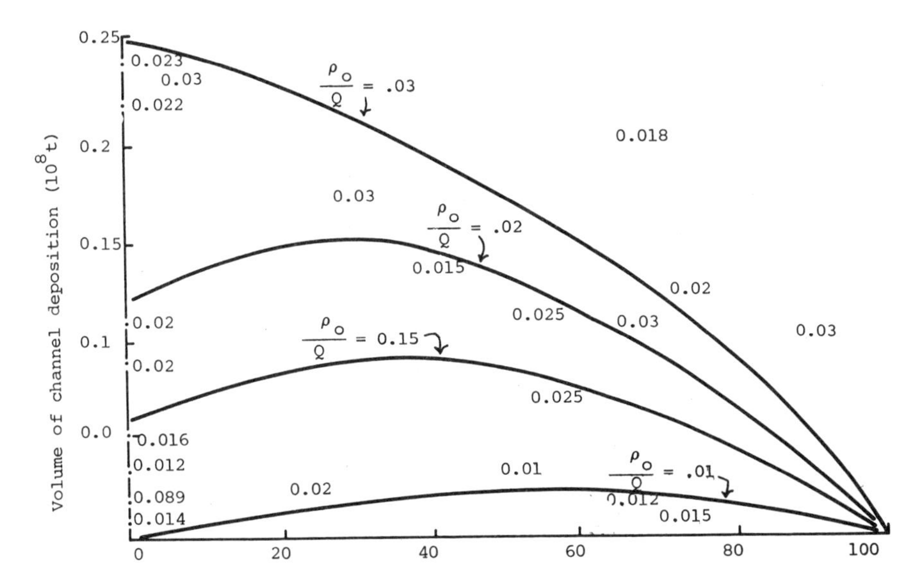

Figure 6. Relation Between Ratio of Coefficient of Sediment and Volume of Channel Deposition.

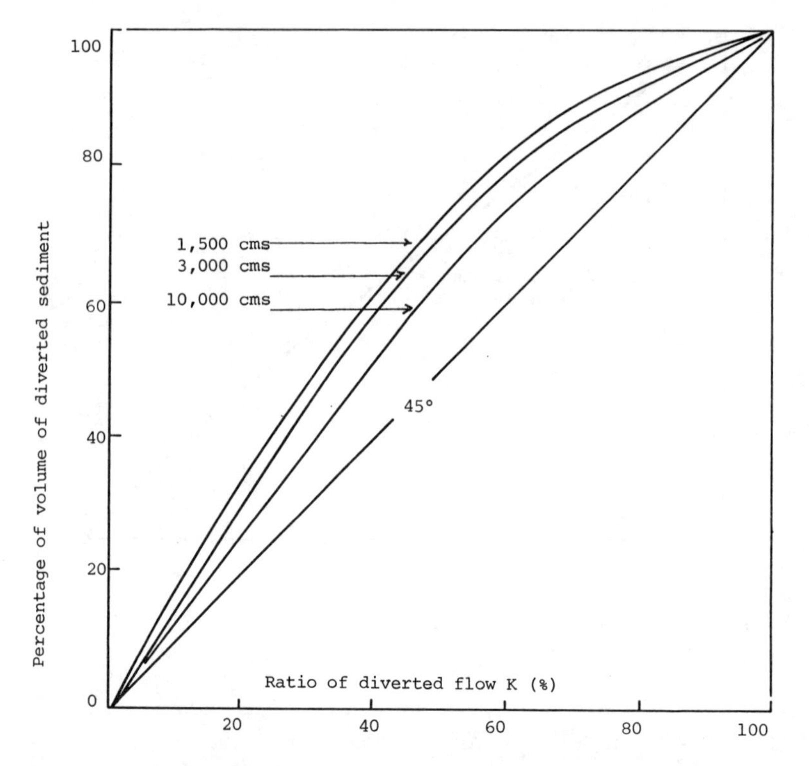

Figure 7. Relation Between ratio of Diverted Flow and Sediment of Discharge During Flooding without Channel Deposition in the Lower Yellow River.

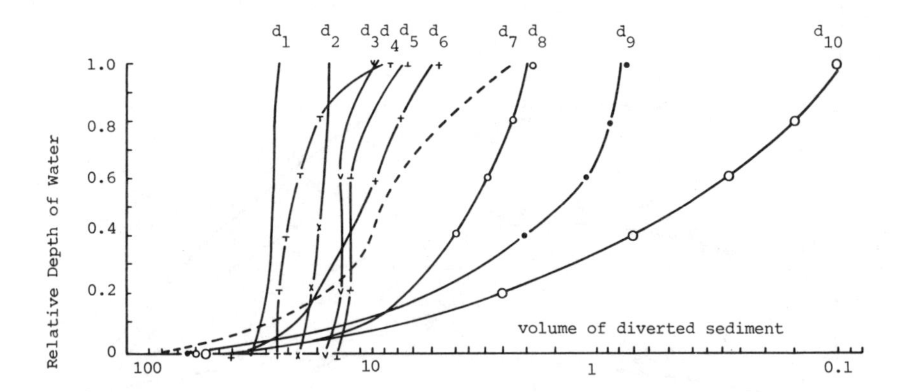

Figure 8. Distribution of Different Grain Sizes of Sediment Along the Depth of Water at Huayuankou Station.

Key: d_1 = < 0.005 mm, d_2 = 0.005-0.01, d_3 = 0.01-0.015, d_4 = 0.015-0.02, d_5 = 0.02-0.03, d_6 = 0.03-0.04, d_7 = 0.04-0.06, d_8 = 0.06-0.08, d_9= 0.08-0.1, d_{10} => 0.1

more than the ratio of diverted flow. If the flood discharge is 10,000 cms with ratio of diverted flow 0.2, the ratio of diverted sediment should be more than 0.26 to prevent deposition in the channel.

ENGINEERING MEASURES RELATED TO THE RATIO OF SEDIMENT WITHDRAWN

The layout of headworks affects the ratio of diverted sediment and will be discussed as follows.

Selection of the location of diversion headworks. Sediment passing through an intake is closely related to the grain size of the sediment in the river and the type of intake. In a cross section of the river channel, the isohyetal lines of sediment concentration are nearly horizontal. However due to secondary circular flow in bends, the average sediment concentration in the vertical near the concave back is relatively small, and it is relatively large at the convex bank. Sediment distributions show fine sediment near the surface and coarse sediment near the bottom of the river. For different mixtures of sediment, the fine sediment is evenly distributed in the vertical, and the coarse sediment concentration is higher in the lower layers.

When sediment laden water is withdrawn on one side of the flow in a straight river channel, the sediment content of the diverted flow is in general more than that in the river channel. It is seen from the measured data that the sediment concentration of the diverted flow is 114%-124% of the sediment content in the river. Water with less sediment content could be withdrawn by a diversion gate built on the concave bank. The optimal location for diversion is the three-fourth arc length along the bend where circular flow is the strongest. Downstream in the Yellow River, sediment grains are relatively small with a grain size less than 0.03 m occupying 40% of the total sediment load during the flood season, and 25%-35% during the non-flood season. These sizes are distributed comparatively evenly along the vertical. As a result, reduction of sediment entering the canal by careful siting is less effective than where coarse bed materials exist. The observed sediment content of the diverted flow and of the river at a few different diversion intakes in the lower Yellow River are shown on Fig.10. In the People's Victory Canal, the relation is scattered, because the main current of river was initially in the vicinity of the intakes located on the concave bank at a diversion angle of 44%. Measured sediment content entering the canal is 78% of the sediment content in the main river. Later on, the thalweg shifted to

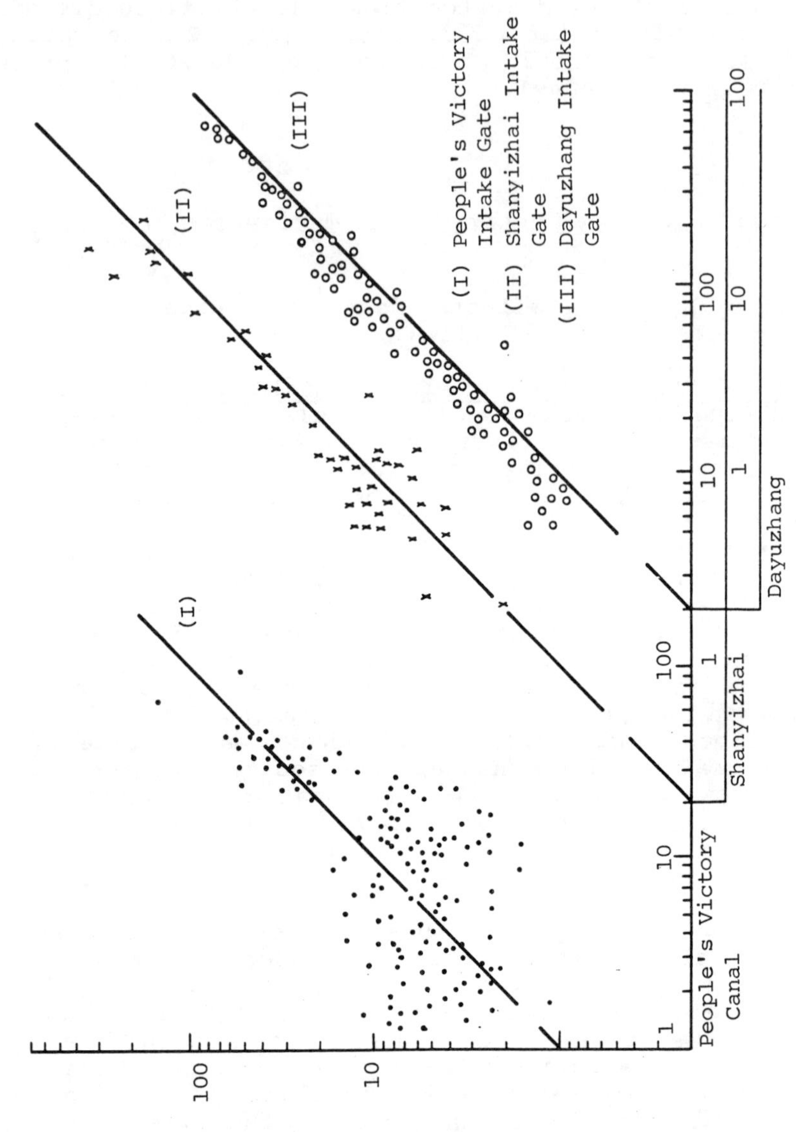

Figure 9. Relation Between Sediment Content of the Yellow River and Sediment Entering Canal.

the other bank and a canal was dug on the flood plain to divert flows during the dry season. The sediment content increased to 147% of that in the Yellow River. Another example is the Sanyizhai diversion intake which is located on the concave side of a sharp bend. The canal is short so sediment content passing through the canal is almost same as the Yellow River sediment content. The Dayuzhang diversion intake is located downstream of the apex to a stable bend with a diversion angle of 40% and was installed with a submerged sill to catch sediment. Sediment concentration of the flow entering the canal is only 85% of that in the Yellow River. The effect of reducing sediment from entering the canal is marked.

Diverting angle for diversional works. The diverting angle denotes the angle of the normal line of diversion works with the thalweg of main river. Due to the wandering nature of the river, its main flow line does not have the same curve as the bank protection works. The angle of diversion is generally set up based on the out-side shape of the bank formed by protection works. If the angle is too large, diversion will not work well, and eddy currents will be formed leading to deposition in front of the intake. On the other hand, when there is a small angle, the headwork of a diversion canal is difficult to layout. For a long intake canal the angle of most diversion gates is set at 30%-75%, with the final angle defined after model tests.

Submerged sill. Submerged sills are installed in front of diversion intakes to catch sediment. This helps to reduce the coarse amount of sediment entering the canal. The People's Victory Canal is an example; the weir is 0.83 m high, top width 1 m, and length 40 m. This worked well during the initial period of operation. Later, because the river bed rose higher than the top of the weir the sill lost its function. The sill at Dayuzhang intake gate is another example. A cofferdam built during construction was used to catch sediment. The reduction of sediment entering the canal is about 12%-15%. From the analysis of sediment grain sizes and suspended sediment concentrations outside and inside of the submerged weirs in diversion intakes at Dayuzhang and Sanyizhai, the sediment content and medium size of sediment beyond the weir are respectively 0.9 and 0.85-0.95 of that in front of the weir.

Due to the deposition of the river bed, the elevation of the gate sill of the Panzhuang intake in Shandong Province is lower than the elevation of the channel bed so that much of bed load entered the intake. In order to reduce the sediment entering the canal, stoplogs of 2 m height were used to raise

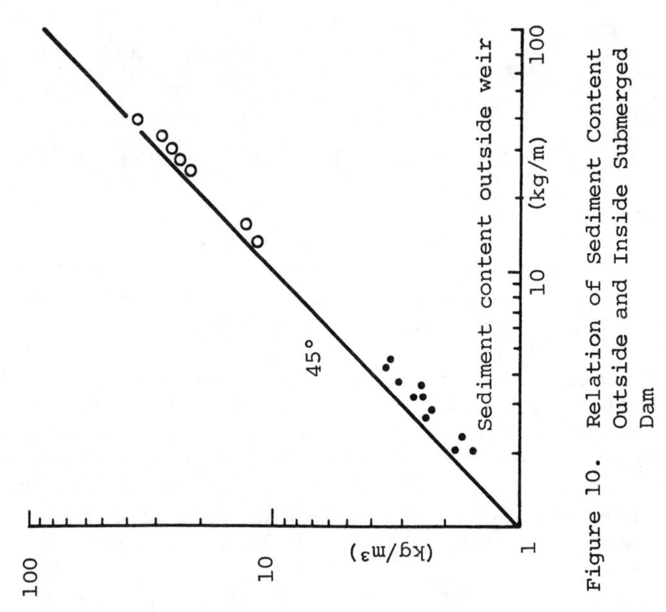

Figure 10. Relation of Sediment Content Outside and Inside Submerged Dam

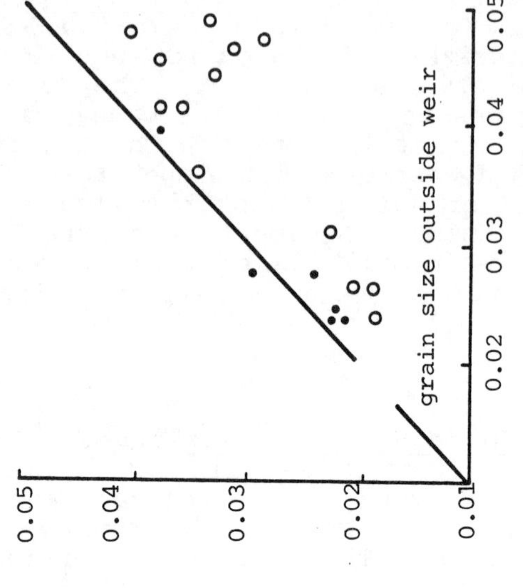

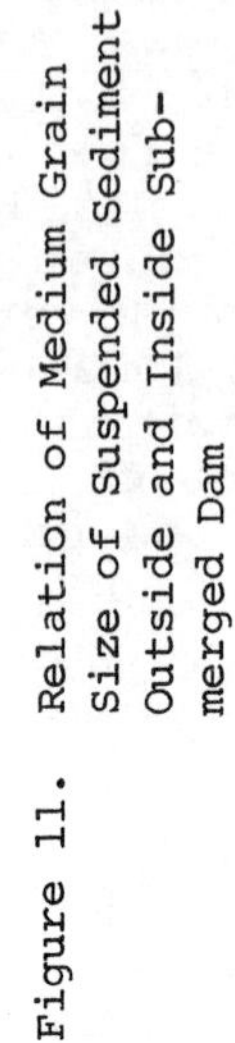

Figure 11. Relation of Medium Grain Size of Suspended Sediment Outside and Inside Sub-merged Dam

the sill elevation. Sediment concentration in the diverted flow amounted to only 75-80% of that in river, whereas previously it was 118% of that in the river.

Floating vanes. Floating vanes are installed in front of the intake. The vanes were made of floating pontoons arranged in two lines and submerged to one-third the depth of water with an angle ususally of 18° to the river flow. More surface water could be diverted to the intake so as to reduce sediment entering the canal, see Fig.11. On the Yellow River such measures had been used for damless diversions, and can reduce by 12%-15% sediment intake and change the grain size from d_{90} = 0.098 mm to d_{90} = 0.074 mm.

Measures for Reducing Deposition in Diversion Canals in Irrigated Areas

MEASURES FOR PREVENTING DEPOSITION IN FRONT OF INTAKE GATES

In the wandering reach of the river, sometimes the intake structure is so far away from the main flow that a canal is needed. The following measures have been taken in order to reduce deposition:

(1) Building sediment-prevention gates to avoid sediment deposition in the diversion canal. When not diverting, the gate is closed to prevent the muddy water from entering the diversion canal. Many sediment-preventing gates made of masonry works are built in front of the diversion intakes in lower Yellow River.

(2) Multi-intakes for diversion. In the wandering reach, there are many permanent bars. New bars also form which interfere with the intakes. In this case, multi-intakes are adopted for diversion. If one of the intakes is filled up or too far from the main flow, the other may be used to ensure diversion.

(3) Using a net-curtain for the reduction of deposition. Stabilized works are not easily built on newly formed bars, but net-curtains may be placed to reduce deposition in the diversion canal. The curtains are made of glass fiber cloth which allows water-penetration. This kind of cloth is easily bought and placed at a low cost and with an efficiency of sediment-prevention of 35%-72% normally.

(4) Utilizing motor-boats for agitation, dredging and sluicing with high-pressure water flows to maintain the clearance of the waterway. Such measures have been used in the People's Victory Canal with a certain degree of success.

(5) Hydraulic flushing to keep water flowing through the canal by proper operation of the gate. When the diverting

operation is stopped, the elevation of deposited sediment in front of the intake is monitored. When the water depth on top is reduced to 0.5 m, the gates are opened to scour the deposited sediment into a settling basin.

MEASURES FOR ENHANCING SEDIMENT TRANSPORT CAPACITY OF CANALS

The water level of the river is increased annually by aggradation. This is an advantage in that it improves the slope of water flow beyond the intake. By increasing the height of the dikes and the bottom of the original canal properly and taking off or rebuilding the structures which severely blocked the water, a certain amount of water-head can be used with great efficiency to improve both the slope of the canal and the capacity for transporting sediment. For instance in Liuzhuang Gate, after rebuilding, the longitudinal slope of canal changed from 1/7,000 to 1/5,000 with an obvious reduction of longitudinal sediment deposition.

In wide, shallow reaches composed of sandy soil, the canal can be trained using willow pile groins to fix the channel to maintain a narrow and deep section. This is simple, the structure inexpensive and construction materials can be found locally. For instance, at Liuzhuang canal, after training by the use of groins with willow piles, the average velocity of flow was increased from 0.85 m/s to 1.14 m/s without appreciable deposition in the canal. The hydraulic radius was increased by 31%, forming a composite cross section with a narrow and deep channel.

Bricks, asphalt soil and plastic cloth materials etc, have been used to line the canals. Concrete boards have also been used in several canals to prevent collapse of the canal banks and to stabilize the section of the canal. The roughness of the concrete is less than the soil, so it improves the capacity for sediment transport. The roughness of the canal after lining will drop from 0.02-0.022 for the original earth canal to 0.013-0.014, and velocity of flow in the canal may reach 1.5 m/s. If anti-percolation is not taken into account, it may be necessary only to line the side slope of canal in order to reduce the amount of investment. For example in the Weishan irrigation area, flow with a discharge of 150 cms annually cleaned out 2 million cu.m of deposits. After lining, the canal was stabilized and there was no need for cleaning the deposits.

From the data analysis of the earth canal in the Yellow River irrigated area, roughness may change with the sediment content of the flow. When the sediment content is 0, $n = 0.013$ approximately, when the content is more than 50 kg/cu.m, $n = 0.01$ or even less than 0.01. It is found in some canals, that deposition occurs when diverting water in the canal with a small sediment content. However, sometimes, no deposition or less deposition may take place even when sediment is large and of fine composition.

In some irrigated areas, a method of flow concentration has been used for the reduction of deposition in canals. From observations in the Weishan irrigation area, diversion of water requires the removal of 1000 cu.m at small discharges, 5.95 cu.m of deposition. However at large discharges, only 2.05 cu.m of deposits need to be cleaned, or a reduction of 60%.

Sediment in the Yellow River is a non-uniform mixture, under the same flow situation. The coarser the sediment, the lower the capacity for transport. In the head reaches of the diversion canals, a majority of the coarse sediment is deposited or treated in combination with warping on the landside of dikes for reinforcement. The settling velocity of the rest of the sediment is appreciably lower and the capability for transport is increased. During the flood season the average settling velocity of the sediment will change from 0.084 to 0.06 cm/s, and the capacity of transport may rise by 40%. During non-flood season, the settling relocity changes from 0.13 cm/s to 0.11 cm/s with an improved capacity of transport of 13%.

Settling Basin for Desilting Sediment in Irrigated Areas

Settling-pools are the basic means of treating sediment from diversions from the Yellow River. The existing alkaline land is usually chosen as a site for desilting. After warping by gravity, it becomes farmland. However, in recent years, because the area of existing land available for warping has become smaller and smaller, pumping the sediment laden water to the settling pool for deposition is sometimes used. Cleaning out the sediment from settling basins may be required for repeated use of a basin.

The pattern of desilting with characteristics of the Yellow River is to use desilting in combination with strengthening of the dikes or improvement of the soil. In some areas with favorable conditions, the backside of an embankment may be heightened through warping by gravity. It is also possible to utilize small suction dredgers to pump the muddy water with heavy concentration of sediment from the basin into the warped area. The desilting basin may be used for a long time. For improvement of soil structure desilting may be carried out on scattered sandy, alkaline lands. A cubic meter of Yellow River sediment contains about 105 g of Nitrogen, 56 g of Phosphorus, and 600 g of Potassium, which increases the productivity of the soil.

Summary

In considering the sediment due to diversion for water supply, a great deal of laboratory experimental work and studies have been undertaken by many relevant agencies. Much experience has been gained regarding diverting water and diverting sediment, the distribution of deposition in the irrigated area, and of the diverted flow and sediment in the irrigated areas. Some effective measures have been adopted to reduce the amount of deposition in canals and to minimize the inflow of sediment. However, many problems remain. Impacts of multi-intakes on the aggradation in lower reaches of the river will have to be studied. Management and operation of existing works have to be improved in the future.

SYNOPSIS OF THE WEST-ROUTE PROJECT OF SOUTH-TO-NORTH WATER TRANSFER FROM THE YANGTZE TO THE YELLOW RIVER

Zhang Weiming
Reconnaissance, Planning and Design Institute
Yellow River Conservancy Commission
People's Republic of China

Necessity and Strategic Location of the West-Route Project

North and Northwest China are arid areas with shortages of water. At present, water resources in these areas cannot meet the demand for industry and agriculture. Water crises have already occurred in some districts. The water shortage will become larger and larger with the growth of the national economy, and the gap between the supply and demand of water will increase in the coming years. Within areas of North and Northwest China there are vast inland areas rich in mineral resources. It is not only a strategic area to the national economy, but also one of the coal and oil bases in China. Deficiencies in water resources are a retarding factor for the development of the economy in the area. In the future, in order to solve the problem of severe shortage of water supply for industry, agriculture and for people's daily use, it is imperative to transfer water over the river basin. This transfer project is an important measure and is aimed at developing natural resources, promoting the economy, opening up Northwest China, and making North China prosperous.

Water and land resources in China are characterized by their uneven distribution. The total area of the Yellow River, the Huai River and the Hai River is 71% of that of the Yangtze River, but the annual average runoff is only 17% of that of the Yangtze River. The farmland and population in all three basins are 1.61 times and 96%, respectively, of that in the Yangtze River basin. The volume of water per capita and per unit land area are only 1/5 and 1/10 as much as that in Yangtze River basin. Owing to the fact that the water is abundant in the South and deficient in the North, the transfer of water from the rich to the poor is a reasonable measure in developing natural resources.

Since the founding of new China, the alternative of a South-to-North water transfer project which diverts water from

L. M. Brush et al. (eds.), Taming the Yellow River: Silt and Floods, 677–686.

the Yangtze River to make up for the deficit of water in Northwest and North China has been studied. Three possibilities have been proposed: transfer of water from upper, middle and lower reaches of the Yangtze River, called the west, middle and east routes. The middle and east routes are mainly to replenish water resources in North China. The west-route project is to complement the shortage of water in Northwest China, especially the Yellow River basin.

At the end of this century and at the beginning of the next, the center of development of the national economy will move toward the northwest. This is a strategic shift in the economic reconstruction. The Yellow River basin lies in the middle belt of the shift. The Northwest is a vast area, abundant in heat energy, sparse in population and rich in mineral resources. Conditions are favorable for developing hydropower, however, there are many reasons accounting for the slow progress in the development of this district. But the shortage of water and dessication of land are two important issues restricting large-scale construction in the future. The west-route project offers hydropower and water resources and provides material basis for the development of industry, agriculture and the economy in Northwest China.

Average annual runoff in the Yellow River is 58 billion cu.m, 720 cu.m per capita and 300 cu.m per mu of cultivated land. Compared with other river basins, the Yellow River is deficient in water resources. The affected irrigation area within the Yellow River basin is 66 million mu. The total irrigation area amounts to nearly 80 million mu if the area outside the basin that is irrigated by Yellow River's water is counted. Total consumption of water resources by agriculture and industry is about 2728 billion cu.m per year including water diverted in the lower reaches of the Yellow River. The actual volume of water diverted from the lower reaches of the Yellow River for irrigation in May and June is generally only 40-50% of the demand.

Estimates by concerned provinces in 1983 showed that by the end of this century 60 billion cu.m of water would be needed to cope with the development of industry and agriculture, of which 48 billion cu.m would be used within the Yellow River basin. Meanwhile 20-24 billion cu.m water must be kept to convey sediment to the ocean to control aggradation in the lower reaches at the present level. However, the maximum volume of water that can be suppliied for use outside the river course is only 34-38 billion cu.m per year. It is obvious that this quantity will not meet the demands for water and the shortage of water will be even larger in the next century. For mitigation of the discrepancy between the supply and demand of water in the Yellow River, it will be necessary to transfer water from the Yangtze valley via west-route as there is no other choice.

Transferrable Quantity and the Impact of Water Transfer

The plan envisioned now is to transfer water from the upper reaches of the Tongtian, Yalong and Dadu Rivers. Near the diversion dam sites of these three rivers, there are three hydrological stations, Zhi-mendai station on Tongtian River, Ganzhi station on Yalong River and Zhumuzhu station on Dadu River. Observations have been carried out there for over 20 years. Annual runoff at these stations amounts to 28 billion cu.m. Perennial regulation may be carried out for these reservoirs to provide for diversion. By computation, 21.7 billion cu.m of water could be diverted from three rivers in normal year at p = 50% (p is the exceeding probability); 18.8 billion cu.m of water could be diverted in dry years at p = 75%. On average 20 billion cms of water could be diverted. The 20 billion cu.m of water to be transferred is only a small portion of annual runoff in the aforesaid three rivers. Ten billion cu.m from Tongtian River constitutes about 25% of the annual runoff above the confluence of Jinshai River and Yalong River; 5 billion cu.m of water from Dadu and another 5 billion cu.m water from Yalong River constitute 10% of the annual runoff respectively of each river. The total amount of diverted flow, 20 billion cu.m diverted from the aforesaid three rivers would take about 5% of the annual runoff of the Yangtze River at Yichang. However, if this quantity of water is diverted to the Yellow River the runoff at Lanzhou in the upper reaches of Yellow River would increase by almost 60%. In addition, the river reach from Longyang Gorge to Guanchang Gorge would be very rich in hydropower. Eighty percent more hydropower could be developed in the reaches from Longyang Gorge to Qintong Gorge.

Because of the large amount of runoff in the Yangtze River, regulation capacity for runoff at power stations on the main stream is rather limited and much water is lost because of inadequate regulation. Ordinary power stations generate power only using the normal discharge. Large amounts of water are wasted during flood seasons. The west-route project is at the upper reaches of Yangtze River where regulation is available. Almost all runoff (including flood runoff) above the dam could be diverted to the Yellow River. The power lost in the Yangtze River by diversion would be very small. After water is transferred to the Yellow River and adjusted by regulation through the reservoirs, the water will be fully utilized. Development of hydropower on the main stem of the Yellow River has occurred earlier than that in the Yangtze. Several hydropower stations have already been built and more will be built in the near future on Yellow River. Due to the availability of high heads, increases in the generation of hydropower at the Yellow River stations would be much larger than reduction of power in the Yangtze River. From the point

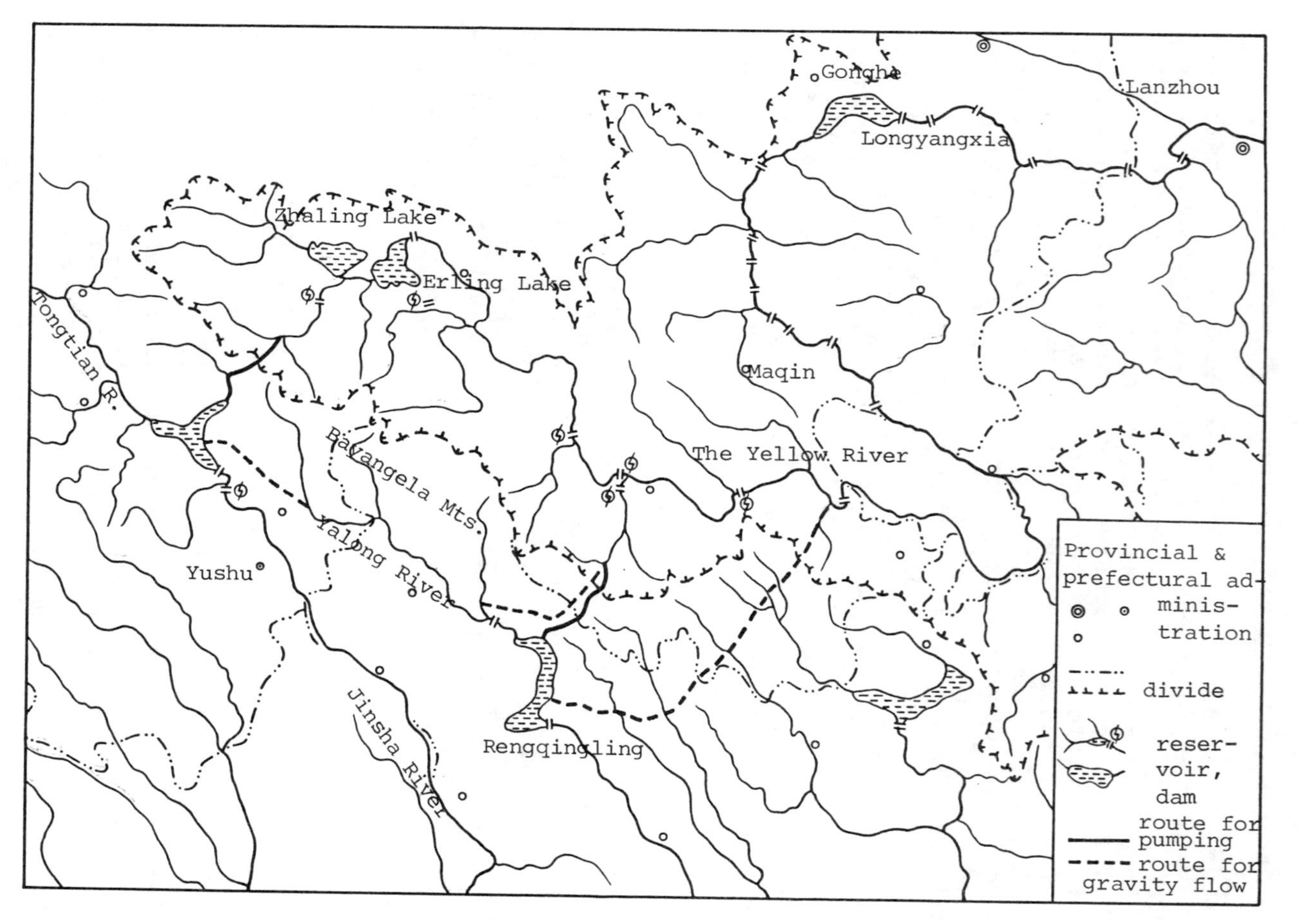

Figure 1. Map showing Main Diversion Routes at the Western Line of South-North Water Transfer.

of view of equitable distribution of water resources, the west-route project is a benefit to China as a whole.

Main Scheme

Since the Fifties reconnaissance and other studies have been carried out to assess the feasibilities of various diversion routes. More tham 30 dam sites and over a dozen river reaches have been examined since 1978. Dam sites at Lianyie and Tunjia on the Tongtian River, Renqinling on the Yalong River and Xie'erga on the Dadu River are considered as the more feasible dam sites for diversion. The proposed routes and sites of diversion have favorable geological and topographic conditions. Earthquake intensity does not exceed 8 degrees and strong earthquake zones are avoided. Difficulties commonly brought about by the height of dams, the size and length of tunnels as well as the lift of pumping could be overcome with our present-day knowledge in science and technology. It is claimed to be technically feasible with obvious economic, environmental and ecological benefits compared with similar water transfer projects through the watersheds. The west-route project is relatively inexpensive. The main route is shown in Figure (I).

River stages at the envisioned diversion sites of each river on the upper reaches of the Yangtze river are 350-500 m below the elevation where the transferred water would empty into the Yellow River. They are 600-860 m lower than the elevation of watershed divide between the Yangtze River and the Yellow River. Therefore a high dam and a long tunnel have to be constructed. To select a technically feasible and economically reasonable route, 50 alternatives have been studied. They may be divided into two groups: One is diversion by gravity and another is by pumping. Schemes of diversion by gravity are characterized by the requirements of a high dam, a long tunnel and canal, a large quantity of work and a high cost of construction. Compared with the gravity plan, the height of the dam may be lower for the pumping scheme, making the selection of damsites easier and requiring smaller construction and a shorter transfer route. The length of the tunnels and the amount of rock and earthwork needed is decreased, the investment lower and the time for construction would be shortened. However, pumping stations and a power supply with high-voltage transmission line must be constructed. In addition, operation and maintenance costs would be higher. After preliminary studies, the pumping alternative is considered more favorable. In-depth studies may show that diversion by gravity is better since there is great potential for saving investment and the quantity of work.

Among the various schemes of diversion by gravity, it seems more favorable to transfer water via the Lian-zhang route. This route would divert water from Lianyie (or Tongjia) reservoir to the Yalong River, then transfer it further to the north from the Renqinling reservoir on the Yalong River to the Yellow River through the Zhanggan River. The height of the Lianyie dam is 410 m (or Tongjia dam, 320 m) and the height of Renqinling dam is 300 m. The total length of the route is 650 km, of which the length of tunnels is 210 km (maximum length of a tunnel is 30 km), the length of canals is 190 km, and the length of river course is 250 km (including length of the route in the reservoirs). To divert 100 cms of water, 2.2 cu.m of earthwork is required.

For diversion from the aforementioned three rivers by pumping, independent schemes have to be worked out for each river. Plans considered favorable are listed as follows.

ROUTE FROM LIANYIE TO DOUQIU AT THE TONGTIAN RIVER

Water would be diverted at the 3,980 m elevation. The volume of water transferred annually would be 10 billion cu.m The length of route is 96 km, of which 29 km is in tunnels, the canal length is 10 km, the length of the river course in the reservoir is 57 km. Seven pumping stations will have to be built with a total pumping lift of 495 m at a consumption of power of 16.8 billion kwhr per year.

ROUTE FROM THE RENQINLING ON YALONG RIVER TO DARI RIVER

Water would be diverted at an elevation of 3,800 m. The volume of water transferred per year would be 5 billion cu.m. The length of the route is 92 km, of which tunnels occupy 30 km and canals 50 km. The length of river course in reservoirs is 12 km. Five pumping stations will be required to lift the water 458 m and the annual consumption of electric power is 8.2 billion kwhr.

ROUTE FROM XIE'ERGA ON THE DADU RIVER TO THE JIAQIU RIVER

Water would be transferred at an elevation of 3,100 m. The amount of water to be transferred annually is 5 billion cu.m. The length of the route is 80 km of which tunnels occupy 24 km and canals 56 km. Five pumping stations will be required to lift the water 358 m and the annual consumption of electric power is 6 billion kwhr.

The volume of water diverted from these three rivers would amount to 20 billion cu.m with an annual consumption of power of 31 billion kwhr. If the pumping plans listed are adopted, power may be supplied by hydropower stations built at Zhijia reservoir on Tongtian River utilizing the high head at Zhijia. This station would make 4.4 billion kwhr annually. The remaining 26.6 billion kwhr may be supplied at hydropower power stations built at suitable places or by a nuclear power station. The quantity of rock and earthwork would amount to 198 million cu.m and the quantity of concrete, 11.1 million cu.m. If pumping plans are adopted, the diversion of each 100 cu.m water requires only 1 cu.m of earthwork.

Staged Development of the Project and Selection of Construction in Earlier Stages

In order to be consistent with the economic conditions of the country, the project has to be developed in stages. Selection of construction sites in the earlier stage depends on which scheme is to be used. If diversion by gravity is used, projects on Yalong River should be done first, of which 4-5 billion cu.m of water could be transferred. Then the project would be expanded to Tongtian River transferring 14-15 billion cu.m of water. If pumping plans are adopted, individual projects should be constructed on the three rivers separately. Benefits and disadvantages of various alternatives for pumping water separatively from the three rivers follows.

THE TONGTIAN RIVER

The Tongtian River has a large annual runoff. The volume of water to be diverted, 5 or 10 billion cu.m, can be decided upon according to the economic capability of the state. Topographical, geological and transportation as well as construction sites are favorable. In the proposed reservoir area, there is little farmland, houses or woodland. There are pieces of grassland and tents of herdsmen on hillsides or in branch ravines. The loss due to inundation would be very small. The Erling Lake which is in the upper reach of the Yellow River could be used as a regulation reservoir. The diverted water joins the Yellow River at a higher elevation which would also promote development on the main stem of the Yellow River above the mouth of Baihe River. This site would supply water to the Chaidamu district in Qinhai province. However, the construction has to be carried out on sites of the Tongtian River lying at high altitudes with a cold climate which would hinder construction. Part of the route passes across a frozen soil region far from any developed area. Construction materials would have to be transported over long distances and the unit price of work would be higher.

THE YALONG RIVER

Compared with the work on Dadu River and Tongtian River,

pumping from Yalong River has advantages of a lower dam height, a shorter tunnel and relatively simple geological conditions. Conditions for transportation are better. It is also favorable with respect to the selection of the dam type and the availability of construction materials.

THE DADU RIVER

Among the three rivers the required height of the dam on the Dadu River is the lowest. Higher temperatures prevail over the site and conditions for constuction would be much better. The proposed route would not cross frozen soil and the project site is much nearer to the developed area so that the hauling distance would be short. However there are large area of woodland in the reservoir proper and the submergence loss would be large. Loss of hydropower in step stations already in existence or planned to be built in Dadu River would be inevitable after diversion.

SUMMARY

In summary, the scale of construction in the earlier stages will be decided according to the plan of national land renovation and economy capability. If the pumping scheme is chosen and 5 billion cu.m of water are required, the Yalong River project should be constructed first. If 10 billion cu.m of water are required, the Tongtian River project should be of high priority. If a combination scheme of Yalong River and Tongtian River diversion by gravity is preferred, the Yalong River project should be constructed first to divert 4-5 billion cu.m of water.

Preliminary proposals for staged development in the west-route project are as follows: The first stage would be to transfer 5-10 billion cu.m of water to replenish the water shortage in the Yellow River basin and nearby districts by 2030. The second stage would be to transfer another 10-15 billion cu.m of water to transfer altogether 20 billion cu.m of water for demand in the Yellow River basin for the long run.

Preliminary Studies of Benefits of the Water Transfer Project

The west-route water transfer project would create large social and economic benefits. It will not only promote industrial and agricultural development in the Northwest area but will also increase power generation by stations already in existence and those to be built in the future.

POWER GENERATION

If 20 billion cu.m of water were diverted and used in the generation of power, an increase of 15 billion kwhr per year would be expected by the existing 8-step hydropower stations on the main stem of the river. This quantity is almost 2.5 times as much as the power generated on the Longyang Gorge power station on the Yellow River. It is planned that other 8-step stations would be built on the main stem of the upper and middle reaches of the river at the time that the transfer of 20 billion cu.m of water is realized. In addition to the existing stations, an increase of 46 billion kwhr per year of power generation could be expected. This is also 2.5 times as much power as is generated by all stations on the Yellow River at present. In the future, after all step stations are built according to the comprehensive plan, an estimate of the increase in power generation is 103-120 billion kwhr per year.

DEVELOPMENT OF INDUSTRY

A part of the diverted 20 billion cu.m of water may be used to develop the Greatwest area (including Chaidamu district) and for coal mines in the relevant provinces. The increase in output from industry alone would be appreciable even if only a part of the diverted water were to be used for industry.

DEVELOPMENT IN AGRICULTURE, FORESTRY, AND STOCK RAISING

Irrigation areas could be enlarged with additional water supply. Reliable water could be provided to promote self-sufficiency of cereal production in Northwest China, and also to promote the growth of forest and grasses for stock raising.

NAVIGATION

After realization of the west-route water tansfer project, discharge in the dry season would increase and with the help of river training works, the navigable length and tonnage of goods would be greatly increased. This would offer chances for the exchange of products along the river and aid in the economic development in the Northwest districts.

REDUCTION OF DEPOSITION IN THE LOWER REACHES

The lower Yellow River aggrades and the bed rises gradually each year, threatening the safety of flood prevention works. It is important to reduce the amount of deposition in the lower reaches of the Yellow River by flushing with the additional amounts of water transferred from the Yangtze River. If 20 billion cu.m of water were used entirely for this purpose, after proper regulation through reservoirs, a reduction of 500-600 million tons of deposition in lower reaches would be expected. Deposition and erosion might be in balance. Flood prevention in the lower reaches of the Yellow River could be ensured for the long run.

DILUTING POLLUTION

At present, nearly 5 million tons of polluted water is discharged into the Yellow River every day. In some reaches the degree of pollution exceeds tolerance limit standards. With the development of industry the quantity of polluted water required to be treated would be larger and larger. Transfer of water in combination with other uses could be used to dilute polluted water.

Not only could many benefits be obtained, but also it would help to solve the problem of water pollution in the Yellow River. Economic benefits discussed above are estimated on an individual basis. Actually it is impossible to use water for only one purpose. Multi-purpose use of water will be made by taking all factors into consideration through regulation on a joint basis with existing reservoirs with a total capacity more than 50 billion cu.m.

RESEARCH NEEDS IN EARLIER STAGES

Comprehensive benefits would be larger than benefits counted on individual items. Studies in depth will be carried out. West- route water transfer is to be carried out over the upper reaches of Yangtze River, where the Yellow River has a cold climate and a high altitude (elevation 2,900-4,500 M). A part of the route may have to pass through frozen soil and the work rate would be slow. The unit price of work would be higher and many difficulties are expected. Several high dams and long tunnels, as well as pumping stations with high heads and large capacities, are required. Environmental and ecological impacts in the area affected by diversion projects need to be evaluated. Research on other problems in science and technology are also needed. Although we have done some reconnaissance work, much remains to be done.

Reference

Zhang Weiming (1986), Preliminary Study of the West Route Project of South to North Water Transfer, "Proceedings IWRA Seminar on Inter-Basin Water Transfer," Beijing, China, 1986.

POSTFACE

Following the presentations during the workshop, the participants were divided, according to their interests in specific topics, into groups. Each group was asked to identify some research needs in their particular areas and to report them to the assembled group. Obviously many of the research topics need not be cooperative but the basic theme of future cooperative research between China and the U.S. was emphasized. In this listing, no attempt was made to set priorities.

CHINA-U.S. BILATERAL SEMINAR ON FLOOD CONTROL MEASURES OF THE LOWER YELLOW RIVER: TENTATIVE PROPOSAL OF TOPICS FOR DISCUSSION AND FOR COOPERATIVE RESEARCH

GROUP 1. REDUCTION OF SEDIMENT DEPOSITION AND CHANNEL STABILITY IN THE LOWER YELLOW RIVER. STUDIES OF RIVER TRAINING, RESERVOIR OPERATION, INTERBASIN TRANSFER AND OTHER MANAGEMENT TECHNIQUES.

Operations

1. Study of effects of clear water transfer from the Yangtze drainage basin and multiple dam construction on the Yellow River system
2. Development of optimization procedure for reservoir releases (effects on river stability and deposition)
3. Test sediment and flood models using Yellow River data
4. Scheduling of diversion and warping to minimize deposition in the main channel

Geomorphology and Geology

1. Study of quantitative geomorphic characteristics of the Yellow River to determine reasons for change of morphology (history of river change)
2. Determination of the effect of active tectonic activity on river morphology, stability and structures
3. Determination of the variability of alluvium (geologic and geotechnical properties) and its effect on structures
4. Description of sedimentology and stratigraphy of Yellow River plain and delta (basic research)
5. Determine effect of base level change on channel morphology and stability including sea level change or channel avulsion)

L. M. Brush et al. (eds.), Taming the Yellow River: Silt and Floods, 687–690.

Hyperconcentrated Flow

1. Determine effect of hyperconcentrated flow on bed forms and channel morphology (scour and fill)
2. Does pulsing flow occur during hyperconcentrated flow? Effect on bank stability?

Instrumentation

1. Development of instrument to measure bed forms during hyperconcentrated flow
2. Development of instrument to detect holes and pipes in levees

GROUP 2. REDUCTION OF SEDIMENT IN THE MIDDLE YELLOW RIVER BY CONTROLLING SOIL EROSION, BANK EROSION AND MASS MOVEMENT AND THEIR ECONOMIC IMPLICATIONS

Fundamental Mechanisms of Gully and Upland Erosion in Loess Area

1. Gully erosion and mass movement and their relation to infiltration, runoff and overland erosion
2. Spatial distribution of sites of severe erosion identified from (1) above
3. Mechanisms for the generation of hyperconcentration of flow
4. Variations in erosional processes and yields from hillslopes and gullies with climatic variations, including those experiences over longer time periods

Evaluation of the Effects of Conservation Control Practices and Land Management

1. Actual reductions (changes) in sediment and runoff produced in different settings over time
2. Relationship of reduction (change) in sediment to changes in runoff
3. Distribution and use of water by plants and their contribution to water loss
4. Delivery of sediment and water to the Yellow River from different sources on the hillslopes and in the valleys or gully bottoms involving extrapolation from small to large areas

Social and Economic Aspects of Water and Sediment Control and Management on the Landscape

1. Development of methods for planning management programs at the watershed and regional level
2. Establishment of policies enhanced by various incentives to encourage farmers to practice conservation including new plant

varieties and techniques which will assure farmers of increased income while promoting conservation
3. Development of methods of evaluating the net benefits not only to farmers but to Chinese Society as a whole of conservation efforts
4. Establishment of data systems, including geographic information systems, to facilitate planning and management

It is suggested that one or several small watersheds might provide appropriate comparative sites for cooperative study. No single watershed is likely to provide the optimum research opportunity for each of the research objectives and a single watershed should not constrain the research objectives. Nevertheless, the watershed provides an important hydrologic unit and a site for a number of process-response studies and may well provide a good beginning if carefully chosen.

GROUP 3. REDUCTION OF FLOOD PEAKS BETWEEN SANMENXIA AND HUAYUANKOU BY STRUCTURAL AND NONSTRUCTURAL MEASURES. EXAMINATION OF RAINFALL ABSTRACTIONS, COORDINATED RESERVOIR OPERATIONS, AND FLOOD WARNING TECHNIQUES

1. Use of radar and remote sensing to measure precipitable cloud water for flood forecasting as a means to increase warning lag time. Other weather forecasting techniques relevant to heavy rain should also be investigated.
2. Catchment runoff modelling which will be useful for more accurate real time flood warning and identification of potential catchment land use changes and management as a means of flood reduction. Suggested emphases are on spatial and temporal variations of abstractions and rainfall.
3. Routing in river channels with irregular floodplains of highly movable bed with high concentration water-sediment mixture
4. Optional operation of a system of reservoirs in parallel and series to reduce flood and to achieve other objectives
5. Flood hazard study to identify not only the level of flooding of different magnitudes, but also the risk and economic, social and environment impact of the flooding

In addition, it has been recognized that studies on the reliability of warning and updating techniques of forecasts are also worth future consideration.

GROUP 4. YELLOW RIVER DELTA MANAGEMENT STRATEGY, TRADEOFFS BETWEEN PETROLEUM DEVELOPMENT AND CHANNEL CONTROL

1. Research on process of sediment disposal by currents and waves in the shallow water around the delta
2. Research on possibility of artificially induced liquifaction and turbidity currents to reduce aggradation. This would

involve laboratory models of liquifaction and density currents and as small scale field test of feasibility.
3. Study of artificial shifts in the position of the distributary channel, in particular management of methods, timing and location and the best criteria for decisionmaking
4. Study of the best methods of filling delta low lands between abandoned distributary channels with constraint of the economic benefits and costs relative to existing oil wells.

MISCELLANEOUS

1. Study of the effect of evapo-transpiration on the flow, particularly low flows in the Yellow River
2. Effect of reduced runoff on the ecosystem of the lower Yellow River

The GeoJournal Library

Bruce Currey and Graeme Hugo (eds.), Famine as Geographical Phenomenon, 1984. ISBN 90–277–1762–1.

S. H. U. Bowie, F.R.S. and I. Thornton (eds.), Environmental Geochemistry and Health, 1985. ISBN 90–277–1879–2.

Leszek A. Kosinski and K. Maudood Elahi (eds.), Population Redistribution and Development in South Asia, 1985. ISBN 90–277–1938–1.

Yehuda Gradus (ed.), Desert Development, 1985. ISBN 90–277–2043–6.

Frank J. Calzonetti and Barry D. Solomon (eds.), Geographical Dimensions of Energy, 1985. ISBN 90–277–2061–4.

Jan Lundqvist, Ulrik Lohm and Malin Falkenmark (eds.), Strategies for River Basin Management, 1985. ISBN 90–277–2111–4.

Andrei Rogers and Frans J. Willekens (eds.), Migration and Settlement. A Multiregional Comparative Study, 1986. ISBN 90–277–2119–X.

Risto Laulajainen, Spatial Strategies in Retailing, 1987. ISBN 90–277–2595–0.

T. H. Lee, H. R. Linden, D. A. Dreyfus and T. Vasko (eds.), The Methane Age, 1988. ISBN 90–277–2745–7.

H. J. Walker (ed.), Artificial Structures and Shorelines, 1988. ISBN 90–277–2746–5.

Aharon Kellerman, Time, Space, and Society: Geographical Societal Perspectives, 1989. ISBN 0–7923–0123–4.

Paolo Fabbri (ed.), Recreational Uses of Coastal Areas, 1989. ISBN 0–7923–0279–6.